Statistical Methods for Handling Incomplete Data

Statistical Methods for Handling Incomplete Data

Second Edition

Jae Kwang Kim

Jun Shao

CRC Press
Taylor & Francis Group
Boca Raton London New York

CRC Press is an imprint of the
Taylor & Francis Group, an **informa** business

A CHAPMAN & HALL BOOK

First edition published 2022
by CRC Press
6000 Broken Sound Parkway NW, Suite 300, Boca Raton, FL 33487-2742

and by CRC Press
2 Park Square, Milton Park, Abingdon, Oxon, OX14 4RN

© 2022 Taylor & Francis Group, LLC

First edition published by CRC Press 2013

CRC Press is an imprint of Taylor & Francis Group, LLC

ISBN: 9780367280543 (hbk)
ISBN: 9781032118130 (pbk)
ISBN: 9780429321740 (ebk)

DOI: 10.1201/9780429321740

Typeset in LM Roman
by KnowledgeWorks Global Ltd.

To Timothy, Jenny, and Jungwoo

and

To Jason, Annie, and Guang

Contents

List of Figures

List of Tables

Preface

Missing data is frequently encountered in statistics. Statistical analysis with missing data is an area of extensive research for the last three decades. Many statistical problems assuming a latent variable can also be viewed as missing data problems. Furthermore, with the advances in statistical computing, there has been a rapid development of techniques and applications in missing data analysis inspired by theoretical findings in this area. This book aims to cover the most up-to-date statistical theories and computational methods for analyzing incomplete data.

The main features of the book can be summarized as follows:

1. Rigorous treatment of statistical theories on likelihood-based inference with missing data.

2. Comprehensive treatment of computational techniques and theories on imputation, including multiple imputation and fractional imputation.

3. Most up-to-date treatment of methodologies involving propensity score weighting, nonignorable missing, longitudinal missing, survey sampling application, and data integration.

This book is developed under the frequentist framework and puts less emphasis on Bayesian methods and nonparametric methods. Apart from some real data examples, many artificial examples are presented to help with the understanding of the methodologies introduced. In the second edition, major changes are made in Chapters 5–7. Also, Chapter 11 and Chapter 12 are newly added.

The book is suitable for use as a textbook for a graduate course in statistics departments. Materials in Chapter 2 to Chapter 8 can be covered systematically in the course. Materials in Chapter 9 to Chapter 12 are more advanced and can perhaps serve for future reference. To be comfortable with the materials the reader should have completed courses in statistical theory and in linear models. This book can also be used as a reference book for those interested in this area. Some of the research ideas introduced in the book can be developed further for specific applications.

We would like to thank those who made enormous contribution to this book. Jae Kwang Kim would like to thank his PhD advisor, professor Wayne Fuller, and his collaborators, including Shu Yang, Zhonglei Wang, Sixia Chen, Kosuke Morikawa, Minsun Riddles, Changbao Wu, Jongho Im, Seunghwan

Park, Seho Park, Chris Skinner, Emily Berg, David Haziza, Dong Wan Shin, Siu-Ming Tam, Danhyang Lee, Gyuhyeong Goh, Cindy Yu, Hejian Sang, Hengfang Wang, Masatoshi Uehara, Myunghee Cho Paik, Youngjo Lee, HaiYing Wang, Youngdeok Hwang and J.N.K. Rao, for their contributions in the research of missing data analysis. His research at Iowa State University was partially supported by a Cooperative Agreement between the US Department of Agriculture Natural Resources Conservation Service and Iowa State University and also by a grant from National Science Foundation (Award No: 1931380). Last but not least, we would like to thank Rob Calver, Vaishali Singh, at Chapman & Hall/CRC Press, and also Meeta Singh for their support during the production of this book. We take full responsibilities for all errors and omissions in the book.

Jae Kwang Kim
Jun Shao

1

Introduction

1.1 Introduction

Missing data, or incomplete data, is frequently encountered in many disciplines. Statistical analysis with missing data has been an area of considerable interest in the statistical community. Many tools, generic or tailor-made, have already been developed, and many more will be forthcoming to handle missing data problems. Missing data is particularly useful because many statistical issues can be treated as special cases of the missing data problem. For example, data with measurement error can be viewed as a special case of missing data where an imperfect measurement is available instead of true measurement. Two-phase sampling can also be viewed as a planned missing data problem where the key items are observed only in the second-phase sample by design. Many statistical problems employing a latent variable can also be viewed as missing data problems. Furthermore, the advances in statistical computing have made the computational aspects of the missing data analysis techniques more feasible. This book aims to cover the most up-to-date statistical theories and computational methods of the missing data analysis.

Generally speaking, let z be the study variable with density function $f(z; \theta)$. We are interested in estimating the parameter θ. If z were observed throughout the sample, then θ would be able to be estimated by the maximum likelihood method. Instead of observing z, however, we only observe $y = T(z, \delta)$ and δ, where $y = T(z, \delta)$ is an incomplete version of z satisfying $T(z, \delta = 1) = z$ and δ is an indicator function that takes either one or zero, depending on the response status. Parameter estimation of θ from the observation of (y, δ) is the core of the problem in missing data analyses.

To handle this problem, the marginal density function of (y, δ) needs to be expressed as a function of the original distribution $f(z; \theta)$. Maximum likelihood estimation can be obtained under some identifying assumptions and statistical theories can be developed for the maximum likelihood estimator obtained from the observed sample. Computational tools for producing the maximum likelihood estimator need to be introduced. How to assess the uncertainty of the resulting maximum likelihood estimator is also an important topic.

When z is a vector, there will be more complications. Because several random variables are subject to missingness, the missing data pattern can

DOI: 10.1201/9780429321740-1

figure in to simply modeling and estimation. The monotone missing pattern refers to the situation where the set of respondents in one variable is always a subset of the set of respondents for another variable, which may host further subsetting. See Table 1.1 for an illustration of the monotone missing pattern.

TABLE 1.1
Monotone Missing Pattern

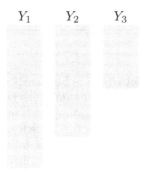

$$Y_1 \qquad Y_2 \qquad Y_3$$

1.2   Outline

Maximum likelihood estimation with missing data serves as the starting point of this book. Chapter 2 is about defining the observed likelihood function from the marginal density of the observed part of the data, finding the maximum of the observed likelihood by solving the mean score equation, and obtaining the observed information matrix from the observed likelihood. Chapter 3 deals with computational tools to arrive at the maximum likelihood estimator, especially the EM algorithm.

Imputation, covered in Chapter 4, is also a popular tool for handling missing data. Imputation can be viewed as a computational technique for the Monte Carlo approximation of the conditional expectation of the original complete-sample estimator given the observed data. As for variance estimation of the imputation estimator, an important subject in missing data analyses, the Taylor linearization or replication method can be used. Multiple imputation, introduced in Chapter 5, has been proposed as a general tool for imputation and simplified variance estimation but it requires some special conditions, called *congeniality* and *self-efficiency*. Fractional imputation is an alternative general-purpose estimation tool for imputation and is covered in Chapter 6.

Propensity score weighting, covered in Chapter 7, is another tool for handling missing data. Basically the responding units are assigned with propensity score weights so that the weighted analysis can lead to valid inference. The propensity score weighting method is often based on an assumption about the response mechanism and the resulting estimator can be made more efficient by properly taking into account of the auxiliary information available from the whole sample. The propensity score weighted estimator can further incorporate the outcome model for imputation and in turn make the resulting estimator doubly protected against the failure of the assumed models.

Nonignorable missing data occurs in the challenging situation where the response mechanism depends on the study variable that is subject to missingness. Nonignorable missing data is also an important area of research that is yet to be thoroughly investigated. Chapter 8 covers some important topics in the analysis of nonignorable missing data. Nonresponse instrumental variables can be used to identify the parameters in some nonignorable missing data and obtain consistent parameter estimates. The generalized method of moments (GMM) and the pseudo likelihood method are useful tools for parameter estimation with nonresponse instrumental variables. Also covered is the exponential tilting model whereby the conditional distribution of the nonrespondents can be expressed via the conditional distribution of the respondents, which eventually enables the derivation of nonparametric maximum likelihood estimators. Instead of using nonresponse instrumental variables, callback samples can also be used to provide consistent parameter estimates for nonignorable missing data.

The rest of the book discusses specific applications of the covered methodology for analyzing missing data. The confluence between missing data and longitudinal analysis, survey sampling, and data integration are the subject of Chapter 9, Chapter 10 and Chapter 11, respectively. Chapter 12 contains some advanced topics.

1.3 How to Use This Book

This book assumes a basic understanding of graduate-level mathematical statistics and linear models. Being slightly more technical than Little and Rubin (2002), it is written at the mathematical level of a second-year Ph.D. course in statistics.

Table 1.2 provides a sample outline for a 15-week semester-long course in the graduate course offering of the Department of Statistics at Iowa State University. Core materials are from Chapter 2 to Chapter 6 and can be covered in about 11–12 weeks. Chapter 7 (excluding Section 7.7 and Section 7.8) can be covered in 2 weeks. Chapter 8 can be covered in 2 weeks.

TABLE 1.2
Outline of a 15-Week Lecture

Weeks	Chapter	Topic
1–2	1–2	Introduction. Likelihood-based approach
3–5	3	Computation
6–7	4	Imputation
8–9	5	Multiple Imputation
10–11	6	Fractional Imputation
12–13	7.1–7.4	Propensity scoring approach
14–15	8.1–8.6	Nonignorable missing data

For readers using this book as reference material, the chapters have been written to be as self-contained as possible. The core theory of maximum likelihood estimation with missing data is covered in Chapter 2 in a concise manner. Chapter 3 may overlap with Little and Rubin (2002) to some extent, but Section 3.2 contains newer materials. Chapter 4 is technically oriented, but Section 4.2 still contains the main theoretical results on the imputation estimator, which was originally covered by Wang and Robins (1998). Chapter 5 presents Bayesian approach to missing data analysis and covers multiple imputation. Chapter 6 covers fractional imputation as a general tool for imputation.

The main novelty of the book comes from the materials in Chapter 7 and after. Topics covered in Chapter 7 through Chapter 12 are either not fully addressed or totally uncharted previously. Researchers in this area may find the materials interesting and useful to their own research.

2

Likelihood-Based Approach

In this chapter, we discuss the likelihood-based approach in the analysis of missing data. To do this, we first review the likelihood-based methods in the case of complete response. The observed likelihood is carefully defined in Section 2.2 from the marginal density of the observed part of the data. Mean score theorem is a fundamental result in missing data analysis and introduced in Section 2.3. Information matrix associated with the observed likelihood is introduced in Section 2.4 and some basic concepts such as missing information principle is also introduced.

2.1 Introduction

In this chapter, we discuss the likelihood-based approach in the analysis of missing data. To do this, we first review the likelihood-based methods in the case of complete response. Let $\mathbf{y} = (y_1, y_2, \ldots y_n)$ be a realization of the random sample from a distribution P with density $f = dP/d\mu$, so that $P(Y \in B) = \int_B f(y)d\mu(y)$ for any measurable set B where $\mu(y)$ is a σ-finite dominating measure. Assume that the true density $f(y)$ belongs to a parametric family of densities $\mathcal{P} = \{f(y;\theta) : \theta \in \Omega\}$ indexed by $\theta \in \Omega \subset \mathbb{R}^p$ for some integer $p \geq 1$. That is, there exists a $\theta_0 \in \Omega$ such that $f(y;\theta_0) = f(y)$ for all y. Once the parametric density is specified, the likelihood function and the maximum likelihood estimator can be defined formally as follows.

Definition 2.1. *The likelihood function of θ, denoted by $L(\theta)$, is defined as the probability density (mass) function of the sample data $\mathbf{y}$ considered as a function of θ. That is,*

$$L(\theta) = f(\mathbf{y};\theta)$$

where $f(\mathbf{y};\theta)$ is the joint density function of $\mathbf{y}$.

Definition 2.2. *Let $\hat{\theta}$ be the maximum likelihood estimator (MLE) of θ_0 if it satisfies*

$$L(\hat{\theta}) = \max_{\theta \in \Omega} L(\theta).$$

DOI: 10.1201/9780429321740-2

If $y_1, y_2, \ldots, y_n$ are independently and identically distributed (IID),

$$L(\theta) = \prod_{i=1}^{n} f(y_i; \theta).$$

Also, if $\hat{\theta}$ is the MLE of θ_0, then $g(\hat{\theta})$ is the MLE of $g(\theta_0)$. The MLE is not necessarily unique.

Definition 2.3. *Let* $\mathcal{P} = \{P_\theta : \theta \in \Omega\}$ *be a statistical model with parameter space* Ω. *Model* $\mathcal{P}$ *is identifiable if the mapping* $\theta \mapsto P_\theta$ *is one-to-one:*

$$P_{\theta_1} = P_{\theta_2} \Rightarrow \theta_1 = \theta_2.$$

If the distributions are defined in terms of the probability density functions (pdfs), then two pdfs should be considered distinct only if they differ on a set of non-zero measure. For example, two functions $f_1(x) = I(0 \le x < 1)$ and $f_2(x) = I(0 \le x \le 1)$ differ only at a single point $x = 1$, a set of measure zero, and thus cannot be considered as distinct pdfs.

Lemma 2.1. *(Shannon-Kolmogorov information inequality) Let* $f_0(y) = f(y; \theta_0)$ *and* $f_1(y) = f(y; \theta_1)$. *The Kullback-Leibler information number defined by*

$$K(f_0, f_1) = E_{\theta_0} \left\{ \ln \frac{f_0(Y)}{f_1(Y)} \right\}$$

satisfies

$$K(f_0, f_1) \ge 0,$$

with equality if and only if $P_{\theta_0} \{f_0(Y) = f_1(Y)\} = 1$.

Proof. Since $\phi(x) = -\ln(x)$ is a strictly convex function of x, we have, using Jensen's inequality,

$$K(f_0, f_1) = E_{\theta_0} \left\{ -\ln \frac{f_1(Y)}{f_0(Y)} \right\} \ge -\ln E_{\theta_0} \left\{ \frac{f_1(Y)}{f_0(Y)} \right\} = 0$$

with equality if and only if $P_{\theta_0} \{f_0(Y) = f_1(Y)\} = 1$. $\qquad\square$

Lemma 2.2. *If* $\mathcal{P} = \{f(y; \theta) : \theta \in \Omega\}$ *is identifiable and* $E\{|\ln f(Y; \theta)|\} < \infty$ *for all* θ, *then*

$$Q(\theta) = E_{\theta_0} \ln [f(Y; \theta)]$$

has a unique maximum at $\theta = \theta_0$.

Proof. By Shannon-Kolmogorov information inequality, we can obtain that

$$Q(\theta) \le Q(\theta_0)$$

for all $\theta \in \Omega$, with equality if and only if

$$P_{\theta_0} \{f(Y; \theta) = f(Y; \theta_0)\} = 1. \tag{2.1}$$

Thus, $Q(\theta)$ has a maximum at θ_0. Furthermore, by identifiability, (2.1) implies $\theta = \theta_0$, which proves the uniqueness of the maximizer of $Q(\theta)$. $\qquad\square$

By Lemma 2.2, we see that model identifiability is an important condition to obtain the uniqueness of the maximizer of $Q(\theta)$ which is the probability limit of

$$Q_n(\theta) = \frac{1}{n} \sum_{i=1}^{n} \ln f(y_i; \theta), \qquad (2.2)$$

where $y_1, \ldots, y_n$ are IID sample from a distribution with density $f(y; \theta_0)$.

The following two theorems present some asymptotic properties of the maximum likelihood estimator (MLE): (weak) consistency and asymptotic normality. To discuss the asymptotic properties, let $\hat{\theta}$ be any solution of

$$Q_n(\hat{\theta}) = \max_{\theta \in \Omega} Q_n(\theta),$$

where $Q_n(\theta)$ is defined in (2.2). Also, define $Q(\theta) = E_{\theta_0}\{\ln f(Y; \theta)\}$ to be the probability limit of $Q_n(\theta)$. By Lemma 2.2, $Q(\theta)$ is maximized at $\theta = \theta_0$. Thus, under some conditions, we may expect that $\hat{\theta} = \arg\max Q_n(\theta)$ converges in probability to $\theta_0 = \arg\max Q(\theta)$. The following theorem presents a formal result.

Theorem 2.1. *Assume the following two conditions:*

1. Identifiability: $Q(\theta)$ is uniquely maximized at θ_0. That is, for any $\epsilon > 0$, there exists a $\delta > 0$ such that $\theta \notin B_\epsilon(\theta_0)$ implies $Q(\theta_0) - Q(\theta) \geq \delta$, where $B_\epsilon(\theta_0) = \{\theta \in \Omega : |\theta - \theta_0| < \epsilon\}$.

2. Uniform weak convergence:

$$\sup_{\theta \in \Omega} |Q_n(\theta) - Q(\theta)| \xrightarrow{p} 0$$

for some nonstochastic function $Q(\theta)$.

Then, $\hat{\theta} \xrightarrow{p} \theta_0$.

Proof. For any $\epsilon > 0$, we can find $\delta > 0$ such that

$$
\begin{aligned}
0 \leq P\left[\hat{\theta} \notin B_\epsilon(\theta_0)\right] &\leq P\left[Q(\theta_0) - Q_n(\hat{\theta}) + Q_n(\hat{\theta}) - Q(\hat{\theta}) \geq \delta\right] \\
&\leq P\left[Q(\theta_0) - Q_n(\theta_0) + Q_n(\hat{\theta}) - Q(\hat{\theta}) \geq \delta\right] \\
&\leq 2P\left[\sup_\theta |Q_n(\theta) - Q(\theta)| \geq \delta/2\right] \to 0.
\end{aligned}
$$

$\square$

In Theorem 2.1, $Q_n(\theta)$ is the log-likelihood of θ obtained from $f(y; \theta)$. The function $Q(\theta)$ is the uniform probability limit of $Q_n(\theta)$ and θ_0 is the unique maximizer of $Q(\theta)$. In Theorem 2.1, it is assumed that $\hat{\theta}$ is not necessarily uniquely determined, but θ_0 is.

Theorem 2.2. *Assume the following regularity conditions:*

1. θ_0 is in the interior of Ω.

2. $Q_n(\theta)$ is twice continuously differentiable on some neighborhood $\Omega_0 (\subset \Omega)$ of θ_0 almost everywhere.

3. The first-order partial derivative of Q_n satisfies

$$\sqrt{n} \frac{\partial}{\partial \theta} Q_n(\theta_0) \xrightarrow{d} N(0, A_0)$$

for some positive definite A_0, where $\xrightarrow{d}$ denotes the convergence in distribution.

4. The second-order partial derivative of Q_n satisfies

$$\sup_{\theta \in \Omega_o} \left\| \frac{\partial^2}{\partial \theta \partial \theta'} Q_n(\theta) - B(\theta) \right\| \xrightarrow{p} 0,$$

for some $B(\theta)$ continuous at θ_0 and $B_0 = B(\theta_0)$ is nonsingular.

Furthermore, assume that $\hat{\theta}$ satisfies

5. $\hat{\theta} \xrightarrow{p} \theta_0$.

6. $\sqrt{n} \partial Q_n(\hat{\theta})/\partial \theta \xrightarrow{p} 0$.

Then

$$\sqrt{n} \left(\hat{\theta} - \theta_0 \right) \xrightarrow{d} N \left(0, B_0^{-1} A_0 B_0^{-1'} \right).$$

Proof. By Assumption 2 and the mean value theorem,

$$\frac{\partial}{\partial \theta} Q_n(\hat{\theta}) = \frac{\partial}{\partial \theta} Q_n(\theta_0) + \left[\frac{\partial^2}{\partial \theta \partial \theta'} Q_n(\theta^*) \right] \left(\hat{\theta} - \theta_0 \right), \tag{2.3}$$

for some θ^* between $\hat{\theta}$ and θ_0. By Assumption 5, $\theta^* \xrightarrow{p} \theta_0$. Hence, by Assumption 4,

$$\frac{\partial^2}{\partial \theta \partial \theta'} Q_n(\theta^*) = B(\theta_0) + o_p(1).$$

Thus, by Assumption 6 and the invertibility of B_0, we can multiply $\sqrt{n} B_0^{-1}$ to both sides of (2.3) to get

$$o_p(1) \quad = \quad B_0^{-1} \sqrt{n} \frac{\partial Q_n(\theta_0)}{\partial \theta} + [1 + o_p(1)] \sqrt{n} \left(\hat{\theta} - \theta_0 \right).$$

By Slutsky's theorem,

$$\sqrt{n} \left(\hat{\theta} - \theta_0 \right) = -B_0^{-1} \sqrt{n} \frac{\partial Q_n(\theta_0)}{\partial \theta} + o_p(1)$$

and the asymptotic normality follows from Assumption 3. $\qquad\square$

In Assumption 3, the partial derivative of the log-likelihood is called the *score function* associated with n complete observation, denoted by

$$S_{\text{com}}(\theta) = \frac{\partial}{\partial \theta} \ln L(\theta).$$

Here, subscript "com" is used to emphasize that it is based on n complete observation. On the other hand, the score function associated with a single observation is denoted by

$$S(\theta; y) = \frac{\partial}{\partial \theta} \ln f(y; \theta).$$

Under the assumption that $y_1, \ldots, y_n \overset{i.i.d.}{\sim} f(y; \theta_0)$, Condition 3 in Theorem 2.2 can be expressed as

$$n^{-1/2} S_{\text{com}}(\theta_0) = n^{-1/2} \sum_{i=1}^{n} S(\theta_0; y_i) \overset{d}{\longrightarrow} N(0, A_0),$$

where $A_0 = E_{\theta_0} \{ S(\theta_0; Y) S(\theta_0; Y)' \}$. In Assumption 4, $B(\theta)$ is essentially the probability limit of the second-order partial derivatives of the log-likelihood. This is the expected Fisher information matrix based on a single observation, defined by

$$\mathcal{I}(\theta) = -E_\theta \left\{ \frac{\partial^2}{\partial \theta \partial \theta'} \ln f(Y; \theta) \right\}$$

and $B_0 = \mathcal{I}(\theta_0)$. We now summarize the definitions associated with the score function.

Definition 2.4. *[1] Score function:*

$$S(\theta; y) = \frac{\partial}{\partial \theta} \ln f(y; \theta).$$

[2] Fisher information (representing the curvature of the log-density function)

$$I(\theta; y) = -\frac{\partial}{\partial \theta'} S(\theta; y).$$

[3] Observed (Fisher) information: $I(\hat{\theta}; y)$, where $\hat{\theta}$ is the MLE.

[4] Expected (Fisher) information: $\mathcal{I}(\theta) = E_\theta \{ I(\theta; Y) \}$.

Because of the definition of the MLE, the observed Fisher information is always positive. The expected information is meaningful as a function of θ across the admissible values of θ, but $I(\theta)$ is only meaningful in the neighborhood of $\hat{\theta}$. The observed information applies to a single dataset. In contrast, the expected information is an average quantity over all possible datasets generated at the true value of the parameter. For exponential families of distributions, we have $I(\hat{\theta}; \mathbf{y}) = \mathcal{I}(\hat{\theta})$. In general, the observed information is preferred for variance estimation of $\hat{\theta}$. The use of the observed information in assessing the accuracy of the MLE is advocated by Efron and Hinkley (1978).

We now present two important equalities of the score function. The equality in (2.5) is often called the (second-order) *Bartlett identity*.

Theorem 2.3. *Under regularity conditions allowing the exchange of the order of integration and differentiation,*

$$E_\theta\{S(\theta; Y)\} = 0 \qquad (2.4)$$

and

$$V_\theta\{S(\theta; Y)\} = \mathcal{I}(\theta). \qquad (2.5)$$

Proof. Note that

$$
\begin{aligned}
E_\theta\{S(\theta; Y)\} &= E_\theta\left\{\frac{\partial}{\partial\theta}\ln f(Y; \theta)\right\} \\
&= E_\theta\left\{\frac{\partial f(Y; \theta)/\partial\theta}{f(Y; \theta)}\right\} \\
&= \int\frac{\partial}{\partial\theta}f(y; \theta)\,d\mu(y).
\end{aligned}
$$

By assumption,

$$\int\frac{\partial}{\partial\theta}f(y; \theta)\,d\mu(y) = \frac{\partial}{\partial\theta}\int f(y; \theta)\,d\mu(y).$$

Since $\int f(y; \theta)\,d\mu(y) = 1$,

$$\frac{\partial}{\partial\theta}\int f(y; \theta)\,d\mu(y) = 0$$

and (2.4) is proven.

To prove (2.5), note that since $E_\theta\{S(\theta; Y)\} = 0$, equality (2.5) is equivalent to

$$E_\theta\{S(\theta; Y)S(\theta; Y)'\} = -E_\theta\left\{\frac{\partial}{\partial\theta'}S(\theta; Y)\right\}. \qquad (2.6)$$

To show (2.6), taking the partial derivative of (2.4) with respect to θ, we get

$$
\begin{aligned}
0 &= \frac{\partial}{\partial\theta'}\int S(\theta; y)f(y; \theta)\,d\mu(y) \\
&= \int\left\{\frac{\partial}{\partial\theta'}S(\theta; y)\right\}f(y; \theta)\,d\mu(y) + \int S(\theta; y)\left\{\frac{\partial}{\partial\theta'}f(y; \theta)\right\}d\mu(y) \\
&= E_\theta\left\{\frac{\partial}{\partial\theta'}S(\theta; Y)\right\} + E_\theta\{S(\theta; Y)S(\theta; Y)'\},
\end{aligned}
$$

and we have shown (2.6). $\square$

Equality (2.5) is a special case of the general equality

$$Cov\{g(Y;\theta), S(\theta;Y)\} = -E\{\partial g(Y;\theta)/\partial\theta'\} \tag{2.7}$$

for any $g(y;\theta)$ such that $E\{g(Y;\theta)\} = 0$. Under the model $y_1,\dots,y_n \overset{i.i.d.}{\sim} f(y;\theta_0)$, Theorem 2.2 states that the limiting distribution of the MLE is

$$\sqrt{n}\left(\hat{\theta} - \theta_0\right) \overset{d}{\longrightarrow} N\left(0, \mathcal{I}(\theta_0)^{-1} A_0 \mathcal{I}(\theta_0)^{-1}\right),$$

where $A_0 = E_{\theta_0}\{S(\theta_0)S(\theta_0)'\}$. By Theorem 2.3, $A_0 = \mathcal{I}(\theta_0)$, and the limiting distribution of the MLE is

$$\sqrt{n}\left(\hat{\theta} - \theta_0\right) \overset{d}{\longrightarrow} N\left(0, \mathcal{I}^{-1}(\theta_0)\right). \tag{2.8}$$

Result (2.8) presents the limiting distribution of the MLE. The asymptotic variance of the MLE is $n^{-1}\mathcal{I}^{-1}(\theta_0)$.

The MLE also satisfies

$$-2\ln\left[\frac{L(\theta_0)}{L(\hat{\theta})}\right] \overset{d}{\longrightarrow} \chi_p^2,$$

which can be used to develop likelihood-ratio (LR) confidence intervals for θ_0. The level $1 - \alpha$ LR confidence intervals (CI) are constructed by

$$\left\{\theta : L(\theta) > k_\alpha \times L(\hat{\theta})\right\}$$

for some k_α which is a function of the upper α quantile of the chi-square distribution with p degrees of freedom. The LR confidence interval is more attractive than the Wald confidence interval in two aspects: (i) A Wald CI can often produce interval estimates beyond the parameter space. (ii) A LR interval is invariant with respect to parameter transformation. For example, if (θ_L, θ_U) is the 95% CI for θ, then $(g(\theta_L), g(\theta_U))$ is the 95% CI for a monotone increasing function $g(\theta)$. For a comprehensive overview of the likelihood-based inference, see Pawitan (2001).

2.2 Observed Likelihood

We now derive the likelihood function under the existence of missing data. Roughly speaking, the likelihood function in the presence of missing data, so-called the *observed likelihood*, is obtained by finding the marginal density for the observed data. To formally define the observed likelihood, let $\mathbf{y} = (y_1, \dots, y_k)'$ be a k-dimensional random vector with probability distribution function $f(\mathbf{y};\theta)$. We are interested in estimating the parameter θ. If

n independent realizations of $\mathbf{y}$ were observed throughout the sample, the maximum likelihood method could be used to estimate θ.

Now suppose that, for unit i in the sample, we observe only a part of $\mathbf{y}_i = (y_{i1}, \ldots, y_{ik})'$. Let δ_{ij} be the response indicator function of y_{ij}, defined by

$$\delta_{ij} = \begin{cases} 1 & \text{if } y_{ij} \text{ is observed} \\ 0 & \text{otherwise.} \end{cases}$$

To estimate the parameter θ, we assume a response probability model $P(\boldsymbol{\delta} \mid \mathbf{y})$, where $\boldsymbol{\delta} = (\delta_1, \ldots, \delta_k)'$.

In general, we have the following two models:

1. Original sample distribution: $f(\mathbf{y}; \theta)$

2. Response mechanism: $P(\boldsymbol{\delta} \mid \mathbf{y})$. Often, it is assumed that $P(\boldsymbol{\delta} \mid \mathbf{y}) = P(\boldsymbol{\delta} \mid \mathbf{y}; \phi)$ with ϕ being the parameter of this response probability model.

Let $(\mathbf{y}_{i,\text{obs}}, \mathbf{y}_{i,\text{mis}})$ be the observed part and missing part of $\mathbf{y}_i$, respectively. For each unit i, we observe $(\mathbf{y}_{i,\text{obs}}, \boldsymbol{\delta}_i)$ instead of observing $\mathbf{y}_i$. Given the above models, we can derive the marginal density of $(\mathbf{y}_{i,\text{obs}}, \boldsymbol{\delta}_i)$ as

$$\tilde{f}(\mathbf{y}_{i,\text{obs}}, \boldsymbol{\delta}_i; \theta, \phi) = \int f(\mathbf{y}_i; \theta) \, P(\boldsymbol{\delta}_i \mid \mathbf{y}_i; \phi) \, d\mu(\mathbf{y}_{i,\text{mis}}). \qquad (2.9)$$

Under the IID assumption, we can write the joint density as

$$\tilde{f}(\mathbf{y}_{\text{obs}}, \boldsymbol{\delta}; \theta, \phi) = \prod_{i=1}^{n} \tilde{f}(\mathbf{y}_{i,\text{obs}}, \boldsymbol{\delta}_i; \theta, \phi), \qquad (2.10)$$

where $\mathbf{y}_{\text{obs}} = (\mathbf{y}_{1,\text{obs}}, \ldots, \mathbf{y}_{n,\text{obs}})$ and $\tilde{f}(\mathbf{y}_{i,\text{obs}}, \boldsymbol{\delta}_i; \theta, \phi)$ is defined in (2.9). The joint density in (2.10) as a function of parameter (θ, ϕ) can be called the *observed likelihood*. The observed likelihood is the marginal density of the observation, expressed as a function of the parameters. To give a formal definition of the observed likelihood function, let

$$\mathcal{R}(\mathbf{y}_{\text{obs}}, \boldsymbol{\delta}) = \{\mathbf{y} : \mathbf{y}_{\text{obs}}(\mathbf{y}_i, \boldsymbol{\delta}_i) = \mathbf{y}_{i,\text{obs}}, \ i = 1, \ldots, n\} \qquad (2.11)$$

be the set of all possible values of $\mathbf{y}$ with the same realized value of $\mathbf{y}_{\text{obs}}$, for a given $\boldsymbol{\delta}$, where $\mathbf{y}_{\text{obs}}(\mathbf{y}_i, \boldsymbol{\delta}_i)$ is a function that gives the value of y_{ij} for $\delta_{ij} = 1$. That is, for given $\mathbf{y}_i = (y_{i1}, \ldots, y_{ik})$, the j-th component of $\mathbf{y}_{\text{obs}}(\mathbf{y}_i, \boldsymbol{\delta}_i)$ is equal to the j-th component of $\mathbf{y}_{i,\text{obs}}$.

Definition 2.5. *Let $f(\mathbf{y}; \theta)$ be the joint density of the complete observation $\mathbf{y} = (\mathbf{y}_1, \ldots, \mathbf{y}_n)$ and let $P(\boldsymbol{\delta} \mid \mathbf{y}; \phi)$ be the conditional probability of $\boldsymbol{\delta} = (\boldsymbol{\delta}_1, \ldots, \boldsymbol{\delta}_n)$ given $\mathbf{y}$. The observed likelihood of (θ, ϕ) based on the realized observation $(\mathbf{y}_{\text{obs}}, \boldsymbol{\delta})$ is given by*

$$L_{\text{obs}}(\theta, \phi) = \int_{\mathcal{R}(\mathbf{y}_{\text{obs}}, \boldsymbol{\delta})} f(\mathbf{y}; \theta) \, P(\boldsymbol{\delta} \mid \mathbf{y}; \phi) \, d\mu(\mathbf{y}). \qquad (2.12)$$

Under the IID setup, the observed likelihood is given by

$$L_{\mathrm{obs}}(\theta,\phi) = \prod_{i=1}^{n} \left[\int f(\mathbf{y}_i;\theta) P(\boldsymbol{\delta}_i \mid \mathbf{y}_i;\phi) \, d\mu(\mathbf{y}_{i,\mathrm{mis}}) \right],$$

where it is understood that, if $\mathbf{y}_i = \mathbf{y}_{i,\mathrm{obs}}$ and $\mathbf{y}_{i,\mathrm{mis}}$ is empty then there is nothing to integrate out.

Parameter ϕ can be viewed as a nuisance parameter in the sense that we are not directly interested in estimating ϕ, but an estimate of ϕ is needed to estimate θ, which is the parameter of interest in $f(\mathbf{y};\theta)$.

In the special case of scalar y, the observed likelihood can be written as

$$L_{\mathrm{obs}}(\theta,\phi) = \prod_{\delta_i=1} [f(y_i;\theta)\pi(y_i;\phi)] \times \prod_{\delta_i=0} \left[\int f(y;\theta)\{1-\pi(y;\phi)\}\, d\mu(y) \right],$$

$$(2.13)$$

where $\pi(y;\phi) = P(\delta = 1 \mid y;\phi)$.

Example 2.1. *Consider the following regression model*

$$y_i = \mathbf{x}_i'\beta + \epsilon_i, \qquad \epsilon_i \sim N(0,\sigma^2)$$

and, instead of y_i, we observe

$$y_i^* = \begin{cases} y_i & \text{if } y_i > 0 \\ 0 & \text{if } y_i \le 0. \end{cases}$$

The observed log-likelihood is

$$l_{\mathrm{obs}}(\beta,\sigma^2) = -\frac{1}{2}\sum_{y_i^*>0}\left[\ln 2\pi + \ln\sigma^2 + \frac{(y_i^* - \mathbf{x}_i'\beta)^2}{\sigma^2}\right] + \sum_{y_i^*=0}\ln\left[1 - \Phi\left(\frac{\mathbf{x}_i'\beta}{\sigma}\right)\right],$$

where $\Phi(x)$ is the cumulative distribution function of the standard normal distribution. This model is sometimes called the Tobit model in the econometric literature (Amemiya, 1985), termed after Tobin (1958).

Example 2.2. *Let $t_1, t_2, \ldots, t_n$ be an IID sample from a distribution with density $f_\theta(t) = \theta e^{-\theta t} I(t > 0)$. Instead of observing t_i, we observe (y_i, δ_i) where*

$$y_i = \begin{cases} t_i & \text{if } \delta_i = 1 \\ c & \text{if } \delta_i = 0 \end{cases}$$

and

$$\delta_i = \begin{cases} 1 & \text{if } t_i \le c \\ 0 & \text{if } t_i > c, \end{cases}$$

where c is a known censoring time. The observed likelihood for θ can be derived as

$$L_{\text{obs}}(\theta) = \prod_{i=1}^{n}\left[\{f_\theta(t_i)\}^{\delta_i}\{P(t_i > c)\}^{1-\delta_i}\right]$$

$$= \theta^{\sum_{i=1}^{n}\delta_i}\exp\left(-\theta\sum_{i=1}^{n}y_i\right).$$

In Example 2.1 and Example 2.2, the missing mechanism is known in the sense that the censoring time is known. In general, the missing mechanism is unknown and it often depends on some unknown parameter ϕ. In this case, the observed likelihood can be expressed as in (2.12) or (2.13). Suppose the observed likelihood can be expressed as a product of two terms:

$$L_{\text{obs}}(\theta,\phi) = L_1(\theta)L_2(\phi). \tag{2.14}$$

In this case, the value of θ that maximizes the observed likelihood does not depend on ϕ. Thus, to compute the MLE of θ, we have only to find $\hat{\theta}$ that maximizes $L_1(\theta)$ and the modeling for the missing mechanism can be avoided in this simple situation. To find a sufficient condition for (2.14), we introduce the concept of missing at random (MAR) of Rubin (1976).

Definition 2.6. *Let the joint density of $\boldsymbol{\delta}$ given $\mathbf{y}$ be $P(\boldsymbol{\delta}\,|\,\mathbf{y})$. Let $\mathbf{y}_{\text{obs}}$ be a random vector that is defined through $\mathbf{y}_{\text{obs}} = \mathbf{y}_{\text{obs}}(\mathbf{y},\boldsymbol{\delta})$, such that*

$$y_{i,\text{obs}} = \begin{cases} y_i & \text{if } \delta_i = 1 \\ * & \text{if } \delta_i = 0. \end{cases}$$

The response mechanism is called missing at random (MAR) *if*

$$P(\boldsymbol{\delta}\,|\,\mathbf{y}_1) = P(\boldsymbol{\delta}\,|\,\mathbf{y}_2), \tag{2.15}$$

for all $\mathbf{y}_1$ and $\mathbf{y}_2$ satisfying $\mathbf{y}_{\text{obs}}(\mathbf{y}_1,\boldsymbol{\delta}) = \mathbf{y}_{\text{obs}}(\mathbf{y}_2,\boldsymbol{\delta})$.

Condition (2.15) can be expressed as

$$P(\boldsymbol{\delta}\,|\,\mathbf{y}) = P(\boldsymbol{\delta}\,|\,\mathbf{y}_{\text{obs}}). \tag{2.16}$$

Thus, the MAR condition means that the response mechanism $P(\boldsymbol{\delta}\,|\,\mathbf{y})$ depends on $\mathbf{y}$ only through $\mathbf{y}_{\text{obs}}$, the observed part of $\mathbf{y}$. By the construction of $\mathbf{y}_{\text{obs}}$, we have $\sigma(\mathbf{y}_{\text{obs}}) \subset \sigma(\mathbf{y})$ and

$$P(\boldsymbol{\delta}\,|\,\mathbf{y},\mathbf{y}_{\text{obs}}) = P(\boldsymbol{\delta}\,|\,\mathbf{y}).$$

Here, $\sigma(\mathbf{y})$ is the sigma-algebra generated by $\mathbf{y}$ (Billingsley, 1986). By Bayes' theorem,

$$P(\mathbf{y}\,|\,\mathbf{y}_{\text{obs}},\boldsymbol{\delta}) = \frac{P(\boldsymbol{\delta}\,|\,\mathbf{y},\mathbf{y}_{\text{obs}})}{P(\boldsymbol{\delta}\,|\,\mathbf{y}_{\text{obs}})}P(\mathbf{y}\,|\,\mathbf{y}_{\text{obs}}).$$

Therefore, the MAR condition (2.16) is equivalent to

$$P(\mathbf{y} \mid \mathbf{y}_{\text{obs}}, \boldsymbol{\delta}) = P(\mathbf{y} \mid \mathbf{y}_{\text{obs}}).$$

The MAR condition is essentially the conditional independence of $\boldsymbol{\delta}$ and $\mathbf{y}$ given $\mathbf{y}_{\text{obs}}$. Using the notation of Dawid (1979), it can be written as

$$\boldsymbol{\delta} \perp \mathbf{y} \mid \mathbf{y}_{\text{obs}}.$$

The MAR is a special case of coarsening at random (Heitjan and Rubin, 1991).

The following theorem, originally proved by Rubin (1976), shows likelihood can be factorized under MAR.

Theorem 2.4. *Let the joint density of $\boldsymbol{\delta}$ given $\mathbf{y}$ be $P(\boldsymbol{\delta} \mid \mathbf{y}; \phi)$ and the joint density of $\mathbf{y}$ be $f(\mathbf{y}; \theta)$. Under the following two conditions*

1. the parameters θ and ϕ are distinct

2. the MAR condition holds,

the observed likelihood can be written as in (2.14) and the MLE of θ can be obtained by maximizing $L_1(\theta)$ only.

Proof. Under MAR, we have (2.16) and the observed likelihood (2.12) can be written as (2.14) where

$$L_1(\theta) = \int_{\mathcal{R}(\mathbf{y}_{\text{obs}}, \boldsymbol{\delta})} f(\mathbf{y}; \theta) \, d\mu(\mathbf{y})$$

and $L_2(\phi) = P(\boldsymbol{\delta} \mid \mathbf{y}_{\text{obs}}; \phi)$. □

Example 2.3. *Consider bivariate data (x_i, y_i) with pdf $f(x, y) = f_1(y \mid x) f_2(x)$, where x_i is always observed and y_i is subject to missingness. Assume that the response status variable δ_i of y_i satisfies*

$$P(\delta_i = 1 \mid x_i, y_i) = \Lambda_1(\phi_0 + \phi_1 x_i + \phi_2 y_i)$$

for $\Lambda_1(x) = 1 - \{1 + \exp(x)\}^{-1}$. Let θ be the parameter of interest in the regression model $f_1(y \mid x; \theta)$. Let α be the parameter in the marginal distribution of x, denoted by $f_2(x_i; \alpha)$. Define $\Lambda_0(x) = 1 - \Lambda_1(x)$. Then, the observed likelihood can be written as

$$
\begin{aligned}
L_{\text{obs}}(\theta, \alpha, \phi) &= \left[\prod_{\delta_i=1} f_1(y_i \mid x_i; \theta) f_2(x_i; \alpha) \Lambda_1(\phi_0 + \phi_1 x_i + \phi_2 y_i) \right] \\
&\quad \times \left[\prod_{\delta_i=0} \int f_1(y_i \mid x_i; \theta) f_2(x_i; \alpha) \Lambda_0(\phi_0 + \phi_1 x_i + \phi_2 y_i) \, dy_i \right] \\
&= L_1(\theta, \phi) \times L_2(\alpha),
\end{aligned}
$$

where

$$L_1(\theta, \phi) = \prod_{\delta_i=1} f_1(y_i \mid x_i; \theta) \Lambda_1(\phi_0 + \phi_1 x_i + \phi_2 y_i)$$

$$\times \prod_{\delta_i=0} \int f_1(y_i \mid x_i; \theta) \Lambda_0(\phi_0 + \phi_1 x_i + \phi_2 y_i) \, dy_i$$

and

$$L_2(\alpha) = \prod_{i=1}^{n} f_2(x_i; \alpha).$$

If $\phi_2 = 0$, then MAR holds and

$$L_1(\theta, \phi) = L_{1a}(\theta) \times L_{1b}(\phi), \tag{2.17}$$

where

$$L_{1a}(\theta) = \prod_{\delta_i=1} f_1(y_i \mid x_i; \theta)$$

and

$$L_{1b}(\phi) = \prod_{\delta_i=1} \Lambda_1(\phi_0 + \phi_1 x_i) \times \prod_{\delta_i=0} \Lambda_0(\phi_0 + \phi_1 x_i).$$

Thus, under MAR, the MLE of θ can be obtained by maximizing $L_{1a}(\theta)$, which is obtained by ignoring the missing part of the data. That is, the MLE of θ is obtained by complete-case (CC) analysis.

In Example 2.3, instead of y_i subject to missingness, if x_i is subject to missingness, then the observed likelihood becomes

$$L_{\text{obs}}(\theta, \phi, \alpha) = \left[\prod_{\delta_i=1} f_1(y_i \mid x_i; \theta) f_2(x_i; \alpha) \Lambda_1(\phi_0 + \phi_1 x_i + \phi_2 y_i) \right]$$

$$\times \left[\prod_{\delta_i=0} \int f_1(y_i \mid x_i; \theta) f_2(x_i; \alpha) \Lambda_0(\phi_0 + \phi_1 x_i + \phi_2 y_i) \, dx_i \right]$$

$$\neq L_1(\theta, \phi) \times L_2(\alpha).$$

If $\phi_1 = 0$, then

$$L_{\text{obs}}(\theta, \alpha, \phi) = L_1(\theta, \alpha) \times L_2(\phi)$$

and MAR holds. Although we are not interested in the marginal distribution of x, we have to specify the model for the marginal distribution of x. If $\phi_1 = \phi_2 = 0$ holds then

$$L_{\text{obs}}(\theta, \alpha, \phi) = L_{1a}(\theta) \times L_{1b}(\phi) \times L_2(\alpha),$$

and this is called *missing completely at random (MCAR)*.

2.3　Mean Score Function

The observed likelihood is simply the marginal likelihood of the observation $(\mathbf{y}_{\text{obs}}, \boldsymbol{\delta})$. Express $\eta = (\theta, \phi)$ and the observed likelihood can be written as follows:

$$
\begin{aligned}
L_{\text{obs}}(\eta) &= \int_{\mathcal{R}(\mathbf{y}_{\text{obs}}, \boldsymbol{\delta})} f(\mathbf{y}; \theta) P(\boldsymbol{\delta} \mid \mathbf{y}; \phi) \, d\mu(\mathbf{y}) \\
&= \int f(\mathbf{y}; \theta) P(\boldsymbol{\delta} \mid \mathbf{y}; \phi) \, d\mu(\mathbf{y}_{\text{mis}}) \\
&= \int f(\mathbf{y}, \boldsymbol{\delta}; \eta) \, d\mu(\mathbf{y}_{\text{mis}}),
\end{aligned}
$$

where $\mathcal{R}(\mathbf{y}_{\text{obs}}, \boldsymbol{\delta})$ is defined in (2.11) and $\mathbf{y}_{\text{mis}}$ is the missing part of $\mathbf{y}$. To find the MLE that maximizes the observed likelihood, we often need to solve

$$
S_{\text{obs}}(\eta) \equiv \frac{\partial}{\partial \eta} \ln L_{\text{obs}}(\eta) = 0. \tag{2.18}
$$

The score equation in (2.18) is called the *observed score equation*, because it is based on the observed likelihood. The function, $S_{\text{obs}}(\eta)$, can be called the *observed score function*. Working with the observed score function can be computationally challenging because the observed likelihood is in integral form.

To overcome this difficulty, we can establish the following theorem, which was originally proposed by Fisher (1922) and also discussed by Louis (1982).

Theorem 2.5. *Under some regularity conditions,*

$$
S_{\text{obs}}(\eta) = \bar{S}(\eta), \tag{2.19}
$$

where $\bar{S}(\eta) = E\{S_{\text{com}}(\eta) \mid \mathbf{y}_{\text{obs}}, \boldsymbol{\delta}\}$,

$$
S_{\text{com}}(\eta) = \partial \ln f(\mathbf{y}, \boldsymbol{\delta}; \eta) / \partial \eta, \tag{2.20}
$$

and

$$
f(\mathbf{y}, \boldsymbol{\delta}; \eta) = f(\mathbf{y}; \theta) P(\boldsymbol{\delta} \mid \mathbf{y}; \phi). \tag{2.21}
$$

Proof.

$$
\frac{\partial}{\partial \eta} \ln \{L_{\text{obs}}(\eta)\} = \frac{\partial L_{obs}(\eta)/\partial \eta}{L_{\text{obs}}(\eta)} = \frac{\int [\partial f(\mathbf{y}, \boldsymbol{\delta}; \eta)/\partial \eta] \, d\mu(\mathbf{y}_{\text{mis}})}{L_{\text{obs}}(\eta)}.
$$

Next, consider the numerator only.

$$
\begin{aligned}
\int [\partial f(\mathbf{y}, \boldsymbol{\delta}; \eta)/\partial \eta] \, d\mu(\mathbf{y}_{\text{mis}}) &= \int \frac{[\partial f(\mathbf{y}, \boldsymbol{\delta}; \eta)/\partial \eta]}{f(\mathbf{y}, \boldsymbol{\delta}; \eta)} \frac{f(\mathbf{y}, \boldsymbol{\delta}; \eta)}{L_{\text{obs}}(\eta)} \, d\mu(\mathbf{y}_{\text{mis}}) \times L_{\text{obs}}(\eta) \\
&= E\{\partial \ln f(\mathbf{y}, \boldsymbol{\delta}; \eta)/\partial \eta \mid \mathbf{y}_{\text{obs}}, \boldsymbol{\delta}\} \times L_{\text{obs}}(\eta) \\
&= E\{S_{\text{com}}(\eta) \mid \mathbf{y}_{\text{obs}}, \boldsymbol{\delta}\} \times L_{\text{obs}}(\eta),
\end{aligned}
$$

we then have the result. □

The function $\bar{S}(\eta)$ is called the *mean score function*. It is computed by taking the conditional expectation of the complete-sample score function given the observation. The mean score function is easier to compute than the observed score function.

Remark 2.1. *An alternative proof for Theorem 2.5 can be made as follows. Since*

$$L_{\text{obs}}(\eta) = f(\mathbf{y}, \boldsymbol{\delta}; \eta) / f(\mathbf{y}_{\text{mis}} \mid \mathbf{y}_{\text{obs}}, \boldsymbol{\delta}; \eta),$$

we have

$$\frac{\partial}{\partial \eta} \ln L_{\text{obs}}(\eta) = \frac{\partial}{\partial \eta} \ln f(\mathbf{y}, \boldsymbol{\delta}; \eta) - \frac{\partial}{\partial \eta} \ln f(\mathbf{y}_{\text{mis}} \mid \mathbf{y}_{\text{obs}}, \boldsymbol{\delta}; \eta), \qquad (2.22)$$

taking conditional expectation of the above equation over the conditional distribution of $(\mathbf{y}, \boldsymbol{\delta})$ given $(\mathbf{y}_{\text{obs}}, \boldsymbol{\delta})$, we have

$$
\begin{aligned}
\frac{\partial}{\partial \eta} \ln L_{\text{obs}}(\eta) &= E\left\{ \frac{\partial}{\partial \eta} \ln L_{\text{obs}}(\eta) \mid \mathbf{y}_{\text{obs}}, \boldsymbol{\delta} \right\} \\
&= E\left\{ S_{\text{com}}(\eta) \mid \mathbf{y}_{\text{obs}}, \boldsymbol{\delta} \right\} - E\left\{ \frac{\partial}{\partial \eta} \ln f(\mathbf{y}_{\text{mis}} \mid \mathbf{y}_{\text{obs}}, \boldsymbol{\delta}; \eta) \mid \mathbf{y}_{\text{obs}}, \boldsymbol{\delta} \right\}.
\end{aligned}
$$

Here, the first equality holds because $L_{\text{obs}}(\eta)$ is a function of $(\mathbf{y}_{\text{obs}}, \boldsymbol{\delta})$ only. The second equality follows from (2.22). The last term is equal to zero by Theorem 2.3 applied to the conditional distribution, which states that the expected value of the score function is zero and the reference distribution in this case is the conditional distribution of $\mathbf{y}_{\text{mis}}$ given $(\mathbf{y}_{\text{obs}}, \boldsymbol{\delta})$.

Example 2.4. *Suppose that the study variable y is randomly distributed as Bernoulli (p_i), where*

$$p_i = p_i(\beta) = \frac{\exp(\mathbf{x}_i'\beta)}{1 + \exp(\mathbf{x}_i'\beta)}$$

for some unknown parameter β and $\mathbf{x}_i$ is a vector of covariates in the logistic regression model for y_i. Let δ_i be the response indicator function for y_i with distribution Bernoulli(π_i), where

$$\pi_i = \frac{\exp(\mathbf{x}_i'\phi_0 + y_i\phi_1)}{1 + \exp(\mathbf{x}_i'\phi_0 + y_i\phi_1)}.$$

We assume that $\mathbf{x}_i$ is always observed, but y_i is missing if $\delta_i = 0$.
    Under complete response, the score function for β is

$$S_1(\beta) = \sum_{i=1}^{n} (y_i - p_i(\beta)) \mathbf{x}_i$$

and the score function for ϕ is

$$S_2(\phi) = \sum_{i=1}^{n} (\delta_i - \pi_i(\phi)) (\mathbf{x}_i', y_i)'.$$

With missing data, the mean score function for β becomes

$$\bar{S}_1(\beta,\phi) = \sum_{\delta_i=1} \{y_i - p_i(\beta)\} \mathbf{x}_i + \sum_{\delta_i=0} \sum_{y=0}^{1} w_i(y;\beta,\phi) \{y - p_i(\beta)\} \mathbf{x}_i, \quad (2.23)$$

where

$$w_i(y;\beta,\phi) = \frac{P(y_i = y \mid \mathbf{x}_i; \beta) P(\delta_i = 0 \mid y_i = y, \mathbf{x}_i; \phi)}{\sum_{z=0}^{1} P(y_i = z \mid \mathbf{x}_i; \beta) P(\delta_i = 0 \mid y_i = z, \mathbf{x}_i; \phi)}$$

Thus, $\bar{S}_1(\beta,\phi)$ is also a function of ϕ. If the response mechanism is MAR so that $\phi_1 = 0$, then

$$w_i(y;\beta,\phi) = \frac{P(y_i = y \mid \mathbf{x}_i; \beta)}{\sum_{z=0}^{1} P(y_i = z \mid \mathbf{x}_i; \beta)} = P(y_i = y \mid \mathbf{x}_i; \beta).$$

Thus,

$$\bar{S}_1(\beta,\phi) = \sum_{\delta_i=1} \{y_i - p_i(\beta)\} \mathbf{x}_i = \bar{S}_1(\beta).$$

Similarly, under missing data, the mean score function for ϕ is

$$\bar{S}_2(\beta,\phi) = \sum_{\delta_i=1} \{\delta_i - \pi(\phi;x_i,y_i)\} (\mathbf{x}_i, y_i)$$

$$+ \sum_{\delta_i=0} \sum_{y=0}^{1} w_i(y;\beta,\phi) \{\delta_i - \pi_i(\phi;\mathbf{x}_i,y)\} (\mathbf{x}_i, y).$$

Thus, $\bar{S}_2(\beta,\phi)$ is also a function of β. In general, finding the solution to $[\bar{S}_1(\beta,\phi), \bar{S}_2(\beta,\phi)] = (0,0)$ is not easy because the weight $w_i(y;\beta,\phi)$ is a function of the unknown parameters.

In the general missing data problem, we have

$$\bar{S}(\theta,\phi) = [E\{S_1(\theta) \mid \mathbf{y}_{\mathrm{obs}}, \boldsymbol{\delta}\}, E\{S_2(\phi) \mid \mathbf{y}_{\mathrm{obs}}, \boldsymbol{\delta}\}] = [\bar{S}_1(\theta,\phi), \bar{S}_2(\theta,\phi)]$$

and, under MAR, we can write

$$S_{\mathrm{obs}}(\theta,\phi) = [\bar{S}_1(\theta), \bar{S}_2(\phi)].$$

Thus, under MAR, the MLE of θ can be obtained by solving $\bar{S}_1(\theta) = 0$.

Example 2.5. *Suppose that the study variable y follows a normal distribution with mean $\mathbf{x}'\beta$ and variance σ^2. The score equations for β and σ^2 under complete response are*

$$S_1(\beta,\sigma^2) = \frac{1}{\sigma^2} \sum_{i=1}^{n} (y_i - \mathbf{x}_i'\beta) \mathbf{x}_i = \mathbf{0}$$

and

$$S_2(\beta,\sigma^2) = -\frac{n}{2\sigma^2} + \frac{1}{2\sigma^4}\sum_{i=1}^{n}\left(y_i - \mathbf{x}_i'\beta\right)^2 = 0.$$

Assume that y_i are observed only for the first r elements and the MAR assumption holds. In this case, the mean score function reduces to

$$\bar{S}_1(\beta,\sigma^2) = \frac{1}{\sigma^2}\sum_{i=1}^{r}\left(y_i - \mathbf{x}_i'\beta\right)\mathbf{x}_i$$

and

$$\bar{S}_2(\beta,\sigma^2) = -\frac{n}{2\sigma^2} + \frac{1}{2\sigma^4}\sum_{i=1}^{r}\left(y_i - \mathbf{x}_i'\beta\right)^2 + (n-r)\frac{1}{2\sigma^2}.$$

The maximum likelihood estimator obtained by solving the mean score equations is

$$\hat{\beta} = \left(\sum_{i=1}^{r}\mathbf{x}_i\mathbf{x}_i'\right)^{-1}\sum_{i=1}^{r}\mathbf{x}_i y_i$$

and

$$\hat{\sigma}^2 = \frac{1}{r}\sum_{i=1}^{r}\left(y_i - \mathbf{x}_i'\hat{\beta}\right)^2.$$

The resulting estimators can also be obtained by simply ignoring the missing part of the sample.

2.4 Observed Information

We discuss some statistical properties of the observed score function in the missing data setup. Before we derive the main theory, it is necessary to give a definition of the information matrix associated with the observed score function. The following definition is an extension of Definition 2.4 to missing data problem.

Definition 2.7. *1. Observed score function:*

$$S_{\text{obs}}(\eta) = \frac{\partial}{\partial\eta}\ln L_{\text{obs}}(\eta).$$

2. Fisher information (representing the curvature of the log-likelihood) from the observed likelihood

$$I_{\text{obs}}(\eta) = -\frac{\partial^2}{\partial\eta\partial\eta'}\ln L_{\text{obs}}(\eta) = -\frac{\partial}{\partial\eta'}S_{\text{obs}}(\eta).$$

3. *Expected (Fisher) information from the observed likelihood:* $\mathcal{I}_{\text{obs}}(\eta) = E_\eta\{I_{\text{obs}}(\eta)\}$.

The following theorem presents the basic properties of the observed score function.

Theorem 2.6. *Under the conditions of Theorem 2.3, we have*

$$E\{S_{\text{obs}}(\eta)\} = 0 \tag{2.24}$$

and

$$V\{S_{\text{obs}}(\eta)\} = \mathcal{I}_{\text{obs}}(\eta), \tag{2.25}$$

where $\mathcal{I}_{\text{obs}}(\eta) = E_\eta\{I_{\text{obs}}(\eta)\}$ *is the expected information from the observed likelihood.*

Proof. Equality (2.24) follows because, by (2.19)

$$\begin{aligned}
E\{S_{\text{obs}}(\eta)\} &= E\left[E\{S_{\text{com}}(\eta) \mid \mathbf{y}_{\text{obs}}, \boldsymbol{\delta}\}\right] \\
&= E\{S_{\text{com}}(\eta)\},
\end{aligned}$$

which is equal to zero by (2.4). Equality (2.25) can be derived using the same argument for proving (2.6), with the observed likelihood at concern this time. $\quad\square$

By Theorem 2.2, under some regularity conditions, the solution to $\bar{S}(\eta) = 0$ is consistent to η_0 and has the asymptotic variance $\mathcal{I}_{\text{obs}}(\eta_0)^{-1}$, where

$$\mathcal{I}_{\text{obs}}(\eta) = E\left\{-\frac{\partial S_{\text{obs}}(\eta)}{\partial \eta'}\right\} = E\{S_{\text{obs}}^{\otimes 2}(\eta)\} = E\{\bar{S}^{\otimes 2}(\eta)\}$$

with $B^{\otimes 2}$ denoting BB'.

For variance estimation of $\hat{\eta}$ obtained by solving $\bar{S}(\eta) = 0$, one can use $\mathcal{I}_{\text{obs}}(\hat{\eta})^{-1}$, the inverse of the expected Fisher information applied to the observed data. Under the IID setup, Redner and Walker (1984) proposed using

$$\left[\hat{H}(\hat{\eta})\right]^{-1} = \left\{\sum_{i=1}^{n} \bar{S}_i^{\otimes 2}(\hat{\eta})\right\}^{-1}$$

to estimate the variance of $\hat{\eta}$, where $\bar{S}_i(\eta) = E\{S_i(\eta) \mid \mathbf{y}_{i,\text{obs}}, \boldsymbol{\delta}_i\}$. Meilijson (1989) termed $\hat{H}(\hat{\eta})$ the empirical Fisher information and touted its computational convenience.

If we define

$$S_{\text{mis}}(\eta) = \frac{\partial}{\partial \eta} \ln f(\mathbf{y}_{\text{mis}} \mid \mathbf{y}_{\text{obs}}, \boldsymbol{\delta}; \eta), \tag{2.26}$$

equation (2.22) can be written as

$$S_{\text{obs}}(\eta) = S_{\text{com}}(\eta) - S_{\text{mis}}(\eta) \tag{2.27}$$

and the result (2.19) is equivalent to

$$E\{S_{\text{mis}}(\eta) \mid \mathbf{y}_{\text{obs}}, \delta\} = 0. \tag{2.28}$$

The score function $S_{\text{mis}}(\eta)$ defined in (2.26) is the score function associated with the conditional distribution $f(\mathbf{y}_{\text{mis}} \mid \mathbf{y}_{\text{obs}}, \delta)$. The expected information derived from $S_{\text{mis}}(\eta)$ is often called the *missing information* and is defined by

$$\mathcal{I}_{\text{mis}}(\eta) = E\left\{-\frac{\partial}{\partial \eta'} S_{\text{mis}}(\eta)\right\}.$$

Using (2.26), the missing information satisfies

$$\mathcal{I}_{\text{mis}}(\eta) = \mathcal{I}_{\text{com}}(\eta) - \mathcal{I}_{\text{obs}}(\eta), \tag{2.29}$$

where

$$\mathcal{I}_{\text{com}}(\eta) = E\left\{-\frac{\partial}{\partial \eta'} S_{\text{com}}(\eta)\right\}$$

is the expected information associated with the complete-sample likelihood. The equality (2.29) is called the *missing information principle* by Orchard and Woodbury (1972).

Remark 2.2. *Using (2.19) and (2.27), we have*

$$Cov\{S_{\text{obs}}(\eta), S_{\text{mis}}(\eta)\} = 0$$

and

$$V\{S_{mis}(\eta)\} = V\{S_{com}(\eta)\} - V\{S_{obs}(\eta)\}.$$

This is an alternative expression of the missing information principle. Since $V\{S_{com}(\eta)\} = \mathcal{I}_{\text{com}}(\eta)$ *and* $V\{S_{obs}(\eta)\} = \mathcal{I}_{\text{obs}}(\eta)$, *we have*

$$\mathcal{I}_{\text{mis}}(\eta) \quad = \quad V\{S_{\text{mis}}(\eta)\}.$$

Thus, Bartlett identity holds for $S_{\text{mis}}(\eta)$.
   Also, the law of total variance for $S_{com}(\eta)$ *can be written as*

$$\begin{aligned}
V\{S_{\text{com}}(\eta)\} &= V[E\{S_{\text{com}}(\eta) \mid \mathbf{y}_{\text{obs}}, \delta\}] + E[V\{S_{\text{com}}(\eta) \mid \mathbf{y}_{\text{obs}}, \delta\}] \\
&= V\{S_{\text{obs}}(\eta)\} + E[V\{S_{\text{com}}(\eta) - S_{\text{obs}}(\eta) \mid \mathbf{y}_{\text{obs}}, \delta\}] \\
&= V\{S_{\text{obs}}(\eta)\} + E[V\{S_{\text{mis}}(\eta) \mid \mathbf{y}_{\text{obs}}, \delta\}] \\
&= V\{S_{\text{obs}}(\eta)\} + V\{S_{\text{mis}}(\eta)\}.
\end{aligned}$$

Thus, the missing information principle is equivalent to the total variance decomposition for $S_{com}(\eta)$.

The following theorem, first proved by Louis (1982), presents a way of computing the observed information obtained from the observed likelihood. The formula in (2.30) is often called *Louis' formula*.

Theorem 2.7. *Let* $L_{\text{com}}(\eta) = f(\mathbf{y}, \boldsymbol{\delta}; \eta)$ *be the complete sample likelihood and let*

$$I_{\text{com}}(\eta) = -\frac{\partial}{\partial \eta'} S_{\text{com}}(\eta) = -\frac{\partial^2}{\partial \eta \partial \eta'} \ln L_{\text{com}}(\eta)$$

be the Fisher information of $L_{\text{com}}(\eta)$.

Under the conditions of Theorem 2.3,

$$I_{\text{obs}}(\eta) = E\{I_{\text{com}}(\eta) \mid \mathbf{y}_{\text{obs}}, \boldsymbol{\delta}\} + \bar{S}(\eta)^{\otimes 2} - E\{S_{\text{com}}^{\otimes 2}(\eta) \mid \mathbf{y}_{\text{obs}}, \boldsymbol{\delta}\} \qquad (2.30)$$

or

$$I_{\text{obs}}(\eta) = E\{I_{\text{com}}(\eta) \mid \mathbf{y}_{\text{obs}}, \boldsymbol{\delta}\} - V\{S_{\text{com}}(\eta) \mid \mathbf{y}_{\text{obs}}, \boldsymbol{\delta}\}, \qquad (2.31)$$

where $I_{\text{obs}}(\eta)$ *is the Fisher information from the observed likelihood and* $\bar{S}(\eta) = E\{S_{\text{com}}(\eta) \mid \mathbf{y}_{\text{obs}}, \boldsymbol{\delta}\}$ *is the mean score function defined in (2.19).*

Proof. By Theorem 2.5, the observed information of $L_{\text{obs}}(\eta)$ can be expressed as

$$I_{\text{obs}}(\eta) = -\frac{\partial}{\partial \eta'} \bar{S}(\eta),$$

where $\bar{S}(\eta) = E\{S_{\text{com}}(\eta) \mid \mathbf{y}_{\text{obs}}, \boldsymbol{\delta}; \eta\}$. Thus, we have

$$
\begin{aligned}
\frac{\partial}{\partial \eta'} \bar{S}(\eta) &= \frac{\partial}{\partial \eta'} \int S_{\text{com}}(\eta; \mathbf{y}) f(\mathbf{y}_{\text{mis}} \mid \mathbf{y}_{\text{obs}}, \boldsymbol{\delta}; \eta) \, d\mu(\mathbf{y}) \\
&= \int \left\{ \frac{\partial}{\partial \eta'} S_{\text{com}}(\eta; \mathbf{y}) \right\} f(\mathbf{y}_{\text{mis}} \mid \mathbf{y}_{\text{obs}}, \boldsymbol{\delta}; \eta) \, d\mu(\mathbf{y}) \\
&\quad + \int S_{\text{com}}(\eta; \mathbf{y}) \left\{ \frac{\partial}{\partial \eta'} f(\mathbf{y}_{\text{mis}} \mid \mathbf{y}_{\text{obs}}, \boldsymbol{\delta}; \eta) \right\} d\mu(\mathbf{y}) \\
&= E\{\partial S_{\text{com}}(\eta)/\partial \eta' \mid \mathbf{y}_{\text{obs}}, \boldsymbol{\delta}\} \\
&\quad + \int S_{\text{com}}(\eta; \mathbf{y}) \left\{ \frac{\partial}{\partial \eta'} \log f(\mathbf{y}_{\text{mis}} \mid \mathbf{y}_{\text{obs}}, \boldsymbol{\delta}; \eta) \right\} f(\mathbf{y}_{\text{mis}} \mid \mathbf{y}_{\text{obs}}, \boldsymbol{\delta}; \eta) \, d\mu(\mathbf{y}).
\end{aligned}
$$

The first term is equal to $-E\{I_{\text{com}}(\eta) \mid \mathbf{y}_{\text{obs}}, \boldsymbol{\delta}\}$ and the second term is equal to

$$
\begin{aligned}
E\{S_{\text{com}}(\eta) S_{\text{mis}}(\eta)' \mid \mathbf{y}_{\text{obs}}, \boldsymbol{\delta}\} &= E\left[\{\bar{S}(\eta) + S_{\text{mis}}(\eta)\} S_{\text{mis}}(\eta)' \mid \mathbf{y}_{\text{obs}}, \boldsymbol{\delta}\right] \\
&= E\{S_{\text{mis}}(\eta) S_{\text{mis}}(\eta)' \mid \mathbf{y}_{\text{obs}}, \boldsymbol{\delta}\}
\end{aligned}
$$

because

$$
\begin{aligned}
E\{\bar{S}(\eta) S_{\text{mis}}(\eta)' \mid \mathbf{y}_{\text{obs}}, \boldsymbol{\delta}\} &= E\left[\bar{S}(\eta) \{S_{\text{com}}(\eta) - \bar{S}(\eta)\}' \mid \mathbf{y}_{\text{obs}}, \boldsymbol{\delta}\right] \\
&= \bar{S}(\eta) E\{S_{\text{mis}}(\eta)' \mid \mathbf{y}_{\text{obs}}, \boldsymbol{\delta}\} = \mathbf{0},
\end{aligned}
$$

by (2.28). Thus, we have

$$I_{\text{obs}}(\eta) = E\{I_{\text{com}}(\eta) \mid \mathbf{y}_{\text{obs}}, \boldsymbol{\delta}\} - E\left\{S_{\text{mis}}(\eta)^{\otimes 2} \mid \mathbf{y}_{\text{obs}}, \boldsymbol{\delta}\right\}.$$

Using (2.22), we have

$$S_{\mathrm{mis}}(\eta) = S_{\mathrm{com}}(\eta) - \bar{S}(\eta).$$

Thus, because $E\{S_{\mathrm{com}}(\eta) \mid \mathbf{y}_{\mathrm{obs}}, \boldsymbol{\delta}\} = \bar{S}(\eta)$, we have

$$E\left\{S_{\mathrm{mis}}(\eta)^{\otimes 2} \mid \mathbf{y}_{\mathrm{obs}}, \boldsymbol{\delta}\right\} = E\left\{S_{\mathrm{com}}(\eta)^{\otimes 2} \mid \mathbf{y}_{\mathrm{obs}}, \boldsymbol{\delta}\right\} - \bar{S}(\eta)^{\otimes 2} \qquad (2.32)$$

and the result follows. □

Example 2.6. *Consider the following bivariate normal distribution:*

$$\begin{pmatrix} y_{1i} \\ y_{2i} \end{pmatrix} \sim N\left[\begin{pmatrix} \mu_1 \\ \mu_2 \end{pmatrix}, \begin{pmatrix} \sigma_{11} & \sigma_{12} \\ \sigma_{12} & \sigma_{22} \end{pmatrix}\right] \qquad (2.33)$$

for $i = 1, 2, \ldots, n$. For simplicity, assume that σ_{11}, σ_{12} and σ_{22} are known constants and $\mu = (\mu_1, \mu_2)'$ is the parameter of interest. The complete sample score function for μ is

$$S_{\mathrm{com}}(\mu) = \sum_{i=1}^{n} S_{\mathrm{com}}^{(i)}(\mu) = \sum_{i=1}^{n} \begin{pmatrix} \sigma_{11} & \sigma_{12} \\ \sigma_{12} & \sigma_{22} \end{pmatrix}^{-1} \begin{pmatrix} y_{1i} - \mu_1 \\ y_{2i} - \mu_2 \end{pmatrix}. \qquad (2.34)$$

The information matrix of μ based on the complete sample is

$$\mathcal{I}_{\mathrm{com}}(\mu) = n \begin{pmatrix} \sigma_{11} & \sigma_{12} \\ \sigma_{12} & \sigma_{22} \end{pmatrix}^{-1}.$$

Suppose that there are some missing values in y_{1i} and y_{2i} and the original sample is partitioned into four sets:

$$
\begin{aligned}
H &= \quad \text{both } y_1 \text{ and } y_2 \text{ are observed,} \\
K &= \quad \text{only } y_1 \text{ is observed,} \\
L &= \quad \text{only } y_2 \text{ is observed,} \\
M &= \quad \text{both } y_1 \text{ and } y_2 \text{ are missing.}
\end{aligned}
$$

Let n_H, n_K, n_L, n_M represent the size of H, K, L, M, respectively. Assume that the response mechanism does not depend on the value of (y_1, y_2) and so it is MAR. In this case, the observed score function of μ based on a single observation in set K is

$$
\begin{aligned}
E\left\{S_{\mathrm{com}}^{(i)}(\mu) \mid y_{1i}, i \in K\right\} &= \begin{pmatrix} \sigma_{11} & \sigma_{12} \\ \sigma_{12} & \sigma_{22} \end{pmatrix}^{-1} \begin{pmatrix} y_{1i} - \mu_1 \\ E(y_{2i} \mid y_{1i}) - \mu_2 \end{pmatrix} \\
&= \begin{pmatrix} \sigma_{11}^{-1}(y_{1i} - \mu_1) \\ 0 \end{pmatrix}.
\end{aligned}
$$

Similarly, we have

$$E\left\{S_{\mathrm{com}}^{(i)}(\mu) \mid y_{2i}, i \in L\right\} = \begin{pmatrix} 0 \\ \sigma_{22}^{-1}(y_{2i} - \mu_2) \end{pmatrix}.$$

Therefore, the expected information matrix of μ from the observed likelihood is

$$\mathcal{I}_{\text{obs}}(\mu) = n_H \begin{pmatrix} \sigma_{11} & \sigma_{12} \\ \sigma_{12} & \sigma_{22} \end{pmatrix}^{-1} + n_K \begin{pmatrix} \sigma_{11}^{-1} & 0 \\ 0 & 0 \end{pmatrix} + n_L \begin{pmatrix} 0 & 0 \\ 0 & \sigma_{22}^{-1} \end{pmatrix}, \quad (2.35)$$

and the asymptotic variance of the MLE of μ can be obtained by the inverse of $\mathcal{I}_{\text{obs}}(\mu)$. In the special case of $n_L = n_M = 0$,

$$\{\mathcal{I}_{\text{obs}}(\mu)\}^{-1} = \left\{ n_H \begin{pmatrix} \sigma_{11} & \sigma_{12} \\ \sigma_{12} & \sigma_{22} \end{pmatrix}^{-1} + n_K \begin{pmatrix} \sigma_{11}^{-1} & 0 \\ 0 & 0 \end{pmatrix} \right\}^{-1}.$$

Using the following Woodbury matrix identity

$$(A + c\mathbf{b}\mathbf{b}')^{-1} = A^{-1} - A^{-1}\mathbf{b}(c^{-1} + \mathbf{b}'A^{-1}\mathbf{b})^{-1}\mathbf{b}'A^{-1}$$

with

$$A = n_H \begin{pmatrix} \sigma_{11} & \sigma_{12} \\ \sigma_{12} & \sigma_{22} \end{pmatrix}^{-1},$$

$\mathbf{b} = (1,0)'$ *and* $c = n_K \sigma_{11}^{-1}$, *we have* $c^{-1} + \mathbf{b}'A^{-1}\mathbf{b} = (1/n_H + 1/n_K)\sigma_{11}$ *and*

$$\{\mathcal{I}_{\text{obs}}(\mu)\}^{-1} = \frac{1}{n_H} \begin{pmatrix} \sigma_{11} & \sigma_{12} \\ \sigma_{12} & \sigma_{22} \end{pmatrix} + \left(\frac{1}{n} - \frac{1}{n_H}\right) \begin{pmatrix} \sigma_{11} & \sigma_{12} \\ \sigma_{12} & \sigma_{12}^2/\sigma_{11} \end{pmatrix}.$$

Thus, the asymptotic variance of the MLE of μ_1 is equal to σ_{11}/n and the asymptotic variance of the MLE of μ_2 is equal to

$$V(\hat{\mu}_2) \doteq \frac{1}{n}\sigma_{22}\rho^2 + \frac{1}{n_H}(1 - \rho^2)\sigma_{22} = \frac{1}{n_H}\sigma_{22} - \rho^2\sigma_{22}\left(\frac{1}{n_H} - \frac{1}{n}\right),$$

where $\rho = \sigma_{12}/\sqrt{\sigma_{11}\sigma_{22}}$. Note that we obtain

$$V(\hat{\mu}_2) = V(\hat{\mu}_{2,H}) - \rho^2\sigma_{22}\left(\frac{1}{n_H} - \frac{1}{n}\right),$$

where $\hat{\mu}_{2,H}$ is the MLE of μ_2 using sample H only. Thus, by incorporating the partial response, the asymptotic variance is reduced by $\rho^2\sigma_{22}(1/n_H - 1/n)$.

Exercises

1. Show that, for a p-dimensional multivariate normal distribution, $\mathbf{y} \sim N(\mu, \Sigma)$, where $\mu = (\mu_1, \ldots, \mu_p)$ and $\Sigma = (\sigma_{ij})$, the Fisher information of $\theta = (\mu', \sigma_{11}, \sigma_{12}, \ldots, \sigma_{pp})'$ is given by

$$\mathcal{I}(\theta) = \begin{pmatrix} \Sigma^{-1} & \mathbf{0} \\ \mathbf{0} & \frac{1}{2}tr\left(\Sigma^{-1}\frac{\partial\Sigma}{\partial\theta}\Sigma^{-1}\frac{\partial\Sigma}{\partial\theta}\right) \end{pmatrix}.$$

2. Show that for the general exponential family model with log-density of the form
$$\log f(x; \theta) = T(x)\eta(\theta) - A(\theta) + c(x)$$
we have
$$\mathcal{I}(\hat{\theta}) = I(\hat{\theta})$$
where $\hat{\theta}$ is the MLE of θ. If θ is the canonical parameter, then $\mathcal{I}(\theta) = I(\theta)$.

3. Prove (2.7).

4. Assume that $t_1, t_2, \ldots, t_n$ are IID with pdf $f(t; \theta) = \theta e^{-\theta t} I(t > 0)$. Instead of observing t_i, we observe (y_i, δ_i), where
$$y_i = \begin{cases} t_i & \text{if } \delta_i = 1 \\ c & \text{if } \delta_i = 0 \end{cases}$$
and
$$\delta_i = \begin{cases} 1 & \text{if } t_i \le c \\ 0 & \text{if } t_i > c, \end{cases}$$
for some c, the known censoring time. Answer the following questions:

   (a) Obtain the observed likelihood and find the MLE for θ.
   (b) Show that the Fisher information for the observed likelihood is $I(\theta) = \sum_{i=1}^{n} \delta_i / \theta^2$ and the expected information from the observed likelihood is $\mathcal{I}(\theta) = n\left(1 - e^{-\theta c}\right) / \theta^2$.
   (c) For variance estimation, we have two candidates:
$$I(\hat{\theta}) = \sum_{i=1}^{n} \delta_i / \hat{\theta}^2$$
   or
$$\mathcal{I}(\hat{\theta}) = n\left(1 - e^{-\hat{\theta}c}\right) / \hat{\theta}^2.$$
   Discuss which one you prefer. Why?

5. Let $f(y; \theta)$ be the density of a distribution and $\theta' = (\theta_1', \theta_2')'$, where θ_1 is a p_1-dimensional vector and θ_2 is a p_2-dimensional vector. Assume that the expected Fisher information matrix $\mathcal{I}(\theta)$ based on n observations can be written as
$$\mathcal{I}(\theta) = \begin{pmatrix} \mathcal{I}_{11} & \mathcal{I}_{12} \\ \mathcal{I}_{21} & \mathcal{I}_{22} \end{pmatrix} = \begin{pmatrix} E(S_1 S_1') & E(S_1 S_2') \\ E(S_2 S_1') & E(S_2 S_2') \end{pmatrix},$$
where $S_1(\theta) = \partial l(\theta) / \partial \theta_1$ and $S_2(\theta) = \partial l(\theta) / \partial \theta_2$. Thus, the asymptotic variance of $\hat{\theta}_{MLE}$ is $\{\mathcal{I}(\theta)\}^{-1}$.

   (a) Derive the inverse of $\mathcal{I}(\theta)$ using the following steps:

i. Consider the following transformation:

$$\tilde{S}(\theta) \equiv \begin{pmatrix} \tilde{S}_1(\theta) \\ \tilde{S}_2(\theta) \end{pmatrix} = \begin{pmatrix} I & -\mathcal{I}_{12}\mathcal{I}_{22}^{-1} \\ 0 & I \end{pmatrix} \begin{bmatrix} S_1(\theta) \\ S_2(\theta) \end{bmatrix} = TS(\theta).$$

ii. Compute $E(\tilde{S}\tilde{S}')$. In particular, show that $E(\tilde{S}_1\tilde{S}_1') = \mathcal{I}_{11} - \mathcal{I}_{12}\mathcal{I}_{22}^{-1}\mathcal{I}_{21}$

iii. Find the inverse of $\mathcal{I}(\theta) = E(SS') = T^{-1}E(\tilde{S}\tilde{S}')(T^{-1})'$.

(b) Show that the asymptotic variance of the MLE of θ_1 (when θ_2 is unknown) is $(\mathcal{I}_{11} - \mathcal{I}_{12}\mathcal{I}_{22}^{-1}\mathcal{I}_{21})^{-1}$.

(c) Note that the asymptotic variance of the MLE of θ_1 when θ_2 is known is $\mathcal{I}_{11}^{-1}$. Compare the two variances. Which one is smaller and why?

6. Let $(x_1, y_1)', \ldots, (x_n, y_n)'$ be a random sample from a bivariate normal distribution with mean $(\mu_x, \mu_y)'$ and variance-covariance matrix $\Sigma = \begin{pmatrix} \sigma_x^2 & \sigma_{xy} \\ \sigma_{xy} & \sigma_y^2 \end{pmatrix}$. We are interested in estimating $\theta = \mu_y$.

(a) Assuming that parameters $\mu_x, \sigma_x^2, \sigma_{xy}, \sigma_y^2$ are known, find the MLE of θ and compute its variance.

(b) If the other parameters are also unknown, derive the MLE of θ. Is the estimator here or the estimator from in (a) more efficient? Explain.

7. Consider a bivariate random variable (Y_1, Y_2), where

$$(Y_1, Y_2) = \begin{cases} (1,1) & \text{with probability } \pi_{11} \\ (1,0) & \text{with probability } \pi_{10} \\ (0,1) & \text{with probability } \pi_{01} \\ (0,0) & \text{with probability } \pi_{00} \end{cases}$$

with $\pi_{00} + \pi_{01} + \pi_{10} + \pi_{11} = 1$. To answer the following questions, it may be helpful to define $\pi_{1+} = P(Y_1 = 1)$, $\pi_{1|1} = P(Y_2 = 1 \mid Y_1 = 1)$ and $\pi_{1|0} = P(Y_2 = 1 \mid Y_1 = 0)$. Note that there is one-to-one correspondence between $\theta_1 = (\pi_{00}, \pi_{01}, \pi_{10})$ and $\theta_2 = (\pi_{1+}, \pi_{1|1}, \pi_{1|0})$. The realized sample observations are presented in Table 2.1.

(a) Compute the observed likelihood and score functions in terms of θ_2.

(b) Obtain the maximum likelihood estimates for θ_1.

(c) Obtain the observed information matrix for θ_1.

8. Assume that $(x_i, y_i)'$, $i = 1, 2, \ldots, n$, are n independent realizations of the random variable $(X, Y)'$ whose distribution follows a canonical exponential family model with log-density of the form

$$\log f(x, y; \theta) = \theta' T(x, y) - A(\theta) + c(x, y).$$

TABLE 2.1

A 2×2 Table with a Supplemental Margin for y_1

Set	y_1	y_2	Count
	1	1	100
H	1	0	50
	0	1	75
	0	0	75
K	1		40
	0		60

(a) Show that the complete-sample score function for θ is

$$S_c(\theta) = \sum_{i=1}^{n} \{T(x_i, y_i) - E(T)\}$$

and the complete-sample Fisher information can be estimated by

$$\sum_{i=1}^{n} (T_i - \bar{T}_n) (T_i - \bar{T}_n)',$$

where $T_i = T(x_i, y_i)$ and $\bar{T}_n = n^{-1} \sum_{i=1}^{n} T_i$.

(b) Suppose that x_i are always observed and y_i are subject to missingness. Assume that y_i is observed for the first r observations and the remaining $n - r$ elements are missing in y. Show that the expected information associated with the observed likelihood is equal to

$$\mathcal{I}_{\text{obs}}(\theta) = nV(T) - (n - r)E\{V(T \mid X)\}.$$

Discuss the missing information principle in this setup.

9. Consider the following regression model

$$y_i = \beta_0 + \beta_1 x_i + e_i,$$

where $x_i \sim N(\mu_x, \sigma_x^2)$, $e_i \sim N(0, \sigma_e^2)$, and e_i is independent of x_i. In addition, a surrogate measurement of x_i, denoted by z_i, is available with distribution

$$z_i = x_i + u_i$$

with $u_i \sim N(0, \sigma_u^2)$. This is often called *measurement error model*. The surrogate variable z_i is conditionally independent of y_i given x_i. That is, we have $f(z \mid x, y) = f(z \mid x)$. Assume that $\mu_x, \sigma_x^2, \sigma_u^2$ are known. Throughout the sample, we observe z_i and y_i. We are interested in estimating $\theta = (\beta_0, \beta_1, \sigma_e^2)$. Answer the following questions.

(a) Under no measurement error ($\sigma_u^2 \equiv 0$), obtain the complete-data likelihood function and its score functions.

(b) Under the existence of measurement error, obtain the observed likelihood function.

(c) Use Bayes' theorem or other technique to derive the conditional distribution of x_i given z_i and y_i.

(d) Compute the mean score function for θ.

10. Under the setup of Example 2.6, use

$$(A+D)^{-1} = A^{-1} - A^{-1}(D^{-1} + A^{-1})^{-1}A^{-1},$$

where A and D are invertible matrices of the same dimension, to compute the inverse of $\mathcal{I}_{\text{obs}}(\mu)$ in (2.35) in terms of $\sigma_{11}, \sigma_{12}, \sigma_{22}$ and n_H, n_K, n_L. Discuss the efficiency gain of the MLEs of μ_1 and μ_2 compared to the case of using only set H.

3

Computation

Maximum likelihood estimation plays a central role in statistical inference. The actual computation to obtain maximum likelihood estimators (MLE) can be challenging in many situations, especially in missing data problems. We first review some of the popular methods of computing maximum likelihood estimators and some issues associated with the computation. EM algorithm, introduced in Section 3.3, is a fundamental computational tool for finding the MLE under missing data. Monte Carlo methods for computing the conditional expectation are also introduced.

3.1 Introduction

In principle, we are interested in finding the solution

$$\hat{\theta} = \arg\max_{\theta} L(\theta).$$

The MLE satisfies the score equation given by

$$S(\theta) \equiv \frac{\partial}{\partial \theta} \log L(\theta) = 0, \tag{3.1}$$

which is generally a system of nonlinear equations. In most cases, the MLE can be obtained by solving the score equation (3.1).

To discuss the computational approaches of solving the score equation (3.1), we first review some methods of solving $g(\theta) = 0$ for θ. If $g(\theta)$ is a scalar function of θ, then the solution to $g(\theta) = 0$ can be easily obtained by the bisection method, which is based on the intermediate value theorem: If g is continuous for all θ in the interval $g(\theta_1)g(\theta_2) < 0$, then a root of $g(\theta)$ lies in the interval (θ_1, θ_2). To describe the bisection method, let (a_0, b_0) be an interval in the domain of $g(\theta)$ such that $g(a_0)g(b_0) < 0$. Hence, the solution θ^* to $g(\theta) = 0$ lies in $[a_0, b_0]$. The bisection method can find the root θ^* for $g(\theta) = 0$ by the following iterative steps:

[Step 1] Set $x_t = (a_t + b_t)/2$.

[Step 2] Evaluate the signs of $g(a_t)g(x_t)$ and $g(x_t)g(b_t)$. If $g(a_t)g(x_t) < 0$ then set $a_{t+1} = a_t$ and $b_{t+1} = x_t$. If $g(b_t)g(x_t) < 0$ then set $a_{t+1} = x_t$ and $b_{t+1} = b_t$.

DOI: 10.1201/9780429321740-3

[Step 3] Stop if the convergence criterion is met. Otherwise, increase t by one and go to step 1.

For the convergence criterion, we can use $|a_{t+1} - b_{t+1}| < \epsilon$ for a sufficiently small $\epsilon > 0$. The bisection method is easy to execute and does not require computing the derivatives. The bisection method is an example of the bracketing method, which can be roughly described as finding a root within a sequence of nested intervals of decreasing length.

Also popular is Newton's method, or the Newton–Raphson method. It uses a linear approximation of $g(\theta)$ at $\theta^{(t)}$

$$g(\theta) \cong g(\theta^{(t)}) + \left\{ \partial g(\theta^{(t)})/\partial\theta' \right\} \left(\theta - \theta^{(t)} \right).$$

We can next make a mental transformation to change $g(\theta)$ to 0, and θ to $\theta^{(t+1)}$. Thus, Newton's method for finding the solution $\hat{\theta}$ to $g(\theta) = 0$ can be described as

$$\theta^{(t+1)} = \theta^{(t)} - \left\{ \partial g(\theta^{(t)})/\partial\theta' \right\}^{-1} g\left(\theta^{(t)}\right). \tag{3.2}$$

Figure 3.1 gives a graphical illustration of the sequence of $x^{(t)}$ obtained from Newton's method. By approximating

$$f(x) \cong f(x^{(t)}) + f'(x^{(t)}) \left(x - x^{(t)} \right) := f_l^{(t)}(x),$$

we can find the solution to $f_l^{(t)}(x) = 0$ easily and iteratively find the solution to $f(x) = 0$.

If $g(\theta)$ is equal to $S(\theta)$, the score function for θ, then (3.2) can be written as

$$\theta^{(t+1)} = \theta^{(t)} + \left[I\left(\theta^{(t)}\right) \right]^{-1} S\left(\theta^{(t)}\right). \tag{3.3}$$

This is the formula for the scoring method. The scoring method can be modified to guarantee that the sequence $\{\theta^{(t)} : t = 1, 2, \ldots\}$ increases the likelihood:

$$\theta^{(t+1)} = \theta^{(t)} + \alpha \left[I\left(\theta^{(t)}\right) \right]^{-1} S\left(\theta^{(t)}\right)$$

for $\alpha \in (0,1]$. If $L(\hat{\theta}^{(t+1)}) < L(\hat{\theta}^{(t)})$, then use $\alpha = \alpha/2$ and compute $\theta^{(t+1)}$ again. Such modification is often called the *ascent method*.

In the scoring method, the behavior of $I(\theta^{(t)})$ can be problematic if $\theta^{(t)}$ is far from the MLE $\hat{\theta}$. Thus, instead of using the observed Fisher information $I(\theta)$ in (3.3), we can use the expected Fisher information to get

$$\theta^{(t+1)} = \theta^{(t)} + \left[\mathcal{I}\left(\theta^{(t)}\right) \right]^{-1} S\left(\theta^{(t)}\right). \tag{3.4}$$

This algorithm is called the *Fisher scoring method*. Generally speaking, the Fisher scoring method can expect rapid computational improvement at the beginning, while Newton's method works better for refinement near the end (Givens and Hoeting, 2005).

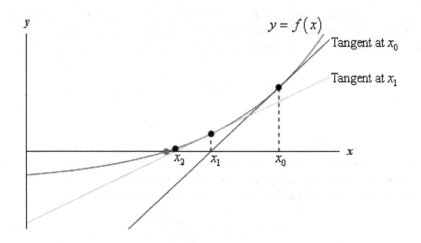

FIGURE 3.1
A graphical illustration of Newton's method for finding the solution to $f(x) = 0$.

Example 3.1. *Consider the logistic regression model*

$$y_i \overset{i.i.d.}{\sim} Bernoulli\,(p_i)$$

with

$$logit(p_i) = \log\left(\frac{p_i}{1-p_i}\right) = \mathbf{x}_i'\beta.$$

The log-likelihood function for β is

$$
\begin{aligned}
l(\beta) &= \sum_{i=1}^{n} \{y_i \log\,(p_i) + (1 - y_i)\log\,(1-p_i)\} \\
&= \sum_{i=1}^{n} \{y_i\,(\mathbf{x}_i'\beta) - \log\,(1+\exp(\mathbf{x}_i'\beta))\}
\end{aligned}
$$

The score function is given by

$$S(\beta) = \frac{\partial}{\partial \beta} l\,(\beta) = \sum_{i=1}^{n} \{y_i - p_i(\beta)\}\,\mathbf{x}_i \tag{3.5}$$

and the Hessian is given by

$$I(\beta) = -\frac{\partial}{\partial \beta'} S\,(\beta) = \sum_{i=1}^{n} p_i(\beta)\,\{1 - p_i(\beta)\}\,\mathbf{x}_i\mathbf{x}_i'.$$

Because the Hessian does not depend on the y-values, the expected Fisher *information evaluated at $\hat{\theta}$ (MLE) is equal to the observed Fisher information. Hence, the Fisher scoring method is equal to the scoring method. The Fisher scoring method can be expressed as*

$$\beta^{(t+1)} = \beta^{(t)} + \left[\sum_{i=1}^{n} p_i^{(t)} (1 - p_i^{(t)}) \mathbf{x}_i \mathbf{x}_i' \right]^{-1} \sum_{i=1}^{n} (y_i - p_i^{(t)}) \mathbf{x}_i$$

where $p_i^{(t)} = p_i(\beta^{(t)})$.

We now discuss the convergence properties of Newton's method. For simplicity, we consider only a scalar parameter θ and a scalar equation, $g(\theta) = 0$. The updating equation for Newton's method can be written as

$$\theta^{(t+1)} = \theta^{(t)} - \frac{g(\theta^{(t)})}{g'(\theta^{(t)})}. \tag{3.6}$$

The convergence of Newton's method depends on the shape of g and the starting value in the iteration. To discuss the convergence properties of Newton's method for $g(\theta) = 0$, suppose that g is twice differentiable with continuous second-order derivatives satisfying $g''(\theta^*) \neq 0$, where θ^* is a solution to $g(\theta) = 0$. A Taylor expansion of $g(\theta^*) = 0$ around $\theta^{(t)}$ leads to

$$0 = g(\theta^*) \quad = \quad g(\theta^{(t)}) + g'(\theta^{(t)}) \left(\theta^* - \theta^{(t)} \right) + 0.5 g''(q) \left(\theta^* - \theta^{(t)} \right)^2,$$

where q is between θ^* and $\theta^{(t)}$. Multiplying both sides of the above equation by $\{g'(\theta^{(t)})\}^{-1}$ and using (3.6), we have

$$\epsilon^{(t+1)} = \left(\epsilon^{(t)} \right)^2 \frac{g''(q)}{2 g'(\theta^{(t)})}, \tag{3.7}$$

where $\epsilon^{(t)} = \theta^{(t)} - \theta^*$. Now, consider a neighborhood of θ^*, $B_\delta(\theta^*) = \{\theta : |\theta - \theta^*| < \delta\}$ for $\delta > 0$. Define

$$c(\delta) = \max_{\theta_1, \theta_2 \in B_\delta(\theta^*)} \left| \frac{g''(\theta_1)}{2 g'(\theta_2)} \right|.$$

Since $c(\delta) \to |g''(\theta^*)/\{2g'(\theta^*)\}|$ as $\delta \to 0$, it follows that $\delta c(\delta) \to 0$ as $\delta \to 0$, so $\delta c(\delta) < 1$ for some $\delta > 0$. For $\theta^{(t)} \in B_\delta(\theta^*)$, by (3.7),

$$\left| c(\delta) \epsilon^{(t+1)} \right| \leq \left\{ c(\delta) \epsilon^{(t)} \right\}^2,$$

which implies

$$\left| c(\delta) \epsilon^{(t+1)} \right| \leq \left\{ c(\delta) \epsilon^{(1)} \right\}^{2^t}. \tag{3.8}$$

Thus, if the initial value $\theta^{(1)}$ is chosen to satisfy $\left|\epsilon^{(1)}\right| < \delta$, then (3.8) implies

$$\left|\epsilon^{(t+1)}\right| \leq \frac{\{c(\delta)\delta\}^{2^t}}{c(\delta)},$$

which converges to zero as $t \to \infty$, because $|c(\delta)\delta| < 1$. Therefore, if g'' is continuous and θ^* is a root of $g(\theta) = 0$, then there exists a neighborhood of x^* for which Newton's method converges to θ^* if $\theta^{(0)}$ is in that neighborhood. We formalize the above result in the following theorem.

Theorem 3.1. *Let $g : (a,b) \to R$ be a differentiable function obeying the following conditions:*

  1. g is twice differentiable with continuous g''.

  2. $|g'|$ is bounded away from 0 on (a,b), and $\inf_{\theta \in (a,b)} |g'(\theta)| > 0$.

  3. There exists a point $\theta^* \in (a,b)$ such that $g(\theta^*) = 0$.

Then, there is $\delta > 0$ such that whenever the initial value $\theta^{(0)} \in (\theta^* - \delta, \theta^* + \delta)$, the sequence $\{\theta^{(t)}\}$ of iterations produced by Newton's method lies in $(\theta^* - \delta, \theta^* + \delta)$, and $\theta^{(t)} \to \theta^*$ as $t \to \infty$.

Theorem 3.1 is a local convergence theorem. It establishes the existence of an interval $(\theta^* - \delta, \theta^* + \delta)$ in which Newton's method converges but gives very little information about that interval. In application, one often wants a global convergence theorem in the sense that the iterative scheme converges regardless of the starting point. The simplest version of the global convergence theorem is as follows: If a convex function g has a root and is twice continuously differentiable, then Newton's method converges to the root from any starting point. For details, see Allen and Issacson (1998).

We now discuss the order of convergence of the iterative solutions.

Definition 3.1. *A sequence $\{\theta^{(t)}\}$ that converges to θ^* is of order p if*

$$\lim_{t \to \infty} \|\theta^{(t)} - \theta^*\| = 0$$

and

$$\lim_{t \to \infty} \frac{\|\theta^{(t+1)} - \theta^*\|}{\|\theta^{(t)} - \theta^*\|^p} = c$$

for some constant $c \neq 0$.

For Newton's method, by (3.7), we have

$$\frac{\theta^{(t+1)} - \theta^*}{\left(\theta^{(t)} - \theta^*\right)^2} = \frac{g''(q)}{2g'\left(\theta^{(t)}\right)}.$$

Then, if Newton's method converges, q also converges to θ^* and

$$\lim_{t\to\infty} \frac{\|\theta^{(t+1)} - \theta^*\|}{\|\theta^{(t)} - \theta^*\|^2} = \left| \frac{g''(\theta^*)}{2g'(\theta^*)} \right| \neq 0.$$

Thus, its convergence is of quadratic order, i.e. $p = 2$. A quadratic order convergence is quite fast. This is regarded as the major strength of Newton's method. Another advantage of Newton's method is that the one-step estimator with a $\sqrt{n}$-consistent initial point is asymptotically efficient.

Remark 3.1. *Newton's method requires computing the Hessian and also updating the Hessian for each iteration, which can be at a huge computational cost, specially when the dimension of θ is large. Instead of (3.2), an alternative iterative method of obtaining $\theta^{(t)}$, called the Quasi-Newton method, can be expressed as*

$$\theta^{(t+1)} = \theta^{(t)} - \left(M^{(t)} \right)^{-1} S(\theta^{(t)}), \tag{3.9}$$

where $M^{(t)}$ is a $p \times p$ matrix approximating the Hessian evaluated at $\theta^{(t)}$. The choice of $M^{(t)} = -\mathcal{I}(\theta^{(t)})$ leads to the Fisher scoring method in (3.3). The secant method finds $M^{(t)}$ by solving

$$S(\theta^{(t)}) - S(\theta^{(t-1)}) = \left(M^{(t)} \right) \left(\theta^{(t)} - \theta^{(t-1)} \right)$$

or even set $M^{(t)}$ to be M, where M is sufficiently close to the Hessian matrix at the solution, $-I(\theta^*)$. In this case, the quadratic convergence of the Newton's method is lost. It can be shown that under some reasonable assumptions, the convergence rate is superlinear in the sense that $\|\theta^{(t+1)} - \theta^*\| < h_t \|\theta^{(t)} - \theta^*\|$ for some h_t that converges to zero as $t \to \infty$. In the IID case, $S(\theta) = \sum_{i=1}^{n} S_i(\theta)$, Redner and Walker (1984) and Meilijson (1989) have suggested using $M^{(t)} = -\hat{H}(\theta^{(t)})$ in (3.9), where

$$\hat{H}(\theta) = \sum_{i=1}^{n} \left\{ S_i(\theta) - n^{-1} \sum_{j=1}^{n} S_j(\theta) \right\}^{\otimes 2}. \tag{3.10}$$

Note that $\hat{H}(\theta)$ is consistent for $\mathcal{I}(\theta) = V_\theta\{S(\theta)\}$ and $\hat{H}(\theta)$ is very easy to compute. The convergence rate is known to be linear.

We now discuss direct computation under the existence of missing data. Under the setup of Section 2.4, the MLE that maximizes the observed likelihood can be obtained as a solution to the observed score equation given by

$$S_{\text{obs}}(\theta) = 0, \tag{3.11}$$

where $S_{\text{obs}}(\theta) = \partial \ln L_{\text{obs}}(\theta)/\partial\theta$. The following example illustrates the direct computation method with missing data.

Example 3.2. *Consider the following bivariate normal distribution*

$$\left(\begin{array}{c} X_i \\ Y_i \end{array} \right) \sim N \left[\left(\begin{array}{c} \mu_x \\ \mu_y \end{array} \right), \left(\begin{array}{cc} \sigma_{xx} & \sigma_{xy} \\ \sigma_{xy} & \sigma_{yy} \end{array} \right) \right], \tag{3.12}$$

where $\sigma_{xx}, \sigma_{xy}, \sigma_{yy}$ are known constants. Suppose that we have complete response in the first $r(< n)$ units $\{(x_i, y_i) : i = 1, 2, \ldots, r\}$ and $n - r$ partial responses from the remaining $n - r$ units $\{x_i : i = r+1, r+2, \ldots, n\}$. In this case, under MAR, the observed score function for $\mu = (\mu_x, \mu_y)'$ can be written as

$$\bar{S}(\mu) = \sum_{i=1}^{r} \left(\begin{array}{cc} \sigma_{xx} & \sigma_{xy} \\ \sigma_{xy} & \sigma_{yy} \end{array} \right)^{-1} \left(\begin{array}{c} x_i - \mu_x \\ y_i - \mu_y \end{array} \right)$$

$$+ \sum_{i=r+1}^{n} \left(\begin{array}{cc} \sigma_{xx} & \sigma_{xy} \\ \sigma_{xy} & \sigma_{yy} \end{array} \right)^{-1} \left(\begin{array}{c} x_i - \mu_x \\ E(y_i \mid x_i) - \mu_y \end{array} \right),$$

where $E(y_i \mid x_i) = \mu_y + (\sigma_{xy}/\sigma_{xx})(x_i - \mu_x)$. The solution to the observed score equation is

$$\hat{\mu}_x = \bar{x}_n \equiv \frac{1}{n} \sum_{i=1}^{n} x_i$$

and

$$\hat{\mu}_y = \bar{y}_r + \frac{\sigma_{xy}}{\sigma_{xx}} (\bar{x}_n - \bar{x}_r)$$

where $(\bar{x}_r, \bar{y}_r) = r^{-1} \sum_{i=1}^{r} (x_i, y_i)$. The variance of $\hat{\mu}_y$ is equal to $\sigma_{yy}\rho^2/n + (1 - \rho^2)\sigma_{yy}/r$, which is consistent with the findings of Example 2.6.

In the above example, the mean score equations are linear and so the computation is straightforward. In general, the mean score equations are nonlinear and we often rely on iterative computation methods. The Fisher-scoring method applied to (3.11) can be written as

$$\hat{\theta}^{(t+1)} = \hat{\theta}^{(t)} + \left\{ \mathcal{I}_{\text{obs}} \left(\hat{\theta}^{(t)} \right) \right\}^{-1} S_{\text{obs}} \left(\hat{\theta}^{(t)} \right). \tag{3.13}$$

Under the IID setup, we can use the empirical Fisher information $\hat{H}(\theta)$ in (3.10) as an estimator of the expected Fisher information $\mathcal{I}_{\text{obs}}(\theta)$.

Example 3.3. *Consider the following normal-theory mixed effect model*

$$y_{ij} = \mathbf{x}'_{ij}\beta + u_i + e_{ij}, \quad u_i \overset{i.i.d.}{\sim} N(0, \sigma_u^2), \quad e_{ij} \overset{i.i.d.}{\sim} N(0, \sigma_e^2)$$

for $i = 1, \ldots, m; j = 1, \ldots, n_i$, and e_{ij} are independent of u_i. We observe $(\mathbf{x}_{ij}, y_{ij})$ for each unit j in cluster i. In this case, the cluster-specific effect u_i can be treated as missing data. The complete sample likelihood for $\theta = (\beta, \sigma_u^2, \sigma_e^2)$ is

$$L_{\text{com}}(\theta) = \prod_{i=1}^{m} \left[\prod_{j=1}^{n_i} \left\{ \frac{1}{\sigma_e} \phi \left(\frac{y_{ij} - \mathbf{x}'_{ij}\beta - u_i}{\sigma_e} \right) \right\} \frac{1}{\sigma_u} \phi \left(\frac{u_i}{\sigma_u} \right) \right]$$

where $\phi(x) = (2\pi)^{-1/2}\exp(-x^2/2)$ is the probability density function of the standard normal distribution. The complete-sample score functions are

$$S_{\text{com},1}(\theta) \equiv \partial\log\{L_{\text{com}}(\theta)\}/\partial\beta = \sum_{i=1}^{m}\sum_{j=1}^{n_i}(y_{ij} - \mathbf{x}'_{ij}\beta - u_i)\,\mathbf{x}_{ij}/\sigma_e^2$$

$$S_{\text{com},2}(\theta) \equiv \partial\log\{L_{\text{com}}(\theta)\}/\partial\sigma_e^2 = \frac{1}{2\sigma_e^4}\sum_{i=1}^{m}\sum_{j=1}^{n_i}\{(y_{ij} - \mathbf{x}'_{ij}\beta - u_i)^2 - \sigma_e^2\}$$

$$S_{\text{com},3}(\theta) \equiv \partial\log\{L_{\text{com}}(\theta)\}/\partial\sigma_u^2 = \frac{1}{2\sigma_u^4}\sum_{i=1}^{m}\sum_{j=1}^{n_i}(u_i^2 - \sigma_u^2).$$

Using Bayes formula, we can obtain

$$f(u_i \mid \mathbf{x}_i, \mathbf{y}_i) = \frac{f_1(\mathbf{y}_i \mid \mathbf{x}_i, u_i; \theta_1)f_2(u_i \mid \mathbf{x}_i; \theta_2)}{\int f_1(\mathbf{y}_i \mid \mathbf{x}_i, u_i; \theta_1)f_2(u_i \mid \mathbf{x}_i; \theta_2)du_i},$$

where

$$f_1(\mathbf{y}_i \mid \mathbf{x}_i, u_i; \theta_1) = \prod_{j=1}^{n_i}\left\{\frac{1}{\sigma_e}\phi\left(\frac{y_{ij} - \mathbf{x}'_{ij}\beta - u_i}{\sigma_e}\right)\right\}$$

and $f_2(u_i \mid \mathbf{x}_i; \theta_2) = \sigma_u^{-1}\phi(u_i/\sigma_u)$. Thus, we obtain

$$u_i \mid (\mathbf{x}_i, \mathbf{y}_i) \sim N\left(\tau_i\left(\bar{y}_i - \bar{\mathbf{x}}'_i\beta\right), \sigma_u^2(1 - \tau_i)\right), \tag{3.14}$$

where $\tau_i = \sigma_u^2/(\sigma_u^2 + \sigma_e^2/n_i)$ and $(\bar{\mathbf{x}}_i, \bar{y}_i) = n_i^{-1}\sum_{j=1}^{n_i}(\mathbf{x}_{ij}, y_{ij})$. Recall that the observed score function can be computed by taking the conditional expectation of the complete-sample score functions with respect to the conditional distribution in (3.14). Thus, the observed score equations can be expressed as

$$\bar{S}_1(\beta) = \sum_{i=1}^{m}\sum_{j=1}^{n_i}\left\{y_{ij} - \mathbf{x}'_{ij}\beta - \tau_i\left(\bar{y}_i - \bar{\mathbf{x}}'_i\beta\right)\right\}\mathbf{x}_{ij}/\sigma_e^2 = 0$$

$$\bar{S}_2(\sigma_e^2) = \frac{1}{2\sigma_e^4}\sum_{i=1}^{m}\sum_{j=1}^{n_i}\left[\left\{y_{ij} - \mathbf{x}'_{ij}\beta - \tau_i\left(\bar{y}_i - \bar{\mathbf{x}}'_i\beta\right)\right\}^2 - (1 - \tau_i/n_i)\sigma_e^2\right] = 0$$

$$\bar{S}_3(\sigma_u^2) = \frac{1}{2\sigma_u^4}\sum_{i=1}^{m}\left\{\tau_i^2\left(\bar{y}_i - \bar{\mathbf{x}}'_i\beta\right)^2 - \tau_i\sigma_u^2\right\} = 0.$$

Now, we can obtain that

$$\sum_{j=1}^{n_i}\left\{y_{ij} - \mathbf{x}'_{ij}\beta - \tau_i\left(\bar{y}_i - \bar{\mathbf{x}}'_i\beta\right)\right\}\mathbf{x}_{ij} = \sum_{j=1}^{n_i}\left\{y_{ij} - \mathbf{x}'_{ij}\beta - c_i\left(\bar{y}_i - \bar{\mathbf{x}}'_i\beta\right)\right\}\left(\mathbf{x}_{ij} - c_i\bar{\mathbf{x}}_{ij}\right)$$

where

$$c_i = 1 - \sqrt{1 - \tau_i} = 1 - \left(\frac{\sigma_e^2/n_i}{\sigma_u^2 + \sigma_e^2/n_i}\right)^{1/2}.$$

If τ_i are known, the resulting $\hat{\beta}$ is obtained by regressing $y_{ij} - c_i \bar{y}_i$ on $(\mathbf{x}_{ij} - c_i \bar{\mathbf{x}}_i)$. Also, if β is known, $\hat{\sigma}_e^2$ and $\hat{\sigma}_u^2$ are easily estimated by solving $\bar{S}_2(\sigma_e^2) = 0$ and $\bar{S}_3(\sigma_u^2) = 0$. Thus, we can solve these equations iteratively until convergence. Fuller and Battese (1973) obtained the same result from the estimated generalized least square method.

3.2 Factoring Likelihood Approach

If the missing mechanism is MAR and the responses are monotone in the sense that the set of respondents for one variable is a proper subset of the set of respondents for another variable, then the computation for the MLE can be simplified using the factoring likelihood approach. To simplify the presentation, consider a bivariate random variable (X, Y) with density $f(x, y; \theta)$. Assume that x_i are completely observed and y_i are subject to missingness. Without loss of generality, assume that y_i are observed only for the first $r < n$ elements.

In this case, assuming MAR, the observed likelihood for θ can be written

$$L_{\text{obs}}(\theta) = \prod_{i=1}^{r} f(x_i, y_i; \theta) \prod_{i=r+1}^{n} \int f(x_i, y_i; \theta) \, dy_i. \tag{3.15}$$

Let $f_X(x; \theta_1)$ be the marginal density of X and $f_{Y|X}(y \mid x; \theta_2)$ be the density for the conditional distribution of Y given $X = x$. Since $f(x_i, y_i; \theta) = f_X(x_i; \theta_1) f_{Y|X}(y_i \mid x_i; \theta_2)$, we can rearrange (3.15) as

$$L_{\text{obs}}(\theta) = \prod_{i=1}^{n} f_X(x_i; \theta_1) \prod_{i=1}^{r} f_{Y|X}(y_i \mid x_i; \theta_2). \tag{3.16}$$

When the observed likelihood is written as (3.16), note that we can write

$$L_{\text{obs}}(\theta) = L_1(\theta_1) \times L_2(\theta_2), \tag{3.17}$$

and the MLEs for each parameter can be obtained by separately maximizing the corresponding likelihood. If parameters θ_1 and θ_2 satisfies (3.17), then the two parameters are called *orthogonal*. If the two parameters, θ_1 and θ_2 are orthogonal, the information matrix for $\theta = (\theta_1, \theta_2)$ becomes a block-diagonal and the MLE of θ_1 is independent of the MLE of θ_2, at least asymptotically (Cox and Reid, 1987).

Example 3.4. *Consider the same setup as in Example 3.2, except that σ_{xx}, σ_{xy}, and σ_{yy} are also unknown parameters. There are now five parameters to identify the distribution. The observed likelihood for $\theta = (\mu_x, \mu_y, \sigma_{xx}, \sigma_{xy}, \sigma_{yy})$ is*

$$L_{\text{obs}}(\theta) = \prod_{i=1}^{r} f(x_i, y_i; \mu_x, \mu_y, \sigma_{xx}, \sigma_{xy}, \sigma_{yy}) \times \prod_{i=r+1}^{n} f(x_i; \mu_x, \sigma_{xx})$$

Finding the MLE of θ would require an iterative computation method. An alternative parametrization is

$$
\begin{aligned}
X_i &\sim N(\mu_x, \sigma_{xx}), \\
Y_i \mid X_i = x &\sim N(\beta_0 + \beta_1 x, \sigma_{ee}),
\end{aligned}
\tag{3.18}
$$

where $\beta_1 = \sigma_{xy}/\sigma_{xx}$, $\beta_0 = \mu_y - \beta_1 \mu_x$, and $\sigma_{ee} = \sigma_{yy} - \sigma_{xy}^2/\sigma_{xx}$. Under this new parametrization,

$$
\begin{aligned}
L_{\text{obs}}(\theta) &= \prod_{i=1}^{n} f(x_i; \mu_x, \sigma_{xx}) \times \prod_{i=1}^{r} f(y_i \mid x_i; \beta_0, \beta_1, \sigma_{ee}) \\
&= L_1(\mu_x, \sigma_{xx}) \times L_2(\beta_0, \beta_1, \sigma_{ee})
\end{aligned}
$$

and the two parameters, $\theta_1 = (\mu_x, \sigma_{xx})$ and $\theta_2 = (\beta_0, \beta_1, \sigma_{ee})$, are orthogonal in the sense of satisfying (3.17). The MLEs are obtained by maximizing each component of $L_{\text{obs}}(\theta)$ separately, which is given by

$$
\begin{aligned}
\hat{\mu}_x &= \bar{x}_n, \\
\hat{\sigma}_{xx} &= S_{xxn},
\end{aligned}
$$

and

$$
\begin{aligned}
\hat{\beta}_1 &= S_{xyr}/S_{xxr}, \\
\hat{\beta}_0 &= \bar{y}_r - \hat{\beta}_1 \bar{x}_r, \\
\hat{\sigma}_{ee} &= S_{yyr} - S_{xyr}^2/S_{xxr},
\end{aligned}
$$

where the subscript r denotes that the statistics are computed from the r complete respondents only, and the subscript n denotes that the statistics are computed from the whole sample of size n. Thus, the MLEs for the original parametrization are

$$
\begin{aligned}
\hat{\mu}_y &= \hat{\beta}_0 + \hat{\beta}_1 \hat{\mu}_x = \bar{y}_r + \hat{\beta}_1(\hat{\mu}_x - \bar{x}_r), \\
\hat{\sigma}_{yy} &= S_{yyr} + \hat{\beta}_1^2(\hat{\sigma}_{xx} - S_{xxr}), \\
\hat{\sigma}_{xy} &= S_{xyr} \frac{\hat{\sigma}_{xx}}{S_{xxr}}.
\end{aligned}
\tag{3.19}
$$

Furthermore, we can compute

$$\hat{\rho} = r_{xy} \times \left(\frac{\hat{\sigma}_{xx}}{S_{xxr}}\right)^{1/2} \left(\frac{\hat{\sigma}_{yy}}{S_{yyr}}\right)^{-1/2},$$

where $r_{xy} = S_{xyr}/(S_{xxr}S_{yyr})^{1/2}$. The MLE of μ_y, $\hat{\mu}_y$ in (3.19), is called the regression estimator and is very popular in sample surveys.

Example 3.5. *We consider the setup of Exercise 7 in Chapter 2. Suppose that we have complete response in the first $r(< n)$ units $\{(y_{1i}, y_{2i}) : i = 1, 2, \ldots, r\}$ and $n - r$ partial responses from the remaining $n - r$ units $\{y_{1i} : i = r+1, r+2, \ldots, n\}$. The observed likelihood can be written as*

$$L_{\text{obs}}(\theta_2) = \prod_{i=1}^{n} \pi_{1+}^{y_{1i}} (1 - \pi_{1+})^{1-y_{1i}}$$

$$\times \prod_{i=1}^{r} \left\{ \pi_{1|1}^{y_{2i}} (1 - \pi_{1|1})^{1-y_{2i}} \right\}^{y_{1i}} \left\{ \pi_{1|0}^{y_{2i}} (1 - \pi_{1|0})^{1-y_{2i}} \right\}^{1-y_{1i}}.$$

Because we can write

$$L_{\text{obs}}(\pi_{1+}, \pi_{1|1}, \pi_{1|0}) = L_1(\pi_{1+}) L_2(\pi_{1|1}) L_3(\pi_{1|0})$$

for some $L_1(\cdot)$, $L_2(\cdot)$, and $L_3(\cdot)$, we can obtain the MLE by separately maximizing each likelihood component. Thus, we have

$$\hat{\pi}_{1+} = \frac{1}{n} \sum_{i=1}^{n} y_{1i},$$

$$\hat{\pi}_{1|1} = \frac{\sum_{i=1}^{r} y_{1i} y_{2i}}{\sum_{i=1}^{r} y_{1i}},$$

$$\hat{\pi}_{1|0} = \frac{\sum_{i=1}^{r} (1 - y_{1i}) y_{2i}}{\sum_{i=1}^{r} (1 - y_{1i})}.$$

The MLE for π_{ij} can then be obtained by $\hat{\pi}_{ij} = \hat{\pi}_{i+} \hat{\pi}_{j|i}$ for $i = 0, 1$ and $j = 0, 1$.

The factoring likelihood approach was first proposed by Anderson (1957), and further discussed by Rubin (1974). It is particularly useful for the monotone missing pattern, where we can relabel the variable in such a way that the set of respondents for each variable is monotonely decreasing:

$$R_1 \supset R_2 \supset \ldots \supset R_p,$$

where R_i denotes the set of respondents for Y_i after relabeling. In this case, under MAR, the observed likelihood can be written as

$$L_{\text{obs}}(\theta) = \prod_{i \in R_1} f(y_{1i}; \theta_1) \times \prod_{i \in R_2} f(y_{2i} \mid y_{1i}; \theta_2) \times \cdots$$

$$\times \prod_{i \in R_p} f(y_{pi} \mid y_{p-1,i}, \ldots, y_{1,i}; \theta_p)$$

and the parameters $\theta_1, \ldots, \theta_p$ are orthogonal. The MLE for each component of the parameters can be obtained by maximizing the corresponding component of the observed likelihood.

We now consider some possible extensions to the non-monotone missing data, discussed by Kim and Shin (2012). In the case of nonmonotone missing data, we have the following steps to apply the factoring likelihood approach:

[Step 1] Partition the original sample into several disjoint sets according to the missing pattern.

[Step 2] Compute the MLEs for the identified parameters separately in each partition of the sample.

[Step 3] Combine the estimators to get a set of final estimates in a generalized least squares (GLS) form.

To simplify the presentation, we describe the proposed method in the bivariate normal setup with a nonmonotone missing pattern. The joint distribution of $(x, y)'$ is parameterized by the five parameters using model (3.12) or (3.18). For the convenience of the factoring method, we use the parametrization in (3.18) and let $\theta = (\beta_0, \beta_1, \sigma_{ee}, \mu_x, \sigma_{xx})'$.

In Step 1, we partition the sample into several disjoint sets according to the pattern of missingness. In the case of a nonmonotone missing pattern with two variables, we have $3 = 2^2 - 1$ types of respondents that contain information about the parameters. The first set H has both x and y observed, the second set K has x observed but y missing, and the third set L has y observed but x missing. See Table 3.1. Let n_H, n_K, n_L be the sample sizes of the sets H, K, L, respectively. The case of both x and y missing can be safely removed from the sample.

TABLE 3.1
An Illustration of the Missing Data Structure under a Bivariate Normal Distribution

Set	x	y	Sample Size	Estimable Parameters
H	Observed	Observed	n_H	$\mu_x, \mu_y, \sigma_{xx}, \sigma_{xy}, \sigma_{yy}$
K	Observed	Missing	n_K	μ_x, σ_{xx}
L	Missing	Observed	n_L	μ_y, σ_{yy}

In Step 2, we obtain the parameter estimates in each set: For set H, we have the five parameters $\eta_H = (\beta_0, \beta_1, \sigma_{ee}, \mu_x, \sigma_{xx})'$ of the conditional distribution of y given x and the marginal distribution of x, with MLEs $\hat{\eta}_H = (\hat{\beta}_{0,H}, \hat{\beta}_{1,H}, \hat{\sigma}_{ee,H}, \hat{\mu}_{x,H}, \hat{\sigma}_{xx,H})'$. For set K, the MLEs $\hat{\eta}_K = (\hat{\mu}_{x,K}, \hat{\sigma}_{xx,K})'$ are obtained for $\eta_K = (\mu_x, \sigma_{xx})'$, the parameters of the marginal distribution of x. For set L, the MLEs $\hat{\eta}_L = (\hat{\mu}_{y,L}, \hat{\sigma}_{yy,L})'$ are obtained for $\eta_L = (\mu_y, \sigma_{yy})'$, where $\mu_y = \beta_0 + \beta_1 \mu_x$ and $\sigma_{yy} = \sigma_{ee} + \beta_1^2 \sigma_{xx}$.

In Step 3, we use the GLS method to combine the three estimators $\hat{\eta}_H, \hat{\eta}_K, \hat{\eta}_L$ to get a final estimator for the parameter θ. Let $\hat{\eta} = (\hat{\eta}_H', \hat{\eta}_K', \hat{\eta}_L')'$. Then

$$\hat{\eta} = \left(\hat{\beta}_{0,H}, \hat{\beta}_{1,H}, \hat{\sigma}_{ee,H}, \hat{\mu}_{x,H}, \hat{\sigma}_{xx,H}, \hat{\mu}_{x,K}, \hat{\sigma}_{xx,K}, \hat{\mu}_{y,L}, \hat{\sigma}_{yy,L} \right)'. \qquad (3.20)$$

The expected value of this estimator is

$$\eta(\theta) = \left(\beta_0, \beta_1, \sigma_{ee}, \mu_x, \sigma_{xx}, \mu_x, \sigma_{xx}, \beta_0 + \beta_1\mu_x, \sigma_{ee} + \beta_1^2\sigma_{xx}\right)' \tag{3.21}$$

and the asymptotic covariance matrix is

$$\mathbf{V} = diag\left\{\frac{\Sigma_{yy.x}}{n_H}, \frac{2\sigma_{ee}^2}{n_H}, \frac{\sigma_{xx}}{n_H}, \frac{2\sigma_{xx}^2}{n_H}, \frac{\sigma_{xx}}{n_K}, \frac{2\sigma_{xx}^2}{n_K}, \frac{\sigma_{yy}}{n_L}, \frac{2\sigma_{yy}^2}{n_L}\right\}, \tag{3.22}$$

where

$$\Sigma_{yy.x} = \begin{pmatrix} \sigma_{ee}\left(1 + \sigma_{xx}^{-1}\mu_x^2\right) & -\sigma_{xx}^{-1}\sigma_{ee}\mu_x \\ -\sigma_{xx}^{-1}\sigma_{ee}\mu_x & \sigma_{xx}^{-1}\sigma_{ee} \end{pmatrix}.$$

Note that

$$\Sigma_{yy.x} = \{E[(1,X)(1,X)']\}^{-1}\sigma_{ee} = \begin{pmatrix} 1 & \mu_x \\ \mu_x & \sigma_{xx} + \mu_x^2 \end{pmatrix}^{-1}\sigma_{ee}.$$

Deriving the asymptotic covariance matrix of the first five estimates in (3.20) is straightforward. Relevant reference can be found, for example, in Subsection 7.2.2 of Little and Rubin (2002). We have a block-diagonal structure of $\mathbf{V}$ in (3.22) because $\hat{\mu}_{1K}$ and $\hat{\sigma}_{11K}$ are independent due to normality, and observations between different sets are independent due to the IID assumptions.

Note that the nine elements in $\eta(\theta)$ are related to each other because they are all functions of the five elements of vector θ. The information contained in the four extra equations has not yet been utilized in constructing the estimators $\hat{\eta}_H, \hat{\eta}_K, \hat{\eta}_L$. The information can be employed to construct a fully efficient estimator of θ by combining $\hat{\eta}_H, \hat{\eta}_K, \hat{\eta}_L$ through a GLS regression of $\hat{\eta} = (\hat{\eta}_H', \hat{\eta}_K', \hat{\eta}_L')'$ on θ as follows:

$$\hat{\eta} - \eta(\hat{\theta}_S) = (\partial\eta/\partial\theta')(\theta - \hat{\theta}_S) + \text{ error},$$

where $\hat{\theta}_S$ is an initial estimator.

The expected value and variance of $\hat{\eta}$ in (3.21) and (3.22) can furnish the specification of a nonlinear model of the five parameters in θ. Using a Taylor series expansion on the nonlinear model, a step of the Gauss–Newton method can be formulated as

$$\mathbf{e}_\eta = \mathbf{X}\left(\theta - \hat{\theta}_S\right) + \mathbf{u}, \tag{3.23}$$

where $\mathbf{e}_\eta = \hat{\eta} - \eta(\hat{\theta}_S)$, $\eta(\hat{\theta}_S)$ is the vector (3.21) evaluated at $\hat{\theta}_S$,

$$\mathbf{X} \equiv \frac{\partial\eta}{\partial\theta'} = \begin{pmatrix} 1 & 0 & 0 & 0 & 0 & 0 & 0 & 1 & 0 \\ 0 & 1 & 0 & 0 & 0 & 0 & 0 & \mu_x & 2\beta_1\sigma_{xx} \\ 0 & 0 & 1 & 0 & 0 & 0 & 0 & 0 & 1 \\ 0 & 0 & 0 & 1 & 0 & 1 & 0 & \beta_1 & 0 \\ 0 & 0 & 0 & 0 & 1 & 0 & 1 & 0 & \beta_1^2 \end{pmatrix}'. \tag{3.24}$$

Approximately, the error term in (3.23) follows

$$\mathbf{u} \sim (\mathbf{0}, \mathbf{V}),$$

where $\mathbf{V}$ is the covariance matrix defined in (3.22). Relations among parameters η, θ, $\mathbf{X}$, and $\mathbf{V}$ are summarized in Table 3.2.

TABLE 3.2
Summary for the Bivariate Normal Case

x	y	Dataset	Size	Estimable Parameters	Asymptotic Variance
O	M	K	n_K	$\eta_K = \theta_1$	$W_K = diag(\sigma_{xx}, 2\sigma_{xx}^2)$
O	O	H	n_H	$\eta_H = (\theta_1, \theta_2)'$	$W_H = diag(W_K, \Sigma_{yy.x}, 2\sigma_{ee}^2)$
M	O	L	n_L	$\eta_L = (\mu_y, \sigma_{yy})'$	$W_L = diag(\sigma_{yy}, 2\sigma_{yy}^2)$

O: observed, M: missing, $\theta_1 = (\mu_x, \sigma_{xx})'$, $\theta_2 = (\beta_0, \beta_1, \sigma_{ee})'$,
$\eta = (\eta_H', \eta_K', \eta_L')'$, $\theta = \eta_H$, $V = diag(W_H/n_H, W_K/n_K, W_L/n_L)$, $\Sigma_{ee} = \{E[(1, x)(1, x)']\}^{-1}\sigma_{ee}$,
$\mathbf{X} = \partial\eta/\partial\theta'$, $\mu_y = \beta_0 + \beta_1\mu_x$, $\sigma_{yy} = \sigma_{ee} + \beta_1^2\sigma_{xx}$

The procedure can be carried out iteratively until convergence. Given the current value $\hat{\theta}^{(t)}$, the solution of the Gauss–Newton method can be obtained iteratively as

$$\hat{\theta}^{(t+1)} = \hat{\theta}^{(t)} + \left(\mathbf{X}'_{(t)}\hat{\mathbf{V}}_{(t)}^{-1}\mathbf{X}_{(t)}\right)^{-1}\mathbf{X}'_{(t)}\hat{\mathbf{V}}_{(t)}^{-1}\left\{\hat{\eta} - \eta\left(\hat{\theta}^{(t)}\right)\right\}, \qquad (3.25)$$

where $\mathbf{X}_{(t)}$ and $\hat{\mathbf{V}}_{(t)}$ are evaluated from $\mathbf{X}$ in (3.24) and $\mathbf{V}$ in (3.22), respectively, using the current value $\hat{\theta}^{(t)}$. The covariance matrix of the estimator in (3.25) can be estimated by

$$\mathbf{C} = \left(\mathbf{X}'_{(t)}\hat{\mathbf{V}}_{(t)}^{-1}\mathbf{X}_{(t)}\right)^{-1}, \qquad (3.26)$$

when the iteration is stopped at the t-th iteration. Gauss–Newton method for the estimation of nonlinear models can be found, for example, in Seber and Wild (1989).

Remark 3.2. *The factoring likelihood method also provides a useful tool for efficiency comparison between different estimators that ignore some part of partial response. In the above bivariate normal example, suppose that we wish to compare the following four types of the estimates:* $\hat{\theta}_H$, $\hat{\theta}_{HK}$, $\hat{\theta}_{HL}$, $\hat{\theta}_{HKL}$ *which are the MLEs obtained from dataset H, $H \cup K$, $H \cup L$, $H \cup K \cup L$, respectively. The Gauss-Newton estimator (3.25) is asymptotically equal to* $\hat{\theta}_{HKL}$. *Write*

$$\mathbf{X}' = \left[\mathbf{X}'_H, \ \mathbf{X}'_K, \mathbf{X}'_L\right],$$

where $\mathbf{X}'_H$ is the left 5×5 submatrix of $\mathbf{X}'$, $\mathbf{X}'_K$ is the 5×2 submatrix in the middle of $\mathbf{X}'$, and $\mathbf{X}'_L$ is the 5×2 submatrix in the right side of $\mathbf{X}'$. Similarly, we can decompose $\hat{\eta}' = (\hat{\eta}'_H, \ \hat{\eta}'_K, \ \hat{\eta}'_L)$ and $\mathbf{V} = diag\{\mathbf{V}_H, \mathbf{V}_K, \mathbf{V}_L\}$.

Note that the asymptotic variance of $\hat{\theta}_H$ is $\left(\mathbf{X}'_H \hat{\mathbf{V}}_H^{-1} \mathbf{X}_H\right)^{-1}$. Similarly, we have

$$Var\left(\hat{\theta}_{HL}\right) = \left(\mathbf{X}'_H \hat{\mathbf{V}}_H^{-1} \mathbf{X}_H + \mathbf{X}'_L \hat{\mathbf{V}}_L^{-1} \mathbf{X}_L\right)^{-1},$$

$$Var\left(\hat{\theta}_{HK}\right) = \left(\mathbf{X}'_H \hat{\mathbf{V}}_H^{-1} \mathbf{X}_H + \mathbf{X}'_K \hat{\mathbf{V}}_K^{-1} \mathbf{X}_K\right)^{-1},$$

and

$$Var\left(\hat{\theta}_{HKL}\right) = \left(\mathbf{X}'_H \hat{\mathbf{V}}_H^{-1} \mathbf{X}_H + \mathbf{X}'_K \hat{\mathbf{V}}_K^{-1} \mathbf{X}_K + \mathbf{X}'_L \hat{\mathbf{V}}_L^{-1} \mathbf{X}_L\right)^{-1}.$$

Using the matrix algebra such as

$$\left(\mathbf{X}'_H \hat{\mathbf{V}}_H^{-1} \mathbf{X}_H + \mathbf{X}'_L \hat{\mathbf{V}}_L^{-1} \mathbf{X}_L\right)^{-1} = \left(\mathbf{X}'_H \hat{\mathbf{V}}_H^{-1} \mathbf{X}_H\right)^{-1}$$
$$- \left(\mathbf{X}'_H \hat{\mathbf{V}}_H^{-1} \mathbf{X}_H\right)^{-1} \mathbf{X}'_L \left[\hat{\mathbf{V}}_L + \mathbf{X}_L \left(\mathbf{X}'_H \hat{\mathbf{V}}_H^{-1} \mathbf{X}_H\right)^{-1} \mathbf{X}'_L\right]^{-1} \mathbf{X}_L \left(\mathbf{X}'_H \hat{\mathbf{V}}_H^{-1} \mathbf{X}_H\right)^{-1},$$

we can derive expressions for the variances of the estimators.

For estimates of the slope parameter, the asymptotic variances are

$$
\begin{aligned}
Var(\hat{\beta}_{1,HK}) &= \frac{\sigma_{ee}}{\sigma_{xx} n_H} = Var(\hat{\beta}_{1,H}) \\
Var(\hat{\beta}_{1,HL}) &= \frac{\sigma_{ee}}{\sigma_{xx} n_H} \left\{1 - 2 p_L \rho^2 \left(1 - \rho^2\right)\right\} \\
Var\left(\hat{\beta}_{1,HKL}\right) &= \frac{\sigma_{ee}}{\sigma_{xx} n_H} \left\{1 - \frac{2 p_L \rho^2 \left(1 - \rho^2\right)}{1 - p_L p_K \rho^4}\right\},
\end{aligned}
$$

where $p_K = n_K/(n_H + n_K)$ and $p_L = n_L/(n_H + n_L)$. Thus, we have

$$Var(\hat{\beta}_{1,H}) = Var(\hat{\beta}_{1,HK}) \geq Var(\hat{\beta}_{1,HL}) \geq Var(\hat{\beta}_{1,HKL}). \qquad (3.27)$$

Here strict inequalities generally hold except for special trivial cases. Note that the asymptotic variance of $\hat{\beta}_{1,HK}$ is the same as the variance of $\hat{\beta}_{1,H}$, which implies that there is no gain of efficiency by adding set K (missing y_2) to H. On the other hand, by comparing $Var(\hat{\beta}_{1,HL})$ with $Var(\hat{\beta}_{1,H})$, we observe an efficiency gain by adding a set L (missing y_1) to H.

Example 3.6. *For a numerical example, we consider the dataset originally presented by Little (1982) and also discussed in Little and Rubin (2002). Table 3.3 presents a partially classified data in a 2×2 table with supplemental margins for both the classification variables. For the orthogonal parametrization, we use $\eta_H = \left(\pi_{1|1}, \pi_{1|2}, \pi_{1+}\right)'$ where $\pi_{1|1} = P(y_2 = 1 \mid y_1 = 1)$, $\pi_{1|2} = P(y_2 = 1 \mid y_1 = 2)$, $\pi_{1+} = P(y_1 = 1)$. We also set $\theta = \eta_H$. Note that the validity of the proposed method does not depend on the choice of the parametrization. This parametrization makes the computation of the information matrix simple.*

TABLE 3.3

A 2×2 Table with Supplemental Margins for Both Variables

Set	y_1	y_2	Count
	1	1	100
H	1	2	50
	2	1	75
	2	2	75
K	1		30
	2		60
L		1	28
		2	60

From the data in Table 3.3, the five observations for the three parameters
are

$$
\hat{\eta} = \left(\hat{\pi}_{1|1,H}, \hat{\pi}_{1|2,H}, \hat{\pi}_{1+,H}, \hat{\pi}_{1+,K}, \hat{\pi}_{+1,L} \right)'
$$
$$
= (100/150, 75/150, 150/300, 30/90, 28/88)'
$$

with the expectations

$$
\eta(\theta) = \left(\pi_{1|1}, \pi_{1|2}, \pi_{1+}, \pi_{1+}, \pi_{1|1}\pi_{1+} + \pi_{1|2} - \pi_{1|2}\pi_{1+} \right)'
$$

and the variance-covariance matrix

$$
\mathbf{V} = diag \left\{ \frac{\pi_{1|1}\left(1 - \pi_{1|1}\right)}{n_H \pi_{1+}}, \frac{\pi_{1|2}\left(1 - \pi_{1|2}\right)}{n_H (1 - \pi_{1+})}, \frac{\pi_{+1}(1 - \pi_{+1})}{n_H}, \frac{\pi_{1+}(1 - \pi_{1+})}{n_K}, \frac{\pi_{+1}(1 - \pi_{+1})}{n_L} \right\},
$$

where $\pi_{+1} = P(y_2 = 1)$. The Gauss-Newton method as described in (3.25) can
be used to solve the nonlinear model of the three parameters, where the initial
estimator of θ is $\hat{\theta}_S = (100/150, 75/150, 180/390)'$ and the X matrix is

$$
\mathbf{X} = \begin{pmatrix} 1 & 0 & 0 & 0 & \pi_{1+} \\ 0 & 1 & 0 & 0 & 1 - \pi_{1+} \\ 0 & 0 & 1 & 1 & \pi_{1|1} - \pi_{1|2} \end{pmatrix}'.
$$

The resulting one-step estimates are $\hat{\pi}_{11} = 0.28, \hat{\pi}_{12} = 0.18, \hat{\pi}_{21} = 0.24$, and
$\hat{\pi}_{22} = 0.31$. The standard errors of the estimated values are computed by (3.26)
and are 0.0205, 0.0174, 0.0195, 0.0211 for $\hat{\pi}_{11}, \hat{\pi}_{12}, \hat{\pi}_{21}, \hat{\pi}_{22}$, respectively.

The proposed method can be shown to be algebraically equivalent to the
Fisher scoring method in (3.13), but avoids the burden of obtaining the observed likelihood. Instead, the MLEs separately computed from each partition
of the marginal likelihoods and the full likelihoods are combined in a natural
way.

3.3   EM Algorithm

When finding the MLE from the observed likelihood $L_{\text{obs}}(\eta)$, where $\eta = (\theta, \phi)$, we often encounter two problems associated with Newton's method:

1. Computing the second order partial derivatives of the log-likelihood can be cumbersome.

2. The likelihood does not always increase for each iteration.

The Expectation-Maximization (EM) algorithm, proposed by Dempster et al. (1977), is an iterative algorithm of finding the MLE without computing the second order partial derivatives. To describe the EM algorithm, suppose that we are interested in finding the MLE $\hat{\eta}$ that maximizes the observed likelihood $L_{\text{obs}}(\eta)$. Let $\eta^{(t)}$ be the current value of the parameter estimate of η. The EM algorithm can be defined by iteratively carrying out the following E-step and M-step:

[E-step] Compute

$$Q(\eta \mid \eta^{(t)}) = E\left\{ \log L_{\text{com}}(\eta) \mid \mathbf{y}_{\text{obs}}, \boldsymbol{\delta}; \eta^{(t)} \right\}, \qquad (3.28)$$

where $L_{\text{com}}(\eta) = f(\mathbf{y}, \boldsymbol{\delta}; \eta)$.

[M-step] Find $\eta^{(t+1)}$ that maximizes $Q(\eta \mid \eta^{(t)})$.

One of the most important properties of the EM algorithm is that its step never decreases the likelihood:

$$L_{\text{obs}}(\eta^{(t+1)}) \geq L_{\text{obs}}(\eta^{(t)}). \qquad (3.29)$$

This makes EM a numerically stable procedure as it "climbs" the likelihood surface. However, Newton's method does not necessarily satisfy (3.29). The following theorem presents this result formally.

Theorem 3.2. *If $Q(\eta^{(t+1)} \mid \eta^{(t)}) \geq Q(\eta^{(t)} \mid \eta^{(t)})$, then (3.29) is satisfied.*

Proof. By the definition of the observed likelihood,

$$\log L_{\text{obs}}(\eta) = \log f(\mathbf{y}, \boldsymbol{\delta}; \eta) - \log f(\mathbf{y}_{\text{mis}} \mid \mathbf{y}_{\text{obs}}, \boldsymbol{\delta}; \eta).$$

Taking the conditional expectation of the above equation given $(\mathbf{y}_{\text{obs}}, \boldsymbol{\delta})$ evaluated at $\eta^{(t)}$, we have

$$\log L_{\text{obs}}(\eta) = Q\left(\eta \mid \eta^{(t)}\right) - H\left(\eta \mid \eta^{(t)}\right), \qquad (3.30)$$

where

$$H\left(\eta \mid \eta^{(t)}\right) = E\left\{\log f\left(\mathbf{y}_{\text{mis}} \mid \mathbf{y}_{\text{obs}}, \boldsymbol{\delta}; \eta\right) \mid \mathbf{y}_{\text{obs}}, \boldsymbol{\delta}, \eta^{(t)}\right\}. \tag{3.31}$$

Using Lemma 2.2, $H(\eta^{(t)} \mid \eta^{(t)}) \geq H(\eta \mid \eta^{(t)})$ for all $\eta \neq \eta^{(t)}$. Therefore, for any $\eta^{(t+1)}$ that satisfies $Q(\eta^{(t+1)} \mid \eta^{(t)}) \geq Q(\eta^{(t)} \mid \eta^{(t)})$, we have $H(\eta^{(t)} \mid \eta^{(t)}) \geq H(\eta^{(t+1)} \mid \eta^{(t)})$ and the result follows. $\qquad\square$

Remark 3.3. *An alternative proof of Theorem 3.2 is made as follows. Writing*

$$
\begin{aligned}
\log L_{\text{obs}}(\eta) &= \log \int_{\mathcal{R}(\mathbf{y}_{\text{obs}}, \boldsymbol{\delta})} f(\mathbf{y}, \boldsymbol{\delta}; \eta) d\mu(\mathbf{y}) \\
&= \log \int_{\mathcal{R}(\mathbf{y}_{\text{obs}}, \boldsymbol{\delta})} \frac{f(\mathbf{y}, \boldsymbol{\delta}; \eta)}{f(\mathbf{y}, \boldsymbol{\delta}; \eta^{(t)})} f\left(\mathbf{y}_{\text{mis}} \mid \mathbf{y}_{\text{obs}}, \boldsymbol{\delta}; \eta^{(t)}\right) L_{\text{obs}}(\eta^{(t)}) d\mu(\mathbf{y}) \\
&= \log E\left\{\frac{f(\mathbf{y}, \boldsymbol{\delta}; \eta)}{f(\mathbf{y}, \boldsymbol{\delta}; \eta^{(t)})} \mid \mathbf{y}_{\text{obs}}, \boldsymbol{\delta}; \eta^{(t)}\right\} + \log L_{\text{obs}}(\eta^{(t)}),
\end{aligned}
$$

we have

$$
\begin{aligned}
\log L_{\text{obs}}(\eta) - \log L_{\text{obs}}(\eta^{(t)}) &= \log E\left\{\frac{f(\mathbf{y}, \boldsymbol{\delta}; \eta)}{f(\mathbf{y}, \boldsymbol{\delta}; \eta^{(t)})} \mid \mathbf{y}_{\text{obs}}, \boldsymbol{\delta}; \eta^{(t)}\right\} \\
&\geq E\left[\log\left\{\frac{f(\mathbf{y}, \boldsymbol{\delta}; \eta)}{f(\mathbf{y}, \boldsymbol{\delta}; \eta^{(t)})}\right\} \mid \mathbf{y}_{\text{obs}}, \boldsymbol{\delta}; \eta^{(t)}\right] \\
&= Q(\eta \mid \eta^{(t)}) - Q(\eta^{(t)} \mid \eta^{(t)}),
\end{aligned}
$$

where the above inequality follows from Lemma 2.1. Therefore, $Q(\eta^{(t+1)} \mid \eta^{(t)}) \geq Q(\eta^{(t)} \mid \eta^{(t)})$ implies (3.29).

We now discuss the convergence of the EM sequence $\{\eta^{(t)}\}$. By (3.29), the sequence $\{L_{\text{obs}}(\eta^{(t)})\}$ is monotonically increasing and it is bounded above if the MLE exists. Thus, the sequence of $L_{\text{obs}}(\eta^{(t)})$ converges to some value L^*. In most cases, L^* is a stationary value in the sense that $L^* = L_{\text{obs}}(\eta^*)$ for some η^* at which $\partial L_{\text{obs}}(\eta)/\partial \eta = 0$. Under fairly weak conditions, such as $Q(\eta \mid \gamma)$ in (3.28) satisfies

$$\partial Q(\eta \mid \gamma)/\partial \eta \text{ is continuous in } \eta \text{ and } \gamma, \tag{3.32}$$

the EM sequence $\{\eta^{(t)}\}$ converges to a stationary point η^*. In particular, if $L_{\text{obs}}(\eta)$ is unimodal in Ω with η^* being the only stationary point, then, under (3.32), any EM sequence $\{\eta^{(t)}\}$ converges to the unique maximizer η^* of $L_{\text{obs}}(\eta)$. Further convergence details can be found in Wu (1983) and McLachlan and Krishnan (2008).

Remark 3.4. *Expression (3.30) gives further insight on the observed information matrix associated with the observed likelihood. Note that (3.30) leads to*

$$0 = \frac{\partial}{\partial \eta_0} Q(\eta \mid \eta_0) - \frac{\partial}{\partial \eta_0} H(\eta \mid \eta_0) \tag{3.33}$$

and

$$I_{\text{obs}}(\eta) = -\frac{\partial^2}{\partial\eta\partial\eta'}\log L_{\text{obs}}(\eta) = -\frac{\partial^2}{\partial\eta\partial\eta'}Q(\eta \mid \eta_0) + \frac{\partial^2}{\partial\eta\partial\eta'}H(\eta \mid \eta_0) \quad (3.34)$$

for any η_0. By applying (2.4) to $S_{\text{mis}}(\eta_0)$, we have

$$E\{S_{\text{mis}}(\eta_0) \mid \mathbf{y}_{\text{obs}}, \boldsymbol{\delta}; \eta_0\} = 0 \quad (3.35)$$

and, by the definition of $S_{\text{mis}}(\eta)$ in (2.26),

$$\begin{aligned}
\frac{\partial^2}{\partial\eta\partial\eta'}H(\eta \mid \eta_0)|_{\eta=\eta_0} &= \frac{\partial}{\partial\eta'}E\{S_{\text{mis}}(\eta) \mid \mathbf{y}_{\text{obs}}, \boldsymbol{\delta}; \eta_0\}|_{\eta=\eta_0} \\
&= -\frac{\partial}{\partial\eta_0'}E\{S_{\text{mis}}(\eta) \mid \mathbf{y}_{\text{obs}}, \boldsymbol{\delta}; \eta_0\}|_{\eta=\eta_0} \\
&= -\frac{\partial^2}{\partial\eta_0\partial\eta'}H(\eta \mid \eta_0)|_{\eta=\eta_0} \\
&= -\frac{\partial^2}{\partial\eta_0\partial\eta'}Q(\eta \mid \eta_0)|_{\eta=\eta_0},
\end{aligned}$$

where the first and the third equalities follow from the definition of $H(\eta \mid \eta_o)$ in (3.31), the second equality follows from (3.35), and the last equality follows from (3.33). Therefore, we have

$$I_{\text{obs}}(\eta_0) = -\left\{\frac{\partial^2}{\partial\eta\partial\eta'}Q(\eta \mid \eta_0) + \frac{\partial^2}{\partial\eta_0\partial\eta'}Q(\eta \mid \eta_0)\right\}|_{\eta=\eta_0}, \quad (3.36)$$

which is another way of computing the observed information using only $Q(\eta \mid \eta_0)$ in the EM algorithm. The formula (3.36) was first derived by Oakes (1999) and is sometimes called Oakes' formula.

To understand Oakes' formula further, note that

$$\begin{aligned}
\frac{\partial}{\partial\eta_0'}E\{S_{\text{mis}}(\eta) \mid \mathbf{y}_{\text{obs}}, \boldsymbol{\delta}; \eta_0\} &= \frac{\partial}{\partial\eta_0'}\int S_{\text{mis}}(\eta)f(\mathbf{y}_{\text{mis}} \mid \mathbf{y}_{\text{obs}}, \boldsymbol{\delta}; \eta_0)\,d\mu(\mathbf{y}_{\text{mis}}) \\
&= \int S_{\text{mis}}(\eta)\left\{\frac{\partial}{\partial\eta_0'}f(\mathbf{y}_{\text{mis}} \mid \mathbf{y}_{\text{obs}}, \boldsymbol{\delta}; \eta_0)\right\}d\mu(\mathbf{y}_{\text{mis}}) \\
&= \int S_{\text{mis}}(\eta)S_{\text{mis}}(\eta_0)'f(\mathbf{y}_{\text{mis}} \mid \mathbf{y}_{\text{obs}}, \boldsymbol{\delta}; \eta_0)\,d\mu(\mathbf{y}_{\text{mis}}) \\
&= E\{S_{\text{mis}}(\eta)S_{\text{mis}}(\eta_0)' \mid \mathbf{y}_{\text{obs}}, \boldsymbol{\delta}; \eta_0\}. \quad (3.37)
\end{aligned}$$

By (3.37), we can express

$$\begin{aligned}
\frac{\partial^2}{\partial\eta_0\partial\eta'}H(\eta \mid \eta_0)|_{\eta=\eta_0} &= \frac{\partial}{\partial\eta_0'}E\{S_{\text{mis}}(\eta) \mid \mathbf{y}_{\text{obs}}, \boldsymbol{\delta}; \eta_0\}|_{\eta=\eta_0} \\
&= E\{S_{\text{mis}}(\eta_0)^{\otimes 2} \mid \mathbf{y}_{\text{obs}}, \boldsymbol{\delta}; \eta_0\}.
\end{aligned}$$

Thus, by (2.32), we have

$$\frac{\partial^2}{\partial \eta_0 \partial \eta'} Q(\eta \mid \eta_0)|_{\eta=\eta_0} = E\left\{ S_{\text{com}}(\eta_0)^{\otimes 2} \mid \mathbf{y}_{\text{obs}}, \boldsymbol{\delta}; \eta_0 \right\} - \bar{S}(\eta_0)^{\otimes 2}$$

and (3.36) is equal to (2.30). This proves equivalence between Oakes' formula and Louis' formula.

The EM algorithm can be expressed as

$$\eta^{(t+1)} = \arg\max_{\eta \in \Omega} Q(\eta \mid \eta^{(t)}), \tag{3.38}$$

and it is often obtained by

$$\eta^{(t+1)} \leftarrow \text{solution to } \bar{S}(\eta \mid \eta^{(t)}) = 0, \tag{3.39}$$

where

$$\bar{S}(\eta \mid \eta^{(t)}) = \frac{\partial}{\partial \eta} Q(\eta \mid \eta^{(t)}) = E\left\{ S_{\text{com}}(\eta) \mid \mathbf{y}_{\text{obs}}, \boldsymbol{\delta}, \eta^{(t)} \right\}.$$

Equivalently,

$$\bar{S}(\eta^{(t+1)} \mid \eta^{(t)}) = 0. \tag{3.40}$$

Now, let η^* be the probability limit point of $\eta^{(t)}$. Applying a Taylor expansion of (3.40) with respect to $(\eta^{(t+1)}, \eta^{(t)})$ around its limiting point (η^*, η^*), we have

$$\begin{aligned}
0 &= \bar{S}(\eta^{(t+1)} \mid \eta^{(t)}) \cong \bar{S}(\eta^* \mid \eta^*) \\
&+ \left\{ \bar{S}_1(\eta^* \mid \eta^*) \right\} \left(\eta^{(t+1)} - \eta^* \right) + \left\{ \bar{S}_2(\eta^* \mid \eta^*) \right\} \left(\eta^{(t)} - \eta^* \right),
\end{aligned} \tag{3.41}$$

where

$$\begin{aligned}
\bar{S}_1(\eta^{(t+1)} \mid \eta^{(t)}) &= \partial \bar{S}(\eta^{(t+1)} \mid \eta^{(t)}) / \partial (\eta^{(t+1)})', \\
\bar{S}_2(\eta^{(t+1)} \mid \eta^{(t)}) &= \partial \bar{S}(\eta^{(t+1)} \mid \eta^{(t)}) / \partial (\eta^{(t)})'.
\end{aligned}$$

Note that

$$\begin{aligned}
\bar{S}(\eta^* \mid \eta^*) &= S_{\text{obs}}(\eta^*) \\
&\cong S_{\text{obs}}(\hat{\eta}_{\text{mle}}) - \mathcal{I}_{\text{obs}}(\hat{\eta}_{\text{mle}})(\eta^* - \hat{\eta}_{\text{mle}}) \\
&\cong \mathcal{I}_{\text{obs}}(\hat{\eta}_{\text{mle}} - \eta^*)
\end{aligned}$$

as $S_{\text{obs}}(\hat{\eta}_{\text{mle}}) = 0$. Also, it can be shown that

$$\plim_n \bar{S}_1(\eta^* \mid \eta^*) = -\mathcal{I}_{\text{com}} \tag{3.42}$$

and

$$\plim_n \bar{S}_2(\eta^* \mid \eta^*) = \mathcal{I}_{\text{mis}}, \tag{3.43}$$

where $p$lim is the notation for "probability limit". Thus, (3.41) reduces to

$$0 \cong \mathcal{I}_{\text{obs}}(\hat{\eta}_{\text{mle}} - \eta^*) - \mathcal{I}_{\text{com}}\left(\eta^{(t+1)} - \eta^*\right) + \mathcal{I}_{\text{mis}}\left(\eta^{(t)} - \eta^*\right),$$

which implies

$$\eta^{(t+1)} - \hat{\eta}_{\text{mle}} \cong \mathcal{I}_{\text{com}}^{-1}\mathcal{I}_{\text{mis}}\left(\eta^{(t)} - \hat{\eta}_{\text{mle}}\right), \tag{3.44}$$

so

$$\eta^{(t+1)} - \eta^{(t)} \cong \mathcal{J}_{\text{mis}}\left(\eta^{(t)} - \eta^{(t-1)}\right).$$

That is, the convergence rate for the EM sequence is linear. The matrix $\mathcal{J}_{\text{mis}} = \mathcal{I}_{\text{com}}^{-1}\mathcal{I}_{\text{mis}}$ is called the *fraction of missing information*. The fraction of missing information may vary across different components of $\eta^{(t)}$, suggesting that certain components of $\eta^{(t)}$ may approach $\hat{\eta}$ rapidly while other components may require many iterations. Roughly speaking, the rate of convergence of a vector sequence $\eta^{(t)}$ from the EM algorithm is given by the largest eigenvalue of the matrix $\mathcal{J}_{\text{mis}}$.

Example 3.7. *Consider the following exponential distribution*

$$z_i \sim f(z; \theta) = \theta \exp(-\theta z), \quad z > 0.$$

Instead of observing z_i, we observe $\{(y_i, \delta_i) : i = 1, 2, \ldots, n\}$, where

$$y_i = \min\{z_i, c_i\}$$

$$\delta_i = \begin{cases} 1 & \text{if } z_i \leq c_i \\ 0 & \text{if } z_i > c_i. \end{cases}$$

The score function for θ, under complete response, is

$$S(\theta) = \sum_{i=1}^{n} \left(\frac{1}{\theta} - z_i\right).$$

Now, we need to evaluate

$$E(z_i \mid y_i, \delta_i) = \begin{cases} y_i & \text{if } \delta_i = 1 \\ E_\theta(z_i \mid z_i > c_i) & \text{if } \delta_i = 0. \end{cases}$$

For given θ,

$$E_\theta(z_i \mid z_i > c_i) = c_i + 1/\theta$$

by a property of the exponential distribution. Thus,

$$\hat{\theta}_{MLE} = \left[\frac{1}{r}\sum_{i=1}^{n}\{\delta_i y_i + (1 - \delta_i)c_i\}\right]^{-1},$$

where $r = \sum_{i=1}^{n} \delta_i$.

Example 3.8. *Consider the following random variable*

$$Y_i = (1 - \delta_i) Z_{1i} + \delta_i Z_{2i}, \quad i = 1, 2, \ldots, n,$$

where $Z_{1i} \sim N\left(\mu_1, \sigma_1^2\right)$, $Z_{2i} \sim N\left(\mu_2, \sigma_2^2\right)$, and $\delta_i \sim Bernoulli\,(\pi)$. Suppose that we do not observe Z_{1i}, Z_{i2}, δ_i but only observe Y_i in the sample. The parameter of interest is $\theta = \left(\mu_1, \mu_2, \sigma_1^2, \sigma_2^2, \pi\right)$. We treat δ_i as a latent variable (i.e., a variable that is always missing) and consider the following complete sample likelihood:

$$L_{\text{com}}(\theta) = \prod_{i=1}^{n} f(y_i, \delta_i \mid \theta),$$

where

$$f(y, \delta \mid \theta) = \left[\phi\left(y \mid \mu_1, \sigma_1^2\right)\right]^{1-\delta} \left[\phi\left(y \mid \mu_2, \sigma_2^2\right)\right]^{\delta} \pi^{\delta} (1 - \pi)^{1-\delta}$$

and $\phi\left(y \mid \mu, \sigma^2\right)$ is the density of $N(\mu, \sigma^2)$ distribution. Hence,

$$
\begin{aligned}
\ln L_{\text{com}}(\theta) \;=\; & \sum_{i=1}^{n} \left\{ (1 - \delta_i) \log \phi\left(y_i \mid \mu_1, \sigma_1^2\right) + \delta_i \log \phi\left(y_i \mid \mu_2, \sigma_2^2\right) \right\} \\
& + \sum_{i=1}^{n} \left\{ \delta_i \log(\pi_i) + (1 - \delta_i) \log(1 - \pi_i) \right\}.
\end{aligned}
$$

To apply the EM algorithm, the E-step can be expressed as

$$
\begin{aligned}
Q(\theta \mid \theta^{(t)}) \;=\; & \sum_{i=1}^{n} \left\{ \left(1 - w_i^{(t)}\right) \log \phi\left(y_i \mid \mu_1, \sigma_1^2\right) + w_i^{(t)} \log \phi\left(y_i \mid \mu_2, \sigma_2^2\right) \right\} \\
& + \sum_{i=1}^{n} \left\{ w_i^{(t)} \log(\pi_i) + \left(1 - w_i^{(t)}\right) \log(1 - \pi_i) \right\},
\end{aligned}
$$

where $w_i^{(t)} = E\left(\delta_i \mid y_i, \theta^{(t)}\right)$ with

$$E(\delta_i \mid y_i, \theta) \;=\; \frac{\pi_i \phi\left(y_i \mid \mu_2, \sigma_2^2\right)}{(1 - \pi_i) \phi\left(y_i \mid \mu_1, \sigma_1^2\right) + \pi_i \phi\left(y_i \mid \mu_2, \sigma_2^2\right)}.$$

The M-step for updating θ is

$$\frac{\partial}{\partial \theta} Q\left(\theta \mid \theta^{(t)}\right) = 0,$$

which can be written as

$$
\begin{aligned}
\mu_j^{(t+1)} \;=\; & \sum_{i=1}^{n} w_{ij}^{(t)} y_i \Big/ \sum_{i=1}^{n} w_{ij}^{(t)}, \\
\sigma_j^{2(t+1)} \;=\; & \sum_{i=1}^{n} w_{ij}^{(t)} \left(y_i - \mu_j^{(t+1)}\right)^2 \Big/ \sum_{i=1}^{n} w_{ij}^{(t)}, \\
\pi^{(t+1)} \;=\; & \sum_{i=1}^{n} w_i^{(t)} / n,
\end{aligned}
$$

for $j = 1, 2$, where $w_{i1}^{(t)} = 1 - w_i^{(t)}$ and $w_{i2}^{(t)} = w_i^{(t)}$.

Remark 3.5. *Consider the exponential family of distributions of the form*

$$f(\mathbf{y}; \theta) = b(\mathbf{y}) \exp\left\{ \theta' \mathbf{T}(\mathbf{y}) - A(\theta) \right\}. \tag{3.45}$$

Under MAR, the E-step of the EM algorithm is

$$Q\left(\theta \mid \theta^{(t)}\right) = \text{constant} + \theta' E\left\{ \mathbf{T}(\mathbf{y}) \mid \mathbf{y}_{\text{obs}}, \theta^{(t)} \right\} - A(\theta), \tag{3.46}$$

and the M-step is

$$\frac{\partial}{\partial \theta} Q\left(\theta \mid \theta^{(t)}\right) = 0 \iff E\left\{ \mathbf{T}(\mathbf{y}) \mid \mathbf{y}_{\text{obs}}; \theta^{(t)} \right\} = \frac{\partial}{\partial \theta} A(\theta).$$

Because $\int f(\mathbf{y}; \theta) \, d\mathbf{y} = 1$, we have

$$\frac{\partial}{\partial \theta} A(\theta) = E\left\{ \mathbf{T}(\mathbf{y}); \theta \right\}.$$

Therefore, the M-step reduces to finding $\theta^{(t+1)}$ as a solution to

$$E\left\{ \mathbf{T}(\mathbf{y}) \mid \mathbf{y}_{\text{obs}}, \theta^{(t)} \right\} = E\left\{ \mathbf{T}(\mathbf{y}); \theta \right\}. \tag{3.47}$$

Navidi (1997) used a graphical illustration for the EM algorithm finding the solution to (3.47). Let $h_1(\theta) = E\{\mathbf{T}(\mathbf{y}) \mid \mathbf{y}_{\text{obs}}, \theta\}$ and $h_2(\theta) = E\{\mathbf{T}(\mathbf{y}); \theta\}$. The EM algorithm finds the solution to $h_1(\theta) = h_2(\theta)$ iteratively by obtaining $\theta^{(t+1)}$ which solves $h_1(\theta^{(t)}) = h_2(\theta)$ for θ, as illustrated in Figure 3.2.

Note that, by (3.46),

$$\frac{\partial^2}{\partial\theta\partial\theta'} Q(\theta \mid \theta_0) = -\frac{\partial^2}{\partial\theta\partial\theta'} A(\theta)$$

and

$$\frac{\partial^2}{\partial\theta\partial\theta_0} Q(\theta \mid \theta_0) = \frac{\partial}{\partial\theta_0} E\left\{ T(\mathbf{y}) \mid \mathbf{y}_{\text{obs}}; \theta_0 \right\}.$$

Thus, by Oakes' formula in (3.36), we have

$$I_{\text{obs}}(\theta) = \frac{\partial^2}{\partial\theta\partial\theta'} A(\theta) - \frac{\partial}{\partial\theta} E\left\{ T(\mathbf{y}) \mid \mathbf{y}_{\text{obs}}; \theta \right\},$$

which can also be obtained by taking the partial derivatives of

$$-S_{\text{obs}}(\theta) = \frac{\partial}{\partial} A(\theta) - E\left\{ T(\mathbf{y}) \mid \mathbf{y}_{\text{obs}}, \theta \right\}$$

with respect to θ'.

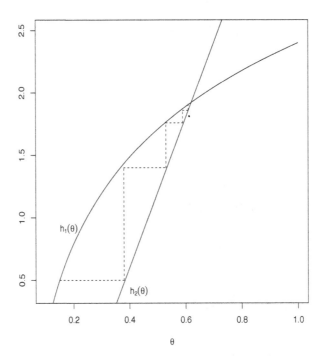

FIGURE 3.2
An illustration of EM algorithm for exponential family.

Example 3.9. *Consider the bivariate normal model in (3.12). Assume that both x_i and y_i are subject to missingness under MAR. Let $\delta_i^{(x)}$ be the response indicator function for x_i and $\delta_i^{(y)}$ be the response indicator function for y_i. The sufficient statistic for $\theta = (\mu_x, \mu_y, \sigma_{xx}, \sigma_{xy}, \sigma_{yy})$ is*

$$T = \left(\sum_{i=1}^n x_i, \sum_{i=1}^n y_i, \sum_{i=1}^n x_i^2, \sum_{i=1}^n x_i y_i, \sum_{i=1}^n y_i^2 \right),$$

and because the distribution belongs to the exponential family, the EM algorithm reduces to solving

$$\sum_{i=1}^n E\left\{ (x_i, y_i, x_i^2, x_i y_i, y_i^2) \mid \mathbf{y}_{i,\text{obs}}; \theta^{(t)} \right\} = \sum_{i=1}^n E\left\{ (x_i, y_i, x_i^2, x_i y_i, y_i^2); \theta \right\}$$

for θ, where $\mathbf{y}_{i,\text{obs}}$ is the observed part of $\mathbf{y}_i = (x_i, y_i)$. The above conditional expectation can be obtained using the usual conditional expectation under normality. For example, if $\delta_i^{(x)} = 1$ and $\delta_i^{(y)} = 0$, then $\mathbf{y}_{i,\text{obs}} = x_i$ and the conditional expectation is equal to

$$\begin{aligned}
&E\left\{ (x_i, y_i, x_i^2, x_i y_i, y_i^2) \mid x_i; \theta \right\} \\
&= \left(x_i, \beta_0 + \beta_1 x_i, x_i^2, x_i (\beta_0 + \beta_1 x_i), (\beta_0 + \beta_1 x_i)^2 + \sigma_{ee} \right),
\end{aligned}$$

where $\beta_1 = \sigma_{xy}/\sigma_{xx}$, $\beta_0 = \mu_y - \beta_1 \mu_x$, and $\sigma_{ee} = \sigma_{yy} - \beta_1 \sigma_{xy}$.

If $\mathbf{y}$ is a categorical variable that takes values in set S_y, then the E-step can be easily computed by a weighted summation

$$E\left\{ \ln f(\mathbf{y}, \boldsymbol{\delta}; \eta) \mid \mathbf{y}_{\text{obs}}, \boldsymbol{\delta}; \eta^{(t)} \right\} = \sum_{\mathbf{y} \in S_y} P(\mathbf{y} \mid \mathbf{y}_{\text{obs}}, \boldsymbol{\delta}, \eta^{(t)}) \ln f(\mathbf{y}, \boldsymbol{\delta}; \eta), \quad (3.48)$$

where the summation is over all possible values of $\mathbf{y}$ and $P(\mathbf{y} \mid \mathbf{y}_{\text{obs}}, \boldsymbol{\delta}, \eta^{(t)})$ is the conditional probability of taking $\mathbf{y}$ given $\mathbf{y}_{\text{obs}}$ and $\boldsymbol{\delta}$ evaluated at $\eta^{(t)}$. The conditional probability $P(\mathbf{y} \mid \mathbf{y}_{\text{obs}}, \boldsymbol{\delta}; \eta^{(t)})$ can be treated as the weight assigned for the categorical variable $\mathbf{y}$. That is, if $S(\eta) = \sum_{i=1}^n S(\eta; \mathbf{y}_i, \delta_i)$ is the score function for η, then the EM algorithm using (3.48) can be obtained by solving

$$\sum_{i=1}^n \sum_{\mathbf{y} \in S_y} P(\mathbf{y}_i = \mathbf{y} \mid \mathbf{y}_{i,\text{obs}}, \boldsymbol{\delta}_i; \eta^{(t)}) S(\eta; \mathbf{y}, \delta_i) = 0$$

for η to get $\eta^{(t+1)}$. Ibrahim (1990) called this approach *EM by weighting.*

Example 3.10. *We now return to Example 3.6. The parameters of interest are $\pi_{ij} = P(Y_1 = i, Y_2 = j)$, for $i = 1, 2, j = 1, 2$. The sufficient statistics for the parameters are $n_{ij}, i = 1, 2; j = 1, 2$, where n_{ij} is the sample size for the set with $Y_1 = i$ and $Y_2 = j$. Let $n_{i+} = n_{i1} + n_{i2}$ and $n_{+j} = n_{1j} + n_{2j}$. Let $n_{ij,H}$ be*

```
setH <- matrix(c(100, 75, 50, 75), nrow = 2, ncol = 2)
setK <- c(30, 60)
setL <- c(28, 60)
th <- prop.table(setH) # initial estimates of pi from setH
nij <- matrix(nrow = 2, ncol = 2)
repeat{
  th0 <- th
  # E step
  for (i in 1:2){
    for (j in 1:2){
      nij[i,j] <- setH[i,j] + setK[i]*th[i,j]/sum(th[i,])
                  + setL[j]*th[i,j]/sum(th[,j])
    }
  }
  # M step
  th <- nij/sum(nij)
  dif <- sum((th0-th)^2)
  if (dif < 1e-8) break
}
round(th, 3)
```

FIGURE 3.3
R program for EM algorithm in Example 3.10.

the number of elements in set H with $Y_1 = i$ and $Y_2 = j$. We define $n_{ij,K}$ and $n_{ij,L}$ in a similar fashion. The E-step computes the conditional expectation of the sufficient statistics. This gives

$$n_{ij}^{(t)} = E\left(n_{ij} \mid data, \pi_{ij}^{(t)}\right) = n_{ij,H} + n_{i+,K}\left(\pi_{ij}^{(t)}/\pi_{i+}^{(t)}\right) + n_{+j,L}\left(\pi_{ij}^{(t)}/\pi_{+j}^{(t)}\right),$$

for $i = 1,2; j = 1,2$. In the M-step, the parameters are updated by $\pi_{ij}^{(t+1)} = n_{ij}^{(t)}/n$.

Example 3.11. *We now return to Example 2.4. In the E-step, the conditional expectations of the score functions are computed by*

$$\bar{S}_1\left(\beta \mid \beta^{(t)}, \phi^{(t)}\right) = \sum_{\delta_i=1}\{y_i - p_i(\beta)\}\mathbf{x}_i + \sum_{\delta_i=0}\sum_{j=0}^{1}w_{ij(t)}\{j - p_i(\beta)\}\mathbf{x}_i,$$

where

$$
\begin{aligned}
w_{ij(t)} &= P\left(Y_i = j \mid \mathbf{x}_i, \delta_i = 0; \beta^{(t)}, \phi^{(t)}\right) \\
&= \frac{P\left(Y_i = j \mid \mathbf{x}_i; \beta^{(t)}\right) P\left(\delta_i = 0 \mid \mathbf{x}_i, j; \phi^{(t)}\right)}{\sum_{y=0}^{1} P\left(Y_i = y \mid \mathbf{x}_i; \beta^{(t)}\right) P\left(\delta_i = 0 \mid \mathbf{x}_i, y; \phi^{(t)}\right)},
\end{aligned}
$$

and

$$\bar{S}_2\left(\phi \mid \beta^{(t)}, \phi^{(t)}\right) = \sum_{\delta_i=1} \{\delta_i - \pi\left(\mathbf{x}_i, y_i; \phi\right)\} \left(\mathbf{x}_i', y_i\right)'$$

$$+ \sum_{\delta_i=0} \sum_{j=0}^{1} w_{ij(t)} \{\delta_i - \pi_i\left(\mathbf{x}_i, j; \phi\right)\} \left(\mathbf{x}_i', j\right)'.$$

In the M-step, the parameter estimates are updated by solving

$$\left[\bar{S}_1\left(\beta \mid \beta^{(t)}, \phi^{(t)}\right), \bar{S}_2\left(\phi \mid \beta^{(t)}, \phi^{(t)}\right)\right] = (0,0)$$

for β and ϕ.

Example 3.12. *Suppose that we have a random sample from a t-distribution with mean μ and scale parameter σ and with ν degrees of freedom such that $x_i = \mu + \sigma e_i$ with $e_i \sim t(\nu)$. Note that we can write*

$$e_i = u_i / \sqrt{w_i},$$

where

$$x_i \mid w_i \sim N\left(\mu, \sigma^2/w_i\right), \quad w_i \sim \chi_\nu^2/\nu. \tag{3.49}$$

In this case, suppose that we are interested in estimating $\theta = (\mu, \sigma)$ by the maximum likelihood method. To apply the EM algorithm, we can treat w_i in (3.49) as a latent variable which is always unobserved.

By Bayes' formula, we have

$$f(w_i \mid x_i) \propto f(w_i) f(x_i \mid w_i)$$

$$\propto (w_i \nu)^{\frac{\nu}{2}-1} \exp\left(-\frac{w_i \nu}{2}\right) \times (\sigma^2/w_i)^{-1/2} \exp\left\{-\frac{w_i}{2}\left(\frac{x_i-\mu}{\sigma}\right)^2\right\}.$$

Note that the above density function is the density for Gamma distribution with parameter $\alpha = (\nu+1)/2$ and $\beta = 2\left\{\nu + (x_i - \mu)^2/\sigma^2\right\}^{-1}$.

Thus, the E-step of EM algorithm can be written as

$$E(w_i \mid x_i, \theta^{(t)}) = \frac{\nu+1}{\nu + \left(d_i^{(t)}\right)^2},$$

where $d_i^{(t)} = (x_i - \mu^{(t)})/\sigma^{(t)}$. The M-step is then

$$\mu^{(t+1)} = \frac{\sum_{i=1}^{n} w_i^{(t)} x_i}{\sum_{i=1}^{n} w_i^{(t)}},$$

$$\sigma^{2(t+1)} = \frac{1}{n} \sum_{i=1}^{n} w_i^{(t)} \left(x_i - \mu^{(t+1)}\right)^2,$$

where $w_i^{(t)} = E(w_i \mid x_i, \theta^{(t)})$.

Next, we briefly discuss how to speed up the convergence of the EM algorithm. Let $\tilde{\theta}^{(t)}$ be the sequence from the usual EM algorithm and $\hat{\theta}^{(t)}$ be the sequence from the Newton method. That is,

$$\hat{\theta}^{(t+1)} = \hat{\theta}^{(t)} + \left\{ \mathcal{I}_{\text{obs}}(\hat{\theta}^{(t)}) \right\}^{-1} S_{\text{obs}}(\hat{\theta}^{(t)})$$

and

$$\tilde{\theta}^{(t+1)} = \tilde{\theta}^{(t)} + \left\{ \bar{I}_{\text{com}}(\tilde{\theta}^{(t)}) \right\}^{-1} S_{\text{obs}}(\tilde{\theta}^{(t)})$$

where $\bar{I}_{\text{com}}(\theta) = -E\left\{ \dot{S}_{\text{com}}(\theta) \mid \mathbf{y}_{\text{obs}}, \boldsymbol{\delta} \right\}$ and $\dot{S}_{\text{com}}(\theta) = \partial S_{\text{com}}(\theta)/\partial\theta'$. Thus, we have

$$\hat{\theta}^{(t+1)} - \hat{\theta}^{(t)} = \left\{ \mathcal{I}_{\text{obs}}(\hat{\theta}^{(t)}) \right\}^{-1} \bar{I}_{\text{com}}(\tilde{\theta}^{(t)}) \left\{ \tilde{\theta}^{(t+1)} - \tilde{\theta}^{(t)} \right\}.$$

Because of the missing information principle, we can write $\mathcal{I}_{\text{obs}}(\theta) = E\{\bar{I}_{\text{com}}(\tilde{\theta})\} - \mathcal{I}_{\text{mis}}(\theta)$ and

$$\hat{\theta}^{(t+1)} - \hat{\theta}^{(t)} = \left\{ I - \mathcal{J}_{\text{mis}}(\hat{\theta}^{(t)}) \right\}^{-1} \left\{ \tilde{\theta}^{(t+1)} - \tilde{\theta}^{(t)} \right\}, \qquad (3.50)$$

where $\mathcal{J}_{\text{mis}}(\theta) = \mathcal{I}_{\text{mis}}(\theta)\{\mathcal{I}_{\text{com}}(\theta)\}^{-1}$ is the fraction of missing information. Setting $\tilde{\theta}^{(t)} = \hat{\theta}^{(t)}$, the procedure in (3.50) can be used to speed up the convergence of the EM algorithm. Such acceleration algorithm is commonly known as *Aitken acceleration* and was also proposed by Louis (1982).

3.4   Monte Carlo Computation

In many situations, the analytic computation methods are not directly applicable because the integral associated with the conditional expectation in the mean score equation is not necessarily of a closed form. In this case, the Monte Carlo method is often used to approximate the expectation.

To explain the Monte Carlo method, suppose that we are interested in computing $\theta = E\{h(X)\}$ where X is a random variable with known density $f(x)$. In this case, a Monte Carlo approximation of θ is

$$\hat{\theta}_m = \frac{1}{m} \sum_{i=1}^{m} h(x_i), \qquad (3.51)$$

where $x_1, \ldots, x_m$ are m independent realizations of random variable X. By the law of large numbers, $\hat{\theta}_m$ in (3.51) converges in probability to θ as $m \to \infty$. Since the variance of $\hat{\theta}_m$ is $m^{-1}\sigma_h^2$, where $\sigma_h^2 = V\{h(X)\}$, we have $\hat{\theta}_m - \theta = O_p(m^{-1/2})$ if σ_h^2 is finite. Here, $X_n = O_p(1)$ denotes that X_n is bounded in probability. Monte Carlo method is attractive because the computation is simple, and it is directly applicable even when X is a random vector.

There are two main issues in the Monte Carlo computation. The first is how to generate samples from a target distribution f. This type of problems, i.e., simulation problem, can be critical when the target density does not follow a familiar parametric density. The second issue is how to reduce the variance of the Monte Carlo estimator for a given Monte Carlo sample size m. Variance reduction techniques in Monte Carlo sampling are not covered here and can be found, for example, in Givens and Hoeting (2005).

Probably, the simplest method of simulation is based on the probability integral transformation. For any continuous distribution function F, if U follows the uniform distribution with support $(0,1)$, then $X = F^{-1}(U)$ has cumulative distribution function equal to F. The probability integral transformation approach is not directly applicable to multivariate distribution.

Rejection sampling is another popular technique for simulation. Given a density of interest f, suppose that there exist a density g and a constant M such that

$$f(x) \leq Mg(x) \tag{3.52}$$

on the support of f. The rejection sampling method proceeds as follows:

1. Sample $Y \sim g$ and $U \sim U(0,1)$, where $U(0,1)$ denotes the uniform $(0,1)$ distribution.

2. Reject Y if

$$U > \frac{f(Y)}{Mg(Y)}. \tag{3.53}$$

   In this case, do not record the value of Y as an element in the target random sample and return to step 1.

3. Otherwise, keep the value of Y. Set $X = Y$, and consider X to be an element of the target random sample.

In the rejection sampling method,

$$
\begin{aligned}
P(X \leq y) &= P\left\{ Y \leq y \mid U \leq \frac{f(Y)}{Mg(Y)} \right\} \\
&= \frac{\int_{-\infty}^{y} \int_{0}^{f(x)/Mg(x)} du \, g(x) \, dx}{\int_{-\infty}^{\infty} \int_{0}^{f(x)/Mg(x)} du \, g(x) \, dx} \\
&= \frac{\int_{-\infty}^{y} f(x) \, dx}{\int_{-\infty}^{\infty} f(x) \, dx}.
\end{aligned}
$$

Note that the rejection sampling method is applicable when the density f is known up to a multiplicative factor, because the above equality still follows even if $f(x) \propto f_1(x)$ with $f_1(x) < Mg(x)$ and the decision rule (3.53) uses $f_1(x)$ instead of $f(x)$. Gilks and Wild (1992) proposed an adaptive rejection algorithm that computes the envelops function $g(x)$ automatically.

Importance sampling is also a very popular Monte Carlo approximation method. Writing

$$\theta \equiv \int h(x) f(x) \, dx = \int h(x) \frac{f(x)}{g(x)} g(x) \, dx$$

for some density $g(x)$, we can approximate θ by

$$\hat{\theta} = \sum_{i=1}^{m} w_i h(x_i),$$

where

$$w_i = \frac{f(x_i)/g(x_i)}{\sum_{j=1}^{m} f(x_j)/g(x_j)}$$

and $x_1, \ldots, x_m$ are IID realizations from a distribution with density $g(x)$. We require that the support of g include the entire support of f. Furthermore, g should have heavier tails than f, or at least the ratio $f(x)/g(x)$ should never grow too large. The distribution g that generates the Monte Carlo sample is called the *proposal distribution,* and the distribution f is called the *target distribution.*

Markov Chain Monte Carlo (MCMC) is also a frequently used Monte Carlo computation tool. A sequence of random variable $\{X_t : t = 0, 1, \ldots\}$ is called a *Markov chain* if the distribution of each element depends only on the previous one. That is, it satisfies

$$P(X_t \mid X_0, X_1, \ldots, X_{t-1}) = P(X_t \mid X_{t-1})$$

for each $t = 1, 2, \ldots$. Markov Chain Monte Carlo (MCMC) refers to a body of methods for generating pseudorandom draws from probability distributions via Markov chains.

To explain the idea of the MCMC methods, let Z be a generic random vector with density $f(Z)$, which is difficult to simulate from directly. In this case, we can construct a Markov chain $\left\{Z^{(t)} : t = 1, 2 \ldots\right\}$ with f as its stationary distribution so that

$$P(Z^{(t)}) \to f \quad \text{as } t \to \infty$$

or

$$\frac{1}{N} \sum_{t=1}^{N} h(Z^{(t)}) \to E_f[h(Z)] = \int h(z) f(z) \, dz \tag{3.54}$$

as $N \to \infty$. A Markov chain that satisfies (3.54) is called *ergodic*. Note that the random variables $Z^{(1)}, Z^{(2)}, \ldots, Z^{(N)}$ are not IID but (3.54) can be satisfied.

Gibbs sampling, introduced by Geman and Geman (1984), is one of the popular MCMC methods. Given $Z^{(t)} = (Z_1^{(t)}, Z_2^{(t)}, \ldots, Z_J^{(t)})$, Gibbs sampling draws $Z^{(t+1)}$ by sampling from the full conditionals of f,

$$Z_1^{(t+1)} \sim P\left(Z_1 \mid Z_2^{(t)}, Z_3^{(t)}, \ldots, Z_J^{(t)}\right)$$
$$Z_2^{(t+1)} \sim P\left(Z_2 \mid Z_1^{(t+1)}, Z_3^{(t)}, \ldots, Z_J^{(t)}\right)$$

$$\vdots$$

$$Z_J^{(t+1)} \sim P\left(Z_J \mid Z_1^{(t+1)}, Z_2^{(t+1)}, \ldots, Z_{J-1}^{(t+1)}\right).$$

Under mild regularity conditions, $P(Z^{(t)}) \to f$ as $t \to \infty$.

Example 3.13. *Suppose* $Z = (Z_1, Z_2)'$ *is bivariate normal,*

$$\begin{pmatrix} Z_1 \\ Z_2 \end{pmatrix} \sim N\left[\begin{pmatrix} 0 \\ 0 \end{pmatrix}, \begin{pmatrix} 1 & \rho \\ \rho & 1 \end{pmatrix}\right].$$

The Gibbs sampler would be

$$Z_1 \sim N\left(\rho Z_2, 1 - \rho^2\right)$$
$$Z_2 \sim N\left(\rho Z_1, 1 - \rho^2\right).$$

After a suitably large "burn-in period" we would find that

$$Z_1^{(t+1)}, Z_1^{(t+2)}, \ldots, Z_1^{(t+n)} \sim N(0,1)$$
$$Z_2^{(t+1)}, Z_2^{(t+2)}, \ldots, Z_2^{(t+n)} \sim N(0,1).$$

Note that, if $\rho \neq 0$ *the samples are dependent.*

Another very popular MCMC method is the *Metropolis–Hastings (MH) algorithm*, proposed by Metropolis et al. (1953) and Hastings (1970). To explain the MH algorithm, let $f(Z)$ be a known distribution on $\mathbb{R}^k$ except for the normalizing constant. The aim is to generate $Z \sim f$. The method starts with the selection of the initial value $Z^{(0)}$ with the requirement that $f(Z^{(0)}) > 0$. Given $Z^{(t)}$, the MH algorithm generating $Z^{(t+1)}$ can be described as follows:

1. Sample a candidate value Z^* from a proposal distribution $g(Z \mid Z^{(t)})$.

2. Compute the ratio

$$R\left(Z^*, Z^{(t)}\right) = \frac{g(Z^{(t)} \mid Z^*)}{g\left(Z^* \mid Z^{(t)}\right)} \frac{f(Z^*)}{f\left(Z^{(t)}\right)}.$$

3. Sample a value $Z^{(t+1)}$ as

$$Z^{(t+1)} = \begin{cases} Z^* & \text{with probability } \rho(Z^*, Z^{(t)}) \\ Z^{(t)} & \text{with probability } 1 - \rho(Z^*, Z^{(t)}) \end{cases}$$

where $\rho(x,y) = \min\{R(x,y), 1\}$.

A rigorous justification for the ergodicity (3.54) of the MH method is beyond the scope of this book and can be found in Robert and Casella (1999). In the independent chain with $g(Z^* \mid Z^{(t)}) = g(Z^*)$, the Metropolis–Hastings ratio is

$$R\left(Z^*, Z^{(t)}\right) = \frac{f(Z^*)/g(Z^*)}{f(Z^{(t)})/g(Z^{(t)})},$$

which is the ratio of the importance weight for Z^* over the importance weight for $Z^{(t)}$. Thus, the Metropolis–Hastings ratio $R(Z^*, Z^{(t)})$ is also called the importance ratio. Roughly speaking, the basic idea of the MH algorithm is

1. From the current position x, move to y according to $g(y \mid x)$,
2. Stay at y with probability $\rho(x, y) = \min\{f(y)/f(x), 1\}$.

Example 3.14. *Suppose that* $Y_1, \ldots, Y_n \overset{i.i.d.}{\sim} N(\theta, 1)$ *and the prior distribution for* θ *is Cauchy (0,1) with density*

$$\pi(\theta) = \frac{1}{\pi(1 + \theta^2)}. \tag{3.55}$$

The posterior distribution can be derived as

$$\pi(\theta \mid y) \quad \propto \quad \exp\left\{-\frac{\sum_{i=1}^{n}(y_i - \theta)^2}{2}\right\} \times \frac{1}{1 + \theta^2}$$

$$\propto \quad \exp\left\{-\frac{n(\theta - \bar{y})^2}{2}\right\} \times \frac{1}{1 + \theta^2},$$

where $\bar{y} = n^{-1}\sum_{i=1}^{n} y_i$. *If we want to generate* $\theta \sim \pi(\theta \mid y)$, *we can apply the MH algorithm as follows:*

1. *Generate* θ^* *from Cauchy (0,1).*
2. *Given* $y_1, \ldots, y_n$, *compute the importance ratio*

$$R\left(\theta^*, \theta^{(t)}\right) = \frac{L(\theta^*)}{L(\theta^{(t)})},$$

where $L(\theta) = \exp\left\{-n(\theta - \bar{y})^2 / 2\right\}$.

3. *Accept* θ^* *as* $\theta^{(t+1)}$ *with probability* $\rho(\theta^*, \theta^{(t)}) = \min\left\{R(\theta^*, \theta^{(t)}), 1\right\}$.

One important special case of the Metropolis-Hasting algorithm is the random-walk Metropolis algorithm, first proposed by Metropolis et al. (1953). In the random-walk Metropolis algorithm, the proposal distribution is $q(Z \mid Z^{(t-1)}) = g(Z - Z^{(t-1)})$ with g being a symmetric distribution. That is,

$$Z^* = Z^{(t-1)} + \epsilon, \quad \epsilon \sim g.$$

In this case, the Metropolis-Hastings ratio becomes

$$R\left(Z^*, Z^{(t)}\right) = \frac{g(Z^{(t-1)} - Z^*)}{g(Z^* - Z^{(t-1)})} \frac{f(Z^*)}{f(Z^{(t-1)})} = \frac{f(Z^*)}{f(Z^{(t-1)})}.$$

3.5 Monte Carlo EM

In the EM algorithm, the E-step involves computing

$$Q\left(\eta \mid \eta^{(t)}\right) = E\left\{\log f\left(\mathbf{y},\boldsymbol{\delta};\eta\right) \mid \mathbf{y}_{\text{obs}},\boldsymbol{\delta};\eta^{(t)}\right\},$$

which involves integration. To avoid the computational challenges in the E-step, Wei and Tanner (1990) proposed the Monte Carlo EM (MCEM) method based on approximating the Q function by a Monte Carlo sampling. That is,

$$Q(\eta \mid \eta^{(t)}) \cong \frac{1}{m}\sum_{j=1}^{m}\log f\left(\mathbf{y}^{*(j)},\boldsymbol{\delta};\eta\right),$$

where $\mathbf{y}^{*(j)} = (\mathbf{y}_{\text{obs}},\mathbf{y}_{mis}^{*(j)})$ and $\mathbf{y}_{mis}^{*(1)},\ldots,\mathbf{y}_{mis}^{*(m)} \overset{i.i.d.}{\sim} f\left(\mathbf{y}_{mis} \mid \mathbf{y}_{\text{obs}},\boldsymbol{\delta};\eta^{(t)}\right)$. That is, Monte Carlo EM uses the Monte Carlo approximation for the E-step, whereas the M-step remains the same. Note that

$$f(\mathbf{y}_{\text{mis}} \mid \mathbf{y}_{\text{obs}},\boldsymbol{\delta};\hat{\eta}) = \frac{f(\mathbf{y};\hat{\theta})P(\boldsymbol{\delta} \mid \mathbf{y};\hat{\phi})}{\int f(\mathbf{y};\hat{\theta})P(\boldsymbol{\delta} \mid \mathbf{y};\hat{\phi})d\mathbf{y}_{\text{mis}}} = \frac{f(\mathbf{y}_{\text{mis}} \mid \mathbf{y}_{\text{obs}};\hat{\theta})P(\boldsymbol{\delta} \mid \mathbf{y};\hat{\phi})}{\int f(\mathbf{y}_{\text{mis}} \mid \mathbf{y}_{\text{obs}};\hat{\theta})P(\boldsymbol{\delta} \mid \mathbf{y};\hat{\phi})d\mathbf{y}_{\text{mis}}}.$$
$$(3.56)$$

Thus, to generated samples from (3.56) given the parameter estimate $\hat{\psi} = (\hat{\theta},\hat{\phi})$, the Metropolis–Hastings algorithm can be implemented as follows:

[Step 1] For a given t-th value $\mathbf{y}^{(t)} = (\mathbf{y}_{\text{obs}},\mathbf{y}_{mis}^{(t)})$, generate $\mathbf{y}^{*} = (\mathbf{y}_{\text{obs}},\mathbf{y}_{\text{mis}}^{*})$ from $f(\mathbf{y}_{\text{mis}} \mid \mathbf{y}_{\text{obs}};\hat{\theta})$ and compute the importance ratio

$$R(\mathbf{y}^{*},\mathbf{y}^{(t)}) = \frac{P(\boldsymbol{\delta} \mid \mathbf{y}^{*};\hat{\phi})}{P(\boldsymbol{\delta} \mid \mathbf{y}^{(t)};\hat{\phi})}.$$

[Step 2] If $R(\mathbf{y}^{*},\mathbf{y}^{*(t)}) > 1$, then $\mathbf{y}^{(t+1)} = \mathbf{y}^{*}$. Otherwise, choose

$$\mathbf{y}^{(t+1)} = \begin{cases} \mathbf{y}^{*} & \text{with probability } R(\mathbf{y}^{*},\mathbf{y}^{(t)}) \\ \mathbf{y}^{(t)} & \text{with probability } 1 - R(\mathbf{y}^{*},\mathbf{y}^{(t)}). \end{cases}$$

Instead of the Metropolis-Hastings algorithm, because $P(\delta \mid \mathbf{y};\hat{\phi})$ is bounded by 1, we can also use the rejection algorithm to generate Monte Carlo samples from the conditional distribution in (3.56). That is, we first generate $\mathbf{y}^{*} = (\mathbf{y}_{\text{obs}},\mathbf{y}_{mis}^{*})$ from $f(\mathbf{y}_{\text{mis}} \mid \mathbf{y}_{\text{obs}};\hat{\theta})$ and then accept it with probability $P(\boldsymbol{\delta} \mid \mathbf{y}^{*};\hat{\phi})$.

Example 3.15. *Consider the conditional distribution*

$$y_i \sim f\left(y_i \mid x_i;\theta\right).$$

Assume that x_i is always observed but we observe y_i only when $\delta_i = 1$ where $\delta_i \sim Bernoulli\{\pi_i(\phi)\}$ and

$$\pi_i(\phi) = \pi(x_i, y_i; \phi) = \frac{\exp(\phi_0 + \phi_1 x_i + \phi_2 y_i)}{1 + \exp(\phi_0 + \phi_1 x_i + \phi_2 y_i)}. \tag{3.57}$$

To implement the MCEM method, we need to generate samples from

$$f(y_i \mid x_i, \delta_i = 0; \hat{\theta}, \hat{\phi}) = \frac{f(y_i \mid x_i; \hat{\theta})\{1 - \pi(x_i, y_i; \hat{\phi})\}}{\int f(y_i \mid x_i; \hat{\theta})\{1 - \pi(x_i, y_i; \hat{\phi})\}dy_i}.$$

We can use the following rejection method

[Step 1] Generate y_i^* from $f\left(y_i \mid x_i; \hat{\theta}\right)$.

[Step 2] Using y_i^* generated from Step 1, compute

$$\pi_i^*(\hat{\phi}) = \frac{\exp(\hat{\phi}_0 + \hat{\phi}_1 x_i + \hat{\phi}_2 y_i^*)}{1 + \exp(\hat{\phi}_0 + \hat{\phi}_1 x_i + \hat{\phi}_2 y_i^*)}. \tag{3.58}$$

Accept y_i^* with probability $1 - \pi_i^*(\hat{\phi})$.

Using the above two steps, we can create m values of y_i, denoted by $y_i^{*(1)}, \ldots, y_i^{*(m)}$, and the M-step can be implemented by solving

$$\sum_{i=1}^{n} \sum_{j=1}^{m} S(\theta; x_i, y_i^{*(j)}) = 0$$

and

$$\sum_{i=1}^{n} \sum_{j=1}^{m} \left\{\delta_i - \pi(\phi; x_i, y_i^{*(j)})\right\}(1, x_i, y_i^{*(j)})' = \mathbf{0},$$

where $S(\theta; x_i, y_i) = \partial \log f(y_i \mid x_i; \theta)/\partial\theta$.

Example 3.16. *In Example 3.15, instead of having y_i missing when $\delta_i = 0$, suppose that a grouped version of y_i, denoted by $\tilde{y}_i = T(y_i)$, is observed. For example, in some surveys, total income is reported in an interval measure for various reasons. Also, age is often reported as groups. In this case, we observe x_i, δ_i, and $\delta_i y_i + (1 - \delta_i)\tilde{y}_i$, for $n = 1, 2, \ldots, n$. The mapping T from y_i to $\tilde{y}_i$ is deterministic in the sense that, if y_i is known, $\tilde{y}_i$ is uniquely determined. To implement the MCEM method, we need to generate samples from*

$$f(y_i \mid x_i, \tilde{y}_i, \delta_i = 0; \hat{\theta}, \hat{\phi}) = \frac{f(y_i \mid x_i; \hat{\theta})\{1 - \pi(x_i, y_i; \hat{\phi})\}I\{y_i \in \mathcal{R}(\tilde{y}_i)\}}{\int_{\mathcal{R}(\tilde{y}_i)} f(y_i \mid x_i; \hat{\theta})\{1 - \pi(x_i, y_i; \hat{\phi})\}dy_i}, \tag{3.59}$$

where $\mathcal{R}(\tilde{y}_i) = \{y : T(y) = \tilde{y}_i\}$. Thus, the rejection method for generating samples from (3.59) is described as follows:

[Step 1] Generate y_i^* from $f(y_i \mid x_i; \hat{\theta})$.

[Step 2] If $y_i^* \in \mathcal{R}(\tilde{y}_i)$, then accept y_i^* with probability $1 - \pi_i^*(\hat{\phi})$ where $\pi_i^*(\hat{\phi})$ is computed by (3.58). Otherwise, go to Step 1.

The observed information matrix can be computed by Louis' formula (2.30) using Monte Carlo computation. Using the alternative formula (2.31), the Monte Carlo approximation of the observed information matrix for the observed likelihood can be computed by

$$\hat{I}_{obs}^*(\eta) = -\frac{1}{m}\sum_{j=1}^{m}\frac{\partial}{\partial \eta'}S_{\mathrm{com}}(\eta; \mathbf{y}^{*(j)}, \boldsymbol{\delta}) - \frac{1}{m}\sum_{j=1}^{m}\left\{S_{\mathrm{com}}(\eta; \mathbf{y}^{*(j)}, \boldsymbol{\delta}) - \bar{S}_{\mathrm{I},m}^*(\eta)\right\}^{\otimes 2},$$

(3.60)

where $\bar{S}_{\mathrm{I},m}^*(\eta) = m^{-1}\sum_{j=1}^{m}S_{\mathrm{com}}(\eta; \mathbf{y}^{*(j)}, \boldsymbol{\delta})$.

Example 3.17. *Suppose that we are interested in estimating parameters in the conditional distribution $f(y \mid x; \theta)$. Instead of observing (x_i, y_i), suppose that we observe (z_i, y_i) in the sample, where z is conditionally independent of y conditional on x and $Cov(x, z) \neq 0$. Such variable z is called instrumental variable of x . We assume that we select a random sample, called validation sample or calibration sample, from the original sample and observe x_i in addition to (z_i, y_i). Thus, we can obtain a consistent estimator for the conditional distribution $g(x \mid z)$ from the validation sample. Such setup is often called external calibration in measurement error model literature (Guo and Little, 2011). The predictive distribution for x_i is, by Bayes' theorem and the conditional independence of z and y given x,*

$$f(x_i \mid y_i, z_i) \propto f(y_i \mid x_i)g(x_i \mid z_i).$$

(3.61)

Thus, we may rely on the Monte Carlo method to compute the E-step of the EM algorithm.

To apply the MCEM approach, suppose that we can estimate $g(x \mid z)$ from the validation sample. We can generate x_i^* from $\hat{g}(x_i \mid z_i)$, an estimated version of $g(x_i \mid z_i)$, and then accept with probability proportional to $f(y_i \mid x_i^*; \hat{\theta})$. If MH algorithm is to be used, the probability of accepting x_i^* is computed by

$$\rho(x_i^*, x_i^{(t-1)}) = min\left\{\frac{f(y_i \mid x_i^*; \hat{\theta})}{f(y_i \mid x_i^{(t-1)}; \hat{\theta})}, 1\right\}.$$

Once m Monte Carlo samples are generated from $f(x_i \mid y_i, z_i)$, the parameters are updated by solving

$$\sum_{i \in V}S(\theta; x_i, y_i) + \sum_{i \in V^c}\sum_{j=1}^{m}S(\theta; x_i^{*(j)}, y_i) = 0$$

for θ, where V is the set of sample indices for the validation sample and $S(\theta; x, y) = \partial \ln f(y \mid x; \theta)/\partial\theta$.

Example 3.18. *We now consider the parameter estimation problem in generalized linear mixed models (GLMM). Let y_{ij} be a binary random variable (that takes values 0 or 1) with probability $p_{ij} = P(y_{ij} = 1 \mid \mathbf{x}_{ij}, a_i)$ and assume that*

$$logit(p_{ij}) = \mathbf{x}'_{ij}\beta + a_i$$

where $\mathbf{x}_{ij}$ is a p-dimension covariate associated with the j-th repetition of unit i, β is the parameter of interest that can represent the treatment effect due to $\mathbf{x}$, and a_i represents the random effect associate with unit i. We assume that a_i are IID from a distribution with $N(0, \sigma_a^2)$. In this case, the latent variable, a variable that is always missing, is a_i, and the observed likelihood is

$$L_{obs}(\beta, \sigma^2) = \prod_i \int \left\{ \prod_j p(x_{ij}, a_i; \beta)^{y_{ij}} \left[1 - p(x_{ij}, a_i; \beta)\right]^{1-y_{ij}} \right\} \frac{1}{\sigma_a} \phi\left(\frac{a_i}{\sigma_a}\right) da_i,$$

where $\phi(\cdot)$ is the pdf of the standard normal distribution. To apply the MCEM approach, we generate a_i^* from

$$f\left(a_i \mid \mathbf{x}_i, \mathbf{y}_i; \hat{\beta}, \hat{\sigma}\right) \propto f_1(\mathbf{y}_i \mid \mathbf{x}_i, a_i; \hat{\beta}) f_2(a_i; \hat{\sigma}_a), \tag{3.62}$$

where

$$f_1(\mathbf{y}_i \mid \mathbf{x}_i, a_i; \beta) = \prod_j p(x_{ij}, a_i; \beta)^{y_{ij}} \left[1 - p(x_{ij}, a_i; \beta)\right]^{1-y_{ij}}$$

and $f_2(a_i; \sigma) = \phi(a_i/\sigma)/\sigma$. To generate a_i^* from (3.62) using the rejection sampling method, we first generate a_i^* from $f_2(a_i; \hat{\sigma})$ and then accept it with probability proportional to $f_1(\mathbf{y}_i \mid \mathbf{x}_i, a_i^*; \hat{\beta})$. We can also use the MH algorithm to generate samples from (3.62).

We use random walk Metropolis algorithm in this example as follows.

1. Starting with any $a_i^{(0)}$, iterate for $t = 1, 2 \ldots$.

2. Draw $\epsilon_i \sim N(0, 0.1)$ and set

$$a_i^* = a_i^{(t-1)} + \epsilon_i.$$

3. Compute

$$R(a_i^*, a_i^{(t-1)}) = \frac{f(a_i^* \mid \mathbf{x}_i, \mathbf{y}_i; \hat{\beta}, \hat{\sigma}_a)}{f(a_i^{(t-1)} \mid \mathbf{x}_i, \mathbf{y}_i; \hat{\beta}, \hat{\sigma}_a)}$$

and

$$\rho(a_i^*, a_i^{(t-1)}) = min\{R(a_i^*, a_i^{(t+1)}), 1\}.$$

4. Set $a_i^{(t)} = a_i^*$ with probability $\rho(a_i^*, a_i^{(t+1)})$. Otherwise, set $a_i^{(t)} = a_i^{(t-1)}$.

In the MCEM method, the convergence of the EM sequence of $\{\theta^{(t)}\}$ is hard to check. Booth and Hobert (1999) discuss some convergence criteria for MCEM.

3.6   Data Augmentation

In this section, we consider an alternative method of generating Monte Carlo samples in the analysis of missing data. Assume that the complete data $\mathbf{y}$ can be written as $\mathbf{y} = (\mathbf{y}_{\text{obs}}, \mathbf{y}_{\text{mis}})$, with $\mathbf{y}_{\text{mis}}$ representing the missing part of $\mathbf{y}$. To make the presentation simple, assume an ignorable missing mechanism. Let $f(\mathbf{y}; \theta)$ be the joint density of the complete data $\mathbf{y}$ with parameter θ. Ideally, the Monte Carlo samples are generated from the conditional distribution of the missing values given the observation evaluated at true parameter value θ_0

$$\mathbf{y}_{\text{mis}}^* \sim f(\mathbf{y}_{\text{mis}} \mid \mathbf{y}_{\text{obs}}; \theta_0).$$

Since the true value is unknown, the MCEM method updates the parameter estimate $\hat{\theta}$ iteratively. In the MCEM method, the parameter estimates are updated deterministically in the sense that the best parameter estimate is chosen based on the current evaluation of the conditional expectation in the MCEM algorithm. Alternatively, one can consider a stochastic update of the parameter estimates by treating the parameter as a random variable as we do in Bayesian analysis. By treating θ as a random variable, we can generate the Monte Carlo samples by averaging over the parameter values. That is,

$$\mathbf{y}_{\text{mis}}^* \sim f_1(\mathbf{y}_{\text{mis}} \mid \mathbf{y}_{\text{obs}}) = \int f(\mathbf{y}_{\text{mis}} \mid \mathbf{y}_{\text{obs}}, \theta) p_{\text{obs}}(\theta \mid \mathbf{y}_{\text{obs}}) d\theta. \qquad (3.63)$$

Here, $f_1(\mathbf{y}_{\text{mis}} \mid \mathbf{y}_{\text{obs}})$ is often called the *posterior predictive distribution* and $p_{\text{obs}}(\theta \mid \mathbf{y}_{\text{obs}})$ is the posterior distribution of θ given the observation $\mathbf{y}_{\text{obs}}$ and is called the *observed posterior distribution*. One can generate Monte Carlo samples from (3.63) by the following steps:

1. Generate θ^* from $p_{\text{obs}}(\theta \mid \mathbf{y}_{\text{obs}})$
2. Given θ^*, generate $\mathbf{y}_{\text{mis}}^*$ from $f(\mathbf{y}_{\text{mis}} \mid \mathbf{y}_{\text{obs}}, \theta^*)$

Often it is hard to generate samples from the observed posterior distribution $p_{\text{obs}}(\theta \mid \mathbf{y}_{\text{obs}})$ directly. Instead, one can consider generating them from

$$\theta^* \sim p_{\text{obs}}(\theta \mid \mathbf{y}_{\text{obs}}) = \int p(\theta \mid \mathbf{y}_{\text{obs}}, \mathbf{y}_{\text{mis}}) f_1(\mathbf{y}_{\text{mis}} \mid \mathbf{y}_{\text{obs}}) d\mathbf{y}_{\text{mis}}, \qquad (3.64)$$

where $p(\theta \mid \mathbf{y}_{\text{obs}}, \mathbf{y}_{\text{mis}}) = p(\theta \mid \mathbf{y})$ is the conditional distribution of θ given $\mathbf{y}$ and $f_1(\mathbf{y}_{\text{mis}} \mid \mathbf{y}_{\text{obs}})$ is the posterior predictive distribution defined in (3.63). The conditional distribution $p(\theta \mid \mathbf{y})$ is often called the *posterior distribution* (under complete response) in Bayesian literature.

To generate samples from the posterior predictive distributions (3.63) and the observed posterior distribution (3.64), we can use the following iterative method:

[Imputation Step (I-step)] Given the parameter value θ^*, generate $\mathbf{y}^*_{\text{mis}}$ from $f(\mathbf{y}_{\text{mis}} \mid \mathbf{y}_{\text{obs}}, \theta^*)$.

[Posterior Step (P-step)] Given the imputed value $\mathbf{y}^* = (\mathbf{y}_{\text{obs}}, \mathbf{y}^*_{\text{mis}})$, generate θ^* from $p(\theta \mid \mathbf{y}^*)$.

The convergence to a stable posterior predictive distribution is very hard to check and often requires tremendous computation time. See Tanner and Wong (1987) for more details.

Example 3.19. *Let $y_1, y_2, \ldots, y_n$ be n independent realizations of a random variable Y following a Bernoulli (θ) distribution, where $\theta = P(Y = 1)$ and the prior distribution for θ is $Beta(\alpha, \beta)$. In this case, the posterior distribution under complete data is*

$$\theta \mid (y_1, y_2, \ldots, y_n) \sim Beta\left(\alpha + \sum_{i=1}^{n} y_i, \beta + n - \sum_{i=1}^{n} y_i\right).$$

Now, suppose that the first r elements are observed in y and the remaining $n - r$ elements are missing in y. Then the observed posterior becomes

$$\theta \mid (y_1, y_2, \ldots, y_r) \sim Beta\left(\alpha + \sum_{i=1}^{r} y_i, \beta + r - \sum_{i=1}^{r} y_i\right). \tag{3.65}$$

We can derive (3.65) using data augmentation. In the I-step, given the current parameter value $\theta^{*(t)}$, we generate $y_i^{*(t)}$ from a Bernoulli $(\theta^{*(t)})$ distribution for $i = r+1, \ldots, n$. For $i = 1, 2, \ldots, r$, we set $y_i^{*(t)} = y_i$. In the P-step, the parameters are generated by

$$\theta^{*(t+1)} \mid (y_1^{*(t)}, \ldots, y_n^{*(t)}) \sim Beta\left(\alpha + \sum_{i=1}^{n} y_i^{*(t)}, \beta + n - \sum_{i=1}^{n} y_i^{*(t)}\right).$$

Note that, for $i > r$,

$$
\begin{aligned}
E\left(y_i^{*(t+1)} \mid Y^{*(t)}\right) &= E\left(\theta^{*(t)} \mid Y^{*(t)}\right) \\
&= \frac{\alpha + \sum_{i=1}^{r} y_i + \sum_{i=r+1}^{n} y_i^{*(t)}}{\alpha + \beta + n} \\
&= \frac{\alpha + \sum_{i=1}^{r} y_i}{\alpha + \beta + r} + \lambda \left(\frac{\sum_{i=r+1}^{n} y_i^{*(t)}}{n - r} - \frac{\alpha + \sum_{i=1}^{r} y_i}{\alpha + \beta + r}\right),
\end{aligned}
$$

where $Y^{*(t)} = (y_1^{*(t)}, \ldots, y_n^{*(t)})$ and $\lambda = (n - r)/(\alpha + \beta + n)$. Writing

$$E\left(y_i^{*(t+1)} \mid Y^{*(t)}\right) = a_0 + \lambda(a^{(t)} - a_0),$$

where $a_0 = (\alpha + \sum_{i=1}^r y_i)/(\alpha + \beta + r)$ and $a^{(t)} = (\sum_{i=r+1}^n y_i^{*(t)})/(n-r)$, we can obtain

$$E\left(y_i^{*(t+1)} \mid Y^{*(1)}\right) = a_0 + \lambda^t \left(a^{(1)} - a_0\right).$$

Thus, as $\lambda < 1$,

$$\lim_{t\to\infty} E\left(y_i^{*(t+1)} \mid y_1, y_2 \ldots, y_r\right) = a_0 = \frac{\alpha + \sum_{i=1}^r y_i}{\alpha + \beta + r},$$

which can also be obtained directly from (3.65).

Example 3.20. *Assume that the scalar outcome variable Y_i follows the model*

$$y_i \quad = \quad \mathbf{x}_i'\beta + e_i, \tag{3.66}$$
$$e_i \quad \overset{i.i.d.}{\sim} \quad N\left(0, \sigma^2\right).$$

The p-dimensional $\mathbf{x}_i$'s are observed on the complete sample and are assumed to be fixed. We assume that the first r units are the respondents. Let $\mathbf{y}_r = (y_1, y_2, \ldots, y_r)'$ and $X_r = (\mathbf{x}_1, \mathbf{x}_2, \ldots, \mathbf{x}_r)'$. Also, let $\mathbf{y}_{n-r} = (y_{r+1}, y_{r+2}, \ldots, y_n)'$ and $X_{n-r} = (\mathbf{x}_{r+1}, \mathbf{x}_{r+2}, \ldots, \mathbf{x}_n)'$.

The suggested Bayesian imputation method for model (3.66) is as follows:

[Posterior Step] Draw

$$\sigma^{*2} \mid \mathbf{y}_r \overset{i.i.d.}{\sim} (r-p)\hat{\sigma}_r^2/\chi_{r-p}^2, \tag{3.67}$$

and

$$\beta^* \mid (\mathbf{y}_r, \sigma^*) \overset{i.i.d.}{\sim} N\left(\hat{\beta}_r, (X_r'X_r)^{-1}\sigma^{*2}\right), \tag{3.68}$$

where $\hat{\sigma}_r^2 = (r-p)^{-1}\mathbf{y}_r'\left[I - X_r(X_r'X_r)^{-1}X_r'\right]\mathbf{y}_r$ and $\hat{\beta}_r = (X_r'X_r)^{-1}X_r'\mathbf{y}_r$.

[Imputation Step] For each missing unit $j = r+1, \ldots, n$, draw

$$e_j^* \mid (\beta^*, \sigma^*) \overset{i.i.d.}{\sim} N\left(0, \sigma^{*2}\right).$$

Then, $y_j^* = \mathbf{x}_j'\beta^* + e_j^*$ is the imputed value associated with unit j.

The above procedure assumes a constant prior for $(\beta, \log\sigma)$ and an ignorable response mechanism in the sense of Rubin (1976). Because of the monotone missing pattern, there is no need to iterate the procedures. The posterior distributions, (3.67) and (3.68), follow from known distributions.
We can also apply Gibbs' sampling approach to generate the imputed values:

[P-step] Given the current imputed data, $\mathbf{y}_n^{*(t)} = (y_1, \ldots, y_r, y_{r+1}^{*(t)}, \ldots, y_n^{*(t)})'$, update the parameters by generating from the posterior distribution

$$\sigma^{*(t)2} \mid \mathbf{y}_n^{*(t)} \overset{i.i.d.}{\sim} (n-p)\hat{\sigma}_n^{*(t)2}/\chi_{n-p}^2$$

and

$$\beta^{*(t)} \mid \left(\mathbf{y}_n^{*(t)}, \sigma^{*(t)}\right) \overset{i.i.d.}{\sim} N\left(\hat{\beta}_n^{(t)}, \left(X_n' X_n\right)^{-1} \sigma^{*(t)2}\right),$$

where $\hat{\sigma}_n^{*(t)2} = (n-p)^{-1} \mathbf{y}_n^{*(t)'} \left[I - X_n \left(X_n' X_n\right)^{-1} X_n'\right] \mathbf{y}_n^{*(t)}$ *and* $\hat{\beta}_n^{(t)} = (X_n' X_n)^{-1} X_n' \mathbf{y}_n^{*(t)}$.

[I-step] Draw

$$e_j^{*(t)} \mid \left(\beta^{*(t)}, \sigma^{*(t)}\right) \overset{i.i.d.}{\sim} N\left(0, \sigma^{*(t)2}\right).$$

Then, the imputed value for unit j is

$$y_j^{*(t)} = \mathbf{x}_j' \beta^{*(t)} + e_j^{*(t)}.$$

There are two main uses of data augmentation. First, data augmentation is used for parameter simulation. That is, it can be used to generate parameters from posterior distributions. The simulated parameters can be used to make Bayesian inference. The second use of data augmentation is to predict the missing values from the posterior predictive distribution. Multiple imputation is a tool for making an inference using the imputed data obtained from data augmentation. More details about multiple imputation inference will be covered in Chapter 5.

Exercises

1. Under the setup of Remark 3.2, prove the following:

    (a) Equations involving the marginal parameters:

    $$Var(\hat{\mu}_{1,HK}) = \frac{\sigma_{11}}{n_H}(1 - p_K),$$

    $$Var(\hat{\mu}_{1,HL}) = \frac{\sigma_{11}}{n_H}(1 - p_L \rho^2),$$

    $$Var(\hat{\mu}_{1,HKL}) = \frac{\sigma_{11}}{n_H}\left\{(1 - p_K) - \frac{p_L \rho^2 (1 - p_K)^2}{(1 - p_L p_K \rho^2)}\right\},$$

    and

    $$Var(\hat{\sigma}_{11,HK}) = \frac{2\sigma_{11}^2}{n_H}(1 - p_K),$$

    $$Var(\hat{\sigma}_{11,HL}) = \frac{2\sigma_{11}^2}{n_H}(1 - p_L \rho^4),$$

    $$Var(\hat{\sigma}_{11,HKL}) = \frac{2\sigma_{11}^2}{n_H}\left\{(1 - p_K) - \frac{p_L \rho^4 (1 - p_K)^2}{1 - p_L p_K \rho^4}\right\}.$$

(b) If $n_K = n_L$, then

$$Var(\hat{\mu}_{1,H}) \geq Var(\hat{\mu}_{1,HL}) \geq Var(\hat{\mu}_{1,HK}) \geq Var(\hat{\mu}_{1,HKL})$$

and

$$Var(\hat{\sigma}_{11,H}) \geq Var(\hat{\sigma}_{11,HL}) \geq Var(\hat{\sigma}_{11,HK}) \geq Var(\hat{\sigma}_{11,HKL}).$$

Give some intuitive explanation on why $\hat{\mu}_{1,HK}$ is more efficient than $\hat{\mu}_{1,HL}$. Discuss when equalities hold.

2. Consider the partially classified categorical data in Table 3.3. Using the EM algorithm, find the maximum likelihood estimates of the parameters and compute standard errors.

3. Consider the data in Table 3.4 which is adapted from Table 1.4-2 of Bishop et al. (1975). Table 3.4 gives the data for a 2^3 table of three categorical variables ($Y_1 = $ Clinic, $Y_2 = $ Parental care, $Y_3 = $ Survival), with one supplemental margin for Y_2 and Y_3 and another supplemental margin for Y_1 and Y_3. In this setup, Y_i are all dichotomous, taking either 0 or 1, and 8 parameters can be defined as $\pi_{ijk} = P(Y_1 = i, Y_2 = j, Y_3 = k)$, $i = 0, 1; j = 0, 1; k = 0, 1$.

For the orthogonal parametrization, we use

$$\eta = \left(\pi_{1|11}, \pi_{1|10}, \pi_{1|01}, \pi_{1|00}, \pi_{+1|1}, \pi_{+1|0}, \pi_{++1}\right)',$$

where $\pi_{i|jk} = P(y_1 = i \mid y_2 = j, y_3 = k)$, $\pi_{+j|k} = P(y_2 = j \mid y_3 = k)$, $\pi_{++k} = P(y_3 = k)$.

(a) Use the factoring likelihood approach to estimate parameter η and estimated standard errors.

(b) Use the EM algorithm to estimate parameter η and use the Louis formula to compute the estimated standard errors.

4. A survey of households in several communities in north-central Iowa was conducted to determine people's views of the community in which they lived. We consider two variables, "Age of Respondent" and "Number of Years Residing in Community". An initial mailing was made to 1,023 households. After two additional mailings a total of 787 eligible units responded. The respondents were divided into seven categories on the basis of age. The age categories and the number of responses to each mailing are given in Table 3.5, which was originally presented in Drew and Fuller (1980).

We assume that the population is partitioned into $K = 7$ age categories. We assume that a proportion $1 - \gamma$ of the population is composed of hard-core nonrespondents who will never answer the survey. We assume that the fraction of hard-core nonrespondents is the same in each age category. Let f_k be the population proportion in category k such that $\sum_{k=1}^{K} f_k = 1$. Let p_k be the conditional

TABLE 3.4

A 2^3 Table with Supplemental Margins

Set	y_1	y_2	y_3	Count
	1	1	1	293
	1	0	1	176
	0	1	1	23
H	0	0	1	197
	1	1	0	4
	1	0	0	3
	0	1	0	2
	0	0	0	17
	1		1	100
K	0		1	82
	1		0	5
	0		0	6
		1	1	90
L		0	1	150
		1	0	5
		0	0	10

probability of response for a unit in category k. Thus, the probability of response at the r-th call for an individual in the k-th category is

$$\pi_{rk} = \gamma(1-p_k)^{r-1}p_k f_k, r = 1,2,3.$$

Also, the probability that an individual in category k will not have responded after $R = 3$ calls is

$$\pi_0 = (1-\gamma) + \gamma \sum_{k=1}^{K}(1-p_k)^3 f_k.$$

The joint distribution of $n_{rk}(k = 1,\ldots,7; r = 1,2,3)$, and n_0 follows from a multinomial distribution with parameter $\pi_{rk}(k = 1,\ldots,7; r = 1,2,3)$, and π_0.

(a) Compute the observed likelihood from the data in Table 3.5 and find the maximum likelihood estimates of the parameters.

(b) Use the EM algorithm to estimate the parameters by applying the following steps:

   i. Write $\pi_0 = \sum_{k=1}^{K}\pi_{0k}$, where $\pi_{0k} = \pi_{0k,M} + \pi_{0k,R}$, $\pi_{0k,M} = (1-\gamma)f_k$, and $\pi_{0k,R} = \gamma(1-p_k)^R f_k$. Let $n_{0k} = n_{0k,M} + n_{0k,R}$ be the decomposition of $n_0 = n\pi_0$ corresponding to $\pi_{0k} =$

TABLE 3.5
Responses by Age for a Community Study

Age	First Mailing	Second Mailing	Third Mailing	No Response after 3 Attempts
15–24	28	17	11	
25–34	63	26	16	
35–44	73	32	23	
45–54	97	36	12	236
55–64	97	32	15	
65–74	72	26	13	
75+	47	28	23	

$\pi_{0k,M} + \pi_{0k,R}$. Find the full-sample likelihood function assuming that $n_{0k,M}, n_{0k,R}, n_{1k}, \ldots n_{Rk}$ are available for $k = 1, \ldots, K$ and follow a multinomial distribution.

ii. Construct the EM algorithm from the full-sample likelihood and compute the MLE of the parameters.

5. Consider bivariate random variable (X, Y) with joint density $f_1(x; \alpha) f_2(y \mid x; \beta)$, where $f_1(x; \alpha)$ is the marginal density of X and $f_2(y \mid x; \beta)$ is the conditional density of Y given $X = x$. Assume that f_1 is a normal distribution with parameter $\alpha = (\mu_x, \sigma_x^2)$, and f_2 is the exponential distribution with mean $\mu_y(x)$, where $1/\mu_y(x) = \beta_0 + \beta_1 x$. Suppose that we have n observations of (x_i, y_i) from the distribution of (X, Y) but some of the items are subject to missingness. We are interested in estimating $\beta = (\beta_0, \beta_1)$ from the data. Assume that the missing data mechanism is ignorable.

(a) Discuss how to obtain the maximum likelihood estimate of β if x_i are always observed and y_i are subject to missingness.

(b) Describe the EM algorithm when y_i are always observed and x_i are subject to missingness.

6. Let $X_1, \ldots, X_n$ be independently and identically distributed random variables with probability density function (pdf)

$$f_X(x; \theta) = \begin{cases} \theta e^{-\theta x} & \text{if } x > 0 \\ 0 & \text{otherwise,} \end{cases}$$

and $\theta > 0$. Now, $Y_1, \ldots, Y_n$ are independently derived by the following steps:

[Step 1] Generate X_i from $f_X(x; \theta)$ above.

[Step 2] Generate U_i from $\text{Exp}(1)$ distribution, independently from X_i, where $\text{Exp}(1)$ is the exponential distribution with mean 1.

[Step 3] If $U_i \leq X_i$, then set $Y_i = X_i$. Otherwise, go to Step 1.

Find the joint density of $Y_1, \ldots, Y_n$ using the Bayes' formula. Find the MLE of θ based on the observations $y_1, \ldots, y_n$. Discuss how to estimate the variance of the MLE.

7. The EM algorithm in Example 3.12 can be further extended to the problem of robust regression, where the error distribution in the regression model is assumed to follow from a t-distribution. Suppose that the model for robust regression can be written as

$$y_i = \beta_0 + \beta_1 x_i + \sigma e_i$$

where $e_i \sim t(\nu)$ with a known ν. Similarly to Example 3.12, we can write $e_i = u_i/\sqrt{w_i}$ where

$$y_i \mid (x_i, w_i) \sim N\left(\beta_0 + \beta_1 x_i, \sigma^2/w_i\right), \quad w_i \sim \chi_\nu^2/\nu.$$

Here, (x_i, y_i) are always observed, and w_i are always missing. Answer the following questions.

(a) Show that the conditional distribution of w_i given x_i and y_i follows from Gamma (α, β) for some α and β. Find the constants α and β.

(b) Show that the E-step can be expressed as

$$E(w_i \mid x_i, y_i, \theta^{(t)}) = \frac{\nu + 1}{\nu + \left(d_i^{(t)}\right)^2},$$

where $d_i^{(t)} = (y_i - \beta_0^{(t)} - \beta_1^{(t)} x_i)/\sigma^{(t)}$.

(c) Show that the M-step can be written as

$$
\begin{aligned}
\left(\mu_x^{(t)}, \mu_y^{(t)}\right) &= \sum_{i=1}^n w_i^{(t)} (x_i, y_i) / (\sum_{i=1}^n w_i^{(t)}), \\
\beta_0^{(t+1)} &= \mu_y^{(t)} - \hat{\beta}_1^{(t+1)} \mu_x^{(t)}, \\
\beta_1^{(t+1)} &= \frac{\sum_{i=1}^n w_i^{(t)} (x_i - \mu_x^{(t)})(y_i - \mu_y^{(t)})}{\sum_{i=1}^n w_i^{(t)} (x_i - \mu_x^{(t)})^2}, \\
\sigma^{2(t+1)} &= \frac{1}{n} \sum_{i=1}^n w_i^{(t)} \left(y_i - \beta_0^{(t)} - \beta_1^{(t)} x_i\right)^2,
\end{aligned}
$$

where $w_i^{(t)} = E(w_i \mid x_i, y_i, \theta^{(t)})$.

8. In repeated measures experiments we typically model the vectors of observations $\mathbf{y}_1, \mathbf{y}_2, \ldots, \mathbf{y}_n$ as

$$\mathbf{y}_i \mid b_i \sim N\left(\mathbf{x}_i' \beta + \mathbf{z}_i b_i, \sigma_e^2 I_{n_i}\right).$$

That is, each $\mathbf{y}_i$ is a vector of length n_i measured from subject i; $\mathbf{x}_i$ and $\mathbf{z}_i$ are known design matrices; β is a p-dimensional vector of parameters; b_i is a latent variable representing the random effect associated with individual i. We never observe b_i and simply assume that b_i' are IID with $N(0, \sigma_b^2)$. We want to derive the EM algorithm to estimate the unknown parameters β, σ_e^2, and σ_b^2.

(a) What is the log-likelihood based on the observed data? (Hint: What is the marginal distribution of $\mathbf{y}_i$?)

(b) Consider $(\mathbf{y}_i, b_i)$ for $i = 1, 2, \ldots, n$ as the "complete data". Write down the log-likelihood of the complete data. Use that and show that the E-step of the EM algorithm involves computing $\hat{b}_i \equiv E(b_i \mid \mathbf{y}_i)$ and $\widehat{b_i^2} \equiv E(b_i^2 \mid \mathbf{y}_i)$.

(c) Given the current values of the unknown parameters, show that

$$\hat{b}_i = \left(\mathbf{z}_i' \mathbf{z}_i + \frac{\sigma_e^2}{\sigma_b^2} \right)^{-1} \mathbf{z}_i' (\mathbf{y}_i - \mathbf{x}_i \beta)$$

$$\widehat{b_i^2} = \left(\hat{b}_i \right)^2 + \sigma_e^2 \left(\mathbf{z}_i' \mathbf{z}_i + \frac{\sigma_e^2}{\sigma_b^2} \right)^{-1}.$$

(d) Obtain the M-step updates for the unknown parameters β, σ_e^2, and σ_b^2.

9. Consider the problem of estimating parameters in the regression model

$$y_i = \beta_0 + \beta_1 x_{1i} + \beta_2 x_{2i} + e_i,$$

where $e_i \sim N(0, \sigma_e^2)$ and

$$\begin{pmatrix} x_{1i} \\ x_{2i} \end{pmatrix} \sim N \left[\begin{pmatrix} \mu_1 \\ \mu_2 \end{pmatrix}, \begin{pmatrix} \sigma_{11} & \sigma_{12} \\ \sigma_{12} & \sigma_{22} \end{pmatrix} \right].$$

Assume, for simplicity, that the parameters $(\mu_1, \mu_2, \sigma_{11}, \sigma_{12}, \sigma_{22})$ are known. Instead of observing (x_{1i}, x_{2i}, y_i), suppose that we observe (x_{1i}, z_i, y_i) in the sample such that $f(y \mid x_1, x_2, z) = f(y \mid x_1, x_2)$ and $Cov(z, x_2) \neq 0$. Assume that the conditional distribution $g(x_2 \mid z)$ is known. In this case, discuss how to implement the EM algorithm to estimate $\theta = (\beta_0, \beta_1, \beta_2, \sigma_e^2)$.

10. Consider the problem of estimating parameter θ in the conditional distribution of y conditional on x, given by $f(y \mid x; \theta)$. Instead of observing (x_i, y_i) throughout the sample, we observe (x_i, y_i) for $\delta_i = 1$ and observe y_i for $\delta_i = 0$. Thus, the covariate in the regression model is subject to missingness. Let $g(x; \alpha)$ be the marginal distribution of x_i in the original sample. Assume further that $P(\delta_i = 1 \mid x_i, y_i)$ does not depend on x_i. Thus, it is MAR. For simplicity, assume that $f(y \mid x; \theta)$ follows from a logistic regression model as in Example 3.1.

(a) Devise an EM algorithm for computing the MLE of θ when α is unknown.

(b) Devise an EM algorithm for computing the MLE of θ when α is known. Discuss the efficiency gain of the MLE of θ over the situation when α is unknown.

11. Consider a simple random effect logistic regression model with dichotomous observation y_{ij} and continuous covariate x_{ij} ($i = 1, 2, \ldots, n; j = 1, 2, \ldots, m$) with conditional probability

$$P(y_{ij} = 1 \mid x_{ij}, u_i, \beta) = \frac{\exp(\beta x_{ij} + u_i)}{1 + \exp(\beta x_{ij} + u_i)},$$

where $u_i \sim N(0, \sigma^2)$ is an unobserved random effect. The vector of random effects $(u_1, \ldots, u_n)$ corresponds to the missing data.

(a) Use the rejection sampling idea to construct a Monte Carlo EM algorithm for obtaining the ML estimates of β and σ^2.

(b) Discuss how to use the Metropolis–Hastings algorithm to construct a MCMC EM algorithm for obtaining the ML estimates of β and σ^2.

(c) Create artificial data from the model with $\sigma^2 = 1$, $\beta = 1$, and $x_{ij} \sim N(5, 1)$ with $n = 100$ and $m = 10$. Compare the above two methods numerically using the artificial data.

12. Consider the problem of estimating parameters in the regression model

$$y_i = \beta_0 + \beta_1 x_i + e_i$$

where $x_i \sim N(\mu_x, \sigma_x^2)$, $e_i \sim N(0, \sigma_e^2)$ and e_i is independent of x_i. Instead of observing (x_i, y_i), suppose that we observe (z_i, y_i) in the sample such that $f(y \mid x, z) = f(y \mid x)$. Assume that the conditional distribution $g(z \mid x)$ is a normal distribution with mean x and variance σ_u^2, where σ_u^2 is known. In this case, discuss how to implement the EM algorithm to estimate $\theta = (\mu_x, \sigma_x^2, \beta_0, \beta_1, \sigma_e^2)$ using the following steps:

(a) Find the conditional distribution of x given z and y.

(b) Describe the E-step for computing $E(x \mid z, y; \theta^{(t)})$ and $E(x^2 \mid z, y; \theta^{(t)})$

(c) Describe the M-step for obtaining $\theta^{(t+1)}$.

(d) Instead of observing (z_i, y_i), suppose that you observe (z_{i1}, z_{i2}, y_i), where (z_{i1}, z_{i2}) satisfies $z_{ij} \mid x_i \overset{indep}{\sim} N(x_i, \sigma_u^2)$, $j = 1, 2$, for unknown σ_u^2. In this case, how can the above EM algorithm be changed?

4

Imputation

Imputation can be viewed as a Monte Carlo approximation of the conditional expectation of the missing part of the complete sample estimator. Unlike the usual Monte Carlo approximation method, the size of Monte Carlo sample (or the imputation size), denoted by m, is not necessarily large. The main motivation for imputation is to provide a completed dataset so that the resulting point estimates are consistent among different users. Because the imputation involves estimated model parameters in the prediction model, its statistical inference is more complicated. We introduce the basic theory for statistical inference with imputed data and discuss variance estimation.

4.1 Introduction

Imputation is a computational tool for handling missing data by using Monte Carlo approximation in Chapter 3. For example, suppose that the parameter of interest is $\mu_g = E\{g(Y)\}$ and the complete sample estimator of μ_g is

$$\hat{\mu}_{g,n} = \frac{1}{n}\sum_{i=1}^{n} g(y_i),$$

where $y_1,\ldots,y_n$ are n independent realizations of a random variable Y with density $f(y)$. If the imputed dataset is provided with the imputed values of y_i for $\delta_i = 0$, then the imputed estimator for μ_g can be computed by

$$\hat{\mu}_{g,I} = \frac{1}{n}\sum_{i=1}^{n} \{\delta_i g(y_i) + (1-\delta_i)g(y_i^*)\},$$

where y_i^* is the imputed value for y_i. Here, the implicit assumption is that

$$E\{g(Y_i) \mid \delta_i = 0\} = E\{g(y_i^*) \mid \delta_i = 0\} \tag{4.1}$$

holds at least approximately for each unit i with $\delta_i = 0$. Thus, one interpretation of imputation is Monte Carlo approximation of the conditional expectation of the unobserved part of the complete sample estimator. A sufficient condition for (4.1) is to generate y_i^* from the conditional distribution of y_i

DOI: 10.1201/9780429321740-4

given $\delta_i = 0$. If the imputed values are not provided then the user needs to obtain a prediction for $g(y_i)$ given $\delta_i = 0$, which can be difficult when little is known about the response mechanism. Further, without imputation, the resulting estimates for μ_g can be different when different users estimate the same parameter under different models. Sometimes it is desirable to avoid this type of inconsistency.

The following example illustrates the statistical issues associated with imputed data.

Example 4.1. *Let $(x,y)'$ be a vector of bivariate random variables. Assume that x_i is always observed and y_i is subject to missingness in the sample, and the probability of missingness does not depend on the value of y_i. That is, it is missing at random. In this case, a consistent estimator of $\theta = E(Y)$ based on a single imputation can be computed by*

$$\hat{\theta}_I = \frac{1}{n} \sum_{i=1}^{n} \left\{ \delta_i y_i + (1-\delta_i) y_i^* \right\}, \tag{4.2}$$

where y_i^* satisfies

$$E\left(y_i^* \mid \delta_i = 0\right) = E\left(y_i \mid \delta_i = 0\right). \tag{4.3}$$

If we have a model for $E(y_i \mid x_i)$, such as

$$y_i \sim N\left(\beta_0 + \beta_1 x_i, \sigma_e^2\right),$$

then we can use

$$y_i^* = \hat{\beta}_0 + \hat{\beta}_1 x_i + e_i^*,$$

where $E(\hat{\beta}_0, \hat{\beta}_1 \mid \mathbf{x}_n, \boldsymbol{\delta}_n) = (\beta_0, \beta_1)$, $E(e_i^* \mid \mathbf{x}_n, \boldsymbol{\delta}_n) = 0$, and $(\hat{\beta}_0, \hat{\beta}_1)$ is the MLE of (β_0, β_1). That is, $(\hat{\beta}_0, \hat{\beta}_1)$ is the solution to

$$\sum_{i=1}^{n} \delta_i \{ y_i - \beta_0 - \beta_1 x_i \}(1, x_i) = (0,0).$$

If $e_i^* \equiv 0$, the imputation method is called the (deterministic) regression imputation. Note that the imputation estimator (4.2) under regression imputation can be written as

$$\hat{\theta}_I = \frac{1}{n} \sum_{i=1}^{n} (\hat{\beta}_0 + \hat{\beta}_1 x_i),$$

which is equivalent to the MLE of $\theta = E(Y)$. See also Example 3.4. Thus, it satisfies

$$E\left(\hat{\theta}_I - \theta\right) = 0$$

and

$$V\left(\hat{\theta}_I\right) = \frac{1}{n}\sigma_y^2 + \left(\frac{1}{r} - \frac{1}{n}\right)\sigma_e^2 = \frac{\sigma_y^2}{r}\left\{1 - \left(1 - \frac{r}{n}\right)\rho^2\right\}. \tag{4.4}$$

If e_i^* are randomly generated from a distribution with mean 0 and variance $\hat{\sigma}_e^2$, which is the MLE of σ_e^2, then

$$V\left(\hat{\theta}_I\right) = \frac{1}{n}\sigma_y^2 + \left(\frac{1}{r} - \frac{1}{n}\right)\sigma_e^2 + \frac{n-r}{n^2}\sigma_e^2,$$

where the third term represents the additional variance due to the stochastic imputation.

Deterministic imputation is unbiased for estimating the population mean but may not be unbiased for estimating proportions. For example, if $\theta = P(Y < c) = E\{I(Y < c)\}$, the imputed estimator

$$\hat{\theta} = n^{-1}\sum_{i=1}^{n}\{\delta_i I(y_i < c) + (1 - \delta_i)I(y_i^* < c)\}$$

is unbiased if $E\{I(Y < c)\} = E\{I(Y^* < c)\}$, which holds only when the marginal distribution of y^* is the same as the marginal distribution of y. In general, under deterministic imputation, we have $E(y) = E(y^*)$ but $V(y^*) = \sigma_y^2(1 - \rho^2) < \sigma_y^2 = V(y)$. Stochastic regression imputation provides an approximate unbiased estimator of the proportions.

Inference with imputed data is challenging because the synthetic observations in the imputed data, $\{\delta_i y_i + (1 - \delta_i)y_i^* : i = 1, 2, \ldots, n\}$, are no longer independent even though the original observations are. For example, suppose that only the first r elements are observed and the imputed values for the remaining $n - r$ elements are all equal to $\bar{y}_r = \sum_{i=1}^{r}\delta_i y_i/r$. In this case, assuming MAR, the correlation between the two imputed values is equal to one. Generally, imputation increases the correlation between the imputed values and, thus, increases the variance of the resulting imputed estimator.

Note that in Example 4.1, the imputed values under the regression estimation is a function of $\hat{\beta} = (\hat{\beta}_0, \hat{\beta}_1)$. Thus, we can express the imputed estimator as $\hat{\theta}_I = \hat{\theta}_I(\hat{\beta})$ to emphasize the fact that it is a function of the estimated parameter $\hat{\beta}$. Here, β can be called a *nuisance parameter* in the sense that we are not directly interested in estimating β, but we need to account for the effect of estimating β when making inferences about θ using $\hat{\theta}_I = \hat{\theta}_I(\hat{\beta})$. How to reflect the uncertainty of the estimated nuisance parameter in the final estimator of θ is a fundamental problem in statistics (Yuan and Jennrich, 2000). It is closely related with the pseudo maximum likelihood estimation of Gong and Samaniego (1981).

4.2 Basic Theory

In this section, we develop basic theories for estimating $\eta = (\theta, \phi)$ from the imputed data, where θ is the parameter in $f(\mathbf{y};\theta)$, and ϕ is the parameter in $P(\boldsymbol{\delta} \mid \mathbf{y}; \phi)$. In other words, θ is related to the complete-sample data, and ϕ is related to the response mechanism. Suppose that $m \geq 1$ imputed values, say $\mathbf{y}_{\text{mis}}^{*(1)}, \ldots, \mathbf{y}_{\text{mis}}^{*(m)}$, are generated from $f(\mathbf{y}_{\text{mis}} \mid \mathbf{y}_{\text{obs}}, \boldsymbol{\delta}; \hat{\eta}_p)$, which is the conditional distribution defined in (3.56) and $\hat{\eta}_p$ is a preliminary estimator of parameter η. Using the m imputed values, we can obtain the imputed score equation as

$$\bar{S}_{\text{I},m}(\eta \mid \hat{\eta}_p) \equiv \frac{1}{m} \sum_{j=1}^{m} S_{\text{com}}(\eta; \mathbf{y}^{*(j)}, \boldsymbol{\delta}) = 0, \qquad (4.5)$$

where $\mathbf{y}^{*(j)} = (\mathbf{y}_{\text{obs}}, \mathbf{y}_{\text{mis}}^{*(j)})$. Let $\hat{\eta}_{\text{I},m}$ be the solution to (4.5). Note that $\hat{\eta}_{\text{I},m}$ is the one-step update of $\hat{\eta}_p$ using the imputed score equation based on $\bar{S}_{\text{I},m}(\eta \mid \hat{\eta}_p)$ in (4.5). We are now interested in the asymptotic properties of $\hat{\eta}_{\text{I},m}$.

Before discussing the asymptotic properties, we first present the following result without proof.

Lemma 4.1. *Let $\hat{\theta}$ be the solution to $\hat{U}(\theta) = 0$, where $\hat{U}(\theta)$ is a function of complete observations $\mathbf{y}_1, \ldots, \mathbf{y}_n$ and parameter θ. Let θ_0 be the solution to $E\{\hat{U}(\theta)\} = 0$. Then, under some regularity conditions,*

$$\hat{\theta} - \theta_0 \cong -\left[E\{\dot{U}(\theta_0)\}\right]^{-1} \hat{U}(\theta_0),$$

where $\dot{U}(\theta) = \partial \hat{U}(\theta)/\partial \theta'$, and the notation $A_n \cong B_n$ means that $B_n^{-1}A_n = 1 + R_n$ for some R_n which converges to zero in probability.

By Lemma 4.1 and by the definition of $\mathcal{I}_{\text{obs}}$, the solution $\hat{\eta}_{\text{mle}}$ to $S_{\text{obs}}(\eta) = 0$ satisfies

$$\hat{\eta}_{\text{mle}} - \eta_0 \cong \mathcal{I}_{\text{obs}}^{-1} S_{\text{obs}}(\eta_0). \qquad (4.6)$$

To discuss the asymptotic properties of $\hat{\eta}_{\text{I},m}^*$ obtained from (4.5), we first consider the asymptotic properties of $\hat{\eta}_{\text{I},\infty} = p\lim_{m \to \infty} \hat{\eta}_{\text{I},m}$, which solves

$$\bar{S}(\eta \mid \hat{\eta}_p) \equiv E\{S_{\text{com}}(\eta) \mid \mathbf{y}_{\text{obs}}, \boldsymbol{\delta}; \hat{\eta}_p\} = 0, \qquad (4.7)$$

where

$$
\begin{aligned}
E\{S_{\text{com}}(\eta) \mid \mathbf{y}_{\text{obs}}, \boldsymbol{\delta}; \hat{\eta}_p\} &= p\lim_{m \to \infty} \frac{1}{m} \sum_{j=1}^{m} S_{\text{com}}\left(\eta; \mathbf{y}^{*(j)}, \boldsymbol{\delta}\right) \\
&= \int S_{\text{com}}(\eta; \mathbf{y}, \boldsymbol{\delta}) f(\mathbf{y}_{\text{mis}} \mid \mathbf{y}_{\text{obs}}, \boldsymbol{\delta}; \hat{\eta}_p) d\mathbf{y}_{\text{mis}}.
\end{aligned}
$$

The following lemma presents the asymptotic properties of $\hat{\eta}_{\text{I},\infty}$. Here, we let η_0 be the true parameter value in the joint density $f(\mathbf{y}, \boldsymbol{\delta}; \eta)$. The first part of Lemma 4.2 is a special case of the general result in Theorem 4.1.

Lemma 4.2. *Let $\hat{\eta}_{I,\infty}$ be the solution to (4.7). Assume that $\hat{\eta}_p$ converges in probability to η_0. Then, under some regularity conditions,*

$$\hat{\eta}_{I,\infty} - \eta_0 \cong \hat{\eta}_{mle} - \eta_0 + \mathcal{I}_{com}^{-1}\mathcal{I}_{mis}\left(\hat{\eta}_p - \hat{\eta}_{mle}\right) \tag{4.8}$$

and

$$V\left(\hat{\eta}_{I,\infty}\right) \doteq \mathcal{I}_{obs}^{-1} + \mathcal{J}_{mis}\left\{V\left(\hat{\eta}_p\right) - V\left(\hat{\eta}_{mle}\right)\right\}\mathcal{J}_{mis}', \tag{4.9}$$

where $\mathcal{J}_{mis} = \mathcal{I}_{com}^{-1}\mathcal{I}_{mis}$ is the fraction of missing information and the notation $\doteq$ denotes approximation.

Proof. Proof of (4.8) will be covered in Theorem 4.1. Now, the asymptotic covariance between $\hat{\eta}_{mle}$ and $\hat{\eta}_p - \hat{\eta}_{MLE}$ is zero. To see this, consider

$$\hat{\eta}^* = \hat{\eta}_{mle} - \frac{Cov(\hat{\eta}_{mle}, \hat{\eta}_p - \hat{\eta}_{mle})}{V(\hat{\eta}_p - \hat{\eta}_{mle})}(\hat{\eta}_p - \hat{\eta}_{mle}).$$

Note that

$$V(\hat{\eta}_{mle}) \geq V(\hat{\eta}^*)$$

with equality hold if and only if $Cov(\hat{\eta}_{mle}, \hat{\eta}_p - \hat{\eta}_{mle}) = 0$. Therefore, since $\hat{\eta}_{mle}$ is asymptotically optimal, the asymptotic covariance between $\hat{\eta}_{mle}$ and $\hat{\eta}_p - \hat{\eta}_{mle}$ should be equal to zero. Therefore, the asymptotic variance is

$$V\left(\hat{\eta}_{I,\infty}\right) \doteq V\left(\hat{\eta}_{mle}\right) + \mathcal{J}_{mis}V\left(\hat{\eta}_p - \hat{\eta}_{mle}\right)\mathcal{J}_{mis}'.$$

Now, since $V\left(\hat{\eta}_{mle}\right) = \mathcal{I}_{obs}^{-1}$ and

$$V(\hat{\eta}_p - \hat{\eta}_{mle}) = V(\hat{\eta}_p) - V(\hat{\eta}_{mle}),$$

we have (4.9). □

Equation (4.8) implies that

$$\hat{\eta}_{I,\infty} = (I - \mathcal{J}_{mis})\,\hat{\eta}_{mle} + \mathcal{J}_{mis}\hat{\eta}_p. \tag{4.10}$$

That is, $\hat{\eta}_{I,\infty}$ is a convex combination of $\hat{\eta}_{mle}$ and $\hat{\eta}_p$. Recall that $\hat{\eta}_{I,\infty}$ is the one-step update of $\hat{\eta}_p$ using the mean score equation in (4.7). Writing $\hat{\eta}^{(t)}$ to be the t-th EM update of η that is computed by solving

$$\bar{S}(\eta \,|\, \hat{\eta}^{(t-1)}) = 0$$

with $\hat{\eta}^{(0)} = \hat{\eta}_p$. Equation (4.10) implies that

$$\hat{\eta}^{(t)} = (I - \mathcal{J}_{mis})\,\hat{\eta}_{mle} + \mathcal{J}_{mis}\hat{\eta}^{(t-1)}.$$

Thus, we can obtain

$$\hat{\eta}^{(t)} = \hat{\eta}_{mle} + (\mathcal{J}_{mis})^{t-1}(\hat{\eta}^{(0)} - \hat{\eta}_{mle}),$$

which justifies $\lim_{t\to\infty} \hat{\eta}^{(t)} = \hat{\eta}_{\mathrm{mle}}$. Therefore, using the same argument for (4.9), we can establish

$$V(\hat{\eta}^{(t)}) \doteq \mathcal{I}_{\mathrm{obs}}^{-1} + (\mathcal{J}_{\mathrm{mis}})^{t-1} \{V(\hat{\eta}_p) - V(\hat{\eta}_{\mathrm{mle}})\} (\mathcal{J}'_{\mathrm{mis}})^{t-1}.$$

Now, we consider the asymptotic properties of $\hat{\eta}_{\mathrm{I},m}$ obtained from (4.5) for finite $m \geq 1$. The following lemma presents a preliminary result.

Lemma 4.3. *Let* $S_I^*(\eta \mid \hat{\eta}_p) = S_{\mathrm{com}}(\eta; \mathbf{y}^*)$ *be the imputed score function evaluated with* $\mathbf{y}^* = (\mathbf{y}_{\mathrm{obs}}, \mathbf{y}_{\mathrm{mis}}^*)$, *where* $\mathbf{y}_{\mathrm{mis}}^*$ *is generated from* $f(\mathbf{y}_{\mathrm{mis}} \mid \mathbf{y}_{\mathrm{obs}}, \boldsymbol{\delta}; \hat{\eta}_p)$. *Assume that* $\hat{\eta}_p$ *converges in probability to* η_0. *Then, under some regularity conditions, the solution* $\hat{\eta}_I^*$ *to* $S_I^*(\eta \mid \hat{\eta}_p) = 0$ *satisfies*

$$\hat{\eta}_I^* - \eta_0 \cong (\hat{\eta}_{\mathrm{mle}} - \eta_0) + \mathcal{J}_{\mathrm{mis}}(\hat{\eta}_p - \hat{\eta}_{\mathrm{mle}}) + \mathcal{I}_{\mathrm{com}}^{-1} Z^*, \tag{4.11}$$

where $Z^* \mid (\mathbf{y}_{\mathrm{obs}}, \boldsymbol{\delta}) \sim (0, \mathcal{I}_{\mathrm{mis}})$ *and* $\mathcal{J}_{\mathrm{mis}} = \mathcal{I}_{\mathrm{com}}^{-1} \mathcal{I}_{\mathrm{mis}}$.

Proof. Taking Taylor expansion of $S_{\mathrm{com}}(\hat{\eta}_I^*; \mathbf{y}_{\mathrm{obs}}, \mathbf{y}_{\mathrm{mis}}^*) = 0$ around $\eta = \hat{\eta}_p$, we can obtain

$$\begin{aligned} 0 &= S_{\mathrm{com}}(\hat{\eta}_I^*; \mathbf{y}_{\mathrm{obs}}, \mathbf{y}_{\mathrm{mis}}^*) \\ &\cong S_{\mathrm{com}}(\hat{\eta}_p; \mathbf{y}_{\mathrm{obs}}, \mathbf{y}_{\mathrm{mis}}^*) + \left\{ \frac{\partial}{\partial \eta'} S_{\mathrm{com}}(\hat{\eta}_p) \right\} (\hat{\eta}_I^* - \hat{\eta}_p). \end{aligned}$$

Note that

$$\begin{aligned} S_{\mathrm{com}}(\hat{\eta}_p; \mathbf{y}_{\mathrm{obs}}, \mathbf{y}_{\mathrm{mis}}^*) &= S_{\mathrm{obs}}(\hat{\eta}_p; \mathbf{y}_{\mathrm{obs}}) + S_{\mathrm{mis}}(\hat{\eta}_p; \mathbf{y}_{mis}^* \mid \mathbf{y}_{\mathrm{obs}}), \\ S_{\mathrm{obs}}(\hat{\eta}_p; \mathbf{y}_{\mathrm{obs}}) &\cong S_{\mathrm{obs}}(\hat{\eta}_{\mathrm{mle}}; \mathbf{y}_{\mathrm{obs}}) \\ &\quad + \left\{ \frac{\partial}{\partial \eta'} S_{\mathrm{obs}}(\hat{\eta}_{\mathrm{mle}}) \right\} (\hat{\eta}_p - \hat{\eta}_{\mathrm{mle}}) \\ &\cong -\mathcal{I}_{\mathrm{obs}}(\hat{\eta}_p - \hat{\eta}_{\mathrm{mle}}) \tag{4.12} \end{aligned}$$

and

$$\frac{\partial}{\partial \eta} S_{\mathrm{com}}(\hat{\eta}_p) \xrightarrow{p} -\mathcal{I}_{\mathrm{com}}.$$

Thus, combining the above results, we have

$$\hat{\eta}_I^* - \hat{\eta}_p \cong \mathcal{I}_{\mathrm{com}}^{-1} \mathcal{I}_{\mathrm{obs}}(\hat{\eta}_{\mathrm{mle}} - \hat{\eta}_p) + \mathcal{I}_{\mathrm{com}}^{-1} S_{\mathrm{mis}}(\hat{\eta}_p; \mathbf{y}_{mis}^* \mid \mathbf{y}_{\mathrm{obs}}).$$

Now, conditional on $(\mathbf{y}_{\mathrm{obs}}, \boldsymbol{\delta})$ and $\hat{\eta}_p$, $Z^* \equiv S_{\mathrm{mis}}(\hat{\eta}_p; \mathbf{y}_{\mathrm{mis}}^* \mid \mathbf{y}_{\mathrm{obs}})$ is a random variable with its mean

$$E\{S_{\mathrm{mis}}(\hat{\eta}_p; \mathbf{y}_{\mathrm{mis}} \mid \mathbf{y}_{\mathrm{obs}}) \mid \mathbf{y}_{\mathrm{obs}}, \hat{\eta}_p\} = 0$$

and its variance

$$E[V\{S_{\mathrm{com}}(\hat{\eta}_p; \mathbf{y}_{\mathrm{mis}} \mid \mathbf{y}_{\mathrm{obs}}) \mid \mathbf{y}_{\mathrm{obs}}, \hat{\eta}_p\}] = \mathcal{I}_{\mathrm{mis}}.$$

Therefore, (4.11) is established. $\square$

Note that the three terms in (4.11) are uncorrelated. The sum of the first two terms is asymptotically equal to $\hat{\eta}_{\mathrm{I},\infty} - \eta_0$ by Lemma 4.2 and so its asymptotic variance is equal to (4.9). The last term has variance

$$V_{\mathrm{imp}} = V(\hat{\eta}_I^* - \hat{\eta}_{\mathrm{I},\infty}) = \mathcal{I}_{\mathrm{com}}^{-1} \mathcal{I}_{\mathrm{mis}} \mathcal{I}_{\mathrm{com}}^{-1}.$$

The variance term V_{imp} is called the *imputation variance* because it is the variance due to random generation of the imputed values. If we have m independent imputed values, then we can expect that

$$V(\hat{\eta}_{\mathrm{I},m} - \hat{\eta}_{\mathrm{I},\infty}) = m^{-1} \mathcal{I}_{\mathrm{com}}^{-1} \mathcal{I}_{\mathrm{mis}} \mathcal{I}_{\mathrm{com}}^{-1}. \tag{4.13}$$

Therefore, we can establish the following theorem, originally proved by Wang and Robins (1998).

Theorem 4.1. *Let $\hat{\eta}_p$ be a preliminary $\sqrt{n}$-consistent estimator of η with variance V_p. Under some regularity conditions, the solution $\hat{\eta}_{\mathrm{I},m}$ to (4.5) has mean η_0 and the asymptotic variance*

$$V\left(\hat{\eta}_{\mathrm{I},m}\right) \doteq \mathcal{I}_{\mathrm{obs}}^{-1} + \mathcal{J}_{\mathrm{mis}} \left\{ V_p - \mathcal{I}_{\mathrm{obs}}^{-1} \right\} \mathcal{J}_{\mathrm{mis}}' + m^{-1} \mathcal{I}_{\mathrm{com}}^{-1} \mathcal{I}_{\mathrm{mis}} \mathcal{I}_{\mathrm{com}}^{-1}, \tag{4.14}$$

where $\mathcal{J}_{\mathrm{mis}} = \mathcal{I}_{\mathrm{com}}^{-1} \mathcal{I}_{\mathrm{mis}}$.
In particular, if we use $\hat{\eta}_p = \hat{\eta}_{\mathrm{mle}}$, then the asymptotic variance is

$$V\left(\hat{\eta}_{\mathrm{I},m}\right) \doteq \mathcal{I}_{\mathrm{obs}}^{-1} + m^{-1} \mathcal{I}_{\mathrm{com}}^{-1} \mathcal{I}_{\mathrm{mis}} \mathcal{I}_{\mathrm{com}}^{-1}. \tag{4.15}$$

Proof. We have only to show that $\hat{\eta}_{I,m}$ satisfies

$$\hat{\eta}_{\mathrm{I},m} \cong \hat{\eta}_{\mathrm{mle}} + \mathcal{J}_{\mathrm{mis}}(\hat{\eta}_p - \hat{\eta}_{\mathrm{mle}}) + \mathcal{I}_{\mathrm{com}}^{-1} \cdot m^{-1} \sum_{k=1}^{m} Z^{*(k)}, \tag{4.16}$$

where

$$Z^{*(k)} \mid (\mathbf{y}_{\mathrm{obs}}, \hat{\eta}_p) \overset{i.i.d}{\sim} (0, \mathcal{I}_{\mathrm{mis}}).$$

Note that, as $m \to \infty$, the last term of (4.16) converges in probability to zero. Thus, (4.16) implies result (4.8) in Lemma 4.2.

The proof for (4.16) is very similar to that of (4.11) in Lemma 4.3. Note that $\hat{\eta}_{I,m}$ satisfies

$$\bar{S}_{I,m}(\eta) \equiv \frac{1}{m} \sum_{j=1}^{m} S_{\mathrm{com}}(\eta; \mathbf{y}_{\mathrm{obs}}, \mathbf{y}_{\mathrm{mis}}^{*(j)}) = 0.$$

Thus, using the same Taylor expansion of $\bar{S}_{I,m}(\hat{\eta}_{I,m}) = 0$ around $\eta = \hat{\eta}_p$, we obtain

$$
\begin{aligned}
0 &\cong m^{-1} \sum_{j=1}^{m} S_{\text{com}}(\hat{\eta}_p; \mathbf{y}_{obs}, \mathbf{y}_{\text{mis}}^{*(j)}) + \left\{ \frac{\partial}{\partial \eta} S_{\text{com}}(\hat{\eta}_p) \right\} (\hat{\eta}_{I,m} - \hat{\eta}_p) \\
&\cong S_{\text{obs}}(\hat{\eta}_p) + m^{-1} \sum_{j=1}^{m} S_{\text{mis}} \left(\hat{\eta}_p; \mathbf{y}_{\text{mis}}^{*(j)} \mid \mathbf{y}_{obs} \right) - \mathcal{I}_{\text{com}} (\hat{\eta}_{I,m} - \hat{\eta}_p) \\
&\cong \mathcal{I}_{\text{obs}}(\hat{\eta}_{\text{mle}} - \hat{\eta}_p) + m^{-1} \sum_{j=1}^{m} S_{\text{mis}} \left(\hat{\eta}_p; \mathbf{y}_{\text{mis}}^{*(j)} \mid \mathbf{y}_{obs} \right) - \mathcal{I}_{\text{com}} (\hat{\eta}_{I,m} - \hat{\eta}_p),
\end{aligned}
$$

where the last approximate equality follows from (4.12). The remaining parts of the proof are essentially the same as that of Lemma 4.3. □

We now consider a more general case of estimating ψ_0, which is defined through an estimating equation $E\{U(\psi; \mathbf{y})\} = 0$. Under complete response, a consistent estimator of ψ can be obtained by solving $U_{\text{com}}(\psi; \mathbf{y}) \equiv \sum_{i=1}^{n} U(\psi; y_i) = 0$. Assume that some part of $\mathbf{y}$, denoted by $\mathbf{y}_{\text{mis}}$, is not observed and m imputed values, say $\mathbf{y}_{\text{mis}}^{*(1)}, \ldots, \mathbf{y}_{\text{mis}}^{*(m)}$, are generated from $f(\mathbf{y}_{\text{mis}} \mid \mathbf{y}_{obs}, \boldsymbol{\delta}; \hat{\eta}_{\text{mle}})$, where $f(\mathbf{y}_{\text{mis}} \mid \mathbf{y}_{obs}, \boldsymbol{\delta}; \hat{\eta}_{\text{mle}})$ is defined in (3.56) and $\hat{\eta}_{\text{mle}}$ is the MLE of η_0. The imputed estimating function using m imputed values is computed as

$$
\bar{U}_{I,m}(\psi) = \frac{1}{m} \sum_{j=1}^{m} U_{\text{com}}(\psi; \mathbf{y}^{*(j)}), \tag{4.17}
$$

where $\mathbf{y}^{*(j)} = (\mathbf{y}_{obs}, \mathbf{y}_{\text{mis}}^{*(j)})$. Let $\hat{\psi}_{I,m}$ be the solution to $\bar{U}_{I,m}(\psi) = 0$. Let $\bar{U}_{I,\infty}(\psi)$ be the probability limit of $\bar{U}_{I,m}(\psi)$ as $m \to \infty$. Note that

$$
\bar{U}_{I,\infty}(\psi) = E\{U_{\text{com}}(\psi) \mid \mathbf{y}_{obs}, \boldsymbol{\delta}; \hat{\eta}_{\text{mle}}\}. \tag{4.18}
$$

Note that, even though $U_{\text{com}}(\eta)$ is a linear function of $U(\psi; y_i)$, $\bar{U}_{I,\infty}(\psi)$ is not a linear function of $(\mathbf{y}_{i,\text{obs}}, \boldsymbol{\delta}_i)$ because of the estimated parameter $\hat{\eta}_{\text{mle}}$. We now introduce the following definition.

Definition 4.1. *Let $X_1, \ldots, X_n$ be IID sample from $f(x; \theta_0), \theta_0 \in \Theta$ and we are interested in estimating $\gamma_0 = \gamma(\theta_0)$, where $\gamma(\cdot) : \Theta \to R^k$. An estimator $\hat{\gamma} = \hat{\gamma}_n$ is called asymptotically linear if there exists a random vector $\psi(x)$ such that*

$$
\sqrt{n}(\hat{\gamma}_n - \gamma_0) = \frac{1}{\sqrt{n}} \sum_{i=1}^{n} \psi(X_i) + o_p(1) \tag{4.19}
$$

with $E_{\theta_0}\{\psi(X)\} = 0$ and $E_{\theta_0}\{\psi(X)\psi(X)'\}$ is finite and non-singular. Here, $Z_n = o_p(1)$ means that Z_n converges to zero in probability.

In this definition, the function $\psi(x)$ is referred to as an *influence function*. The phrase influence function was used by Hampel (1974) and is motivated by the fact that to the first order $\psi(x)$ is the influence of a single observation on the estimator $\hat{\gamma} = \hat{\gamma}(X_1, \ldots, X_n)$. The asymptotic properties of an asymptotically linear estimator, $\hat{\gamma}_n$, can be summarized by only considering its influence function. Since $\psi(X)$ has zero mean, the central limit theorem (CLT) tells us that

$$\frac{1}{\sqrt{n}} \sum_{i=1}^{n} \psi(X_i) \xrightarrow{\mathcal{L}} N\left[0, E_{\theta_0}\{\psi(X)\psi(X)'\}\right]. \tag{4.20}$$

Thus, combining (4.19) with (4.20) and applying Slutsky's theorem, we have

$$\sqrt{n}\,(\hat{\gamma}_n - \gamma_0) \xrightarrow{\mathcal{L}} N\left[0, E_{\theta_0}\{\psi(X)\psi(X)'\}\right].$$

The Taylor method is often used to obtain the linearization form in (4.19). The following theorem, originally proved by Robins and Wang (2000), presents some asymptotic properties of the estimator that is a solution to $\bar{U}_{I,\infty}(\psi) = 0$.

Theorem 4.2. *Suppose that the parameter of interest ψ_0 is defined as the solution to $E\{U(\psi; \mathbf{y})\} = 0$. Let $\hat{\psi}_{I,\infty}$ be the solution to $\bar{U}_{I,\infty}(\psi) = 0$, where $\bar{U}_{I,\infty}(\psi)$ is defined in (4.18). Then, under some regularity conditions, we can express*

$$\sqrt{n}\left(\hat{\psi}_{I,\infty} - \psi_0\right) = \frac{1}{\sqrt{n}} \sum_{i=1}^{n} D(\mathbf{y}_{i,\mathrm{obs}}, \delta_i) + o_p(1), \tag{4.21}$$

where

$$D(\mathbf{y}_{i,\mathrm{obs}}, \delta_i; \psi_0, \eta_0) = \tau^{-1} E\{U(\psi_0; \mathbf{y}_i) + \kappa S(\eta_0; \mathbf{y}_i, \boldsymbol{\delta}_i) \mid \mathbf{y}_{i,\mathrm{obs}}, \boldsymbol{\delta}_i\},$$

$\tau = -E\{\partial U(\psi_0)/\partial \psi'\}$, *and*

$$\kappa = E\left\{U_{\mathrm{com}}(\psi_0) S_{\mathrm{mis}}(\eta_0)'\right\} \mathcal{I}_{\mathrm{obs}}^{-1}. \tag{4.22}$$

Proof. Note that, in (4.18), $\bar{U}_{I,\infty}(\psi)$ depends on $\hat{\eta}_{\mathrm{mle}}$. To emphasize its dependency on $\hat{\eta}_{\mathrm{mle}}$, define

$$\bar{U}(\psi \mid \hat{\eta}_{\mathrm{mle}}) = E\left\{U_{\mathrm{com}}(\psi; \mathbf{y}) \mid \mathbf{y}_{\mathrm{obs}}, \boldsymbol{\delta}; \hat{\eta}_{\mathrm{mle}}\right\}.$$

By a Taylor expansion of $\bar{U}(\psi \mid \hat{\eta}_{\mathrm{mle}})$ around $\hat{\eta}_{\mathrm{mle}} = \eta_0$, we have

$$\bar{U}(\psi \mid \hat{\eta}_{\mathrm{mle}}) \cong \bar{U}(\psi \mid \eta_0) + \left\{\partial \bar{U}(\psi \mid \eta_0)/\partial \eta'\right\}(\hat{\eta}_{\mathrm{mle}} - \eta_0). \tag{4.23}$$

Since

$$\begin{aligned}
\frac{\partial}{\partial \eta'} \bar{U}(\psi \mid \eta) &= \int U_{\mathrm{com}}(\psi; \mathbf{y}) \frac{\partial}{\partial \eta'} f(\mathbf{y}_{\mathrm{mis}} \mid \mathbf{y}_{\mathrm{obs}}, \boldsymbol{\delta}; \eta) \, d\mu(\mathbf{y}_{\mathrm{mis}}) \\
&= E\left\{U_{\mathrm{com}}(\psi) S_{\mathrm{mis}}(\eta)' \mid \mathbf{y}_{\mathrm{obs}}, \boldsymbol{\delta}; \eta\right\},
\end{aligned}$$

using (4.6), the expansion in (4.23) reduces to

$$\bar{U}(\psi \mid \hat{\eta}_{\mathrm{mle}}) \cong \bar{U}(\psi \mid \eta_0) + E\left\{U_{\mathrm{com}}(\psi) S_{\mathrm{mis}}(\eta_0)'\right\} \mathcal{I}_{\mathrm{obs}}^{-1} S_{\mathrm{obs}}(\eta_0). \tag{4.24}$$

Writing

$$\bar{U}_l(\psi \mid \eta_0) \equiv \bar{U}(\psi \mid \eta_0) + E\{U_{\mathrm{com}}(\psi) S_{\mathrm{mis}}(\eta_0)'\} \mathcal{I}_{\mathrm{obs}}^{-1} S_{\mathrm{obs}}(\eta_0),$$

with subscript l denoting linearization, we can express, by Lemma 4.1,

$$\hat{\psi}_{\mathrm{I},\infty} - \psi_0 \cong -\left[E\left\{\frac{\partial}{\partial \psi'}\bar{U}_l(\psi_0 \mid \eta_0)\right\}\right]^{-1} \bar{U}_l(\psi_0 \mid \eta_0).$$

Since $E\{S_{\mathrm{obs}}(\eta_0)\} = 0$, we have

$$E\left\{\frac{\partial}{\partial \psi'}\bar{U}_l(\psi_0 \mid \eta_0)\right\} = E\left\{\frac{\partial}{\partial \psi'}\bar{U}(\psi_0 \mid \eta_0)\right\} = nE\left\{\frac{\partial}{\partial \psi'}U(\psi_0)\right\},$$

and we can write

$$
\begin{aligned}
\hat{\psi}_{\mathrm{I},\infty} - \psi_0 &\cong n^{-1}\tau^{-1}\left\{\bar{U}(\psi_0 \mid \eta_0) + \kappa S_{\mathrm{obs}}(\eta_0)\right\} &(4.25)\\
&= n^{-1}\tau^{-1}E\left\{U_{\mathrm{com}}(\psi_0) + \kappa S_{\mathrm{com}}(\eta_0) \mid \mathbf{y}_{\mathrm{obs}}, \boldsymbol{\delta}; \eta_0\right\}\\
&= n^{-1}\tau^{-1}\sum_{i=1}^{n} E\left\{U(\psi_0; \mathbf{y}_i) + \kappa S(\eta_0; \mathbf{y}_i, \boldsymbol{\delta}_i) \mid \mathbf{y}_{i,\mathrm{obs}}, \boldsymbol{\delta}_i; \eta_0\right\},
\end{aligned}
$$

where κ is as defined in (4.22), and the result follows.   □

By (4.21), we can obtain

$$\sqrt{n}\left(\hat{\psi}_{\mathrm{I},\infty} - \psi_0\right) \xrightarrow{\mathcal{L}} N(0, \Omega_0), \qquad (4.26)$$

where

$$\Omega_0 = \lim \frac{1}{n}\sum_{i=1}^{n} E\{D(\mathbf{y}_{i,\mathrm{obs}}, \boldsymbol{\delta}_i; \psi_0, \eta_0)^{\otimes 2}\}. \qquad (4.27)$$

To discuss the variance of $\hat{\psi}_{\mathrm{I},m}$ for finite $m \geq 1$, note that we can write

$$\hat{\psi}_{\mathrm{I},m} = \hat{\psi}_{\mathrm{I},\infty} + \left(\hat{\psi}_{\mathrm{I},m} - \hat{\psi}_{\mathrm{I},\infty}\right)$$

and the two terms are independent. Thus, we have

$$V(\hat{\psi}_{\mathrm{I},m}) = V(\hat{\psi}_{\mathrm{I},\infty}) + V(\hat{\psi}_{\mathrm{I},m} - \hat{\psi}_{\mathrm{I},\infty}).$$

By the same argument for (4.25), we have

$$\hat{\psi}_{\mathrm{I},m} - \psi_0 \cong n^{-1}\tau^{-1}\left\{\bar{U}_{I,m}(\psi_0 \mid \eta_0) + \kappa S_{\mathrm{obs}}(\eta_0)\right\}$$

and so

$$\hat{\psi}_{\mathrm{I},m} - \hat{\psi}_{\mathrm{I},\infty} \cong n^{-1}\tau^{-1}\left\{\bar{U}_{I,m}(\psi_0 \mid \eta_0) - \bar{U}_{I,\infty}(\psi_0 \mid \eta_0)\right\}.$$

Thus, writing $V_{\text{imp}}(\bar{U}) = E\{V(n^{-1}U_{\text{com}} \mid \mathbf{y}_{\text{obs}}, \boldsymbol{\delta})\}$, we can obtain

$$V(\hat{\psi}_{\text{I},m} - \hat{\psi}_{\text{I},\infty}) \cong m^{-1}\tau^{-1}V_{\text{imp}}(U)\tau^{-1'}$$

and

$$V(\hat{\psi}_{\text{I},m}) \cong n^{-1}\Omega_0 + m^{-1}\tau^{-1}V_{\text{imp}}(U)\tau^{-1'}, \qquad (4.28)$$

where Ω_0 is defined in (4.27).

Example 4.2. *We now go back to Example 4.1. In this case, we can write*

$$U(\theta) = \sum_{i=1}^{n}(y_i - \theta)/\sigma_e^2$$

and the score function for β under complete response is, under joint normality,

$$S_{\text{com}}(\beta) = \frac{1}{\sigma_e^2}\sum_{i=1}^{n}(y_i - \beta_0 - \beta_1 x_i)(1, x_i)'.$$

Assuming that the response mechanism is ignorable, we have

$$S_{\text{obs}}(\beta) = \frac{1}{\sigma_e^2}\sum_{i=1}^{n}\delta_i(y_i - \beta_0 - \beta_1 x_i)(1, x_i)'.$$

Let $\hat{\beta} = (\hat{\beta}_0, \hat{\beta}_1)$ be the solution to $S_{\text{obs}}(\beta) = 0$. Note that the imputed estimator (4.2) with $y_i^* = \hat{\beta}_0 + \hat{\beta}_1 x_i$ can be written as the solution to

$$E\left\{U(\theta) \mid \mathbf{y}_{\text{obs}}, \boldsymbol{\delta}; \hat{\beta}\right\} = 0$$

since

$$E\left\{U(\theta) \mid \mathbf{y}_{\text{obs}}, \boldsymbol{\delta}; \beta\right\} \equiv \bar{U}(\theta \mid \beta) = \sum_{i=1}^{n}\{\delta_i y_i + (1 - \delta_i)(\beta_0 + \beta_1 x_i) - \theta\}/\sigma_e^2.$$

Thus, using the linearization formula (4.24), we have

$$\bar{U}_l(\theta \mid \beta) = \bar{U}(\theta \mid \beta) + (\kappa_0, \kappa_1)S_{\text{obs}}(\beta),$$

where

$$(\kappa_0, \kappa_1)' = \mathcal{I}_{\text{obs}}^{-1}E\left\{S_{\text{mis}}(\beta)U(\theta)\right\}.$$

In this example, we have

$$\begin{pmatrix} \kappa_0 \\ \kappa_1 \end{pmatrix} = \left[E\left\{\sum_{i=1}^{n}\delta_i(1, x_i)'(1, x_i)\right\}\right]^{-1}E\left\{\sum_{i=1}^{n}(1 - \delta_i)(1, x_i)'\right\}. \qquad (4.29)$$

Thus,

$$\bar{U}_l(\theta \mid \beta)\sigma_e^2 = \sum_{i=1}^{n} \delta_i(y_i - \theta) + \sum_{i=1}^{n}(1 - \delta_i)(\beta_0 + \beta_1 x_i - \theta)$$

$$+ \sum_{i=1}^{n} \delta_i (y_i - \beta_0 - \beta_1 x_i)(\kappa_0 + \kappa_1 x_i).$$

Note that the solution to $\bar{U}_l(\theta \mid \beta) = 0$ *leads to*

$$\hat{\theta}_l = \frac{1}{n} \sum_{i=1}^{n} \{\beta_0 + \beta_1 x_i + \delta_i(1 + \kappa_0 + \kappa_1 x_i)(y_i - \beta_0 - \beta_1 x_i)\} = \frac{1}{n} \sum_{i=1}^{n} d_i. \quad (4.30)$$

For variance estimation, one can just apply the standard variance estimation formula to $\hat{d}_i = d_i(\hat{\beta})$, *the pseudo values for variance estimation. We will cover this topic further in Section 4.3. Under a uniform response mechanism,* $1 + \kappa_0 + \kappa_1 x_i \cong n/r$ *and the asymptotic variance of* $\hat{\theta}_l$ *is equal to*

$$\frac{1}{n}\beta_1^2 \sigma_x^2 + \frac{1}{r}\sigma_e^2 = \frac{1}{n}\sigma_y^2 + \left(\frac{1}{r} - \frac{1}{n}\right)\sigma_e^2$$

which is consistent with the result in (4.4).

4.3   Variance Estimation after Imputation

We now discuss variance estimation of the imputed point estimators. Variance estimation can be implemented using either a linearization method or a replication method. We first discuss the linearization method. The replication method shall be discussed in Section 4.4.

In the linearization method, we first decompose the imputed estimator into a sum of two components, a deterministic part and a stochastic part. The linearization method is applied to the deterministic component. For example, in Example 4.1, the imputed estimator based on the regression imputation can be written as a function of $\hat{\beta} = (\hat{\beta}_0, \hat{\beta}_1)$. That is, we can write

$$\hat{\theta}_I(\hat{\beta}) = \hat{\theta}_{Id} = \frac{1}{n} \sum_{i=1}^{n} \left\{ \delta_i y_i + (1 - \delta_i)(\hat{\beta}_0 + \hat{\beta}_1 x_i) \right\}. \quad (4.31)$$

By the linearization result in (4.30), we can find $d_i = d_i(\beta)$ such that

$$\hat{\theta}_I(\hat{\beta}) \cong \frac{1}{n} \sum_{i=1}^{n} d_i(\beta)$$

where

$$d_i(\beta) = \beta_0 + \beta_1 x_i + \delta_i \left(1 + \kappa_0 + \kappa_1 x_i\right)\left(y_i - \beta_0 - \beta_1 x_i\right)$$

and (κ_0, κ_1) is defined in (4.29). Note that, if (x_i, y_i, δ_i) are IID, then $d_i = d(x_i, y_i, \delta_i)$ are also IID. Thus, the variance of $\bar{d}_n = n^{-1} \sum_{i=1}^{n} d_i$ is unbiasedly estimated by

$$\hat{V}(\bar{d}_n) = \frac{1}{n} \frac{1}{n-1} \sum_{i=1}^{n} \left(d_i - \bar{d}_n\right)^2. \tag{4.32}$$

Unfortunately, we cannot compute $\hat{V}(\bar{d}_n)$ in (4.32) since $d_i = d_i(\beta)$ is a function of an unknown parameter. Thus, we use $\hat{d}_i = d_i(\hat{\beta})$ in (4.32) to get a consistent variance estimator of the imputed estimator.

Instead of the deterministic imputation, suppose that a stochastic imputation is used such that

$$\hat{\theta}_I = \frac{1}{n} \sum_{i=1}^{n} \left\{ \delta_i y_i + (1 - \delta_i)\left(\hat{\beta}_0 + \hat{\beta}_1 x_i + \hat{e}_i^*\right)\right\},$$

where $\hat{e}_i^*$ are the additional noise terms in the stochastic imputation. Often $\hat{e}_i^*$ are randomly selected from the empirical distribution of the sample residuals in the respondents. That is, $\hat{e}_i^*$ is randomly selected among the set $\hat{\mathbf{e}}_R = \{\hat{e}_i : \delta_i = 1\}$, where $\hat{e}_i = y_i - \hat{\beta}_0 - \hat{\beta}_1 x_i$. The variance of the imputed estimator can be decomposed into two parts:

$$V\left(\hat{\theta}_I\right) = V\left(\hat{\theta}_{Id}\right) + V\left(\hat{\theta}_I - \hat{\theta}_{Id}\right), \tag{4.33}$$

where the first part is the deterministic part and the second part is the additional increase in the variance due to stochastic imputation. The first part can be estimated by the linearization method discussed above. The second part is called the *imputation variance*. If we require the imputation mechanism to satisfy

$$\sum_{i=1}^{n} (1 - \delta_i) \hat{e}_i^* = 0,$$

then the imputation variance is equal to zero. Also, if we impute more than one imputed values so that

$$\hat{\theta}_I = n^{-1} m^{-1} \sum_{i=1}^{n} \sum_{j=1}^{m} \left\{ \delta_i y_i + (1 - \delta_i)\left(\hat{\beta}_0 + \hat{\beta}_1 x_i + \hat{e}_i^{*(j)}\right)\right\},$$

then the imputation variance is reduced by increasing the imputation size m. Often the variance of $\hat{\theta}_I - \hat{\theta}_{Id} = n^{-1} \sum_{i=1}^{n} (1 - \delta_i) \hat{e}_i^*$ can be computed under the known imputation mechanism. For example, if simple random sampling without replacement is used and $n < 2r$, then

$$V\left(\hat{\theta}_I - \hat{\theta}_{Id}\right) = E\left\{V\left(\hat{\theta}_I \mid \mathbf{y}_{obs}, \boldsymbol{\delta}\right)\right\} = \frac{n-r}{n^2}\left(2 - \frac{n}{r}\right)\frac{1}{r-1}\sum_{i=1}^{n} \delta_i \hat{e}_i^2.$$

If we can write $\hat{e}_i^* = \sum_{k=1}^{n} d_{ki}\delta_k\hat{e}_k$ for some d_{ji}, where $d_{ji} = 1$ if $\hat{e}_j$ is used for $\hat{e}_i^*$ and $d_{ji} = 0$ otherwise, then $\hat{\theta}_I - \hat{\theta}_{Id} = \sum_{i=1}^{n} \delta_i d_i \hat{e}_i$ where $d_i = \sum_{j=1}^{n}(1-\delta_j)d_{ij}$ is the number of times that $\hat{e}_i$ is used as a donor for random imputation and

$$\hat{V}_{imp} = \frac{1}{n^2}\sum_{i=1}^{n} \delta_i d_i^2 \hat{e}_i^2$$

can be used to estimate the imputation variance.

Instead of the above tailer-made estimation method for imputation variance, we may consider an alternative approach when the imputed values are independently generated and m is greater than one. Writing

$$\hat{\theta}_I = \frac{1}{m}\sum_{j=1}^{m} \hat{\theta}_I^{(j)} = \bar{\theta}_I^{(\cdot)},$$

where

$$\hat{\theta}_I^{(j)} = \frac{1}{n}\sum_{i=1}^{n}\{\delta_i y_i + (1-\delta_i)(\hat{\beta}_0 + \hat{\beta}_1 x_i + \hat{e}_i^{*(j)})\},$$

we can show that the imputation variance is unbiasedly estimated by $m^{-1}B_m$, where

$$B_m = \frac{1}{m-1}\sum_{j=1}^{m} \left(\hat{\theta}_I^{(j)} - \bar{\theta}_I^{(\cdot)}\right)^2. \tag{4.34}$$

The following lemma presents the properties of B_m in (4.34).

Lemma 4.4. *Let* $X_1,\dots,X_m$ *be identically distributed (not necessarily independent) with mean* $\theta = E(X)$ *and covariance*

$$Cov(X_i, X_j) = \begin{cases} c_{11} & \text{if } i = j \\ c_{12} & \text{otherwise.} \end{cases}$$

Let $\bar{X}_m = m^{-1}\sum_{i=1}^{m} X_i$ *and* $B_m = (m-1)^{-1}\sum_{i=1}^{m}(X_i - \bar{X}_m)^2$. *Then, we have*

$$V(\bar{X}_m) = c_{12} + m^{-1}(c_{11} - c_{12}), \tag{4.35}$$

and

$$E(B_m) = c_{11} - c_{12}. \tag{4.36}$$

Proof. Result (4.35) can be easily proved by

$$V(\bar{X}_m) = \frac{1}{m^2}\left\{\sum_{i=1}^{m} V(X_i) + \sum_{i=1}^{m}\sum_{j\neq i} Cov(X_i, X_i)\right\}.$$

Result (4.36) can be proved using the equality

$$\sum_{i=1}^{m}\left(X_i - \bar{X}_m\right)^2 = \sum_{i=1}^{m}\left(X_i - c\right)^2 - m\left(\bar{X}_m - c\right)^2$$

for any constant c. Choosing $c = E(X) = \theta$ and taking the expectation on both sides of the above equality, we have

$$(m-1)E\left(B_m\right) = mc_{11} - m\left\{c_{12} + m^{-1}\left(c_{11} - c_{12}\right)\right\},$$

and (4.36) follows. □

If we apply Lemma 4.4 to the imputed estimator with $X_j = \hat{\theta}_I^{(j)} - \hat{\theta}_{Id}$, then

$$c_{12} = Cov\left\{n^{-1}\sum_{i=1}^{n}(1-\delta_i)\hat{e}_i^{*(1)}, n^{-1}\sum_{i=1}^{n}(1-\delta_i)\hat{e}_i^{*(2)}\right\}.$$

If $\hat{e}_i^*$ are randomly selected among $\hat{\mathbf{e}}_R$, then the above covariance term is zero and $c_{12} = 0$. Thus, we have

$$E\left(m^{-1}B_m\right) = m^{-1}V\left(\bar{\theta}_I^{(1)} - \hat{\theta}_{Id}\right) = V\left(\bar{\theta}_I^{(\cdot)} - \hat{\theta}_{Id}\right) \qquad (4.37)$$

and the imputation variance is unbiasedly estimated by $m^{-1}B_m$.

We now discuss a general case of parameter estimation when the parameter of interest ψ is estimated by the solution $\hat{\psi}_n$ to

$$\sum_{i=1}^{n}U(\psi;\mathbf{y}_i) = 0 \qquad (4.38)$$

under complete response of $\mathbf{y}_1,\ldots,\mathbf{y}_n$. The complete-sample variance estimator of $\hat{\psi}_n$ is

$$\hat{V}(\hat{\psi}_n) = \hat{\tau}_u^{-1}\hat{\Omega}_u(\hat{\tau}_u^{-1})', \qquad (4.39)$$

where

$$\hat{\tau}_u = n^{-1}\sum_{i=1}^{n}\dot{U}(\hat{\psi}_n;\mathbf{y}_i)$$

$$\hat{\Omega}_u = n^{-1}(n-1)^{-1}\sum_{i=1}^{n}(\hat{u}_i - \bar{u}_n)^{\otimes 2},$$

$\dot{U}(\psi;\mathbf{y}) = \partial U(\psi;\mathbf{y})/\partial\psi'$, $\bar{u}_n = n^{-1}\sum_{i=1}^{n}\hat{u}_i$, and $\hat{u}_i = U(\hat{\psi}_n;\mathbf{y}_i)$. The variance estimator in (4.39) is often called the *sandwich variance estimator*.

Under the existence of missing data, let $\mathbf{y}_{i,\text{obs}}$ be the observed part of $\mathbf{y}_i$ and $\mathbf{y}_{i,\text{mis}}$ the missing part of $\mathbf{y}_i$. Assume that m imputed values of $\mathbf{y}_{i,\text{mis}}$, denoted by $\mathbf{y}_{i,\mathrm{mis}}^{*(1)},\ldots,\mathbf{y}_{i,\mathrm{mis}}^{*(m)}$, are randomly generated from the conditional distribution $f\left(\mathbf{y}_{i,\mathrm{mis}}\mid\mathbf{y}_{i,\mathrm{obs}},\boldsymbol{\delta}_i;\hat{\eta}_p\right)$ where $\hat{\eta}_p$ is estimated by solving

$$\hat{U}_p\left(\eta\right)\equiv\sum_{i=1}^n U_p\left(\eta;\mathbf{y}_{i,\mathrm{obs}}\right)=0. \tag{4.40}$$

If we apply the m imputed values to (4.38), we can get the imputed estimating equation

$$\bar{U}_m^*\left(\psi\mid\hat{\eta}_p\right)\equiv m^{-1}\sum_{i=1}^n\sum_{j=1}^m U(\psi;\mathbf{y}_i^{*(j)})=0, \tag{4.41}$$

where $\mathbf{y}_i^{*(j)}=(\mathbf{y}_{i,\mathrm{obs}},\mathbf{y}_{i,\mathrm{mis}}^{*(j)})$. To apply the linearization method, we first compute the conditional expectation of $U(\psi;\mathbf{y}_i)$ given $(\mathbf{y}_{i,\mathrm{obs}},\boldsymbol{\delta}_i)$ evaluated at $\hat{\eta}_p$. That is, compute

$$\bar{U}\left(\psi\mid\hat{\eta}_p\right)=\sum_{i=1}^n\bar{U}_i\left(\psi\mid\hat{\eta}_p\right)=\sum_{i=1}^n E\left\{U(\psi;\mathbf{y}_i)\mid\mathbf{y}_{i,\mathrm{obs}},\boldsymbol{\delta}_i;\hat{\eta}_p\right\}, \tag{4.42}$$

which is the probability limit of $\bar{U}_m^*\left(\psi\mid\hat{\eta}_p\right)$ in (4.41) as $m\to\infty$. Let $\hat{\psi}_R$ be the solution to $\bar{U}\left(\psi\mid\hat{\eta}_p\right)=0$. Using the linearization technique, we have

$$\bar{U}\left(\psi\mid\hat{\eta}_p\right)\cong\bar{U}\left(\psi\mid\eta_0\right)+E\left\{\frac{\partial}{\partial\eta'}\bar{U}\left(\psi\mid\eta_0\right)\right\}(\hat{\eta}_p-\eta_0) \tag{4.43}$$

and

$$0=\hat{U}_p(\hat{\eta}_p)=\hat{U}_p(\eta_0)+E\left\{\frac{\partial}{\partial\eta'}\hat{U}_p(\eta_0)\right\}(\hat{\eta}_p-\eta_0). \tag{4.44}$$

Thus, combining (4.43) and (4.44), we have

$$\bar{U}\left(\psi\mid\hat{\eta}_p\right)\cong\bar{U}\left(\psi\mid\eta_0\right)+\kappa(\psi)\hat{U}_p\left(\eta_0\right)=\sum_{i=1}^n\left\{\bar{U}_i\left(\psi\mid\eta_0\right)+\kappa(\psi)U_p\left(\eta_0;\mathbf{y}_{i,\mathrm{obs}}\right)\right\}, \tag{4.45}$$

where

$$\kappa(\psi)=-E\left\{\frac{\partial}{\partial\eta'}\bar{U}\left(\psi\mid\eta_0\right)\right\}\left[E\left\{\frac{\partial}{\partial\eta'}\hat{U}_p\left(\eta_0\right)\right\}\right]^{-1}.$$

Write

$$\bar{U}_l\left(\psi\mid\eta_0\right)=\sum_{i=1}^n\left\{\bar{U}_i\left(\psi\mid\eta_0\right)+\kappa(\psi)U_p\left(\eta_0;\mathbf{y}_{i,\mathrm{obs}}\right)\right\}=\sum_{i=1}^n q_i\left(\psi\mid\eta_0\right), \tag{4.46}$$

and $q_i\left(\psi\mid\eta_0\right)=\bar{U}_i\left(\psi\mid\eta_0\right)+\kappa(\psi)U_p\left(\eta_0;\mathbf{y}_{i,\mathrm{obs}}\right)$, and the variance of $\bar{U}\left(\psi\mid\hat{\eta}_p\right)$ is asymptotically equal to the variance of $\bar{U}_l\left(\psi\mid\eta_0\right)$. Thus, the sandwich-type variance estimator for $\hat{\psi}_R$ is

$$\hat{V}\left(\hat{\psi}_R\right)=\hat{\tau}_q^{-1}\hat{\Omega}_q(\hat{\tau}_q^{-1})', \tag{4.47}$$

where

$$\hat{\tau}_q = n^{-1} \sum_{i=1}^{n} E\left\{ \dot{U}(\hat{\psi}_R; \mathbf{y}_i) \mid \mathbf{y}_{i,\mathrm{obs}}, \boldsymbol{\delta}_i; \hat{\eta}_p \right\}$$

$$\hat{\Omega}_q = n^{-1}(n-1)^{-1} \sum_{i=1}^{n} (\hat{q}_i - \bar{q}_n)^{\otimes 2},$$

$\bar{q}_n = n^{-1} \sum_{i=1}^{n} \hat{q}_i$, and $\hat{q}_i = q_i(\hat{\psi}_R \mid \hat{\eta}_p)$.

Remark 4.1. *The variance estimator (4.47) can be understood as a sandwich formula based on the joint estimating equations (4.40) and (4.42). Because* $(\hat{\psi}_R, \hat{\eta}_p)$ *is the solution to*

$$\mathbf{U}(\psi, \eta) \equiv \left[\begin{array}{c} U_1(\psi, \eta) \\ U_2(\eta) \end{array} \right] = 0,$$

where $U_1(\psi, \eta) = \bar{U}(\psi \mid \eta)$ *and* $U_2(\eta) = \hat{U}_p(\eta)$, *we can apply the Taylor expansion to get*

$$\left(\begin{array}{c} \hat{\psi}_R \\ \hat{\eta}_p \end{array} \right) \cong \left(\begin{array}{c} \psi_0 \\ \eta_0 \end{array} \right) - \left(\begin{array}{cc} B_{11} & B_{12} \\ B_{21} & B_{22} \end{array} \right)^{-1} \left[\begin{array}{c} U_1(\psi_0, \eta_0) \\ U_2(\eta_0) \end{array} \right],$$

where

$$\left(\begin{array}{cc} B_{11} & B_{12} \\ B_{21} & B_{22} \end{array} \right) = \left[\begin{array}{cc} E\left(\partial U_1 / \partial \psi' \right) & E\left(\partial U_1 / \partial \eta' \right) \\ E\left(\partial U_2 / \partial \psi' \right) & E\left(\partial U_2 / \partial \eta' \right) \end{array} \right].$$

Since $B_{21} = 0$, *we have*

$$\left(\begin{array}{cc} B_{11} & B_{12} \\ 0 & B_{22} \end{array} \right)^{-1} = \left(\begin{array}{cc} B_{11}^{-1} & -B_{11}^{-1} B_{12} B_{22}^{-1} \\ 0 & B_{22}^{-1} \end{array} \right)$$

and

$$\hat{\psi}_R \cong \psi_0 - B_{11}^{-1} \left\{ U_1(\psi_0, \eta_0) - B_{12} B_{22}^{-1} U_2(\eta_0) \right\}.$$

Thus, the result in (4.47) follows directly.

Finally, we consider the variance estimation of the imputed estimator $\hat{\psi}_{\mathrm{I},m}$, which is the solution to the imputed estimating equation (4.41). Writing

$$\bar{U}_m^*(\psi \mid \hat{\eta}_p) = \bar{U}(\psi \mid \hat{\eta}_p) + \left\{ \bar{U}_m^*(\psi \mid \hat{\eta}_p) - \bar{U}(\psi \mid \hat{\eta}_p) \right\},$$

we can express

$$V\left\{ \bar{U}_m^*(\psi \mid \hat{\eta}_p) \right\} = V\left\{ \bar{U}(\psi \mid \hat{\eta}_p) \right\} + V_{\mathrm{imp}}\left(\bar{U}_m^* \right),$$

where $V_{\mathrm{imp}}\left(\bar{U}_m^* \right)$ is the imputation variance of $\bar{U}_m^* = \bar{U}_m^*(\psi \mid \hat{\eta}_p)$. The imputation variance can be estimated, for example, by applying Lemma 4.3 with $X_j = U^{*(j)} = \sum_{i=1}^{n} U(\hat{\psi}_{\mathrm{I},m}; \mathbf{y}_i^{*(j)})$. That is, we can estimate $V_{\mathrm{imp}}\left(\bar{U}_m^* \right)$ by

$$\hat{V}_{\mathrm{imp}} = \frac{1}{m} \cdot \frac{1}{m-1} \sum_{j=1}^{m} \left(U^{*(j)} - \bar{U}_m^* \right)^{\otimes 2},$$

and $\bar{U}_m^* = m^{-1} \sum_{j=1}^m U^{*(j)}$. Thus, the sandwich-type variance estimator of $\hat{\psi}_m^*$ is

$$\hat{V}\left(\hat{\psi}_m^*\right) = \left(\hat{\tau}_q^*\right)^{-1} \left\{\hat{\Omega}_q^* + \hat{V}_{\text{imp}}\right\} \left(\hat{\tau}_q^{*-1}\right)', \tag{4.48}$$

where

$$\hat{\tau}_q^* = (nm)^{-1} \sum_{i=1}^n \sum_{j=1}^m \dot{U}(\hat{\psi}_{I,m}; \mathbf{y}_i^{*(j)}),$$

$$\hat{\Omega}_q^* = \{n(n-1)\}^{-1} \sum_{i=1}^n (\hat{q}_i^* - \bar{q}_n^*)^{\otimes 2},$$

and $\hat{q}_i^* = q_i^*(\hat{\psi}_{I,m} \mid \hat{\eta}_p)$ by (4.46).

Example 4.3. *Assume that the original sample is decomposed into G disjoint groups (often called* imputation cells*) and the sample observations are independently and identically distributed within the same cell. That is,*

$$y_i \mid i \in A_g \overset{i.i.d.}{\sim} \left(\mu_g, \sigma_g^2\right), \tag{4.49}$$

where A_g is the set of sample indices in cell g. Assume that there are n_g sample elements in cell g and r_g elements are observed. Assume that the response mechanism is MAR. The parameter of interest is $\theta = E(Y)$. Model (4.49) is often called the cell mean model.

In this case, a deterministic imputation can be used with $\hat{\eta} = (\hat{\mu}_1, \ldots, \hat{\mu}_G)$. Let $\hat{\mu}_g = r_g^{-1} \sum_{i \in A_g} \delta_i y_i$ be the g-th cell mean of y among respondents. The imputed estimator of θ is

$$\hat{\theta}_{Id} = \frac{1}{n} \sum_{g=1}^G \sum_{i \in A_g} \{\delta_i y_i + (1 - \delta_i)\hat{\mu}_g\} = \frac{1}{n} \sum_{g=1}^G n_g \hat{\mu}_g. \tag{4.50}$$

By the linearization technique in (4.45), the imputed estimator can be expressed as

$$\hat{\theta}_{Id} \cong \frac{1}{n} \sum_{g=1}^G \sum_{i \in A_g} \left\{\mu_g + \frac{n_g}{r_g}\delta_i(y_i - \mu_g)\right\} \tag{4.51}$$

and the plug-in variance estimator can be expressed as

$$\hat{V}(\hat{\theta}_{Id}) = \frac{1}{n} \frac{1}{n-1} \sum_{g=1}^G \sum_{i \in A_g} \left(\hat{d}_{gi} - \bar{d}_n\right)^2, \tag{4.52}$$

where

$$\hat{d}_{gi} = \hat{\mu}_g + \frac{n_g}{r_g}\delta_i(y_i - \hat{\mu}_g)$$

and $\bar{d}_n = n^{-1} \sum_{g=1}^G \sum_{i \in A_g} \hat{d}_{gi}$.

If a stochastic imputation is used, where an imputed value is randomly selected from the set of respondents in the same cell, then we can write

$$\hat{\theta}_{Is} = \frac{1}{n} \sum_{g=1}^{G} \sum_{i \in A_g} \{\delta_i y_i + (1 - \delta_i) y_i^*\}. \tag{4.53}$$

Such an imputation method is often called hot deck imputation. *Writing*

$$\hat{\theta}_{Is} = \hat{\theta}_{Id} + \frac{1}{n} \sum_{g=1}^{G} \sum_{i \in A_g} (1 - \delta_i)(y_i^* - \hat{\mu}_g),$$

the variance of the first term can be estimated by (4.52), and the variance of the second term in (4.53) can be estimated by

$$n^{-2} \sum_{g=1}^{G} \sum_{i \in A_g} (1 - \delta_i)(y_i^* - \hat{\mu}_g)^2,$$

if the imputed values are generated independently, conditional on the respondents.

An extension of Example 4.3 can be made by considering a general model for imputation

$$y_i \mid \mathbf{x}_i \overset{i.i.d.}{\sim} \{E(y_i \mid \mathbf{x}_i), V(y_i \mid \mathbf{x}_i)\}. \tag{4.54}$$

If $E(y_i \mid \mathbf{x}_i)$ is a known function of unknown parameters, such as $E(y_i \mid \mathbf{x}_i) = m(\mathbf{x}_i; \eta)$, then we can use the linearization technique as discussed in Kim and Rao (2009). If $E(y_i \mid \mathbf{x}_i)$ is unknown, then we can use a nonparametric regression technique as in Wang and Chen (2009). Some of the nonparametric approach to imputation will be introduced in Section 6.2.

4.4 Replication Variance Estimation

Replication variance estimation is a simulation-based method using replicates of the given point estimator. Let $\hat{\theta}_n$ be the complete-sample estimator of θ. The replication variance estimator of $\hat{\theta}_n$ takes the form of

$$\hat{V}_{rep}(\hat{\theta}_n) = \sum_{k=1}^{L} c_k \left(\hat{\theta}_n^{(k)} - \hat{\theta}_n \right)^2, \tag{4.55}$$

where L is the number of replicates, c_k is the replication factor associated with replication k, and $\hat{\theta}_n^{(k)}$ is the k-th replicate of $\hat{\theta}_n$. If $\hat{\theta}_n = \sum_{i=1}^{n} y_i/n$, then we can write $\hat{\theta}_n^{(k)} = \sum_{i=1}^n w_i^{(k)} y_i$ for some replication weights $w_1^{(k)}, w_2^{(k)}, \ldots, w_n^{(k)}$. For example, in the jackknife method, we have $L = n$, $c_k = (n-1)/n$, and

$$w_i^{(k)} = \begin{cases} (n-1)^{-1} & \text{if } i \neq k \\ 0 & \text{if } i = k. \end{cases}$$

If we use the above jackknife method to $\hat{\theta}_n = \sum_{i=1}^n y_i/n$, the resulting jackknife variance estimator in (4.55) is algebraically equivalent to $n^{-1}(n-1)^{-1}\sum_{i=1}^n (y_i - \bar{y}_n)^2$, which is unbiased for $V(\hat{\theta}_n)$. Furthermore, if we apply the jackknife to $\hat{\theta}_n = \sum_{i=1}^n y_i/(\sum_{i=1}^n x_i)$, then

$$\hat{V}_{rep}(\hat{\theta}_n) = \frac{1}{n}\frac{1}{n-1}\sum_{k=1}^n \left(\frac{1}{\bar{x}_n^{(k)}}\right)^2 \left(y_k - \hat{\theta}_n x_k\right)^2,$$

which is close to the linearized variance estimator

$$\hat{V}_l(\hat{\theta}_n) = \frac{1}{(\bar{x}_n)^2}\frac{1}{n}\frac{1}{n-1}\sum_{k=1}^n \left(y_k - \hat{\theta}_n x_k\right)^2.$$

In general, under some regularity conditions, for $\hat{\theta}_n = g(\bar{y}_n)$ which is a smooth function of $\bar{y}_n$, the replication variance estimator of $\hat{\theta}_n$, defined by

$$\hat{V}_{rep}\left(\hat{\theta}_n\right) = \sum_{k=1}^L c_k \left(\hat{\theta}_n^{(k)} - \hat{\theta}_n\right)^2, \tag{4.56}$$

where $\hat{\theta}_n^{(k)} = g(\bar{y}_n^{(k)})$, satisfies

$$\hat{V}_{rep}\left(\hat{\theta}_n\right) \cong \left\{g'(\bar{y}_n)\right\}^2 \hat{V}_{rep}(\bar{y}_n).$$

That is, the replication variance estimator is asymptotically equivalent to the linearized variance estimator.

We now look at parameters other than regression parameters such as β_0, β_1 and σ_e^2, often denoted by θ. Denote one such nonregression parameter, an example of which could be a proportion parameter, as ψ, and estimate it by $\hat{\psi}_n$ obtained by solving an estimating equation $\sum_{i=1}^n U(\psi; y_i) = 0$. A consistent variance estimator can be obtained by the sandwich formula in (4.39). If we want to use the replication method of the form (4.55), we can construct the replication variance estimator of $\hat{\psi}_n$ by

$$\hat{V}_{rep}(\hat{\psi}_n) = \sum_{k=1}^L c_k \left(\hat{\psi}_n^{(k)} - \hat{\psi}_n\right)^2, \tag{4.57}$$

where $\hat{\psi}_n^{(k)}$ is computed by

$$\hat{U}^{(k)}(\psi) \equiv \sum_{i=1}^n w_i^{(k)} U(\psi; y_i) = 0. \tag{4.58}$$

The replication variance estimator (4.57) is asymptotically equivalent to the sandwich-type variance estimator. Note that the replication variance estimator does require computing partial derivatives in variance estimation. In some cases, finding the solution to (4.58) can be computationally challenging. In this case, the one-step approximation method can be used. The one-step approximation method is based on Taylor expansion, as described below.

$$\begin{aligned} 0 &= \hat{U}^{(k)}(\hat{\psi}^{(k)}) \\ &\cong \hat{U}^{(k)}(\hat{\psi}) + \dot{U}^{(k)}\left(\hat{\psi}\right)\left(\hat{\psi}^{(k)} - \hat{\psi}\right), \end{aligned}$$

where $\dot{U}^{(k)}(\psi) = \partial \hat{U}^{(k)}(\psi)/\partial \psi'$. Thus, the one-step approximation of $\hat{\psi}^{(k)}$ is to use

$$\hat{\psi}_1^{(k)} = \hat{\psi} - \left\{\dot{U}^{(k)}(\hat{\psi})\right\}^{-1} \hat{U}^{(k)}(\hat{\psi}) \tag{4.59}$$

or, even more simply, use

$$\hat{\psi}_1^{(k)} = \hat{\psi} - \left\{\dot{U}(\hat{\psi})\right\}^{-1} \hat{U}^{(k)}(\hat{\psi}). \tag{4.60}$$

The replication variance estimator of (4.60) is algebraically equivalent to

$$\left\{\dot{U}(\hat{\psi})\right\}^{-1} \left[\sum_{k=1}^{n} c_k \left\{\hat{U}^{(k)}(\hat{\psi}) - \hat{U}(\hat{\psi})\right\}^{\otimes 2}\right] \left\{\dot{U}(\hat{\psi})\right\}^{-1},$$

which is very close to the sandwich variance formula in (4.39).

We now discuss replication variance estimation after a deterministic imputation. The replication method can be applied to the deterministic part naturally. For example, in the regression imputation estimator of the form (4.31), the replication variance estimator can be computed by

$$\hat{V}_{rep}\left(\hat{\theta}_{Id}\right) = \sum_{k=1}^{L} c_k \left(\hat{\theta}_{Id}^{(k)} - \hat{\theta}_{Id}\right)^2, \tag{4.61}$$

where the k-th replicate of the imputed estimator is

$$\hat{\theta}_{Id}^{(k)} = \sum_{i=1}^{n} w_i^{(k)} \left\{\delta_i y_i + (1 - \delta_i)\left(\hat{\beta}_0^{(k)} + \hat{\beta}_1^{(k)} x_i\right)\right\},$$

and $(\hat{\beta}_0^{(k)}, \hat{\beta}_1^{(k)})$ is the solution to

$$\sum_{i=1}^{n} w_i^{(k)} \delta_i (y_i - \beta_0 - \beta_1 x_i)(1, x_i) = (0, 0). \tag{4.62}$$

Note that, by (4.62), the k-th replicate of $\hat{\theta}_{Id}$ can be written as

$$\hat{\theta}_{Id}^{(k)} = \sum_{i=1}^{n} w_i^{(k)} \left(\hat{\beta}_0^{(k)} + \hat{\beta}_1^{(k)} x_i\right),$$

which is a smooth function of $\hat{\beta}_0^{(k)}, \hat{\beta}_1^{(k)}$ and $\bar{x}_n^{(k)} = \sum_{i=1}^{n} w_i^{(k)} x_i$.

Note that we can express

$$\hat{\theta}_{Id}^{(k)} - \hat{\theta}_{Id} = g(\hat{\beta}_0^{(k)}, \hat{\beta}_1^{(k)}, \bar{x}_n^{(k)}) - g(\hat{\beta}_0, \hat{\beta}_1, \bar{x}_n)$$

for some smooth function $g(\cdot)$ with continuous partial derivatives. Thus, writing $\hat{\eta} = (\hat{\beta}_0, \hat{\beta}_1, \bar{x}_n)'$ and $\hat{\eta}^{(k)} = (\hat{\beta}_0^{(k)}, \hat{\beta}_1^{(k)}, \bar{x}_n^{(k)})'$, we can obtain

$$
\begin{aligned}
\sum_{k=1}^{L} c_k \left(\hat{\theta}_{Id}^{(k)} - \hat{\theta}_{Id} \right)^2 &= \sum_{k=1}^{L} c_k \left\{ g(\hat{\eta}^{(k)}) - g(\hat{\eta}) \right\}^2 \\
&\cong \sum_{k=1}^{L} c_k \left\{ \dot{g}(\hat{\eta})(\hat{\eta}^{(k)} - \hat{\eta}) \right\}^2 \\
&= \dot{g}(\hat{\eta}) \hat{V}(\hat{\eta}) \left\{ \dot{g}(\hat{\eta}) \right\}' \cong \hat{V}\{g(\hat{\eta})\},
\end{aligned}
$$

where $\dot{g}(\eta) = \partial g(\eta)/\partial \eta'$. Because we can write $\hat{\theta}_{Id} = g(\hat{\eta})$, the above result establishes the asymptotic equivalence of the jackknife variance estimator with the linearization variance estimator.

If a stochastic imputation is used such that the imputed estimator can be written as

$$\hat{\theta}_{I,s} = \frac{1}{n} \sum_{i=1}^{n} \left\{ \delta_i y_i + (1 - \delta_i)(\hat{\beta}_0 + \hat{\beta}_1 x_i + \hat{e}_i^*) \right\}$$

then the k-th replicate of $\hat{\theta}_{Is}$ can be computed by

$$\hat{\theta}_{Is}^{(k)} = \sum_{i=1}^{n} w_i^{(k)} \left\{ \delta_i y_i + (1 - \delta_i) \left(\hat{\beta}_0^{(k)} + \hat{\beta}_1^{(k)} x_i + \hat{e}_i^* \right) \right\}.$$

The replication variance estimator defined by

$$\sum_{i=1}^{L} c_k \left(\hat{\theta}_{Is}^{(k)} - \hat{\theta}_{Is} \right)^2$$

can be shown to be consistent for the variance of the imputed estimator $\hat{\theta}_{Is}$. For details, see Rao and Shao (1992) and Rao and Sitter (1995).

Example 4.4. *We now return to the setup of Example 3.11. In this case, the deterministically imputed estimator of $\theta = E(Y)$ is constructed by*

$$\hat{\theta}_{Id} = n^{-1} \sum_{i=1}^{n} \left\{ \delta_i y_i + (1 - \delta_i) \hat{p}_{0i} \right\}, \tag{4.63}$$

where $\hat{p}_{0i}$ is the predictor of y_i given $\mathbf{x}_i$ and $\delta_i = 0$. That is,

$$\hat{p}_{0i} = \frac{p(\mathbf{x}_i; \hat{\beta})\{1 - \pi(\mathbf{x}_i, 1; \hat{\phi})\}}{\{1 - p(\mathbf{x}_i; \hat{\beta})\}\{1 - \pi(\mathbf{x}_i, 0; \hat{\phi})\} + p(\mathbf{x}_i; \hat{\beta})\{1 - \pi(\mathbf{x}_i, 1; \hat{\phi})\}},$$

where $\hat{\beta}$ and $\hat{\phi}$ are jointly estimated by the EM algorithm described in Example 3.11. For replication variance estimation, we can use (4.61) with

$$\hat{\theta}_{Id}^{(k)} = \sum_{i=1}^{n} w_i^{(k)} \left\{ \delta_i y_i + (1 - \delta_i)\hat{p}_{0i}^{(k)} \right\}. \tag{4.64}$$

In the above formula,

$$\hat{p}_{0i}^{(k)} = \frac{p(\mathbf{x}_i; \hat{\beta}^{(k)})\{1 - \pi(\mathbf{x}_i, 1; \hat{\phi}^{(k)})\}}{\{1 - p(\mathbf{x}_i; \hat{\beta}^{(k)})\}\{1 - \pi(\mathbf{x}_i, 0; \hat{\phi}^{(k)})\} + p(\mathbf{x}_i; \hat{\beta}^{(k)})\{1 - \pi(\mathbf{x}_i, 1; \hat{\phi}^{(k)})\}},$$

and $(\hat{\beta}^{(k)}, \hat{\phi}^{(k)})$ is obtained by solving the mean score equations with original weights replaced by replication weights $w_i^{(k)}$. That is, $(\hat{\beta}^{(k)}, \hat{\phi}^{(k)})$ is the solution to

$$
\begin{aligned}
\bar{S}_1^{(k)}(\beta, \phi) &\equiv \sum_{\delta_i=1} w_i^{(k)} \{y_i - p(\mathbf{x}_i; \beta)\} \mathbf{x}_i \\
&+ \sum_{\delta_i=0} w_i^{(k)} \sum_{y=0}^{1} w_{iy}^*(\beta, \phi)\{y - p(\mathbf{x}_i; \beta)\}\mathbf{x}_i = 0 \\
\bar{S}_2^{(k)}(\beta, \phi) &\equiv \sum_{\delta_i=1} w_i^{(k)} \{\delta_i - \pi(\mathbf{x}_i, y_i; \phi)\} (\mathbf{x}_i', y_i)' \\
&+ \sum_{\delta_i=0} w_i^{(k)} \sum_{y=0}^{1} w_{iy}^*(\beta, \phi)\{\delta_i - \pi(\mathbf{x}_i, y; \beta)\}(\mathbf{x}_i', y)' = 0
\end{aligned}
$$

and

$$w_{iy}^*(\beta, \phi) = \frac{p(\mathbf{x}_i; \beta)\{1 - \pi(\mathbf{x}_i, 1; \phi)\}}{\{1 - p(\mathbf{x}_i; \beta)\}\{1 - \pi(\mathbf{x}_i, 0; \phi)\} + p(\mathbf{x}_i; \beta)\{1 - \pi(\mathbf{x}_i, 1; \phi)\}}.$$

Thus, we may apply the same EM algorithm to compute $(\hat{\beta}^{(k)}, \hat{\phi}^{(k)})$ iteratively. Under MAR, $\hat{p}_{0i}^{(k)} = p(\mathbf{x}_i; \hat{\beta}^{(k)})$ and $\hat{\beta}^{(k)}$ is computed by solving

$$\sum_{i=1}^{n} w_i^{(k)} \delta_i \{y_i - p_i(\beta)\}\mathbf{x}_i = \mathbf{0} \tag{4.65}$$

with respect to β. Instead of solving (4.65), by (4.59), one can use an one-step approximation

$$\hat{\beta}^{(k)} = \hat{\beta} + \left\{ \sum_{i=1}^{n} w_i^{(k)} \delta_i \hat{p}_i (1 - \hat{p}_i)\mathbf{x}_i\mathbf{x}_i' \right\}^{-1} \sum_{i=1}^{n} w_i^{(k)} \delta_i (y_i - \hat{p}_i)\mathbf{x}_i.$$

Exercises

1. Let $\hat{\theta}$ be an unbiased estimator of θ and $\hat{a}$ be an unbiased estimator of a constant a with known a. Now, consider the following class of estimators of the form

$$\hat{\theta}_k = \hat{\theta} + k(\hat{a} - a)$$

where k is a constant to be determined.

   (a) Show that $\hat{\theta}_k$ is unbiased for all $k \in \mathbb{R}$.
   (b) Find the value of k, say k^*, that minimizes the variance of $\hat{\theta}_k$.
   (c) Show that if $\hat{\theta}$ has the minimum variance among the class of unbiased estimator of θ then $Cov(\hat{\theta}, \hat{a}) = 0$.
   (d) Let $\hat{\theta}_p$ be a preliminary estimator of θ and assume that $\hat{\theta}_p$ is unbiased for θ. Let $\hat{\theta}_{MLE}$ be the maximum likelihood estimator. Show that

$$Cov(\hat{\theta}_{MLE}, \hat{\theta}_p - \hat{\theta}_{MLE}) = 0.$$

2. Under the setup of Example 4.1, let $\hat{y}_i = \hat{\beta}_0 + \hat{\beta}_1 x_i$ be the (deterministically) imputed value of y_i. Let $y_i^* = \hat{y}_i + \hat{e}_i^*$, where $\hat{e}_i^*$ is randomly selected from $\hat{e}_i = y_i - \hat{y}_i$ among the respondents of y.

   (a) Prove that the estimator of σ_y^2 using the deterministic imputation

$$\hat{\sigma}_{y,d}^2 = \frac{1}{n} \sum_{i=1}^{n} \left\{ \delta_i y_i^2 + (1 - \delta_i)\hat{y}_i^2 - (\bar{y}_{Id})^2 \right\},$$

   where $\bar{y}_{Id} = n^{-1} \sum_{i=1}^{n} \{ \delta_i y_i + (1 - \delta_i)\hat{y}_i \}$, satisfies

$$E\left(\hat{\sigma}_{y,d}^2 \right) = \sigma_y^2 - \frac{n-r}{n} \sigma_y^2 \left(1 - \rho^2 \right) < \sigma_y^2.$$

   (b) Prove that the estimator of σ_y^2 using the stochastic imputation

$$\hat{\sigma}_{y,s}^2 = \frac{1}{n} \sum_{i=1}^{n} \left\{ \delta_i y_i^2 + (1 - \delta_i)y_i^{*2} - (\bar{y}_{Is})^2 \right\},$$

   where $\bar{y}_{Is} = n^{-1} \sum_{i=1}^{n} \{ \delta_i y_i + (1 - \delta_i)y_i^* \}$, satisfies $E\left(\hat{\sigma}_{y,s}^2 \right) \doteq \sigma_y^2$.

3. Let $y_1, \dots, y_n$ be IID samples from a Bernoulli distribution with probability of $y_i = 1$ being equal to

$$p_i = \frac{\exp(\beta_0 + \beta_1 x_i)}{1 + \exp(\beta_0 + \beta_1 x_i)}$$

and let δ_i be the response indicator function of y_i. Assume that x_i is always observed and the response mechanism for y is MAR. In this case, the maximum likelihood estimator of the probability p_i is $\hat{p}_i = p_i(\hat{\beta})$ with $\hat{\beta} = (\hat{\beta}_0, \hat{\beta}_1)$ computed by

$$\sum_{i=1}^{n} \delta_i \{y_i - p_i(\beta)\}(1, x_i)' = (0, 0)'.$$

The parameter of interest is $\theta = E(Y)$.

(a) If the imputation method is deterministic in the sense that we use $\hat{y}_i = \hat{p}_i$ as the imputed value for missing y_i, derive the variance estimation formula for the imputed estimator of θ.

(b) If the imputation method is stochastic in the sense that we use $y_i^* \sim Bernoulli(\hat{p}_i)$ as the imputed value for missing y_i, derive the variance estimation formula for the imputed estimator of θ.

4. A random sample of size $n = 100$ is drawn from an infinite population composed of two cells, where the distribution of y is

$$y_i \sim \begin{cases} N\left(\mu_1, \sigma_1^2\right), & \text{if } i \in \text{cell 1} \\ N\left(\mu_2, \sigma_2^2\right), & \text{if } i \in \text{cell 2}, \end{cases}$$

where μ_g and σ_g^2 are unknown parameters and subscript g denotes cells, $g = 1, 2$. In this sample, we have n_1 elements from cell 1 and n_2 elements from cell 2. Among the n_1 elements, r_1 elements are observed and $m_1 = n_1 - r_1$ elements are missing in y. In cell 2, we have r_2 respondents and $m_2 = n_2 - r_2$ nonrespondents. We assume that the response mechanism is missing at random in the sense that the respondents are independently and identically distributed within the same cell. We use a with-replacement hot deck imputation in the sense that, for each cell g, m_g imputed values are randomly selected from r_g respondents with replacement. Let y_i^* be the imputed value of y_i obtained from the hot deck imputation. The parameter of interest is the population mean of y.

(a) Under the above setup, compute the variance of the imputed point estimator.

(b) Show that the expected value of the naive variance estimator $\hat{V}_I$ (that is obtained by treating the imputed values as if observed and applying the standard variance estimation formula to the imputed data) is equal to

$$E\{\hat{V}_I\} = V\left\{n^{-1}\sum_{g=1}^{2} n_g \mu_g\right\}$$
$$+ \frac{1}{n(n-1)} E\left\{\sum_{g=1}^{2}\left(n_g - \frac{n_g}{n} - 2\frac{m_g}{n} - \frac{m_g(m_g-1)}{nr_g}\right)\right\}\sigma_g^2.$$

(c) We use the following bias-corrected estimator

$$\hat{V} = \hat{V}_I + \sum_{g=1}^{2} c_g S_{Rg}^2$$

to estimate the variance of the imputed point estimator. Find c_1 and c_2.

(d) Show that the Rao and Shao (1992) jackknife variance estimator defined below is approximately unbiased.

$$\hat{V}_{JK} = \frac{n-1}{n} \sum_{k=1}^{n} \left(\hat{\theta}_I^{(k)} - \hat{\theta}_I \right)^2,$$

where

$$w_i^{(k)} = \begin{cases} 1/(n-1) & \text{if } i \neq k \\ 0 & \text{if } i = k, \end{cases}$$

$$\hat{\theta}_I^{(k)} = \sum_{i=1}^{n} w_i^{(k)} \left(\delta_i y_i + (1 - \delta_i) y_i^{*(k)} \right),$$

and

$$y_i^{*(k)} = \bar{y}_g^{(k)} - \bar{y}_g + y_i^*$$

for $i \in A_g$ and A_g is the index set of samples in cell g. Here, $\bar{y}_g$ is the sample mean of y_i (with $\delta_i = 1$) in cell g and $\bar{y}_g^{(k)} = \left(\sum_{i \in A_c} w_i^{(k)} \delta_i y_i \right) / \left(\sum_{i \in A_c} w_i^{(k)} \delta_i \right)$.

5. Assume that we have a random sample of $(\mathbf{x}_i, y_i)$, where only y_i is subject to missingness. Let $\hat{y}_i = m(\mathbf{x}_i; \hat{\beta})$ be the imputed value of y_i, with $m(\cdot)$, $q_i(\cdot)$ being known functions, and $\hat{\beta}$ obtained by solving

$$\sum_{i=1}^{n} \delta_i \{ y_i - m(\mathbf{x}_i; \beta) \} q_i(\beta) \mathbf{x}_i = 0 \qquad (4.66)$$

for β. In this case, the deterministic imputation for $\theta = E(Y)$ is given by

$$\hat{\theta}_{dI} = n^{-1} \sum_{i=1}^{n} \left\{ \delta_i y_i + (1 - \delta_i) m(\mathbf{x}_i; \hat{\beta}) \right\}.$$

(a) Under the assumption of $y_i \mid \mathbf{x}_i \sim (m(\mathbf{x}_i; \beta_0), v_i)$ for some $v_i > 0$ and MAR, discuss the optimal choice of q_i in (4.66) that minimizes the variance of $\hat{\beta}$.

(b) Find the linearization variance estimator of $\hat{\theta}_{dI}$.

6. Assume that (x_1, x_2, y) is a vector of random variables from a multivariate normal distribution. Let the sample be partitioned into three sets, H_1, H_2 and H_3, where (x_{1i}, x_{2i}, y_i) are observed in H_1, (x_{1i}, x_{2i}) are observed in H_2, and only x_{1i} are observed in H_3. That is, x_{1i} are observed throughout the sample, x_{2i} are observed in $H_1 \cup H_2$, and y_i are observed only in H_1. Suppose that we are interested in estimating $\theta = E(Y)$ from the sample. We consider the following imputation estimator of θ:

$$\hat{\theta}_1 = n^{-1} \left\{ \sum_{i \in H_1} y_i + \sum_{i \in H_2} \hat{y}_{2i} + \sum_{i \in H_3} \hat{y}_{1i} \right\},$$

where $\hat{y}_{2i} = \hat{\beta}_0 + \hat{\beta}_1 x_{1i} + \hat{\beta}_2 x_{2i}$, $\hat{y}_{1i} = \hat{\beta}_0 + \hat{\beta}_1 x_{1i} + \hat{\beta}_2 \hat{x}_{2i}$, and $\hat{x}_{2i} = \hat{\gamma}_0 + \hat{\gamma}_1 x_{1i}$. The regression coefficients are computed by

$$\begin{pmatrix} \hat{\beta}_0 \\ \hat{\beta}_1 \\ \hat{\beta}_2 \end{pmatrix} = \left\{ \sum_{i \in H_1} \begin{pmatrix} 1 \\ x_{1i} \\ x_{2i} \end{pmatrix} \begin{pmatrix} 1 \\ x_{1i} \\ x_{2i} \end{pmatrix}' \right\}^{-1} \sum_{i \in H_1} \begin{pmatrix} 1 \\ x_{1i} \\ x_{2i} \end{pmatrix} y_i$$

and

$$\begin{pmatrix} \hat{\gamma}_0 \\ \hat{\gamma}_1 \end{pmatrix} = \left\{ \sum_{i \in H_1 \cup H_2} \begin{pmatrix} 1 \\ x_{1i} \end{pmatrix} \begin{pmatrix} 1 \\ x_{1i} \end{pmatrix}' \right\}^{-1} \sum_{i \in H_1 \cup H_2} \begin{pmatrix} 1 \\ x_{1i} \end{pmatrix} y_i.$$

(a) Compute the variance of $\hat{\theta}$ in terms of the model parameters in the joint distribution of (x_1, x_2, y).

(b) Derive the linearized variance estimator of $\hat{\theta}$.

5

Multiple Imputation

Multiple imputation (MI), first proposed by Rubin (1978), is one popular method of imputation motivated from a Bayesian perspective. Bayesian prediction tools can be used to develop MI for various situations. The MI also have a frequentist justification for the parameters in the imputation model. The frequentist theory behind multiple imputation has been unknown until the work of Wang and Robins (1998). We present the frequentist justification in Section 5.3. Some recent findings on the use of MI for general purpose estimation are presented in Section 5.5.

5.1 Review of Bayesian Inference

In Bayesian analysis, a parameter is treated as random, and the inference is based on the conditional probability of a parameter given observed data, called the posterior distribution:

$$p(\theta \mid \text{data}) = \frac{L(\theta)\pi(\theta)}{\int L(\theta)\pi(\theta)d\theta},$$

(5.1)

where $L(\theta)$ is the likelihood function of θ, and $\pi(\theta)$ is a prior distribution of θ. The likelihood function reflects the information about the parameter contained in the data, and the prior distribution reflects the information about the parameter before observing the current data. The prior information may include the prior knowledge or belief. With large sample sizes, reasonable choices of prior distributions will have minor effects on posterior inferences.

Example 5.1. *Let $y_1, \ldots, y_n$ be a realization of random sample from $N(\theta, \sigma^2)$ with known σ^2, where θ is the parameter of interest. The likelihood function is given as*

$$L(\theta) = \prod_{i=1}^{n} \left[\frac{1}{\sqrt{2\pi\sigma^2}} \exp\left\{ -\frac{1}{2\sigma^2}(y_i - \theta)^2 \right\} \right] \propto \exp\left\{ -\frac{n}{2\sigma^2}(\theta - \bar{y})^2 \right\},$$

where $\bar{y} = n^{-1}\sum_{i=1}^{n} y_i$. For the prior distribution, we may use $\theta \sim N(\mu_0, \tau_0^2)$, where μ_0 and τ_0 are pre-specified constants, called hyper-parameters. Using

DOI: 10.1201/9780429321740-5

Bayes formula (5.1), the density of the posterior distribution is given as

$$p(\theta \mid y_1, \ldots, y_n) \quad \propto \quad L(\theta)\pi(\theta)$$

$$\propto \quad \exp\left\{-\frac{n}{2\sigma^2}(\theta - \bar{y})^2\right\} \exp\left\{-\frac{1}{2\tau_0^2}(\theta - \mu_0)^2\right\}$$

$$\propto \quad \exp\left\{-\frac{1}{2\sigma_1^2}(\theta - \mu_1)^2\right\},$$

where

$$\mu_1 = \frac{\tau_0^2}{n^{-1}\sigma^2 + \tau_0^2}\bar{y} + \frac{n^{-1}\sigma^2}{n^{-1}\sigma^2 + \tau_0^2}\mu_0,$$

and

$$\frac{1}{\sigma_1^2} = \frac{n}{\sigma^2} + \frac{1}{\tau_0^2}.$$

Thus, the posterior distribution is $N(\mu_1, \sigma_1^2)$. If the prior is flat in the sense that $\tau_0^2 \to \infty$, the posterior distribution is $N(\bar{y}, n^{-1}\sigma^2)$, and the resulting confidence interval is the same as the frequentist one. Also, for large sample sizes, the posterior distribution is also $N(\bar{y}, n^{-1}\sigma^2)$, and the choice of the hyperparameters does not make any difference in the posterior inference.

Example 5.2. *Let $y_1, \ldots, y_n$ be a random sample from a Bernoulli distribution with parameter $\theta \in (0, 1)$. That is,*

$$y_i = \begin{cases} 1 & \text{with probability } \theta \\ 0 & \text{with probability } 1 - \theta. \end{cases}$$

Let $n = n_1 + n_0$, where $n_1 = \sum_{i=1}^{n} y_i$. For the prior for θ, we may use Beta distribution with parameter (α, β). Using this prior, the posterior distribution is

$$p(\theta \mid y_1, \ldots, y_n) \quad \propto \quad L(\theta)\pi(\theta)$$

$$\propto \quad \theta^{n_1}(1-\theta)^{n_0} \times \theta^{\alpha-1}(1-\theta)^{\beta-1}$$

$$= \quad \theta^{n_1+\alpha-1}(1-\theta)^{n_0+\beta-1},$$

which is the density of $Beta(n_1 + \alpha, n_0 + \beta)$ distribution. Note that

$$E(\theta \mid y_1, \ldots, y_n) = \frac{n_1 + \alpha}{n + \alpha + \beta} \cong \frac{n_1}{n}$$

$$V(\theta \mid y_1, \ldots, y_n) = \frac{(n_1 + \alpha)(n_0 + \beta)}{(n + \alpha + \beta)^2(n + \alpha + \beta + 1)} \cong \frac{1}{n}\frac{n_1}{n}\left(1 - \frac{n_1}{n}\right)$$

for sufficiently large n. Thus, the parameters of the prior distribution have no influence on the posterior distribution as $n \to \infty$.

In the above two examples, the posterior distribution follows the same parametric form as the prior distribution. If $\mathcal{F}$ is a class of distributions $f(y \mid$

θ), and $\mathcal{P}$ is a class of prior distributions $\pi(\theta)$, then the class $\mathcal{P}$ is conjugate for $\mathcal{F}$ if the posterior distribution satisfies $p(\theta \mid y) \in \mathcal{P}$. If the prior belongs to the conjugate for $\mathcal{F}$, it is called conjugate prior. Conjugate prior distributions have the practical advantage, in addition to computational convenience, of being interpretable as additional data.

In many statistical problems, parameter θ can be vector-valued, and we may be interested in only a part of the parameter. Let $\theta = (\theta_1, \theta_2)$ and suppose that we are only interested in making inference for θ_1. In this case, we are not directly interested in θ_2, which is often called a nuisance parameter. In Bayesian analysis, handling nuisance parameter can be easily solved by integrating it out in the posterior distribution. That is, the posterior distribution of θ_1 can be easily obtained from

$$p(\theta_1 \mid y_1, \ldots, y_n) = \int p(\theta_1, \theta_2 \mid y_1, \ldots, y_n) d\theta_2.$$

Example 5.3. *Consider n independent observations from $N(\mu, \sigma^2)$, where μ and σ^2 are unknown. We assume a flat prior for $(\mu, \log \sigma^2)$ so that*

$$\pi(\mu, \sigma^2) \propto (\sigma^2)^{-1}.$$

The joint posterior is

$$
\begin{aligned}
p(\mu, \sigma^2 \mid y_1, \ldots, y_n) \;\; &\propto \;\; L(\mu, \sigma^2) \pi(\mu, \sigma^2) \\
&\propto \;\; \left[\prod_{i=1}^{n} (\sigma^2)^{-1/2} \exp\left\{ -\frac{1}{2\sigma^2} (y_i - \mu)^2 \right\} \right] (\sigma^2)^{-1} \\
&\propto \;\; (\sigma^2)^{-n/2-1} \exp\left[-\frac{1}{2\sigma^2} \left\{ \sum_{i=1}^{n} (y_i - \bar{y})^2 + n(\bar{y} - \mu)^2 \right\} \right] \\
&\propto \;\; (\sigma^2)^{-n/2-1} \exp\left[-\frac{1}{2\sigma^2} \left\{ (n-1)s^2 + n(\bar{y} - \mu)^2 \right\} \right],
\end{aligned}
$$

where $s^2 = (n-1)^{-1} \sum_{i=1}^{n} (y_i - \bar{y})^2$. Since

$$
\begin{aligned}
p(\mu, \sigma^2 \mid y_1, \ldots, y_n) \;\; &\propto \;\; (\sigma^2/n)^{-1/2} \exp\left\{ -\frac{n}{2\sigma^2} (\mu - \bar{y})^2 \right\} \\
&\quad \times (\sigma^2)^{-(n+1)/2} \exp\left\{ -\frac{1}{2\sigma^2} (n-1)s^2 \right\} \\
&\propto \;\; p(\mu \mid y_1, \ldots, y_n, \sigma^2) \cdot p(\sigma^2 \mid y_1, \ldots, y_n),
\end{aligned}
$$

we obtain

$$\mu \mid (\sigma^2, y_1, \ldots, y_n) \;\; \sim \;\; N(\bar{y}, \sigma^2/n) \tag{5.2}$$

$$\sigma^2 \mid (y_1, \ldots, y_n) \;\; \sim \;\; \text{Inv-Gamma}\left(\frac{n-1}{2}, \frac{(n-1)s^2}{2} \right). \tag{5.3}$$

Recall that the density of the inverse-Gamma distribution with parameter (α, β) *is*

$$f(\sigma^2; \alpha, \beta) = \frac{\beta^\alpha}{\Gamma(\alpha)} \left(\sigma^2\right)^{-(\alpha+1)} \exp\left(-\frac{\beta}{\sigma^2}\right)$$

for $\sigma^2 > 0$.

Now, if we are only interested in obtaining the posterior distribution of μ, *we can derive the marginal posterior density for* μ *by integrating the joint posterior density over* σ^2. *That is,*

$$
\begin{aligned}
p(\mu|y_1,\ldots,y_n) &= \int_0^\infty p(\mu, \sigma^2|y_1,\ldots,y_n)d\sigma^2 \\
&\propto \int_0^\infty (\sigma^2)^{-n/2-1} \exp\left[-\frac{1}{2\sigma^2}\left\{(n-1)s^2 + n(\bar{y}-\mu)^2\right\}\right]d\sigma^2 \\
&\propto \left\{(n-1)s^2 + n(\bar{y}-\mu)^2\right\}^{-n/2} \\
&\propto \left\{1 + \frac{1}{(n-1)}\frac{(\mu-\bar{y})^2}{s^2/n}\right\}^{-\frac{(n-1)+1}{2}},
\end{aligned}
$$

which is proportional to $t_{n-1}(\bar{y}, s^2/n)$ *density, called non-standardized Student's t distribution with* $n-1$ *degrees of freedom. Recall that the standard T-distribution with p* ν *degrees of freedom, denoted as* T_ν, *has the density of*

$$f_T(t; \nu) \propto \left(1 + \frac{t^2}{\nu}\right)^{-\frac{\nu+1}{2}}$$

for $-\infty < t < \infty$.

From Example 5.3, we derive

$$\frac{\mu - \bar{y}}{s/\sqrt{n}} \mid (y_1, \ldots, y_n) \sim T_{n-1}.$$

It is interesting to note that the same conclusion can be made from the frequentist perspective, where the reference distribution is the sampling distribution $f(y_1, \ldots, y_n \mid \mu)$ treating μ as fixed. In the Bayesian inference, the reference distribution is the opposite, $p(\mu \mid y_1, \ldots, y_n)$.

Example 5.4. *We now consider a normal regression model*

$$y_i = \mathbf{x}_i'\beta + \epsilon_i, \quad i = 1, \ldots, n,$$

where $\mathbf{x}_i = (x_{i1}, \ldots, x_{ip})'$, $\beta = (\beta_1, \ldots, \beta_p)'$, *and* $\epsilon_i \overset{i.i.d.}{\sim} N(0, \sigma^2)$. *Let* $\mathbf{y} = (y_1, \ldots, y_n)'$ *and* $X = (\mathbf{x}_1, \ldots, \mathbf{x}_n)'$. *Suppose that* X *has full rank and* $p < n$. *Then the likelihood is given as*

$$L(\beta, \sigma^2) = \frac{1}{(2\pi\sigma^2)^{n/2}} \exp\left(-\frac{1}{2\sigma^2}\|\mathbf{y} - X\beta\|^2\right),$$

where $\|\cdot\|$ *denote the Euclidean norm, i.e.,* $\|y\| = \sqrt{y'y}$.
If we use a noninformative prior for (β, σ^2),

$$\pi(\beta, \sigma^2) \propto 1/\sigma^2,$$

the posterior is

$$
\begin{aligned}
p(\beta, \sigma^2 \mid \mathbf{y}) \quad &\propto \quad (\sigma^2)^{-n/2-1} \exp\left(-\frac{1}{2\sigma^2}\|\mathbf{y} - X\beta\|^2\right) \\
&= \quad (\sigma^2)^{-(n-p)/2-1} \exp\left(-\frac{1}{2\sigma^2}\|\mathbf{y} - X\hat{\beta}\|^2\right) \\
&\quad \times (\sigma^2)^{-p/2} \exp\left(-\frac{1}{2\sigma^2}(\beta - \hat{\beta})'(X'X)(\beta - \hat{\beta})\right),
\end{aligned}
$$

where $\hat{\beta} = (X'X)^{-1}X'\mathbf{y}$. *This implies that*

$$
\begin{aligned}
\beta | \mathbf{y}, \sigma^2 \quad &\sim \quad N\left(\hat{\beta}, \sigma^2(X'X)^{-1}\right), \\
\sigma^2 | \mathbf{y} \quad &\sim \quad Inv\text{-}Gamma\left(\frac{n-p}{2}, \frac{\|\mathbf{y} - X\hat{\beta}\|^2}{2}\right).
\end{aligned}
$$

If the dimension of X is high, then the $p \times p$ *matrix* $(X'X)$ *can be singular. In this case, one can consider an informative prior such as the Bayesian Lasso (Park and Casella, 2008) as follows:*

$$
p(\beta|\sigma^2) \quad = \quad \prod_{j=1}^{p} \frac{\lambda}{2\sqrt{\sigma^2}} \exp\left(-\frac{\lambda}{\sqrt{\sigma^2}}|\beta_j|\right),
$$

where $\lambda(>0)$ *is a deterministic hyperparameter.*

We now briefly introduce the asymptotic normality of the posterior distribution. Let $\hat{\theta}$ be the posterior mode, that is, $\hat{\theta} = \arg\max_\theta p(\theta|y_1, \ldots, y_n)$. A Taylor expansion of $\log p(\theta|y_1, \ldots, y_n)$ centered at $\hat{\theta}$ gives

$$
\begin{aligned}
\log p(\theta|y_1, \ldots, y_n) \quad &\cong \quad \log p(\hat{\theta}|y_1, \ldots, y_n) \\
&\quad + \frac{1}{2}(\theta - \hat{\theta})' \left[\frac{\partial^2}{\partial\theta\partial\theta'} \log p(\theta|y_1, \ldots, y_n)\right]_{\theta=\hat{\theta}} (\theta - \hat{\theta}).
\end{aligned}
$$

Under some regularity conditions, it can be shown that the posterior distribution converges to a point mass at the true parameter value, θ_0, as $n \to \infty$. That is, for every $\epsilon > 0$,

$$
\lim_{n \to \infty} P(|\theta - \theta_0| > \epsilon|y_1, \ldots, y_n) = 0, \tag{5.4}
$$

P_{θ_0}-almost surely as $n \to \infty$.

From the posterior consistency in (5.4), we observe that as $n \to \infty$, the mass of the posterior distribution $p(\theta|y_1,\ldots,y_n))$ becomes concentrated in smaller and smaller neighborhoods of θ_0. The posterior consistency in (5.4) implies the posterior mode is consistent for θ_0, the true parameter value of θ.

Note that

$$\left[\frac{\partial^2}{\partial\theta\partial\theta'}\log p(\theta|y_1,\ldots,y_n)\right]_{\theta=\hat{\theta}} = \left[\frac{\partial^2}{\partial\theta\partial\theta'}\log p(\theta)\right]_{\theta=\hat{\theta}} + \sum_{i=1}^{n}\left[\frac{\partial^2}{\partial\theta\partial\theta'}\log f(y_i|\theta)\right]_{\theta=\hat{\theta}}.$$

Under some regularity conditions, the first term is negligible and

$$\left[\frac{\partial^2}{\partial\theta\partial\theta'}\log p(\theta|y_1,\ldots,y_n)\right]_{\theta=\hat{\theta}} = -nI(\hat{\theta}),$$

where $I(\theta)$ is the observed Fisher information. Therefore, under some regularity conditions, as $n \to \infty$, we obtain

$$p(\theta|y_1,\ldots,y_n) \to C(y_1,\ldots,y_n)\exp\left\{-\frac{1}{2}(\theta-\hat{\theta})'[nI(\hat{\theta})](\theta-\hat{\theta})\right\}$$

which is the density of $N(\hat{\theta},[nI(\hat{\theta})]^{-1})$ distribution. This is the asymptotic normality of the posterior distribution, often called Berstein-von Mises Theorem (van der Vaart, 1998, Ch. 10).

5.2 Multiple Imputation: Bayesian Justification

Imputation is essentially a prediction problem. Thus, we can apply Bayesian approach to prediction to handle imputation. Given the current observations $y_1,\ldots,y_n$, the posterior predictive distribution of a future observation, y_{n+1}, can be calculated directly by integration as follows:

$$f(y_{n+1} \mid y_1,\ldots,y_n) = \int f(y_{n+1} \mid \theta)p(\theta \mid y_1,\ldots,y_n)d\theta. \tag{5.5}$$

Under the missing data setup with $\mathbf{y}_{\text{com}} = (\mathbf{y}_{\text{obs}},\mathbf{y}_{\text{mis}})$, the posterior predictive distribution of $\mathbf{y}_{\text{mis}}$ can be calculated by

$$f(\mathbf{y}_{\text{mis}} \mid \mathbf{y}_{\text{obs}},\boldsymbol{\delta}) = \int f(\mathbf{y}_{\text{mis}} \mid \mathbf{y}_{\text{obs}},\boldsymbol{\delta};\eta)p(\eta \mid \mathbf{y}_{\text{obs}},\boldsymbol{\delta})d\eta, \tag{5.6}$$

where $\eta = (\theta,\phi)$, $f(\mathbf{y}_{\text{mis}} \mid \mathbf{y}_{\text{obs}},\boldsymbol{\delta};\eta)$ is the prediction distribution defined in (3.56), and $p(\eta \mid \mathbf{y}_{\text{obs}},\boldsymbol{\delta})$ is the posterior distribution of η given the observation $(\mathbf{y}_{\text{obs}},\boldsymbol{\delta})$. The posterior distribution is often computed by

$$p(\eta \mid \mathbf{y}_{\text{obs}},\boldsymbol{\delta}) = \int p(\eta \mid \mathbf{y}_{\text{com}})f(\mathbf{y}_{\text{mis}} \mid \mathbf{y}_{\text{obs}},\boldsymbol{\delta})d\mathbf{y}_{\text{mis}}, \tag{5.7}$$

which involves the posterior predictive distribution in (5.6). Thus, the data augmentation algorithm in Section 3.6 applied to (5.6) and (5.7) can be used iteratively to obtain the Monte Carlo samples of $(\mathbf{y}_{\text{mis}}, \eta)$ given $(\mathbf{y}_{\text{obs}}, \boldsymbol{\delta})$. Or, if $p(\eta \mid \mathbf{y}_{\text{obs}}, \boldsymbol{\delta})$ is directly available, then the following two-step procedure for imputation can be made:

[Step 1] Generate η^* from $p(\eta \mid \mathbf{y}_{\text{obs}}, \boldsymbol{\delta})$.

[Step 2] Given η^* from [Step 1], generate $\mathbf{y}_{\text{mis}}^*$ from $f(\mathbf{y}_{\text{mis}} \mid \mathbf{y}_{\text{obs}}, \boldsymbol{\delta}; \eta^*)$.

Multiple imputation, proposed by Rubin (1978) and further developed by Rubin (1987), is a Bayesian approach of generating imputed values with simplified variance estimation. In this procedure, Bayesian methods of generating imputed values are considered, where $m(> 1)$ multiply imputed data $\mathbf{y}_{\text{com}}^{*(j)} = (\mathbf{y}_{\text{obs}}, \mathbf{y}_{\text{mis}}^{*(j)})$ are generated for $j = 1, \ldots, m$, and $\mathbf{y}_{\text{mis}}^{*(1)}, \ldots, \mathbf{y}_{\text{mis}}^{*(m)}$ are m independent Monte Carlo samples from (5.6).

Using the m imputed values, $\mathbf{y}_{\text{com}}^{*(1)}, \ldots, \mathbf{y}_{\text{com}}^{*(m)}$, the multiple imputation (MI) estimator of η, denoted by $\hat{\eta}_{\text{mi}}$, can be obtained by

$$\hat{\eta}_{\text{mi}} = \frac{1}{m} \sum_{j=1}^{m} \hat{\eta}_I^{(j)}, \tag{5.8}$$

where $\hat{\eta}_I^{(j)} = \hat{\eta}(\mathbf{y}_{\text{com}}^{*(j)})$ is obtained by applying the complete sample estimator of η. For variance estimation, a simple formula was proposed by Rubin (1987), and it is given by

$$\hat{V}_{\text{mi}} = W_m + \left(1 + \frac{1}{m}\right) B_m, \tag{5.9}$$

where

$$W_m = \frac{1}{m} \sum_{j=1}^{m} \hat{V}(\mathbf{y}_{\text{com}}^{*(j)}),$$

with $\hat{V}(\mathbf{y}_{\text{com}}^{*(j)})$ being the imputed version of the complete-sample variance estimator of $\hat{\eta}$ applied to $\mathbf{y}_{\text{com}}^{*(j)}$, and

$$B_m = \frac{1}{m-1} \sum_{j=1}^{m} \left(\hat{\eta}_I^{(j)} - \hat{\eta}_{\text{mi}}\right)^{\otimes 2}.$$

To justify (5.8), we first assume that

$$\hat{\eta}(\mathbf{y}_{\text{com}}) = E(\eta \mid \mathbf{y}_{\text{com}}). \tag{5.10}$$

Now, as $\mathbf{y}_{\text{com}}^{*(j)}$ are generated from $f(\mathbf{y}_{\text{com}} \mid \mathbf{y}_{\text{obs}}, \boldsymbol{\delta})$, we have

$$\lim_{m \to \infty} m^{-1} \sum_{j=1}^{m} \hat{\eta}(\mathbf{y}_{\text{com}}^{*(j)}) = E\{\hat{\eta}(\mathbf{y}_{\text{com}}) \mid \mathbf{y}_{\text{obs}}, \boldsymbol{\delta}\}$$

$$= E\{E(\eta \mid \mathbf{y}_{\text{com}}) \mid \mathbf{y}_{\text{obs}}, \boldsymbol{\delta}\}$$

$$= E(\eta \mid \mathbf{y}_{\text{obs}}, \boldsymbol{\delta}),$$

where the second equality follows from (5.10). Thus, the MI point estimator in (5.8) is a Monte Carlo approximation of the posterior expectation $E(\eta \mid \mathbf{y}_{\text{obs}}, \boldsymbol{\delta})$.

To justify (5.9), we also assume that

$$\hat{V}(\mathbf{y}_{\text{com}}) = V(\eta \mid \mathbf{y}_{\text{com}}). \tag{5.11}$$

Now, under (5.11), we have

$$\lim_{m \to \infty} W_m = E\{V(\eta \mid \mathbf{y}_{\text{com}}) \mid \mathbf{y}_{\text{obs}}, \boldsymbol{\delta}\} \tag{5.12}$$

as $\mathbf{y}_{\text{com}}^{*(j)}$ are generated from $f(\mathbf{y}_{\text{com}} \mid \mathbf{y}_{\text{obs}}, \boldsymbol{\delta})$. Also, under (5.10), we have

$$\lim_{m \to \infty} B_m = V\{E(\eta \mid \mathbf{y}_{\text{com}}) \mid \mathbf{y}_{\text{obs}}, \boldsymbol{\delta}\}. \tag{5.13}$$

Therefore, combining (5.12) and (5.13), we have

$$
\begin{aligned}
\lim_{m \to \infty} \hat{V}_{\text{mi}} &= \lim_{m \to \infty} W_m + \lim_{m \to \infty} B_m \\
&= E\{V(\eta \mid \mathbf{y}_{\text{com}}) \mid \mathbf{y}_{\text{obs}}, \boldsymbol{\delta}\} + V\{E(\eta \mid \mathbf{y}_{\text{com}}) \mid \mathbf{y}_{\text{obs}}, \boldsymbol{\delta}\} \\
&= V(\eta \mid \mathbf{y}_{\text{obs}}, \boldsymbol{\delta}).
\end{aligned} \tag{5.14}
$$

Thus, the Rubin's variance estimator is an approximation of the posterior variance of η given the observed data (i.e., $\mathbf{y}_{\text{obs}}$ and $\boldsymbol{\delta}$).

5.3   Multiple Imputation: Frequentist Justification

We now study the frequentist justification of the multiple imputation procedure. Unlike the Bayesian justification in the previous section, we assume that the complete sample estimator of η is the MLE of η by solving $S_{\text{com}}(\eta; \mathbf{y}, \boldsymbol{\delta}) = 0$, where $S_{\text{com}}(\eta; \mathbf{y}, \boldsymbol{\delta})$ is the complete-data score function defined in (2.20).

In multiple imputation, m imputed values are generated from the posterior predictive distribution in (5.6), which can be described as a two-step procedure as follows:

1. Posterior Step: m independent realizations of η, denoted by $\eta^{*(1)}, \ldots, \eta^{*(m)}$, are first generated from the posterior distribution $p(\eta \mid \mathbf{y}_{\text{obs}}, \boldsymbol{\delta})$.

2. Imputation Step: For each $j = 1, \ldots, m$, $\mathbf{y}_{\text{mis}}^{*(j)}$ is independently generated from the predictive distribution $f(\mathbf{y}_{\text{mis}} \mid \mathbf{y}_{\text{obs}}, \boldsymbol{\delta}; \eta)$ in (3.56) evaluated at $\eta = \eta^{*(j)}$.

The MI estimator $\hat{\eta}_{\text{mi}}$ is computed by

$$\hat{\eta}_{\text{mi}} = \frac{1}{m} \sum_{j=1}^{m} \hat{\eta}_I^{(j)}, \tag{5.15}$$

where $\hat{\eta}_I^{(j)}$ is the solution to $S_{\text{com}}(\eta; \mathbf{y}_{\text{obs}}, \mathbf{y}_{\text{mis}}^{*(j)}) = 0$ for η.

To discuss the asymptotic properties of $\hat{\eta}_{\text{mi}}$ obtained from (5.15), we present the following result, which was established by Wang and Robins (1998) and also by Nielsen (2003).

Theorem 5.1. *Assume that the posterior distribution $p(\eta \mid \mathbf{y}_{\text{obs}}, \boldsymbol{\delta})$ is asymptotically normal with mean $\hat{\eta}_{\text{mle}}$ and variance $\{I_{\text{obs}}(\hat{\eta}_{\text{mle}})\}^{-1}$, where $I_{\text{obs}}(\eta)$ is the observed Fisher information derived from the marginal density of $(\mathbf{y}_{\text{obs}}, \boldsymbol{\delta})$. Under some regularity conditions, the MI estimator $\hat{\eta}_{\text{mi}}$ is approximately unbiased for η_0 and has the asymptotic variance*

$$V(\hat{\eta}_{\text{mi}}) \cong \mathcal{I}_{\text{obs}}^{-1} + m^{-1} \mathcal{J}_{\text{mis}} \mathcal{I}_{\text{obs}}^{-1} \mathcal{J}_{\text{mis}}' + m^{-1} \mathcal{I}_{\text{com}}^{-1} \mathcal{I}_{\text{mis}} \mathcal{I}_{\text{com}}^{-1}, \tag{5.16}$$

where $\mathcal{J}_{\text{mis}} = \mathcal{I}_{\text{com}}^{-1} \mathcal{I}_{\text{mis}}$ is the fraction of missing information.

Proof. By the assumption of the posterior distribution, the j-th draw $\eta^{*(j)}$ from $p(\eta \mid \mathbf{y}_{\text{obs}}, \boldsymbol{\delta})$ satisfies

$$(\eta^{*(j)} - \hat{\eta}_{\text{mle}}) \mid (\mathbf{y}_{\text{obs}}, \boldsymbol{\delta}) \sim N \left[0, \{I_{\text{obs}}(\hat{\eta}_{\text{mle}})\}^{-1} \right] \tag{5.17}$$

and

$$\hat{\eta}_{\text{mle}} - \eta_0 \sim N \left(0, \mathcal{I}_{\text{obs}}^{-1} \right), \tag{5.18}$$

for sufficiently large n.

Now, $\hat{\eta}_I^{(j)}$ is the solution to $S_{\text{com}}(\eta; \mathbf{y}_{\text{obs}}, \mathbf{y}_{\text{mis}}^{*(j)}) = 0$, we can apply Lemma 4.3 to obtain

$$\hat{\eta}_I^{(j)} - \eta_0 \cong \hat{\eta}_{\text{mle}} - \eta_0 + \mathcal{J}_{\text{mis}}(\eta^{*(j)} - \hat{\eta}_{\text{mle}}) + \mathcal{I}_{\text{com}}^{-1} Z^{*(j)}, \tag{5.19}$$

where $Z^{*(j)} \mid (\mathbf{y}_{\text{obs}}, \boldsymbol{\delta}) \sim (0, \mathcal{I}_{\text{mis}})$ and the three terms are mutually independent. The first term has variance $\mathcal{I}_{\text{obs}}^{-1}$ by (5.18), the second term has variance $\mathcal{J}_{\text{mis}} \mathcal{I}_{\text{obs}}^{-1} \mathcal{J}_{\text{mis}}'$ by (5.17), and the third term has variance $\mathcal{I}_{\text{com}}^{-1} \mathcal{I}_{\text{mis}} \mathcal{I}_{\text{com}}^{-1}$ by (4.13). Furthermore, by (5.15), we have

$$\hat{\eta}_{\text{mi}} \cong \hat{\eta}_{\text{mle}} + m^{-1} \sum_{j=1}^{m} \mathcal{J}_{\text{mis}} \left(\eta^{*(j)} - \hat{\eta}_{\text{mle}} \right) + m^{-1} \sum_{j=1}^{m} \mathcal{I}_{\text{com}}^{-1} Z^{*(j)}. \tag{5.20}$$

Since the m values of $(\eta^{*(j)}, \mathbf{y}_{\text{mis}}^{*(j)})$ are independently generated, we obtain (5.16). $\qquad\square$

Comparing (4.15) with (5.16), the MI estimator has greater variance than the imputation estimator using the MLE. The first additional variance term, $m^{-1}\mathcal{J}_{\text{mis}}\mathcal{I}_{\text{obs}}^{-1}\mathcal{J}_{\text{mis}}'$, comes from the posterior step, as the parameters $\eta^{*(1)},\ldots,\eta^{*(m)}$ are generated from the observed posterior distribution. The second additional variance term, $m^{-1}\mathcal{I}_{\text{com}}^{-1}\mathcal{I}_{\text{mis}}\mathcal{I}_{\text{com}}^{-1}$, represents the variance due to the imputation step given the realized parameter values.

The following theorem presents the conditions for the asymptotic unbiasedness of Rubin's variance estimator.

Theorem 5.2. *Assume that m posterior values of η, denoted by $\eta^{*(1)},\ldots,\eta^{*(m)}$, are independently generated from a normal distribution with mean $\hat{\eta}_{\text{mle}}$ and variance matrix $\{I_{\text{obs}}(\hat{\eta}_{\text{mle}})\}^{-1}$. Assume that the complete sample variance estimator $\hat{V}$ satisfies*

$$E\left\{\hat{V}_I^{(j)}\right\} \cong \mathcal{I}_{\text{com}}^{-1}, \qquad (5.21)$$

where $\hat{V}_I^{(j)}$ is the naive variance estimator computed by applying $\hat{V}$ to the j-th imputed data $\mathbf{y}^{*(j)}$. Then, Rubin's variance estimator (5.9) is asymptotically unbiased for the variance of the MI estimator $\hat{\eta}_{\text{mi}}$.

Proof. By (4.36) and (5.19), we have

$$E(B_m) = V(\hat{\eta}_I^{(1)}) - Cov(\hat{\eta}_I^{(1)}, \hat{\eta}_I^{(2)}) = \mathcal{J}_{\text{mis}}\mathcal{I}_{\text{obs}}^{-1}\mathcal{J}_{\text{mis}}' + \mathcal{I}_{\text{com}}^{-1}\mathcal{I}_{\text{mis}}\mathcal{I}_{\text{com}}^{-1}. \quad (5.22)$$

Thus, by assumption (5.21), we have

$$E\left\{\hat{V}_{\text{mi}}(\hat{\eta}_{\text{mi}})\right\} \cong \mathcal{I}_{\text{com}}^{-1} + \left(1+m^{-1}\right)\left(\mathcal{J}_{\text{mis}}\mathcal{I}_{\text{obs}}^{-1}\mathcal{J}_{\text{mis}}' + \mathcal{I}_{\text{com}}^{-1}\mathcal{I}_{\text{mis}}\mathcal{I}_{\text{com}}^{-1}\right). \quad (5.23)$$

Using Woodbury matrix identity, we have

$$\left(A+BCB'\right)^{-1} = A^{-1} - A^{-1}B\left(C^{-1}+B'A^{-1}B\right)^{-1}B'A^{-1}$$

and

$$\left(C^{-1}+B'A^{-1}B\right)^{-1} = C - CB'\left(A+BCB'\right)^{-1}BC,$$

which leads to

$$\left(A+BCB'\right)^{-1} = A^{-1} - A^{-1}BCB'A^{-1} + A^{-1}BCB'\left(A+BCB'\right)^{-1}BCB'A^{-1}.$$

Applying the above equality to $A = \mathcal{I}_{\text{com}}$, $B = I$, and $C = -\mathcal{I}_{\text{mis}}$, we have

$$\mathcal{I}_{\text{obs}}^{-1} = \mathcal{I}_{\text{com}}^{-1} + \mathcal{J}_{\text{mis}}\mathcal{I}_{\text{obs}}^{-1}\mathcal{J}_{\text{mis}}' + \mathcal{I}_{\text{com}}^{-1}\mathcal{I}_{\text{mis}}\mathcal{I}_{\text{com}}^{-1} \qquad (5.24)$$

and (5.23) reduces to

$$E\left\{\hat{V}_{\text{mi}}(\hat{\eta}_{\text{mi}})\right\} \cong \mathcal{I}_{\text{obs}}^{-1} + m^{-1}\mathcal{J}_{\text{mis}}\mathcal{I}_{\text{obs}}^{-1}\mathcal{J}_{\text{mis}}' + m^{-1}\mathcal{I}_{\text{com}}^{-1}\mathcal{I}_{\text{mis}}\mathcal{I}_{\text{com}}^{-1}, \qquad (5.25)$$

which shows the asymptotic unbiasedness of Rubin's variance estimator by (5.16). $\square$

Example 5.5. *(Univariate Normal distribution)*
    Let $y_1, \ldots, y_n$ be IID observations from $N(\mu, \sigma^2)$ and only the first r elements are observed and the remaining $n - r$ elements are missing. Assume that the response mechanism is ignorable. To summarize, we have

$$y_1, \ldots, y_r \overset{i.i.d.}{\sim} N(\mu, \sigma^2). \tag{5.26}$$

In this case, the j-th posterior values of (μ, σ^2) are generated from

$$\sigma^{*(j)2} \mid \mathbf{y}_r \sim r\hat{\sigma}_r^2 / \chi_{r-1}^2 \tag{5.27}$$

and

$$\mu^{*(j)} \mid (\mathbf{y}_r, \sigma^{*(j)2}) \sim N\left(\bar{y}_r, r^{-1}\sigma^{*(j)2}\right), \tag{5.28}$$

where $\mathbf{y}_r = (y_1, \ldots, y_r)$, $\bar{y}_r = r^{-1} \sum_{i=1}^r y_i$, and $\hat{\sigma}_r^2 = r^{-1} \sum_{i=1}^r (y_i - \bar{y}_r)^2$. Given the posterior sample $(\mu^{*(j)}, \sigma^{*(j)2})$, the imputed values are generated from

$$y_i^{*(j)} \mid \left(\mathbf{y}_r, \mu^{*(j)}, \sigma^{*(j)2}\right) \sim N\left(\mu^{*(j)}, \sigma^{*(j)2}\right) \tag{5.29}$$

independently for $i = r+1, \ldots, n$. The m imputed values are generated by independently repeating (5.27)-(5.29) m times.
    Let $\theta = E(Y)$ be the parameter of interest and the MI estimator of θ can be expressed as

$$\hat{\theta}_{\mathrm{mi}} = \frac{1}{m} \sum_{j=1}^m \hat{\theta}_I^{(j)},$$

where

$$\hat{\theta}_I^{(j)} = \frac{1}{n} \left\{ \sum_{i=1}^r y_i + \sum_{i=r+1}^n y_i^{*(j)} \right\}.$$

Then,

$$\hat{\theta}_{\mathrm{mi}} = \bar{y}_r + \frac{n-r}{nm} \sum_{j=1}^m \left(\mu^{*(j)} - \bar{y}_r\right) + \frac{1}{nm} \sum_{i=r+1}^n \sum_{j=1}^m \left(y_i^{*(j)} - \mu^{*(j)}\right). \tag{5.30}$$

Asymptotically, the first term has mean μ and variance $r^{-1}\sigma^2$, the second term has mean zero and variance $(1 - r/n)^2 \sigma^2/(mr)$, the third term has mean zero and variance $\sigma^2(n-r)/(n^2 m)$, and the three terms are mutually independent. Thus, the variance of $\hat{\theta}_{\mathrm{mi}}$ is

$$V\left(\hat{\theta}_{\mathrm{mi}}\right) = \frac{1}{r}\sigma^2 + \frac{1}{m}\left(\frac{n-r}{n}\right)^2 \left(\frac{1}{r}\sigma^2 + \frac{1}{n-r}\sigma^2\right), \tag{5.31}$$

which is consistent with the general result in (5.16).

For variance estimation, note that

$$
\begin{aligned}
V(y_i^{*(j)}) &= V(\bar{y}_r) + V(\mu^{*(j)} - \bar{y}_r) + V(y_i^{*(j)} - \mu^{*(j)}) \\
&= \frac{1}{r}\sigma^2 + \frac{1}{r}\sigma^2\left(\frac{r+1}{r-1}\right) + \sigma^2\left(\frac{r+1}{r-1}\right) \\
&\cong \sigma^2.
\end{aligned}
$$

Writing

$$
\begin{aligned}
\hat{V}_I^{(j)}(\hat{\theta}) &= n^{-1}(n-1)^{-1}\sum_{i=1}^{n}\left\{\tilde{y}_i^{*(j)} - \frac{1}{n}\sum_{k=1}^{n}\tilde{y}_k^{*(j)}\right\}^2 \\
&= n^{-1}(n-1)^{-1}\left\{\sum_{i=1}^{n}\left(\tilde{y}_i^{*(j)} - \mu\right)^2 - n\left(\frac{1}{n}\sum_{k=1}^{n}\tilde{y}_k^{*(j)} - \mu\right)^2\right\}
\end{aligned}
$$

where $\tilde{y}_i^* = \delta_i y_i + (1-\delta_i)y_i^{*(j)}$, *we have*

$$
\begin{aligned}
E\left\{\hat{V}_I^{(j)}(\hat{\theta})\right\} &= n^{-1}(n-1)^{-1}\left\{\sum_{i=1}^{n}E\left(\tilde{y}_i^{*(j)} - \mu\right)^2 - nV\left(\frac{1}{n}\sum_{k=1}^{n}\tilde{y}_k^{*(j)}\right)\right\} \\
&\cong n^{-1}(n-1)^{-1}\left[n\sigma^2 - n\left\{\frac{1}{r}\sigma^2 + \left(\frac{n-r}{n}\right)^2\left(\frac{1}{r}\sigma^2 + \frac{1}{n-r}\sigma^2\right)\right\}\right] \\
&\cong n^{-1}\sigma^2,
\end{aligned}
$$

which satisfies (5.21). By (4.36) and (5.31), we have

$$
\begin{aligned}
E(B_m) &= V\left(\hat{\theta}_I^{*(1)}\right) - Cov\left(\hat{\theta}_I^{*(1)}, \hat{\theta}_I^{*(2)}\right) \\
&= V\left\{\frac{n-r}{n}\left(\mu^{*(1)} - \bar{y}_r\right) + \frac{1}{n}\sum_{i=r+1}^{n}\left(y_i^{*(1)} - \mu^{*(1)}\right)\right\} \\
&\cong \left(\frac{n-r}{n}\right)^2\left(\frac{1}{r} + \frac{1}{n-r}\right)\sigma^2 \\
&= \left(\frac{1}{r} - \frac{1}{n}\right)\sigma^2.
\end{aligned}
$$

Thus, Rubin's variance estimator satisfies

$$
E\left\{\hat{V}_{mi}(\hat{\theta}_{mi})\right\} \cong \frac{1}{r}\sigma^2 + \frac{1}{m}\left(\frac{n-r}{n}\right)^2\left(\frac{1}{r}\sigma^2 + \frac{1}{n-r}\sigma^2\right) \cong V\left(\hat{\theta}_{mi}\right),
$$

which is consistent with the general result in (5.25).

Example 5.6. *Multiple imputation can be implemented nonparametrically using the Bayesian bootstrap of Rubin (1981), in which we first assume that an element of the population takes one of the values $d_1, \ldots, d_K$ with probability $p_1, \ldots, p_K$, respectively. That is, we assume*

$$P(Y = d_k) = p_k, \quad \sum_{k=1}^{K} p_k = 1. \tag{5.32}$$

Let $y_1, \ldots, y_n$ be an IID sample from (5.32) and n_k be the number of y_i equal to d_k. The parameter is a vector of probabilities $\mathbf{p} = (p_1, \ldots, p_K)$, such that $\sum_{i=1}^{K} p_i = 1$. In this case, the population mean $\theta = E(Y)$ can be expressed as $\theta = \sum_{i=1}^{K} p_i d_i$, and we only need to estimate $\mathbf{p}$. If the improper Dirichlet prior with density proportional to $\prod_{k=1}^{K} p_k^{-1}$ is placed on the vector $\mathbf{p}$, then the posterior distribution of $\mathbf{p}$ is proportional to

$$\prod_{k=1}^{K} p_k^{n_k - 1},$$

which is a Dirichlet distribution with parameter $(n_1, \ldots, n_K)$. This posterior distribution can be simulated using $n - 1$ independent uniform random numbers. Let $u_1, \ldots, u_{n-1}$ be IID $U(0,1)$, and let $g_i = u_{(i)} - u_{(i-1)}, i = 1, 2, \ldots, n-1$ where $u_{(k)}$ is the k-th order statistic of $u_1, \ldots, u_{n-1}$ with $u_{(0)} = 0$ and $u_{(n)} = 1$. Partition the $g_1, \ldots, g_n$ into K collections, with the k-th one having n_k elements, and let p_k be the sum of the g_i in the k-th collection. Then, the realized value of $p_1, \ldots, p_k$ follows a $(K-1)$-variate Dirichlet distribution with parameter $(n_1, \ldots, n_K)$. In particular, if $K = n$, then $(g_1, \ldots, g_n)$ is the vector of probabilities to attach to the data values $y_1, \ldots, y_n$ in that Bayesian bootstrap replication.

To implement Rubin's Bayesian bootstrap to multiple imputation, assume that the first r elements are observed and the remaining $n - r$ elements are missing. The imputed values can be generated with the following steps:

[Step 1] From $\mathbf{y}_r = (y_1, \ldots, y_r)$, generate $\mathbf{p}_r^* = (p_1^*, \ldots, p_r^*)$ from the posterior distribution using the Bayesian bootstrap as follows.

  1. *Generate $u_1, \ldots, u_{r-1}$ independently from $U(0,1)$, and sort them to get $0 = u_{(0)} < u_{(1)} < \ldots < u_{(r-1)} < u_{(r)} = 1$.*
  2. *Compute $p_i^* = u_{(i)} - u_{(i-1)}$, $i = 1, 2, \ldots, r - 1$ and $p_r^* = 1 - \sum_{i=1}^{r-1} p_i^*$.*

[Step 2] Select the imputed value of y_i by

$$y_i^* = \begin{cases} y_1 & \text{with probability } p_1^* \\ \ldots & \ldots \\ y_r & \text{with probability } p_r^* \end{cases}$$

independently for each $i = r+1, \ldots, n$.

Using the above Bayesian bootstrap imputation m times independently, we can compute the MI point estimator and the MI variance estimator. For estimating $\theta = E(Y)$, we can also establish (5.30) with $\mu^{*(j)} = \sum_{i=1}^{r} p_i^{*(j)} y_i$, where $p_i^{*(j)}$ is the realized value of selection probability p_i^* in [Step1] obtained from the j-th application of Rubin's Bayesian bootstrap method. Using a property of the Dirichlet distribution and the multinomial distribution, we can establish the same variance formula as in (5.31). Thus, the above Bayesian bootstrap imputation is asymptotically equivalent to the normal imputation. Also, it can be shown that the MI variance estimator is asymptotically unbiased.

Rubin and Schenker (1986) proposed an approximation of this Bayesian bootstrap method, called the approximate Bayesian boostrap (ABB) method, which provides an alternative approach of generating imputed values from the empirical distribution. The ABB method can be described as follows:

[Step 1] From $\mathbf{y}_r = (y_1, \ldots, y_r)$, generate a donor set $\mathbf{y}_r^* = (y_1^*, \ldots, y_r^*)$ by bootstrapping. That is, we select

$$
y_i^* = \begin{cases} y_1 & \text{with probability } 1/r \\ \ldots & \ldots \\ y_r & \text{with probability } 1/r \end{cases}
$$

independently for each $i = 1, \ldots, r$.

[Step 2] From the donor set $\mathbf{y}_r^* = (y_1^*, \ldots, y_r^*)$, select an imputed value of y_i by

$$
y_i^{**} = \begin{cases} y_1^* & \text{with probability } 1/r \\ \ldots & \ldots \\ y_r^* & \text{with probability } 1/r \end{cases}
$$

independently for each $i = r+1, \ldots, n$.

Using the above ABB imputation m times independently, we can compute the MI point estimator and the MI variance estimator. For the estimation of $\theta = E(Y)$, we can also establish (5.31) and the asymptotic unbiasedness of the MI variance estimator.

Example 5.7. *(Regression model imputation)*
Under the linear regression model setup of Example 3.20, multiple imputation can be implemented by applying the steps [P-step]-[I-step] of Example 3.20 independently m times. At each repetition of the imputation ($j = 1, \ldots, m$), we can calculate the imputed version of the full sample estimators

$$
\hat{\beta}_I^{(j)} = \left(\sum_{i=1}^{n} \mathbf{x}_i \mathbf{x}_i' \right)^{-1} \left\{ \sum_{i=1}^{r} \mathbf{x}_i y_i + \sum_{i=r+1}^{n} \mathbf{x}_i y_i^{*(j)} \right\}
$$

and

$$
\hat{V}_I^{(j)} = \left(\sum_{i=1}^{n} \mathbf{x}_i \mathbf{x}_i' \right)^{-1} \hat{\sigma}_I^{(j)2},
$$

where

$$\hat{\sigma}_I^{(j)2} = (n-p)^{-1} \left\{ \sum_{i=1}^{r} \left(y_i - \mathbf{x}_i' \hat{\beta}_I^{(j)} \right)^2 + \sum_{i=r+1}^{n} \left(y_i^{*(j)} - \mathbf{x}_i' \hat{\beta}_I^{(j)} \right)^2 \right\}.$$

The proposed point estimator for the regression coefficient based on m repeated imputations is

$$\hat{\beta}_{\mathrm{mi}} = \frac{1}{m} \sum_{j=1}^{m} \hat{\beta}_I^{(j)}, \tag{5.33}$$

and the proposed estimator for the variance of $\hat{\beta}_{\mathrm{mi}}$ is given by $\hat{V}_{\mathrm{mi}}$ in (5.9). Since we can express

$$\hat{\beta}_{\mathrm{mi}} = \hat{\beta}_r + \frac{1}{m} \sum_{j=1}^{m} \sum_{i=r+1}^{n} \mathbf{h}_i \mathbf{x}_i' \left(\beta^{*(j)} - \hat{\beta}_r \right) + \frac{1}{m} \sum_{j=1}^{m} \sum_{i=r+1}^{n} \mathbf{h}_i e_i^{*(j)}, \tag{5.34}$$

where $\hat{\beta}_r = (X_r'X_r)^{-1} X_r' \mathbf{y}_r$, $\mathbf{h}_i = (X_n'X_n)^{-1} \mathbf{x}_i$ and $\beta^{*(j)}$ is the j-th realization of the parameter values generated from posterior distribution (3.68). The total variance is

$$\begin{aligned}
V\left(\hat{\beta}_{\mathrm{mi}}\right) &\cong \left(X_r'X_r\right)^{-1} \sigma^2 \\
&+ m^{-1} \left(X_n'X_n\right)^{-1} \left(X_{n-r}'X_{n-r}\right) \left(X_r'X_r\right)^{-1} \left(X_{n-r}'X_{n-r}\right) \left(X_n'X_n\right)^{-1} \sigma^2 \\
&+ m^{-1} \left(X_n'X_n\right)^{-1} \left(X_{n-r}'X_{n-r}\right) \left(X_n'X_n\right)^{-1} \sigma^2,
\end{aligned}$$

which is a special case of the general result in (5.16). Using some matrix algebra similar to (5.24),

$$\begin{aligned}
\left(X_r'X_r\right)^{-1} &= \left(X_n'X_n - X_{n-r}'X_{n-r}\right)^{-1} \\
&= \left(X_n'X_n\right)^{-1} + \left(X_n'X_n\right)^{-1} \left(X_{n-r}'X_{n-r}\right) \left(X_n'X_n\right)^{-1} \\
&+ \left(X_n'X_n\right)^{-1} \left(X_{n-r}'X_{n-r}\right) \left(X_r'X_r\right)^{-1} \left(X_{n-r}'X_{n-r}\right) \left(X_n'X_n\right)^{-1},
\end{aligned}$$

we can write

$$V(\hat{\beta}_{\mathrm{mi}}) \cong (X_r'X_r)^{-1} \sigma^2 + m^{-1} \left\{ (X_r'X_r)^{-1} - (X_n'X_n)^{-1} \right\} \sigma^2.$$

Also, it can be shown that

$$E(W_m) \cong (X_n'X_n)^{-1} \sigma^2,$$

and using an argument similar to that in (5.22),

$$\begin{aligned}
E(B_m) &\cong \left(X_n'X_n\right)^{-1} \left(X_{n-r}'X_{n-r}\right) \left(X_n'X_n\right)^{-1} \\
&+ \left(X_n'X_n\right)^{-1} \left(X_{n-r}'X_{n-r}\right) \left(X_r'X_r\right)^{-1} \left(X_{n-r}'X_{n-r}\right) \left(X_n'X_n\right)^{-1} \\
&= \left\{ \left(X_r'X_r\right)^{-1} - \left(X_n'X_n\right)^{-1} \right\} \sigma^2.
\end{aligned}$$

Thus, the asymptotic unbiasedness of the Rubin's variance estimator can be established. Kim (2004) showed that, instead of (3.67), if one uses

$$\sigma^{*2} \mid \mathbf{y}_r \overset{i.i.d.}{\sim} (r-p)\hat{\sigma}_r^2/\chi_{r-p+1}^2, \tag{5.35}$$

then the resulting MI variance estimator can have smaller bias in small sample sizes.

We now discuss an extension under the setup of Example 5.7. Suppose that the parameter of interest is not the regression coefficient β. Let $\hat{\theta}_n = \sum_{i=1}^n \alpha_i y_i$ be the complete sample estimator of a parameter θ for some coefficients α_i. The MI estimator of θ is computed by

$$\hat{\theta}_{\mathrm{mi}} = \frac{1}{m} \sum_{j=1}^m \hat{\theta}_I^{(j)},$$

and, for the case of $\hat{\theta}_n = \sum_{i=1}^n \alpha_i y_i$, we can write

$$\hat{\theta}_{\mathrm{mi}} = \hat{\theta}_{I,\infty} + \sum_{i=r+1}^n \alpha_i \mathbf{x}_i' \left(\bar{\beta}_m^* - \hat{\beta}_r \right) + \sum_{i=r+1}^n \alpha_i \bar{e}_i^*,$$

where $\hat{\theta}_{I,\infty} = \sum_{i=1}^r \alpha_i y_i + \sum_{i=r+1}^n \alpha_i \mathbf{x}_i' \hat{\beta}_r$, $\bar{\beta}_m = m^{-1} \sum_{j=1}^m \beta^{*(j)}$, and $\bar{e}_i^* = m^{-1} \sum_{j=1}^m e_i^{*(j)}$. Note that $\hat{\theta}_{I,\infty} = \mathrm{plim}_{m \to \infty} \hat{\theta}_{\mathrm{mi}}$. Thus, the total variance of $\hat{\theta}_{\mathrm{mi}}$ is

$$V(\hat{\theta}_{\mathrm{mi}}) = V(\hat{\theta}_{I,\infty}) + \frac{1}{m} \left\{ \alpha_{n-r}' X_{n-r} \left(X_r' X_r \right)^{-1} X_{n-r}' \alpha_{n-r} + \alpha_{n-r}' \alpha_{n-r} \right\} \sigma^2,$$

where $\alpha_{n-r} = (\alpha_{r+1}, \ldots, \alpha_n)'$.

To discuss variance estimation, first note that, by the same argument for (5.22),

$$E(B_m) = \left\{ \alpha_{n-r}' X_{n-r} \left(X_r' X_r \right)^{-1} X_{n-r}' \alpha_{n-r} + \alpha_{n-r}' \alpha_{n-r} \right\} \sigma^2.$$

The following theorem presents the conditions for the asymptotic unbiasedness of the MI variance estimator.

Theorem 5.3. *Assume that $E(\hat{V}_I^{(j)}) \cong V(\hat{\theta}_n)$ holds for each $j = 1, 2, \ldots, m$. Also, assume that*

$$V(\hat{\theta}_{I,\infty}) = V(\hat{\theta}_n) + V(\hat{\theta}_{I,\infty} - \hat{\theta}_n) \tag{5.36}$$

holds. Then, under a linear regression model, the MI variance estimator is asymptotically unbiased for the variance of the MI point estimator.

Proof. By (5.36), the variance of MI point estimator is decomposed into three terms:

$$V\left(\hat{\theta}_{\text{mi}}\right) = V(\hat{\theta}_n) + V(\hat{\theta}_{I,\infty} - \hat{\theta}_n) + V(\hat{\theta}_{\text{mi}} - \hat{\theta}_{I,\infty}).$$

The first term is estimated by W_m by assumption. The third term,

$$V\left(\hat{\theta}_{\text{mi}} - \hat{\theta}_{I,\infty}\right) = m^{-1}\left\{\alpha'_{n-r}X_{n-r}\left(X'_r X_r\right)^{-1}X'_{n-r}\alpha_{n-r} + \alpha'_{n-r}\alpha_{n-r}\right\}\sigma^2,$$

is estimated by $m^{-1}B_m$. It remains to be shown that the second term is estimated by B_m. Since $\hat{\theta}_{I,\infty} - \hat{\theta}_n = \alpha'_{n-r}(X_{n-r}\hat{\beta}_r - \mathbf{y}_{n-r})$, we have

$$V\left(\hat{\theta}_{I,\infty} - \hat{\theta}_n\right) = \left\{\alpha'_{n-r}X_{n-r}\left(X'_r X_r\right)^{-1}X'_{n-r}\alpha_{n-r} + \alpha'_{n-r}\alpha_{n-r}\right\}\sigma^2 = E(B_m),$$

and so the MI variance estimator is asymptotically unbiased. □

Condition (5.36) is crucial for the asymptotic unbiasedness of the MI variance estimator. Meng (1994) called the condition *congeniality*. The congeniality condition is not always achieved. Kim et al. (2006a) discussed sufficient conditions for the congeniality under the linear regression models. More details will be covered in Section 5.5.

5.4 Multiple Imputation Using Mixture Models

Mixture distributions arise in practical problems when the measurements of a random variable are taken under different conditions. For example, the distribution of heights in a population of adults reflects the mixture of males and females in the population, that is,

$$y \sim \pi_{\text{male}}N(\mu_{\text{male}},\sigma^2_{\text{male}}) + \pi_{\text{female}}N(\mu_{\text{female}},\sigma^2_{\text{female}}),$$

where π_{male} and π_{female} respectively denote proportions of male and female in the population.

Generally speaking, finite mixture models are expressed as

$$p(y|\theta,\pi) = \sum_{k=1}^{K}\pi_k f(y \mid z = k, \theta_k), \tag{5.37}$$

where $\pi = (\pi_1,\ldots,\pi_K)$, $\theta = (\theta_1,\ldots,\theta_K)$, $\pi_k = P(z = k)$ satisfies $0 < \pi_1 < \ldots < \pi_K < 1$ and $\sum_{k=1}^{K}\pi_k = 1$. Here, $z \in \{1,\ldots,K\}$ is the latent variable for the group information. The joint density of (y,z) is

$$p(y,z \mid \theta,\pi) = \prod_{k=1}^{K}\left\{\pi_k f(y \mid z = k, \theta_k)\right\}^{\mathbb{I}(z=k)}. \tag{5.38}$$

A Bayesian inference for the mixture model requires a joint prior distribution for (π, θ), where $\pi = (\pi_1, \ldots, \pi_K)$ and $\theta = (\theta_1, \ldots, \theta_K)$. Prior distribution is

$$(\pi_1, \ldots, \pi_K) \sim \text{Dirichlet}(a_1, \ldots, a_K)$$
$$\theta_k \overset{iid}{\sim} P_k(\cdot), \quad k = 1, \ldots, K,$$

where P_k is a distribution.

Under complete response of $\mathbf{y} = (y_1, \ldots, y_n)$, we can apply Gibbs sampler to generate Monte Carlo samples from the full posterior $p(\theta, \pi | \mathbf{y})$ by iterating the following sampling:

$$\mathbf{z} \sim p(\mathbf{z} | \mathbf{y}, \theta, \pi),$$
$$\theta \sim p(\theta | \mathbf{y}, \pi, \mathbf{z}),$$
$$\pi \sim p(\pi | \mathbf{y}, \theta, \mathbf{z}).$$

Specifically, we use

$$P(z_i = k \mid y_i, \theta, \pi) = \frac{\pi_k f(y_i \mid z_i = k; \theta_k)}{\sum_{h=1}^{K} \pi_h f(y_i \mid z_i = h; \theta_h)}, \tag{5.39}$$

$$P(\theta_k \mid \mathbf{y}, \mathbf{z}, \pi) = \frac{\{\prod_{z_i=k} f(y_i \mid z_i = k, \theta_k)\} P_k(\theta_k)}{\int \{\prod_{z_i=k} f(y_i \mid z_i = k, \theta_k)\} P_k(\theta_k) d\theta_k}, \tag{5.40}$$

$$P(\pi_k \mid \mathbf{y}, \mathbf{z}, \theta) \sim \text{Dirichlet}(a_1 + n_1, \ldots, a_k + n_k), \tag{5.41}$$

where $n_k = \sum_{i=1}^{n} I(z_i = k)$.

Example 5.8. *(Gaussian mixture model)*
Assume the joint model of (z, y):

$$y_i \mid (z_i = k) \sim N(\mu_k, \tau_k^2),$$
$$P(z_i = k) = \pi_k, \quad k = 1, \ldots, K,$$

for $i = 1, \ldots, n$. The prior distribution is

$$(\pi_1, \ldots, \pi_K) \sim \text{Dirichlet}(a_0, \ldots, a_0),$$
$$(\mu_k, \tau_k^2) \overset{iid}{\sim} N(\mu_k | \mu_0, \kappa_0 \tau_k^2) \times \text{Inv-Gamma}(\tau_k^2 | \alpha_0, \beta_0);$$

where μ_0, κ_0, α_0 and β_0 are hyperparameters.
Using (5.39)-(5.41), we can show that

$$Pr(z_i = k | others) = \frac{\pi_k N(y_i | \mu_k, \tau_k^2)}{\sum_{h=1}^{K} \pi_h N(y_i | \mu_h, \tau_h^2)}$$

for $k = 1, 2, \ldots, K$.

$$\pi | others \sim \text{Dirichlet}(a_0 + n_1, \ldots, a_0 + n_K),$$

where $n_k = \sum_{i=1}^{n} \mathbb{I}\{z_i = k\}$. *Also,*

$$\mu_k, \tau_k^2 | \text{others} \overset{ind}{\sim} N(\mu_k | \hat{\mu}_k, \hat{\kappa}_k \tau_k^2) \times \text{Inv-Gamma}(\tau_k^2 | \hat{\alpha}_k, \hat{\beta}_k),$$

where $\hat{\kappa}_k = (\kappa_0^{-1} + n_k)^{-1}$, $\hat{\mu}_k = \hat{\kappa}_k(\kappa_0^{-1}\mu_0 + n_k \bar{y}_k)$, $\hat{\alpha}_k = \alpha_0 + n_k/2$,

$$\hat{\beta} = \beta_0 + \frac{1}{2}\left\{ \sum_{i:z_i=k} (y_i - \bar{y}_k)^2 + \frac{n_k}{1 + \kappa_0 n_k}(\bar{y}_h - \mu_0)^2 \right\},$$

and $\bar{y}_k = n_k^{-1} \sum_{i:z_i=k} y_i$.

For handling missing data, we have only to add one more step (imputation step) in the Gibbs sampling. That is, we add

$$y_{i,\text{mis}} | \text{others} \sim \sum_{k=1}^{K} \mathbb{I}(z_i^* = k) f(y_{i,\text{mis}} | y_{i,\text{obs}}, z_i = k, \theta_k^*),$$

where (θ_k^*, z_i^*) is the realized value of (θ_k, z_i) in the previous step of the Gibbs sampling.

Example 5.9. *Consider a bivariate random vector* (X, Y) *with density*

$$f(x, y) = \sum_{k=1}^{K} \pi_k \phi_2(x, y; \mu_k, \Sigma_k) \tag{5.42}$$

where $\phi_2(x, y; \mu_k, \Sigma_k)$ *is the density of bivariate normal distribution with parameter* μ_k *and* Σ_k, $\mu_k = (\mu_{x,k}, \mu_{y,k})'$,

$$\Sigma_k = \begin{pmatrix} \sigma_{xx,k} & \sigma_{xy,k} \\ \sigma_{xy,k} & \sigma_{yy,k} \end{pmatrix}.$$

Thus,

$$\phi_2(x, y; \mu_k, \Sigma_k) \propto \exp\left[-\frac{1}{2} \begin{pmatrix} x - \mu_{x,k} \\ y - \mu_{y,k} \end{pmatrix}' \Sigma_k^{-1} \begin{pmatrix} x - \mu_{x,k} \\ y - \mu_{y,k} \end{pmatrix} \right]$$

and $\int \phi_2(x, y; \mu, \Sigma) d(x, y) = 1$.

Instead of observing (x_i, y_i), we observe $(x_i, \delta_i y_i)$, where δ_i is the response indicator function of y_i and satisfies MAR assumption in the sense that

$$f(y | x, \delta = 1) = f(y | x).$$

Let $\theta_k = (\mu_k, \Sigma_k)$ and assume a prior $P(\theta_k)$ for θ_k. The following data augmentation algorithm can be developed for the mixture model with missing data.

[I-step] Given the current parameter values θ^* and z_i^*, for unit i with $\delta_i = 0$ and $z_i^* = k$, generate

$$y_i^* \mid (x_i, z_i^* = k, \theta_k^*) \sim N(\beta_{0,k}^* + \beta_{1,k}^* x_i, \sigma_{e,k}^{2*}),$$

where $\beta_{1,k}^* = \sigma_{xy,k}^* / \sigma_{xx,k}^*$, $\beta_{0,k}^* = \mu_{y,k}^* - \beta_{1,k}^* \mu_{x,k}^*$, and $\sigma_{ee,k}^* = \sigma_{yy,k}^* - \beta_{1,k}^* \sigma_{xy,k}^*$.

[P-step]

1. *Generate z_i^* from $\{1, \ldots, K\}$ with probability*

$$P(z_i^* = k \mid others) = \frac{\pi_k^* \{\phi(x_i, y_i; \theta_k^*)\}^{\delta_i} \{\phi(x_i, y_i^*; \theta_k^*)\}^{(1-\delta_i)}}{\sum_{k=1}^{K} \pi_k^* \{\phi(x_i, y_i; \theta_k^*)\}^{\delta_i} \{\phi(x_i, y_i^*; \theta_k^*)\}^{(1-\delta_i)}}.$$

2. *Generate*

$$\theta_k^* \sim p(\theta_k \mid others) = \frac{\prod_{\{z_i^* = k\}} \{\phi(x_i, y_i; \theta_k)\}^{\delta_i} \{\phi(x_i, y_i^*; \theta_k)\}^{(1-\delta_i)} P(\theta_k)}{\int \prod_{\{z_i = k\}} \{\phi(x_i, y_i; \theta_k)\}^{\delta_i} \{\phi(x_i, y_i^*; \theta_k)\}^{(1-\delta_i)} P(\theta_k) d\theta_k}.$$

3. *Generate*

$$\pi \mid others \quad \sim \quad Dirichlet(a_0 + n_1^*, \ldots, a_0 + n_K^*)$$

where $n_k^* = \sum_{i=1}^{n} \mathbb{I}\{z_i^* = k\}$.

So far, we have assumed that K is known. Note that K is the model parameter that determines the level of complexity in the class of finite mixture models. In practice, K is unknown and can be determined by using cross validation or model selection criteria such as Bayesian Information Criterion (BIC) of Schwartz (Schwarz, 1978). Another approach is to employ the Dirichlet process (DP) mixture distribution. In practice, we choose a sufficiently large value of K and apply the stick-breaking representation of a truncated Dirichlet process (Sethuraman, 1994; Ishwaran and James, 2011), given by

$$\pi_k \quad = \quad V_k \prod_{g<k} (1 - V_g), \text{ for } k = 1, \ldots, K,$$
$$V_k \quad \sim \quad Beta(1, \alpha), \text{ for } k = 1, \ldots, K-1; V_K = 1.$$

If α is very small, most of the probability in $\pi = (\pi_1, \ldots, \pi_K)$ is allocated to the first few components. To choose α in Bayesian way, we may use a hyperparameter distribution

$$\alpha \sim \Gamma(a_\alpha, b_\alpha),$$

where a_α and b_α are generally chosen to be small values so that the gamma prior becomes relatively non-informative. For Example 5.8, the posterior step is changed to

$$P(z_i = k | \text{others}) = \frac{\pi_k N(y_i | \mu_k, \tau_k^2)}{\sum_{h=1}^K \pi_h N(y_i | \mu_h, \tau_h^2)}, \quad k = 1, \ldots, K,$$

$$V_k | \text{others} \sim \text{Beta}\left(1 + n_k, \alpha + \sum_{h=k+1}^K n_h\right),$$

$$\mu_k, \tau_k^2 | \text{others} \sim N(\mu_k | \hat{\mu}_k, \hat{\kappa}_k \tau_k^2) \times \text{Inv-Gamma}(\tau_k^2 | \hat{\alpha}_k, \hat{\beta}_k),$$

$$\alpha \, | \, \text{others} \sim \text{Gamma}(a_\alpha + K - 1, b_\alpha - \log(\pi_K)).$$

Kim et al. (2014) used the DP mixture of multivariate normal model for a flexible multiple imputation. Murray and Reiter (2016) extend the idea further to handle missing categorical and continuous values in multiple imputation.

5.5 Multiple Imputation for General Purpose Estimation

We now consider estimating ψ_0, which is defined through an estimating equation $E\{U(\psi; \mathbf{y})\} = 0$ using multiple imputation. Under complete response, a consistent estimator of ψ can be obtained by solving $U_{\text{com}}(\psi; \mathbf{y}) \equiv \sum_{i=1}^n U(\psi; \mathbf{y}_i) = 0$. Assume that m imputed values, say $\mathbf{y}_{\text{mis}}^{*(1)}, \ldots, \mathbf{y}_{\text{mis}}^{*(m)}$, are generated from the posterior predictive distribution in (5.6). Let $\hat{\psi}_I^{(j)}$ be the solution to $U_{\text{com}}(\psi; \mathbf{y}^{*(j)}) = 0$, where $\mathbf{y}^{*(y)} = (\mathbf{y}_{\text{obs}}, \mathbf{y}_{\text{mis}}^{*(j)})$. The MI estimator of ψ is then given by

$$\hat{\psi}_{\text{mi}} = \frac{1}{m} \sum_{j=1}^m \hat{\psi}_I^{(j)}. \tag{5.43}$$

The MI point estimator is asymptotically unbiased, and its asymptotic distribution for $m \to \infty$ is given in Theorem 4.2. Unfortunately, the Rubin's variance estimator does not work in this general case. The following example gives an illustration of the phenomenon.

Example 5.10. *Consider the bivariate data (x_i, y_i) of size $n = 200$, where x_i is always observed, and y_i is subject to missingness. The sampling distribution of (x_i, y_i) is $x_i \sim N(3, 1)$ and $y_i = -2 + x_i + e_i$ with $e_i \sim N(0, 1)$. Multiple imputation can be used to estimate $\theta_1 = E(Y)$ and $\theta_2 = P(Y < 1)$. To test the performance, a small simulation study is performed. The response mechanism is uniform with response rate 0.6. In estimating θ_2, we used a method-of-moment estimator (MME) $\hat{\theta}_2 = n^{-1} \sum_{i=1}^n I(y_i < 1)$ under complete response. An unbiased estimator for the variance of $\hat{\theta}_2$ is then $\hat{V}_2 = (n-1)^{-1} \hat{\theta}_2(1-$*

$\hat{\theta}_2$). *Multiple imputation with size $m = 50$ is used. After multiple imputation, Rubin's variance formula (5.9) is used to obtain variance estimation.*

Table 5.1 presents the simulation results for the multiple imputation point estimators. For comparison, we have also computed the complete-sample point estimators. Table 5.2 presents the performance of the multiple imputation variance estimators. The t-statistic is computed to test the significance of the Monte Carlo bias of the variance estimator. The MI variance estimator shows significant bias for estimating the variance of $\hat{\theta}_{2,\mathrm{mi}}$.

TABLE 5.1
Simulation Results of the MI Point Estimators

Parameter	Mean	$V(\hat{\theta}_n)$	$V(\hat{\theta}_{\mathrm{mi}})$	$V(\hat{\theta}_{\mathrm{mi}} - \hat{\theta}_n)$	$Cov(\hat{\theta}_n, \hat{\theta}_{\mathrm{mi}} - \hat{\theta}_n)$
θ_1	1.00	0.0100	0.0134	0.0035	0.0000
θ_2	0.50	0.00129	0.00137	0.00046	−0.00019

TABLE 5.2
Simulation Results of the MI Variance Estimators

Parameter	$E(W_m)$	$E(B_m)$	Rel. Bias (%)	t-statistics
$V(\hat{\theta}_1)$	0.0100	0.0033	−0.24	−0.08
$V(\hat{\theta}_2)$	0.00125	0.000436	23.08	7.48

For $\theta = \theta_1$,

$$V(\hat{\theta}_{\mathrm{mi}}) = V(\hat{\theta}_n) + V(\hat{\theta}_{\mathrm{mi}} - \hat{\theta}_n) = \frac{\sigma_y^2}{n} + \left(\frac{1}{r} - \frac{1}{n}\right)\sigma_e^2 = \frac{2}{200} + \left(\frac{1}{120} - \frac{1}{200}\right)\cdot 1 = 0.0133,$$

which is roughly equal to $E(W_m) + E(B_m)$ in the simulation result. However, for $\theta = \theta_2$, we have

$$
\begin{aligned}
V(\hat{\theta}_{\mathrm{mi}}) &= V(\hat{\theta}_n) + V(\hat{\theta}_{\mathrm{mi}} - \hat{\theta}_n) + 2Cov(\hat{\theta}_n, \hat{\theta}_{\mathrm{mi}} - \hat{\theta}_n) \\
&\doteq 0.00129 + 0.00046 + 2 \cdot (-0.00019) = 0.00137,
\end{aligned}
$$

while

$$E(\hat{V}_{\mathrm{mi}}) = E(W_m) + (1 + m^{-1})E(B_m) = 0.00125 + 1.02 \cdot 0.000436 = 0.00169 > 0.00137.$$

Thus, the MI variance estimator overestimates the variance because it ignores the covariance term between $\hat{\theta}_n$ and $\hat{\theta}_{\mathrm{mi}} - \hat{\theta}_n$.

We now explain the theory behind Example 5.10. To discuss asymptotic results, we assume a scalar y with density $f(y \mid x; \theta)$ and further assume that the posterior distribution in $p(\theta \mid \mathbf{x}_r, \mathbf{y}_r)$ is asymptotically normal in the sense that

$$\theta \mid (\mathbf{x}_n, \mathbf{y}_r) \sim N\left[\hat{\theta}_{\text{mle}}, \hat{V}(\hat{\theta}_{\text{mle}})\right] \tag{5.44}$$

where $\hat{\theta}_{\text{mle}}$ is the MLE of θ, $\hat{V}(\hat{\theta}_{\text{mle}}) = \{\sum_{i=1}^{r} I(\hat{\theta}_{\text{mle}}; x_i, y_i)\}^{-1}$ and

$$I(\theta; x, y) = -\frac{\partial}{\partial \theta'} S(\theta; x, y) = -\frac{\partial^2}{\partial \theta \partial \theta'} \log f(y \mid x; \theta).$$

Also, define $\mathcal{I}_\theta = E\{I(\theta; X, Y)\}$ and $p = E(r/n)$.

The parameter of interest ψ is defined through $E\{U(\psi; Y)\} = 0$. Define

$$g(x; \psi, \theta) = \frac{\partial}{\partial \theta'} E\{U(\psi; Y) \mid x; \theta\} = E\{U(\psi; Y)S(\theta; x, Y)' \mid x\},$$

and $\hat{U}(\psi) = n^{-1} \sum_{i=1}^{n} U(\psi; y_i)$. We first establish the following theorem.

Theorem 5.4. *Under MAR and the multiple imputation using (5.44) in the posterior step, we have*

$$\sqrt{n}\left(\hat{\psi}_{mi} - \psi\right) \to N\left(0, (\tau)^{-1}\Sigma_m(\tau')^{-1}\right) \tag{5.45}$$

where $\tau = E\{\dot{U}(\psi; Y)\}$,

$$\Sigma_m = V_1 + m^{-1}V_2 + m^{-1}V_3,$$

$$
\begin{aligned}
V_1 &= V(\bar{U}) + pE_1\{V(U \mid X)\} + p^{-1}(1-p)^2 g_0(\psi \mid \theta)\mathcal{I}_\theta^{-1} g_0(\psi \mid \theta)' \\
&\quad + 2(1-p)g_1(\psi \mid \theta)\mathcal{I}_\theta^{-1} g_0(\psi \mid \theta)' \\
V_2 &= (1-p)^2 p^{-1} g_0(\psi \mid \theta)\mathcal{I}_\theta^{-1} g_0(\psi \mid \theta)' \\
V_3 &= (1-p)E_0\{V(U \mid X)\}
\end{aligned}
$$

and $g_k(\eta \mid \theta) = E_k\{g(X; \psi, \theta)\}$ *with* $E_k(\cdot) = E(\cdot \mid \delta = k)$ *for* $k = 0, 1$.

Proof. Using the argument similar to Lemma 4.3, we can obtain

$$\hat{U}^{*(j)}(\psi) \cong \bar{U}(\psi \mid \mathbf{x}_n, \mathbf{y}_r; \hat{\theta}_{\text{mle}}) + (1-p)g_0(\psi \mid \hat{\theta}_{\text{mle}})(\theta_p^{*(j)} - \hat{\theta}_{\text{mle}}) + Z^{*(j)}, \tag{5.46}$$

where

$$
\begin{aligned}
\bar{U}(\psi \mid \mathbf{x}_n, \mathbf{y}_r; \hat{\theta}_{\text{mle}}) &= E\left\{\hat{U}(\psi) \mid \mathbf{x}_n, \mathbf{y}_r; \hat{\theta}_{\text{mle}}\right\} \\
&= n^{-1}\left[\sum_{i=1}^{r} U(\psi; y_i) + \sum_{i=r+1}^{n} E\{U(\psi; Y_i) \mid x_i; \hat{\theta}_{\text{mle}}\}\right],
\end{aligned}
$$

$$
\begin{aligned}
E\left\{\frac{\partial}{\partial \theta'}\bar{U}(\psi \mid \mathbf{x}_n, \mathbf{y}_r; \theta)\right\} &= E\left[n^{-1}\sum_{i=r+1}^{n} \frac{\partial}{\partial \theta'} E\{U(\psi; Y_i) \mid x_i, \theta\}\right] \\
&= (1-p)E\{g(X; \psi, \theta) \mid \delta = 0\} = g_0(\psi|\theta),
\end{aligned}
$$

and

$$Z^{*(j)} = \hat{U}^{*(j)}(\psi) - E\left\{\hat{U}(\psi) \mid \mathbf{x}_n, \mathbf{y}_r; \theta^{*(j)}\right\}.$$

The three terms in (5.46) are independent.

By the first order Taylor expansion, we have, for $i > r$,

$$E\{U(\psi; Y) \mid x_i; \hat{\theta}_{\mathrm{mle}}\} = E\{U(\psi; Y) \mid x_i; \theta\} + g(x_i; \psi, \theta)(\hat{\theta}_{\mathrm{mle}} - \theta) + o_p(n^{-1/2}), \tag{5.47}$$

where

$$\hat{\theta}_{\mathrm{mle}} - \theta = \mathcal{I}_\theta^{-1} r^{-1} \sum_{i=1}^{r} S(\theta; x_i, y_i) + o_p(n^{-1/2}).$$

Thus, by (5.47),

$$\sum_{i=r+1}^{n} E\{U(\psi; Y_i) \mid x_i; \hat{\theta}_{\mathrm{mle}}\} = \sum_{i=r+1}^{n} E\{U(\psi; Y_i) \mid x_i; \theta\}$$

$$+ \quad (n-r) g_0(\psi \mid \theta) r^{-1} \mathcal{I}_\theta^{-1} \sum_{i=1}^{r} S(\theta; x_i, y_i) + o_p(n^{1/2})$$

and, writing $\kappa = g_0(\psi \mid \theta) \mathcal{I}_\theta^{-1}$, $\bar{U}_i(\psi \mid \theta) = E\{U(\psi; Y_i) \mid x_i; \theta\}$, we have

$$\bar{U}(\psi \mid \mathbf{x}_n, \mathbf{y}_r; \hat{\theta}_{\mathrm{mle}})$$

$$= \quad n^{-1}\left[\sum_{i=1}^{r} U(\psi; y_i) + \sum_{i=r+1}^{n} E\{U(\psi; Y_i) \mid x_i; \hat{\theta}_{\mathrm{mle}}\}\right]$$

$$= \quad n^{-1}\left[\sum_{i=1}^{r} U(\psi; y_i) + \sum_{i=r+1}^{n} E\{U(\psi; Y_i) \mid x_i; \theta\}\right] + (1-p)\kappa r^{-1} \sum_{i=1}^{r} S(\theta; x_i, y_i)$$

$$= \quad n^{-1}\sum_{i=1}^{n} \bar{U}_i(\psi \mid \theta) + (1-p)\kappa r^{-1} \sum_{i=1}^{r} S(\theta; x_i, y_i) + n^{-1}\sum_{i=1}^{r} \{U(\psi; y_i) - \bar{U}_i(\psi \mid \theta)\}$$

$$:= \quad P_n + Q_n + R_n + o_p(n^{-1/2}),$$

where $Cov(P_n, R_n) = 0$ and $Cov(P_n, Q_n) = 0$. Now, ignoring the smaller order terms, the variance of the first term of (5.46) is

$$V\left\{\bar{U}(\psi \mid \mathbf{x}_n; \hat{\theta}_{\mathrm{mle}})\right\} = V(P_n) + V(Q_n) + V(R_n) + 2Cov(Q_n, R_n).$$

Now,

$$V(P_n) = n^{-1}V\{\bar{U}(\psi \mid \theta)\}$$
$$V(Q_n) = r^{-1}(1-p)^2 g_0(\psi \mid \theta) \mathcal{I}_\theta^{-1} g_0(\psi \mid \theta)'$$
$$V(R_n) = n^{-1} p E_1[V\{U \mid X\}]$$

and

$$Cov(Q_n, R_n) = n^{-1}(1-p) g_1(\psi \mid \theta) \mathcal{I}_\theta^{-1} g_0(\psi \mid \theta)',$$

where $E_k(\cdot) = E(\cdot \mid \delta = k)$ for $k = 0, 1$.

Thus,

$$V\left\{\bar{U}(\psi \mid \mathbf{x}_n, \mathbf{y}_r, \hat{\theta}_{\text{mle}})\right\}$$

$$= n^{-1}\left[V\{\bar{U}(\psi \mid \theta)\} + pE\{V_1(U \mid X)\} + p^{-1}(1-p)^2 g_0(\psi \mid \theta)\mathcal{I}_\theta^{-1} g_0(\psi \mid \theta)'\right]$$

$$+ 2n^{-1}(1-p)g_1(\psi \mid \theta)\mathcal{I}_\theta^{-1} g_0(\eta \mid \theta)'. \tag{5.48}$$

The variance of the second term of (5.46) is

$$V\{(1-p)g_0(\eta|\hat{\theta}_{\text{mle}})(\theta_p^{*(j)} - \hat{\theta}_{\text{mle}})\} = (1-p)^2 r^{-1} g_0(\eta \mid \theta)\mathcal{I}_\theta^{-1} g_0(\eta \mid \theta)'.$$

Finally, the variance of the third term of (5.46) is

$$V(Z^{*(j)}) = n^{-1}(1-p)E_0[V\{U \mid X\}].$$

Therefore, combining the results, we have established (5.45). □

Now, let's investigate the bias of MI variance estimator. We first assume that W_m is unbiased for $V(\hat{\psi}_n)$.

Theorem 5.5. *Under some regularity assumptions, the asymptotic bias of* $\hat{V}_{mi}$ *is*

$$Bias\left(\hat{V}_{mi}\right) = 2n^{-1}(1-p)\tau^{-1}[E\{V(U \mid X)\}$$

$$- g_1(\psi \mid \theta)\mathcal{I}_\theta^{-1} g_0(\psi \mid \theta)'](\tau')^{-1}. \tag{5.49}$$

Proof. In Rubin's formula, the W_m will essentially estimate $\tau^{-1}V\{\hat{U}(\psi)\}(\tau')^{-1}$ term. Thus, ignoring the smaller order terms,

$$E(W_m) = n^{-1}\tau^{-1}V(U)(\tau')^{-1}.$$

Also, using Lemma 5.2 and (5.46), we can obtain

$$E(B_m) = n^{-1}\tau^{-1}\left\{B_1 + n^{-1}(1-p)E_0[V\{U \mid X\}]\right\}(\tau')^{-1},$$

where

$$B_1 = (1-p)^2 p^{-1} g_0(\psi \mid \theta)\mathcal{I}_\theta^{-1} g_0(\psi \mid \theta)'.$$

The bias of MI variance estimator is

$$E(\hat{V}_{mi}) - V(\hat{U}_{mi})$$

$$= n^{-1}\tau^{-1}[V(U) + (1-p)E_0\{V(U \mid X)\} + B_1](\tau')^{-1}$$

$$- n^{-1}\tau^{-1}[V(\bar{U}) + pE_1\{V(U \mid X)\} + B_1 + 2(1-p)g_1(\psi \mid \theta)\mathcal{I}_\theta^{-1} g_0(\psi \mid \theta)'](\tau')^{-1}$$

$$= 2n^{-1}(1-p)\tau^{-1}\left\{E_0\{V(U \mid X)\} - g_1(\psi \mid \theta)\mathcal{I}_\theta^{-1} g_0(\psi \mid \theta)'\right\}(\tau')^{-1}.$$

□

Yang and Kim (2016) cover a more rigorous theory on the bias of the MI variance estimator for MME under MAR assumption. Xie and Meng (2017) also give a comprehensive discussion of the congeniality issue in multiple imputation.

6

Fractional Imputation

Fractional imputation (FI) was originally proposed by Kalton and Kish (1984) as an imputation method with a reduced variance. In FI, m imputed values are generated for each missing component $\mathbf{y}_{i,\mathrm{mis}}$ of the complete observation $\mathbf{y}_i = (\mathbf{y}_{i,\mathrm{obs}}, \mathbf{y}_{i,\mathrm{mis}})$, and m fractional weights are assigned to the imputed values so that the mean score function can be approximated by a weighted sum of the imputed score functions. FI is motivated from a frequentist framework. Table 6.1 presents a comparison of two popular imputation approaches, MI and FI. The main difference is that FI treats parameter θ as fixed and the prediction model is used for imputation.

TABLE 6.1

Comparison of Two Popular Imputation Approaches

	Bayesian	Frequentist
Model	Posterior distribution $f(\text{latent}, \theta \mid \text{data})$	Prediction model $f(\text{latent} \mid \text{data}, \theta)$
Computation Prediction Parameter update	Data augmentation I-step P-step	EM algorithm E-step M-step
Parameter est'n	Posterior mode	ML estimation
Imputation	Multiple imputation	Fractional imputation
Variance estimation	Rubin's formula	Linearization or Bootstrap

Fractional imputation is introduced using a fully parametric model in Section 6.1. In Sections 6.2 and 6.3, the parametric model assumptions are relaxed. Semiparametric fractional imputation using mixture models is introduced in Section 6.4. Fractional imputation for multivariate categorical data is introduced in Section 6.5. Model selection is discussed in Section 6.6.

DOI: 10.1201/9780429321740-6

6.1   Parametric Fractional Imputation

Let $\mathbf{y}_{ij}^* = (\mathbf{y}_{i,\mathrm{obs}}, \mathbf{y}_{i,\mathrm{mis}}^{*(j)})$ be the j-th imputed value of $\mathbf{y}_i$ and let w_{ij}^* be the fractional weight assigned to $\mathbf{y}_{ij}^*$. The fractional weights are constructed to satisfy

$$\sum_{j=1}^m w_{ij}^* = 1 \tag{6.1}$$

for each $i = 1, 2, \ldots, n$. If we have a parametric model with a joint density $f(\mathbf{y}, \boldsymbol{\delta}; \eta)$, the fractional weights should also satisfy

$$\sum_{i=1}^n \sum_{j=1}^m w_{ij}^* S(\hat{\eta}_{\mathrm{mle}}; \mathbf{y}_{ij}^*) = 0, \tag{6.2}$$

where $S(\eta; \mathbf{y})$ is the score function of η and $\hat{\eta}_{\mathrm{mle}}$ is the MLE of η.

Let ψ be the parameter of interest that is consistently estimated by solving

$$\sum_{i=1}^n U(\psi; \mathbf{y}_i) = 0$$

for ψ under complete response of $\mathbf{y}_i$. In fractional imputation, fractional weights $w_{i1}^*, \ldots, w_{im}^*$ are assigned to $\mathbf{y}_{i1}^*, \ldots, \mathbf{y}_{im}^*$, respectively, such that

$$\sum_{i=1}^n \sum_{j=1}^m w_{ij}^* U(\psi; \mathbf{y}_{ij}^*) \cong \sum_{i=1}^n E\left\{ U(\psi; \mathbf{y}_i) \mid \mathbf{y}_{i,\mathrm{obs}}, \boldsymbol{\delta}_i; \hat{\eta}_{\mathrm{mle}} \right\}. \tag{6.3}$$

If $\mathbf{y}_i$ is categorical, then (6.3) can be easily achieved by choosing

$$w_{ij}^* = P\left(\mathbf{y} = \mathbf{y}_{ij}^* \mid \mathbf{y}_{i,\mathrm{obs}}, \boldsymbol{\delta}_i; \hat{\eta} \right).$$

For continuous $\mathbf{y}_i$, we use the following iterative procedure to achieve (6.3) as closely as possible:

[Step 1] Generate m imputed values from some density $h_m(\mathbf{y}_{i,\mathrm{mis}})$ which has the same support as $f(\mathbf{y}_{i,\mathrm{mis}} \mid \mathbf{y}_{i,\mathrm{obs}}, \boldsymbol{\delta}_i; \eta)$ in (3.56). Often, the choice of the proposal density is $h_m(\mathbf{y}_{i,\mathrm{mis}}) = f(\mathbf{y}_{i,\mathrm{mis}} \mid \mathbf{y}_{i,\mathrm{obs}}, \boldsymbol{\delta}_i; \hat{\eta}_p)$, where $\hat{\eta}_p$ is a preliminary estimator of η.

[Step 2] Given $\hat{\eta}_{(t)}$, compute the fractional weights by

$$w_{ij(t)}^* \propto \frac{f(\mathbf{y}_{i,\mathrm{mis}}^{*(j)} \mid \mathbf{y}_{i,\mathrm{obs}}, \boldsymbol{\delta}_i; \hat{\eta}_{(t)})}{h_m(\mathbf{y}_{i,\mathrm{mis}}^{*(j)})} \tag{6.4}$$

with $\sum_{j=1}^m w_{ij(t)}^* = 1$.

[Step 3] Given the fractional weights computed from [Step 2], update the parameter $\hat{\eta}_{(t+1)}$ by maximizing

$$Q^*(\eta \mid \hat{\eta}_{(t)}) = \sum_{i=1}^{n} \sum_{j=1}^{m} w^*_{ij(t)} \ln\{f(\mathbf{y}^*_{ij}, \boldsymbol{\delta}_i; \eta)\} \tag{6.5}$$

over η, where $f(\mathbf{y}_i, \boldsymbol{\delta}_i; \eta)$ is the joint density of $(\mathbf{y}_i, \boldsymbol{\delta}_i)$.

[Step 4] Go to [Step 2] until convergence.

Step 1 can be called the *imputation step*, Step 2 can be called the *weighting step*, and Step 3 can be called the *maximization step* (M-step). The imputation and weighting steps can be combined to implement the E-step of the EM algorithm. Unlike the MCEM method, imputed values are not changed for each EM iteration - only the fractional weights are changed. Thus, the FI method has some computational advantage over the MCEM method.

Note that the fractional weights of the form (6.4) can be written as

$$w^*_{ij(t)} = \frac{f(\mathbf{y}^{*(j)}_{i,\text{mis}} \mid \mathbf{y}_{i,\text{obs}}, \boldsymbol{\delta}_i; \hat{\eta}^{(t)})/h_m(\mathbf{y}^{*(j)}_{i,\text{mis}})}{\sum_{k=1}^{m} f(\mathbf{y}^{*(k)}_{i,\text{mis}} \mid \mathbf{y}_{i,\text{obs}}, \boldsymbol{\delta}_i; \hat{\eta}^{(t)})/h_m(\mathbf{y}^{*(k)}_{i,\text{mis}})}.$$

Since the conditional distribution can be written as

$$f(\mathbf{y}_{i,\text{mis}} \mid \mathbf{y}_{i,\text{obs}}, \boldsymbol{\delta}_i; \hat{\eta}) = \frac{f(\mathbf{y}_i, \boldsymbol{\delta}_i; \hat{\eta})}{\int f(\mathbf{y}_i, \boldsymbol{\delta}_i; \hat{\eta})\, d\mathbf{y}_{i,\text{mis}}} = \frac{f(\mathbf{y}_i, \boldsymbol{\delta}_i; \hat{\eta})}{f_{\text{obs}}(\mathbf{y}_{i,\text{obs}}, \boldsymbol{\delta}_i; \hat{\eta})},$$

where $f_{\text{obs}}(\mathbf{y}_{i,\text{obs}}, \boldsymbol{\delta}_i; \hat{\eta}) = \int f(\mathbf{y}_i, \boldsymbol{\delta}_i; \hat{\eta})\, d\mathbf{y}_{i,\text{mis}}$ is the marginal density of $(\mathbf{y}_{i,\text{obs}}, \boldsymbol{\delta}_i)$, we can express

$$w^*_{ij(t)} = \frac{f(\mathbf{y}^*_{ij}, \boldsymbol{\delta}_i; \hat{\eta}_{(t)})/h_m(\mathbf{y}^{*(j)}_{i,\text{mis}})}{\sum_{k=1}^{m} f(\mathbf{y}^*_{ik}, \boldsymbol{\delta}_i; \hat{\eta}_{(t)})/h_m(\mathbf{y}^{*(k)}_{i,\text{mis}})}. \tag{6.6}$$

Thus, the marginal density in computing the conditional distribution is not needed in computing the fractional weights. Only the joint density is needed.

Given the m imputed values, $\mathbf{y}^*_{i1}, \ldots, \mathbf{y}^*_{im}$, generated from $h_m(\mathbf{y}_{i,\text{mis}})$, the sequence of estimators $\{\hat{\eta}_{(1)}, \hat{\eta}_{(2)}, \ldots\}$ can be constructed using [Step 2]-[Step 3]. The following theorem presents some convergence properties of the sequence of estimators.

Theorem 6.1. *Let $Q^*(\eta \mid \hat{\eta}_{(t)})$ be the weighted log likelihood function (6.5) based on fractional imputation. If*

$$Q^*(\hat{\eta}_{(t+1)} \mid \hat{\eta}_{(t)}) \geq Q^*(\hat{\eta}_{(t)} \mid \hat{\eta}_{(t)}), \tag{6.7}$$

then

$$l^*_{\text{obs}}(\hat{\eta}_{(t+1)}) \geq l^*_{\text{obs}}(\hat{\eta}_{(t)}), \tag{6.8}$$

where

$$l^*_{\text{obs}}(\eta) = \sum_{i=1}^{n} \ln\{f^*_{obs(i)}(\mathbf{y}_{i,\text{obs}}, \boldsymbol{\delta}_i; \eta)\} \qquad (6.9)$$

is the observed log-likelihood constructed from the fractional imputation and

$$f^*_{obs(i)}(\mathbf{y}_{i,\text{obs}}, \boldsymbol{\delta}_i; \eta) = \frac{\sum_{j=1}^{m} f(\mathbf{y}^*_{ij}, \boldsymbol{\delta}_i; \eta)/h_m(\mathbf{y}^{*(j)}_{i,\text{mis}})}{\sum_{j=1}^{m} 1/h_m(\mathbf{y}^{*(j)}_{i,\text{mis}})}.$$

Proof. By (6.6) and using Jensen's inequality,

$$
\begin{aligned}
l^*_{\text{obs}}(\hat{\eta}_{(t+1)}) - l^*_{\text{obs}}(\hat{\eta}_{(t)}) &= \sum_{i=1}^{n} \ln \left\{ \sum_{j=1}^{m} w^*_{ij(t)} \frac{f(\mathbf{y}^*_{ij}, \boldsymbol{\delta}_i; \hat{\eta}_{(t+1)})}{f(\mathbf{y}^*_{ij}, \boldsymbol{\delta}_i; \hat{\eta}_{(t)})} \right\} \\
&\geq \sum_{i=1}^{n} \sum_{j=1}^{m} w^*_{ij(t)} \ln \left\{ \frac{f(\mathbf{y}^*_{ij}, \boldsymbol{\delta}_i; \hat{\eta}_{(t+1)})}{f(\mathbf{y}^*_{ij}, \boldsymbol{\delta}_i; \hat{\eta}_{(t)})} \right\} \\
&= Q^*(\hat{\eta}_{(t+1)} \mid \hat{\eta}_{(t)}) - Q^*(\hat{\eta}_{(t)} \mid \hat{\eta}_{(t)}).
\end{aligned}
$$

Therefore, (6.7) implies (6.8). $\square$

Note that $l^*_{\text{obs}}(\eta)$ is an imputed version of the observed log-likelihood based on the m imputed values, $\mathbf{y}^*_{i1}, \ldots, \mathbf{y}^*_{im}$. By Theorem 6.1, the sequence $l^*_{\text{obs}}(\hat{\eta}_{(t)})$ is monotonically increasing and, under some conditions, the convergence of $\hat{\eta}_{(t)}$ to a stationary point $\hat{\eta}^*_m$ follows for fixed m. The stationary point $\hat{\eta}^*_m$ converges to the MLE of η as $m \to \infty$. Theorem 6.1 does not hold for the sequence obtained from the Monte Carlo EM method for fixed m, because the imputed values are re-generated for each E-step of the Monte Carlo EM method.

In some case, it is desired to create fractional imputation with a small imputation size, say $m = 10$. If the MLE of η is obtained analytically or computed from a fractional imputation with sufficiently large m, then we can create the final weights using smaller m with constraints (6.2). With this further constraint, the solution to the imputed score equation is equal to the MLE of η even for a small m. The fractional imputation satisfying constraints such as (6.2) is called *calibration fractional imputation*.

Finding the fractional weights for calibration fractional imputation can be achieved by the regression weighting technique, by which the fractional weights that satisfy (6.2) and $\sum_{j=1}^{m} w^*_{ij} = 1$ are constructed by

$$w^*_{ij} = w^*_{ij0} + w^*_{ij0}\Delta \left(S^*_{ij} - \bar{S}^*_i \right), \qquad (6.10)$$

where w^*_{ij0} is the initial fractional weights defined in (6.4), $S^*_{ij} = S(\hat{\eta}; \mathbf{y}^*_{ij}, \boldsymbol{\delta}_i)$,

$\bar{S}_{i.}^* = \sum_{j=1}^m w_{ij0}^* S_{ij}^*,$

$$\Delta = -\left\{\sum_{i=1}^n \sum_{j=1}^m w_{ij0}^* S_{ij}^*\right\}' \left[\sum_{i=1}^n \sum_{j=1}^m w_{ij0}^* \left(S_{ij}^* - \bar{S}_{i.}^*\right)^{\otimes 2}\right]^{-1}.$$

Note that some of the fractional weights computed by (6.10) can take negative values. In this case, some alternative algorithm other than the regression weighting should be used. For example, the fractional weights of the form

$$w_{ij}^* = \frac{w_{ij0}^* \exp\left(\Delta S_{ij}^*\right)}{\sum_{k=1}^m w_{ik0}^* \exp\left(\Delta S_{ik}^*\right)}$$

is approximately equal to the regression fractional weight in (6.10) and is always positive.

Now consider the special case of the exponential family of distributions in (3.45) with $\sum_{i=1}^n \mathbf{T}(\mathbf{y}_i)$ being the complete sufficient statistic for θ. Recall that, under MAR, the M-step of the EM algorithm is implemented by solving (3.47). In this setup, the weighting step in the calibration fractional imputation at the t-th step of the EM algorithm can be expressed as

$$\sum_{i=1}^n \sum_{j=1}^m w_{ij(t)}^* \mathbf{T}\left(\mathbf{y}_{ij}^*\right) = \sum_{i=1}^n E\left\{\mathbf{T}(\mathbf{y}_i) \mid \mathbf{y}_{i,\text{obs}}, \delta_i; \hat{\theta}^{(t)}\right\},$$

with $\sum_{j=1}^m w_{ij(t)}^* = 1$. The M-step remains the same. That is, the parameter is updated by solving

$$\sum_{i=1}^n \sum_{j=1}^m w_{ij(t)}^* \mathbf{T}\left(\mathbf{y}_{ij}^*\right) = \sum_{i=1}^n E_\theta\left\{\mathbf{T}(\mathbf{y}_i)\right\}$$

for θ to get $\hat{\theta}^{(t+1)}$.

Asymptotic properties of the FI estimator are derived as a special case of the general theory in Section 4.2. Variance estimation after fractional imputation is also discussed in Kim (2011).

Example 6.1. *We consider the bivariate normal distribution (2.33) with MAR. Let δ_{1i} and δ_{2i} be the response indicator functions of y_{1i} and y_{2i}, respectively. A set of sufficient statistics for $\theta = (\mu_1, \mu_2, \sigma_{11}, \sigma_{12}, \sigma_{22})$ is $T = \sum_{i=1}^n \left(y_{1i}, y_{2i}, y_{1i}^2, y_{1i}y_{2i}, y_{2i}^2\right)$. By a property of the normal distribution, and the fractional weights for $\delta_{1i} = 0$ and $\delta_{2i} = 1$ can be constructed by*

$$\sum_{j=1}^m w_{ij(t)}^* \left\{1, y_{1i}^{*(j)}, \left(y_{1i}^{*(j)}\right)^2\right\} = \left\{1, E\left(y_{1i} \mid y_{2i}, \hat{\theta}_{(t)}\right), E\left(y_{1i}^2 \mid y_{2i}, \hat{\theta}_{(t)}\right)\right\},$$

where

$$E\left(y_{1i} \mid y_{2i}, \hat{\theta}\right) = \hat{\mu}_1 + \frac{\hat{\sigma}_{12}}{\hat{\sigma}_{22}}(y_{2i} - \hat{\mu}_2)$$

and

$$E\left(y_{1i}^2 \mid y_{2i}, \hat{\theta}\right) = \left\{\hat{\mu}_1 + \frac{\hat{\sigma}_{12}}{\hat{\sigma}_{22}}\left(y_{2i} - \hat{\mu}_2\right)\right\}^2 + \hat{\sigma}_{11} - \hat{\sigma}_{12}^2/\hat{\sigma}_{22}.$$

Similarly, the fractional weights for $\delta_{i1} = 1$ *and* $\delta_{2i} = 0$ *are constructed by*

$$\sum_{j=1}^m w_{ij(t)}^* \left\{1, y_{2i}^{*(j)}, \left(y_{2i}^{*(j)}\right)^2\right\} = \left\{1, E\left(y_{2i} \mid y_{1i}, \hat{\theta}_{(t)}\right), E\left(y_{2i}^2 \mid y_{1i}, \hat{\theta}_{(t)}\right)\right\},$$

and the fractional weights for $\delta_{i1} = 0$ *and* $\delta_{2i} = 0$ *are constructed by*

$$\sum_{j=1}^m w_{ij(t)}^* \left\{1, y_{1i}^{*(j)}, y_{2i}^{*(j)}, \left(y_{1i}^{*(j)}\right)^2, \left(y_{2i}^{*(j)}\right)^2, y_{1i}^{*(j)} y_{2i}^{*(j)}\right\}$$

$$= \left\{1, \hat{\mu}_{1(t)}, \hat{\mu}_{2(t)}, \hat{\mu}_{1(t)}^2 + \sigma_{1(t)}^2, \hat{\mu}_{2(t)}^2 + \sigma_{2(t)}^2, \hat{\mu}_{1(t)}\hat{\mu}_{2(t)} + \hat{\sigma}_{12(t)}\right\}.$$

In the M-step, the parameters are updated by

$$\hat{\mu}_{1(t+1)} = n^{-1}\sum_{i=1}^n\sum_{j=1}^m w_{ij(t)}^* y_{1i}^{*(j)}$$

$$\hat{\mu}_{2(t+1)} = n^{-1}\sum_{i=1}^n\sum_{j=1}^m w_{ij(t)}^* y_{2i}^{*(j)}$$

$$\hat{\sigma}_{1(t+1)}^2 = n^{-1}\sum_{i=1}^n\sum_{j=1}^m w_{ij(t)}^* \left(y_{1i}^{*(j)}\right)^2 - \hat{\mu}_{1(t+1)}^2$$

$$\hat{\sigma}_{2(t+1)}^2 = n^{-1}\sum_{i=1}^n\sum_{j=1}^m w_{ij(t)}^* \left(y_{2i}^{*(j)}\right)^2 - \hat{\mu}_{2(t+1)}^2$$

$$\hat{\sigma}_{12(t+1)} = n^{-1}\sum_{i=1}^n\sum_{j=1}^m w_{ij(t)}^* y_{1i}^{*(j)} y_{2i}^{*(j)} - \hat{\mu}_{1(t+1)}\hat{\mu}_{2(t+1)}.$$

Note that the parameter estimates are computed by the standard formula for the MLE using fractional weights.

Example 6.2. *We now consider fractional imputation for partially classified categorical data. Let* $\mathbf{y} = (y_1, \ldots, y_p)$ *be the vector of study variables that take categorical values. Let* $\mathbf{y}_i = (y_{i1}, \ldots, y_{ip})$ *be the i-th realization of* $\mathbf{y}$*. Let* δ_{ij} *be the response indicator function for* y_{ij}*. Assume that the response mechanism is missing at random. Based on the realization of* $\boldsymbol{\delta}_i = (\delta_{i1}, \ldots, \delta_{ip})$*, the original observation* $\mathbf{y}_i$ *can decompose into* $(\mathbf{y}_{i,\text{obs}}, \mathbf{y}_{i,\text{mis}})$*. Let* $D_i = \{\mathbf{y}_{i,\text{mis}}^{*(1)}, \ldots, \mathbf{y}_{i,\text{mis}}^{*(M_i)}\}$ *be the set of all possible values of* $\mathbf{y}_{i,\text{mis}}$*. Such enumeration is possible as* $\mathbf{y}_i$ *is categorical with known categories. In this case, the fractional imputation consists of taking all of* M_i *possible values as the imputed values and then assigning them with fractional weights. The fractional* *weight assigned to* $\mathbf{y}_{i,\mathrm{mis}}^{*(j)}$ *is*

$$w_{ij}^* = \frac{\pi(\mathbf{y}_{i,\mathrm{obs}}, \mathbf{y}_{i,\mathrm{mis}}^{*(j)})}{\sum_k \pi(\mathbf{y}_{i,\mathrm{obs}}, \mathbf{y}_{i,\mathrm{mis}}^{*(k)})}, \tag{6.11}$$

where $\pi(\tilde{\mathbf{y}})$ *is the joint probability of* $\tilde{\mathbf{y}}$ *composed of observed and imputed* y_i *'s. If the joint probability is nonparametrically modeled, it is computed by*

$$\pi(\tilde{\mathbf{y}}) = \frac{1}{n} \sum_{i=1}^n \sum_{j \in D_i} w_{ij}^* I\left\{ (\mathbf{y}_{i,\mathrm{obs}}, \mathbf{y}_{i,\mathrm{mis}}^{*(j)}) = \tilde{\mathbf{y}} \right\}. \tag{6.12}$$

Note that (6.11) and (6.12) correspond to the E-step and M-step of the EM algorithm, respectively. The M-step (6.12) can be changed if there is a parametric model for the joint probability. For example, if the joint probability can be modeled by a multinomial distribution with parameter θ*, then rather than solving the mean score equation, the M step becomes solving the imputed score equation. The initial values of fractional weights in the EM algorithm can be* $w_{ij}^* = 1/M_i$.

Example 6.3. *We now consider fractional imputation under the setup of Example 3.15. Instead of the rejection method, we can consider the following fractional imputation method:*

[Step 1] Generate $y_i^{*(1)}, \ldots, y_i^{*(m)}$ *from* $f\left(y_i \mid x_i; \hat{\theta}_{(0)}\right)$.

[Step 2] Using the m imputed values generated from Step 1, compute the fractional weights by

$$w_{ij(t)}^* \propto \frac{f\left(y_i^{*(j)} \mid x_i; \hat{\theta}_{(t)}\right)}{f\left(y_i^{*(j)} \mid x_i; \hat{\theta}_{(0)}\right)} \left\{ 1 - \pi(x_i, y_i^{*(j)}; \hat{\phi}_{(t)}) \right\} \tag{6.13}$$

where

$$\pi(x_i, y_i; \hat{\phi}) = \frac{\exp\left(\hat{\phi}_0 + \hat{\phi}_1 x_i + \hat{\phi}_2 y_i\right)}{1 + \exp\left(\hat{\phi}_0 + \hat{\phi}_1 x_i + \hat{\phi}_2 y_i\right)}.$$

Use the imputed data and the fractional weights and implement the M-step by solving

$$\sum_{i=1}^n \sum_{j=1}^m w_{ij(t)}^* S\left(\theta; x_i, y_i^{*(j)}\right) = 0 \tag{6.14}$$

and

$$\sum_{i=1}^n \sum_{j=1}^m w_{ij(t)}^* \left\{ \delta_i - \pi(\phi; x_i, y_i^{*(j)}) \right\} \left(1, x_i, y_i^{*(j)}\right) = 0, \tag{6.15}$$

where $S\left(\theta; x_i, y_i\right) = \partial \log f(y_i \mid x_i; \theta)/\partial \theta$.

Example 6.4. *We now return to the setup of Example 3.18. The random effects, a_i, are generated from the conditional distribution in (3.62). Instead of using the Metropolis–Hastings algorithm, which can be computationally heavy, we can use fractional imputation which starts with the generation of m values of a_i, denoted by $a_i^{*(1)}, \ldots, a_i^{*(m)}$, from some proposal distribution $h(a_i)$. The choice of the proposal distribution $h(a)$ is somewhat arbitrary, but t-distribution with four degrees of freedom seems to work well in many cases. Then, compute the fractional weights by*

$$w_{ik}^* \propto f_1(\mathbf{y}_i \mid \mathbf{x}_i, a_i^{*(k)}; \hat{\beta}) f_2(a_i^{*(k)}; \hat{\sigma})/h(a_i^{*(k)}),$$

where $f_1(\cdot)$ and $f_2(\cdot)$ are defined in (3.62). Given the current parameter values, the M-step updates the parameter estimates by solving

$$\sum_{i=1}^{n} \sum_{k=1}^{m} w_{ik}^* \sum_j S_1(\beta; x_{ij}, y_{ij}, a_i^{*(k)}) = 0$$

and

$$\sum_{i=1}^{n} \sum_{k=1}^{m} w_{ik}^* S_2(\sigma; a_i^{*(k)}) = 0,$$

where $S_1(\cdot)$ and $S_2(\cdot)$ are the score functions derived from $f_1(\cdot)$ and $f_2(\cdot)$, respectively.

Example 6.5. *Suppose that we are interested in estimating parameter θ in the conditional distribution $f(y \mid x; \theta)$. Instead of observing (x_i, y_i), suppose that we observe (z_i, y_i), where z is conditionally independent of y given x and the joint distribution of (x, z) is known (or estimable from a calibration sample). In this case, we want to create x_i^* from the observed values of (z_i, y_i) in order to perform regression analysis. This setup is related to the problem of inference with linked data (Lahiri and Larsen, 2005) or measurement error model problem discussed in Example 3.17.*

To apply the FI method in this setup, we first generate m values of x_i, denoted by $x_i^{*(1)}, \ldots, x_i^{*(m)}$, from the conditional distribution $g(x \mid z_i)$ obtained from another source and then assign fractional weights computed by

$$w_{ij}^* \propto f\left(x_i^{*(j)} \mid y_i, z_i\right)/g\left(x_i^{*(j)} \mid z_i\right). \tag{6.16}$$

Using (3.61), the above fractional weights can be written

$$w_{ij}^* \propto f\left(y_i \mid x_i^{*(j)}\right),$$

which depends on unknown parameters $\theta = (\beta_0, \beta_1, \sigma)$. Thus, an EM algorithm can be developed for solving

$$\sum_{i=1}^{n} \sum_{j=1}^{m} w_{ij}^*(\theta) S(\theta; x_i^{*(j)}, y_i) = 0,$$

where $S(\theta; x, y) = \partial \ln f(y \mid x; \theta) / \partial \theta$ and

$$w_{ij}^*(\theta) = \frac{f\left(y_i \mid x_i^{*(j)}; \theta\right)}{\sum_{k=1}^m f\left(y_i \mid x_i^{*(k)}; \theta\right)}.$$

Write

$$Q^*(\eta \mid \eta_0) = \sum_{i=1}^n \sum_{j=1}^m w_{ij}^*(\eta_0) \log f\left(y_{ij}^*, \boldsymbol{\delta}_i; \eta\right), \tag{6.17}$$

where $w_{ij}^*(\eta)$ is the fractional weight associated with y_{ij}^*, denoted by

$$w_{ij}^*(\eta) = \frac{f(\mathbf{y}_{ij}^*, \boldsymbol{\delta}_i; \eta) / h_m(\mathbf{y}_{i,\text{mis}}^{*(j)})}{\sum_{k=1}^m f(\mathbf{y}_{ik}^*, \boldsymbol{\delta}_i; \eta) / h_m(\mathbf{y}_{i,\text{mis}}^{*(k)})}, \tag{6.18}$$

the EM algorithm for fractional imputation can be expressed as

$$\hat{\eta}^{(t+1)} \leftarrow \operatorname{argmax} Q^*(\eta \mid \hat{\eta}^{(t)}).$$

Instead of the EM algorithm, a Newton-type algorithm can also be used. Using the Oakes' formula in (3.36), we can obtain the observed information from the Q^* function (6.17) alone, without having to know the observed likelihood function. That is, we have

$$I_{obs}^*(\eta) = -\sum_{i=1}^n \sum_{j=1}^m w_{ij}^*(\eta) \dot{S}(\eta; \mathbf{y}_{ij}^*, \boldsymbol{\delta}_i) - \sum_{i=1}^n \sum_{j=1}^m w_{ij}^*(\eta) \left\{ S(\eta; \mathbf{y}_{ij}^*, \boldsymbol{\delta}_i) - \bar{S}_i^*(\eta) \right\}^{\otimes 2},$$

$$\tag{6.19}$$

where $S(\eta; \mathbf{y}, \boldsymbol{\delta}) = \partial \log f(\mathbf{y}, \boldsymbol{\delta}; \eta) / \partial \eta$, $\dot{S}(\eta; \mathbf{y}, \boldsymbol{\delta}) = \partial S(\eta; \mathbf{y}, \boldsymbol{\delta}) / \partial \eta'$ and

$$\bar{S}_i^*(\eta) = \sum_{j=1}^m w_{ij}^*(\eta) S(\eta; \mathbf{y}_{ij}^*, \boldsymbol{\delta}_i).$$

Note that, for $m \to \infty$, (6.19) converges to

$$-\sum_{i=1}^n E\left\{ \dot{S}(\eta; \mathbf{y}_i, \boldsymbol{\delta}_i) \mid \mathbf{y}_{i,obs}, \boldsymbol{\delta}_i \right\} - \sum_{i=1}^n V\left\{ S(\eta; \mathbf{y}_i, \boldsymbol{\delta}_i) \mid \mathbf{y}_{i,obs}, \boldsymbol{\delta}_i \right\},$$

which is equal to the observed information matrix discussed in (2.30). Thus, a Newton-type algorithm for computing the MLE from the fractionally imputed data is given by

$$\hat{\eta}^{(t+1)} = \hat{\eta}^{(t)} + \left\{ I_{obs}^*(\hat{\eta}^{(t)}) \right\}^{-1} \bar{S}^*(\hat{\eta}^{(t)}), \tag{6.20}$$

with $I_{obs}^*(\hat\eta)$ defined in (6.19) and

$$\bar{S}^*(\eta) = \sum_{i=1}^{n}\sum_{j=1}^{m} w_{ij}^*(\eta) S(\eta;\mathbf{y}_{ij}^*,\boldsymbol{\delta}_i).$$

We now briefly discuss estimating general parameters Ψ that can be estimated by (4.38) under complete response. The FI estimator of Ψ is then computed by solving

$$\sum_{i=1}^{n}\sum_{j=1}^{m} w_{ij}^*(\hat\eta) U(\Psi;\mathbf{y}_{ij}^*) = 0, \tag{6.21}$$

where $\hat\eta$ is the solution to

$$\sum_{i=1}^{n}\sum_{j=1}^{m} w_{ij}^*(\hat\eta) S\left(\hat\eta;\mathbf{y}_{ij}^*\right) = 0.$$

We can use either the linearization method or the replication method for variance estimation. For the former, we can use (4.47) with $\hat{q}_i^* = \bar{U}_i^* + \hat\kappa \bar{S}_i^*$, where $(\bar{U}_i^*,\bar{S}_i^*) = \sum_{j=1}^{m} w_{ij}^*(\hat\eta)(U_{ij}^*,S_{ij}^*)$, $U_{ij}^* = U(\hat\Psi;\mathbf{y}_{ij}^*)$, $S_{ij}^* = S(\hat\eta;\mathbf{y}_{ij}^*)$,

$$\hat\kappa = \sum_{i=1}^{n}\sum_{j=1}^{m} w_{ij}^*(\hat\eta)\left(U_{ij}^* - \bar{U}_i^*\right) S_{ij}^* \{I_{obs}^*(\hat\eta)\}^{-1}, \tag{6.22}$$

and $I_{obs}^*(\eta)$ is defined in (6.19). For linearization method, we can use sandwich formula

$$\hat{V}(\hat\Psi) = \hat\tau_q^{-1}\hat\Omega_q\hat\tau_q^{-1'} \tag{6.23}$$

where

$$\hat\tau_q = n^{-1}\sum_{i=1}^{n}\sum_{j=1}^{m} w_{ij}^*(\hat\eta)\dot{U}(\hat\Psi;\mathbf{y}_{ij}^*),$$

$$\hat\Omega_q = n^{-1}(n-1)^{-1}\sum_{i=1}^{n}(\hat{q}_i^* - \bar{q}_n^*)^{\otimes 2},$$

with $\hat{q}_i^* = \bar{U}_i^* + \hat\kappa\bar{S}_i^*$, where $(\bar{U}_i^*,\bar{S}_i^*) = \sum_{j=1}^{m} w_{ij}^*(U_{ij}^*,S_{ij}^*)$, $U_{ij}^* = U(\hat\Psi;\mathbf{y}_{ij}^*)$, $S_{ij}^* = S(\hat\eta;\mathbf{y}_{ij}^*)$, and

$$\hat\kappa = \sum_{i=1}^{n}\sum_{j=1}^{m} w_{ij}^*(\hat\eta)\left(U_{ij}^* - \bar{U}_i^*\right) S_{ij}^* \{I_{obs}^*(\hat\eta)\}^{-1}.$$

Justification of (6.23) is given in Kim (2011).

For the replication method, we first obtain the k-th replicate $\hat{\eta}^{(k)}$ of $\hat{\eta}$ by solving

$$\bar{S}^{*(k)}(\eta) \equiv \sum_{i=1}^{n} \sum_{j=1}^{m} w_i^{(k)} w_{ij}^*(\eta) S\left(\eta; \mathbf{y}_{ij}^*\right) = 0. \tag{6.24}$$

Once $\hat{\eta}^{(k)}$ is obtained, then the k-th replicate $\hat{\Psi}^{(k)}$ of $\hat{\Psi}$ is obtained by solving

$$\sum_{i=1}^{n} \sum_{j=1}^{m} w_i^{(k)} w_{ij}^*(\hat{\eta}^{(k)}) U(\Psi; \mathbf{y}_{ij}^*) = 0 \tag{6.25}$$

for ψ and the replication variance estimator of $\hat{\Psi}$ from (6.21) is obtained by

$$\hat{V}_{rep}(\hat{\Psi}) = \sum_{k=1}^{L} c_k \left(\hat{\Psi}^{(k)} - \hat{\Psi}\right)^2.$$

Note that the imputed values are not changed. Only the fractional weights are changed for each replication. Finding the solution $\hat{\eta}^{(k)}$ to (6.24) can require some iterative computation such as the EM algorithm. To avoid iterative computation, one may consider a one-step approximation by applying the following Taylor expansion:

$$\begin{aligned} 0 &= \bar{S}^{*(k)}(\hat{\eta}^{(k)}) \\ &\cong \bar{S}^{*(k)}(\hat{\eta}) + \left\{ \frac{\partial}{\partial \eta'} \bar{S}^{*(k)}(\hat{\eta}) \right\} \left(\hat{\eta}^{(k)} - \hat{\eta}\right). \end{aligned}$$

Now, writing $I_{obs}^{*(k)}(\eta) = \partial \bar{S}^{*(k)}(\eta)/\partial \eta'$, we can obtain, using the argument for (6.19),

$$I_{obs}^{*(k)}(\eta) = -\sum_{i=1}^{n} w_i^{(k)} \sum_{j=1}^{m} w_{ij}^*(\eta) \dot{S}(\eta; \mathbf{y}_{ij}^*, \boldsymbol{\delta}_i) \tag{6.26}$$

$$-\sum_{i=1}^{n} w_i^{(k)} \sum_{j=1}^{m} w_{ij}^*(\eta) \left\{ S(\eta; \mathbf{y}_{ij}^*, \boldsymbol{\delta}_i) - \bar{S}_i^*(\eta) \right\}^{\otimes 2}.$$

Thus, the one-step approximation of $\hat{\eta}^{(k)}$ to (6.24) is given by

$$\hat{\eta}^{(k)} \cong \hat{\eta} + \left\{ I_{obs}^{*(k)}(\hat{\eta}) \right\}^{-1} \bar{S}^{*(k)}(\hat{\eta}).$$

One-step replicate $\hat{\eta}^{(k)}$ can be used in (6.25) to compute $\hat{\Psi}^{(k)}$.

Example 6.6. *Consider the problem of estimating the p-th quantile $\Psi = F^{-1}(p)$, where $F(y)$ is the marginal cumulative distribution function (CDF) of the study variable Y. Under complete response, the estimating equation for Ψ is*

$$U(\Psi) \equiv \sum_{i=1}^{n} \{I(Y_i \leq \Psi) - p\} = 0. \tag{6.27}$$

We may need an interpolation technique to solve (6.27) from the realized sample. Now, using fractionally imputed data, we can solve $\hat{F}_{FI}(\Psi) = p$ for Ψ, where

$$\hat{F}_{FI}(\Psi) \equiv n^{-1}\sum_{i=1}^{n}\left\{\delta_i I\left(y_i < \Psi\right) + (1-\delta_i)\sum_{j=1}^{M} w_{ij}^* I\left(y_{ij}^* < \Psi\right)\right\},$$

to obtain FI estimator of Ψ. To estimate the variance of $\hat{\Psi}$ from the FI data, we can simply apply the sandwich formula (6.23) with $U(\Psi;y) = I(y \leq \Psi) - p$ and $\hat{\tau}_q = f(\hat{\Psi};\hat{\eta})$, which is the value of marginal density of Y at $y = \hat{\Psi}$ with parameter value $\hat{\eta}$. If the density is unknown, one can use

$$\hat{f}_h(y) = \frac{1}{2h}\left\{\hat{F}_{FI}(\hat{\Psi}+h) - \hat{F}_{FI}(\hat{\Psi}-h)\right\}$$

to estimate the density f around $y = \hat{\Psi}$, where h is the bandwidth. The choice of $h = n^{-1/2}$ can be used.

We now consider the *fractional hot deck imputation* in which m imputed values are taken from the set of respondents. For simplicity, we consider a bivariate data structure (x_i, y_i) with x_i fully observed. Let $\{y_1, \ldots, y_r\}$ be the set of respondents, and we want to obtain m imputed values, $y_i^{*(1)}, \ldots, y_i^{*(m)}$, for $i = r+1, \ldots, n$ from the respondents. Let w_{ij}^* be the fractional weights assigned to $y_i^{*(j)}$ for $j = 1, 2, \ldots, m$. We consider the special case of $m = r$. In this case, the fractional weight represents the point mass assigned to each responding y_i. Thus, it is desirable to compute the fractional weights $w_{i1}^*, \ldots, w_{ir}^*$ such that $\sum_{j=1}^{r} w_{ij}^* = 1$ and

$$\sum_{j=1}^{r} w_{ij}^* I(y_j < y) \cong Pr\left(y_i < y \mid x_i, \delta_i = 0\right). \qquad (6.28)$$

Note that for the special case of $w_{ij}^* = 1/r$, the left side of the above equality estimates $P(y_i < y \mid \delta_i = 1)$.

If we can assume a parametric model $f(y \mid x; \theta)$ for the conditional distribution of y on x and the response probability model is given by $P(\delta_i = 1 \mid x_i, y_i) = \pi(x_i, y_i; \phi)$, then the fractional weights satisfying (6.28) are given by

$$w_{ij}^* \propto f\left(y_j \mid x_i, \delta_i = 0; \hat{\theta}, \hat{\phi}\right)/f(y_j \mid \delta_j = 1)$$

$$\propto f\left(y_j \mid x_i, \hat{\theta}\right)\left\{1 - \pi\left(x_i, y_j; \hat{\phi}\right)\right\}/f(y_j \mid \delta_j = 1).$$

Since

$$f(y_j \mid \delta_j = 1) \propto \int \pi(x, y_j) f(y_j \mid x) f(x)dx$$

$$\cong \frac{1}{n}\sum_{i=1}^{n} \pi(x_i, y_j) f(y_j \mid x_i),$$

we can compute

$$w_{ij}^* \propto \frac{f\left(y_j \mid x_i, \hat{\theta}\right)\left\{1 - \pi\left(x_i, y_j; \hat{\phi}\right)\right\}}{\sum_{k=1}^n \pi\left(x_k, y_j; \hat{\phi}\right) f\left(y_j \mid x_k; \hat{\theta}\right)}.$$

Under MAR, the fractional weight is

$$w_{ij}^* \propto \frac{f(y_j \mid x_i; \hat{\theta})}{\sum_{k; \delta_k = 1} f(y_j \mid x_k; \hat{\theta})}$$

with $\sum_{j; \delta_j = 1} w_{ij}^* = 1$. Once the fractional imputation is created, the imputed estimating equation for η is computed by

$$\sum_{i=1}^n \left\{ \delta_i U(\eta; x_i, y_i) + (1 - \delta_i) \sum_{j; \delta_j = 1} w_{ij}^* U(\eta; x_i, y_j) \right\} = 0, \qquad (6.29)$$

where

$$w_{ij}^* \propto \frac{f(y_j \mid x_i; \hat{\theta})}{\sum_{k; \delta_k = 1} f(y_j \mid x_k; \hat{\theta})} \qquad (6.30)$$

with $\sum_{j=1}^n \delta_j w_{ij}^* = 1$. The fractional weights in (6.29) leads to robust estimation in the sense that certain level of misspecification in $f(y \mid x)$ can still provide a consistent estimator of η. See Yang and Kim (2014) for more details. The fractional weights can be further adjusted to satisfy

$$\sum_{i=1}^n \{\delta_i S(\hat{\theta}; x_i, y_i) + (1 - \delta_i) \sum_{j; \delta_j = 1} w_{ij}^* S(\hat{\theta}; x_i, y_j)\} = 0, \qquad (6.31)$$

where $S(\theta; x_i, y_i)$ is the score function of θ, and $\hat{\theta}$ is the MLE of θ.

Example 6.7. *Consider the setup of Example 5.10, except that $e_i = u_i - 1$ where u_i is the exponential distribution with mean 1. Suppose that the imputation model for the error term is $e_i \sim N(0, \sigma^2)$. Thus, the imputation model is not correct because the true sampling distribution is $e_i \sim Exp(1) - 1$. We are interested in estimating $\theta_1 = E(Y)$ and $\theta_2 = P(Y < 1)$. A simulation study was performed to compare the three imputation methods: multiple imputation, parametric fractional imputation of Kim (2011), and fractional hot deck imputation, with $m = 50$ for all methods. Table 6.2 presents the result from the simulation study. For estimation of $\theta_1 = E(Y)$, all imputation methods provide unbiased point estimators because $E(Y_i^*) = E(Y_i)$, where the expectation is taken under the true model. However, the imputation methods considered here provide biased estimates for θ_2 because $E\{I(Y_i^* < 1)\} \neq E\{I(Y_i < 1)\}$. The simulation results in Table 6.2 show that the bias of the FHDI estimator is the smallest for θ_2 in the setup considered.*

TABLE 6.2
Simulation Results of the Point Estimators

Parameter	Method	Bias	Standard Error
θ_1	MI	0.00	0.084
	PFI	0.00	0.084
	FHDI	0.00	0.084
θ_2	MI	−0.014	0.026
	PFI	−0.014	0.026
	FHDI	−0.001	0.029

6.2   Nonparametric Approach

In this section, we briefly introduce a nonparametric approach to fractional imputation. Specifically, we consider a nonparametric regression imputation using a kernel-based method. We assume a bivariate data structure (x_i, y_i) with x_i being fully observed. Assume that the response mechanism is missing at random. Let $K_h(x_i, x_j) = K((x_i - x_j)/h)$ be the kernel function with bandwidth h such that $K(x) \geq 0$ and

$$\int K(x)dx = 1, \quad \int xK(x)dx = 0, \quad \sigma_K^2 \equiv \int x^2 K(x)dx > 0.$$

Examples of the kernel functions include the following:

- Boxcar kernel: $K(x) = \frac{1}{2}I(|x| \leq 1)$,

- Gaussian kernel: $K(x) = \frac{1}{\sqrt{2\pi}}\exp(-\frac{1}{2}x^2)$,

- Epanechnikov kernel: $K(x) = \frac{3}{4}(1 - x^2)I(|x| \leq 1)$,

- Tricube kernel: $K(x) = \frac{70}{81}(1 - |x|^3)^3 I(|x| \leq 1)$.

Define

$$\hat{f}(x) = \frac{1}{n}\sum_{i=1}^{n} K_h(x_i - x)$$

to be the kernel-based estimator of the density of X, where $K_h(x) = h^{-1}K(x/h)$, h is the bandwidth, and $K(\cdot)$ is the kernel function. For simplicity, assume $\dim(x) = 1$. Let $h = h_n$ be a smoothing bandwidth such that $h_n \to 0$ and $nh_n \to \infty$ as $n \to \infty$. It is well known that

$$E\{\hat{f}(x)\} = f(x) + O(h^2)$$

and

$$V\{\hat{f}(x)\} = O((nh)^{-1}),$$

for each x, where $f(x)$ is the true density function. Thus,

$$MSE\{\hat{f}(x)\} = O(h^4 + (nh)^{-1}).$$

The optimal choice of the bandwidth is $h^* = c(x)n^{-1/5}$, and the MSE is $O(n^{-4/5})$.

The following lemma is a useful result to derive the asymptotic properties of the kernel-based nonparametric regression estimator. The proof can be found, for example, in Li and Wang (2007).

Lemma 6.1. *Assume for simplicity that x is a scalar. Under some regularity conditions,*

$$n^{-1} \sum_{i=1}^{n} K_h(x_i, x)y_i = f(x) \cdot E(Y \mid x) + O_p(h^2 + (nh)^{-1/2}). \tag{6.32}$$

We now consider nonparametric regression imputation under MAR. A nonparametric regression estimator of $m(x) = E(y \mid x)$ can be obtained by finding $\hat{m}(x)$ that minimizes

$$\sum_{i=1}^{n} K_h(x_i, x)\delta_i \{y_i - m(x)\}^2. \tag{6.33}$$

The minimizer of (6.33) is

$$\hat{m}(x) = \sum_{j=1}^{n} \delta_j w_{j1}(x)y_j, \tag{6.34}$$

where

$$w_{j1}(x) = \frac{K_h(x_j, x)}{\sum_{i=1}^{n} \delta_i K_h(x_i, x)}.$$

The weight $w_{i1}(x)$ in (6.34) represents the point mass assigned to y_i, and $E(Y \mid x)$ is approximated by $\sum_{j=1}^{n} \delta_j w_{j1}(x)y_j$. The nonparametric regression imputation estimator of $\theta = E(Y)$ is

$$\hat{\theta}_{NP} = \frac{1}{n} \sum_{i=1}^{n} \{\delta_i y_i + (1 - \delta_i)\hat{m}(x_i)\} := \frac{1}{n} \sum_{i=1}^{n} \left\{ \delta_i y_i + (1 - \delta_i) \sum_{j=1}^{n} \delta_j w_{j1}(x_i)y_j \right\}, \tag{6.35}$$

where $w_{ij}^* = \delta_j w_{j1}(x_i)$ is essentially the fractional weights assigned to the j-th imputed value for missing unit i. For the choice of the bandwidth, we can use 10-fold cross validation from the respondents that minimizes the mean squared prediction error.

The following theorem, first proved by Cheng (1994) and further proved by Wang and Chen (2009), presents the $\sqrt{n}$-consistency of the nonparametric regression imputation estimator.

Theorem 6.2. *Assume that h satisfies* $nh \to \infty$ *and* $nh^4 \to 0$. *Under some regularity conditions,*

$$\sqrt{n}\left(\hat{\theta}_{NP} - \theta\right) \to N\left(0, \sigma_1^2\right), \tag{6.36}$$

where $\sigma_1^2 = V\left\{m\left(X\right)\right\} + E\left[\left\{\pi\left(X\right)\right\}^{-1}V\left(Y \mid X\right)\right]$ *and* $\pi(X) = P(\delta = 1 \mid X)$.

Proof. First, write

$$\hat{\theta}_{NP} = \frac{1}{n}\sum_{i=1}^{n}\delta_i\{y_i - m(x_i)\} + \frac{1}{n}\sum_{i=1}^{n}(1-\delta_i)\{\hat{m}(x_i) - m(x_i)\} + \frac{1}{n}\sum_{i=1}^{n}m(x_i)$$

$$:= S_n + T_n + R_n.$$

Also, we can express

$$\hat{m}(x) = \frac{C_n(x)}{D_n(x)},$$

where

$$C_n(x) = n^{-1}\sum_{j=1}^{n}\delta_j K_h(x_j - x)y_j,$$

$$D_n(x) = n^{-1}\sum_{j=1}^{n}\delta_j K_h(x_j - x).$$

Using (6.32), we can obtain

$$C_n(x) = E(\delta \mid x)E(Y \mid x)f(x) + O_p(h^2 + (nh)^{-1/2})$$
$$D_n(x) = E(\delta \mid x)f(x) + O_p(h^2 + (nh)^{-1/2}).$$

Now, by Taylor expansion,

$$\frac{C_n(x)}{D_n(x)} - m(x) = -\frac{1}{E(\delta \mid x)f(x)}\{C_n(x) - m(x)D_n(x)\} + O_p(h^2 + (nh)^{-1/2}).$$

Thus, writing $\pi(x) = E(\delta \mid x)$,

$$T_n = \frac{1}{n^2}\sum_{i=1}^{n}(1-\delta_i)\frac{1}{\pi(x_i)f(x_i)}\sum_{j=1}^{n}\delta_j K_h(x_j - x_i)\{y_j - m(x_i)\} + O_p(h^2 + n^{-1}h^{-1/2}).$$

Since h satisfies $nh \to \infty$ and $nh^4 \to 0$, we have

$$h^2 + n^{-1}h^{-1/2} = o(n^{-1/2}).$$

Thus,

$$T_n = \frac{1}{n^2}\sum_{i=1}^{n}(1-\delta_i)\frac{1}{\pi(x_i)f(x_i)}\sum_{j=1}^{n}\delta_j K_h(x_j - x_i)\{y_j - m(x_i)\} + o_p(n^{-1/2}).$$

Ignoring the smaller order terms, we can express

$$T_n = \frac{1}{n(n-1)} \sum_i \sum_{j \neq i} h(z_i, z_j)$$

where $z_i = (x_i, \delta_i, y_i)$ and

$$
\begin{aligned}
h(z_i, z_j) &= \frac{1}{2} \Bigg[(1-\delta_i)\delta_j \frac{1}{\pi(x_i)f(x_i)} K_h(x_i, x_j)\{y_j - m(x_i)\} \\
&\quad + (1-\delta_j)\delta_i \frac{1}{\pi(x_j)f(x_j)} K_h(x_j, x_i)\{y_i - m(x_j)\} \Bigg] \\
&:= \frac{1}{2}(\zeta_{ij} + \zeta_{ji}).
\end{aligned}
$$

Now,

$$
\begin{aligned}
E(\zeta_{ij} \mid z_i) &= (1-\delta_i)\frac{1}{\pi(x_i)f(x_i)}\frac{1}{h}\int K\left(\frac{x_i - x_j}{h}\right)\pi(x_j)\left\{m(x_j) - m(x_i)\right\}f(x_j)dx_j \\
&= (1-\delta_i)\frac{1}{\pi(x_i)f(x_i)}\int K(s)\pi(x_i + hs)\left\{m(x_i + hs) - m(x_i)\right\}f(x_i + hs)ds \\
&= O(h^2),
\end{aligned}
$$

and

$$
\begin{aligned}
E(\zeta_{ji} \mid z_i) &= \delta_i\frac{1}{h}\int \frac{1-\pi(x_j)}{\pi(x_j)f(x_j)}K\left(\frac{x_j - x_i}{h}\right)\{y_i - m(x_j)\}f(x_j)dx_j \\
&= \delta_i\int \frac{1-\pi(x_i + hs)}{\pi(x_i + hs)f(x_i + hs)}K(s)\{y_i - m(x_i + hs)\}f(x_i + hs)ds \\
&= \delta_i\left\{\frac{1}{\pi(x_i)} - 1\right\}\{y_i - m(x_i)\} + O(h^2).
\end{aligned}
$$

Thus, using the theory of U-statistics (van der Vaart, 1998; Ch. 12), we have

$$T_n = \frac{2}{n}\sum_{i=1}^n E\{h(z_i, z_j) \mid z_i\} + o_p(n^{-1/2}),$$

where

$$\frac{2}{n}\sum_{i=1}^n E\{h(z_i, z_j) \mid z_i\} = \frac{2}{n}\sum_{i=1}^n \delta_i\left\{\frac{1}{\pi(x_i)} - 1\right\}\{y_i - m(x_i)\} + O(h^2).$$

Therefore, we can obtain

$$T_n = \frac{1}{n}\sum_{i=1}^n \delta_i\left\{\frac{1}{\pi(x_i)} - 1\right\}\{y_i - m(x_i)\} + o_p(n^{-1/2}). \tag{6.37}$$

Finally, using (6.37), we establish

$$\hat{\theta}_{NP} = \tilde{\theta}_{NP} + o_p(n^{-1/2}). \tag{6.38}$$

where

$$\tilde{\theta}_{NP} = \frac{1}{n}\sum_{i=1}^{n}\left[m(x_i) + \delta_i\frac{1}{\pi(x_i)}\{y_i - m(x_i)\}\right].$$

Note that $\tilde{\theta}_{NP}$ is asymptotically linear with influence function

$$d(x_i, y_i, \delta_i) = m(x_i) + \delta_i\frac{1}{\pi(x_i)}\{y_i - m(x_i)\}.$$

The variance of $\tilde{\theta}_{NP}$ is equal to $n^{-1}\sigma_1^2$, where σ_1^2 is defined after (6.36).   □

For variance estimation of the nonparametric regression imputation (NRI) estimator $\hat{\theta}_{NP}$ in (6.35), we can use (6.38) as the linearization of the NRI estimator. We can use

$$\hat{g}(x) = \sum_{j=1}^{n}\left\{\frac{K_h(x_j, x)}{\sum_{k=1}^{n}\delta_k K_h(x_j, x_k)}\right\}$$

as an estimator of $1/\pi(x)$. Note that $\hat{g}(x)$ satisfies

$$\sum_{i=1}^{n}\delta_i\hat{g}(x_i)y_i = \sum_{i=1}^{n}\hat{m}(x_i),$$

where $\hat{m}(x)$ is defined in (6.34). Thus, using (6.38), we can express

$$\hat{\theta}_{NP} \cong n^{-1}\sum_{i=1}^{n}\hat{d}_i,$$

where

$$\hat{d}_i = \hat{m}(x_i) + \delta_i\hat{g}(x_i)\{y_i - m(x_i)\}$$

Using the above $\hat{d}_i$, we can apply the standard variance estimation formula in (4.32) to estimate the asymptotic variance of the nonparametric regression imputation estimator in (6.35). That is, we use

$$\hat{V}(\hat{\theta}_{NP}) = \frac{1}{n}\frac{1}{n-1}\sum_{i=1}^{n}\left(\hat{d}_i - \bar{\hat{d}}_n\right)^2,$$

where $\bar{\hat{d}}_n = n^{-1}\sum_{i=1}^{n}\hat{d}_i$.

6.3   Semiparametric Fractional Imputation

The nonparametric regression imputation using kernel-based method is not applicable when the dimension of x is large, due to the curse of dimensionality. In this case, one may consider a semiparametric approach to imputation, where the parameter in the model consists of two parts; one is a parametric part with a fixed dimension, and the other is a nonparametric part. For example, one may use the following regression model

$$Y = m(x; \beta) + \epsilon, \tag{6.39}$$

where $m(x; \cdot)$ is a known function with unknown parameter β of fixed dimension, and ϵ satisfies $E(\epsilon \mid x) = 0$. No parametric model assumptions on ϵ is made.

Suppose that the parameter of interest is the solution to $E\{U(\psi; Y)\} = 0$. Under MAR assumption, a consistent estimator of ψ is obtained by solving the following expected estimating equation:

$$\sum_{i=1}^{n} [\delta_i U(\psi; y_i) + (1 - \delta_i) E\{U(\psi; Y_i) \mid x_i\}] = 0. \tag{6.40}$$

To compute the conditional expectation in (6.40), note that

$$E\{U(\psi; Y_i) \mid x_i\} = E\{U(\psi; m(x_i; \beta) + \epsilon_i) \mid x_i\} = \int U(\psi; m(x_i; \beta) + \epsilon) f(\epsilon \mid x_i) d\epsilon, \tag{6.41}$$

where $f(\epsilon \mid x)$ is the density function for the conditional distribution of ϵ given x. Note that $f(\epsilon \mid x)$ is unknown except that $E(\epsilon \mid x) = 0$ holds.

To approximate the conditional expectation in (6.41) under the restriction $E(\epsilon \mid x) = 0$, Müller (2009) proposed a novel application of the empirical likelihood method of Owen (2001). The proposed method is further extended by Chen and Kim (2017) to fractional imputation. To explain the idea, assume for now that the true parameter β is known. In this case, ϵ_i are observed when $\delta_i = 1$. In the empirical likelihood approach, under MAR, the density function $f(\epsilon \mid x)$ belongs to the class

$$f_w(\epsilon \mid x) = \sum_{i=1}^{n} \delta_i w_i I(\epsilon = \epsilon_i), \tag{6.42}$$

where $w_i \in (0, 1)$ and $\sum_{i=1}^{n} \delta_i w_i = 1$. That is, the density function has positive support only in the observed data. Note that, under the model (6.42), the moment condition $E(\epsilon \mid x) = 0$ is equivalent to

$$\sum_{i=1}^{n} \delta_i w_i \epsilon_i = 0. \tag{6.43}$$

In practice, we cannot observe ϵ_i as β is unknown. Thus, we use $\hat{\epsilon}_i = y_i - m(x_i; \hat{\beta})$ instead of ϵ_i, where $\hat{\beta}$ is a consistent estimator of β obtained by complete-case (CC) analysis method.

The proposed empirical likelihood (EL) method is to find $\hat{w}_i$ that maximize

$$l(w) = \sum_{\delta_i=1} \log(w_i)$$

subject to $\sum_{i=1}^{n} \delta_i w_i = 1$ and $\sum_{i=1}^{n} \delta_i w_i \hat{\epsilon}_i = 0$. Using Lagrange multiplier method, the solution to the optimization problem can be written as

$$\hat{w}_i = \frac{1}{r} \frac{1}{1 + \hat{\lambda} \hat{\epsilon}_i}, \tag{6.44}$$

where $r = \sum_{i=1}^{n} \delta_i$, and $\hat{\lambda}$ satisfies $\sum_{i=1}^{n} \delta_i \hat{w}_i \hat{\epsilon}_i = 0$. Using (6.41) and $\hat{w}_i$ in (6.44), we can obtain

$$E\{U(\psi; Y_i) \mid x_i\} \cong \sum_{j=1}^{n} \delta_j \hat{w}_j U(\psi; m(x_i; \hat{\beta}) + \hat{\epsilon}_j).$$

Note that we can express the above approximation in the form of fractional imputation

$$E\{U(\psi; Y_i) \mid x_i\} \cong \sum_{j=1}^{n} w_{ij}^* U(\psi; y_{ij}^*),$$

where $w_{ij}^* = \delta_j \hat{w}_j$ and $y_{ij}^* = m(x_i; \hat{\beta}) + \hat{\epsilon}_j$. Thus, it is a semiparametric fractional imputation using empirical likelihood method. Chen and Kim (2017) presented further asymptotic properties of the proposed semiparametric fractional imputation and also discussed variance estimation.

Another semiparametric fractional imputation can be developed using the semiparametric generalized linear model (SGLM). In SGLM, the finite population follows a superpopulation model with moment condition only: $E(Y \mid \mathbf{x}) = \mu(\mathbf{x}) = g^{-1}(\mathbf{x}'\beta)$, where $g(\cdot)$ is a known link function. In this case, we can express

$$f(y \mid x; \beta, f_0) = \frac{f_0(y) \exp(\theta y)}{\int f_0(y) \exp(\theta y) dy}, \tag{6.45}$$

where θ satisfies $\int y f(y \mid \mathbf{x}; \beta, f_0) dy = \mu(\mathbf{x})$ and the baseline density f_0 is an infinite dimensional nuisance parameter. The pseudo log-likelihood function can be written as

$$l_p(\beta, f_0) = \sum_{i=1}^{n} \log f(y_i \mid x_i; \beta, f_0) = \sum_{i=1}^{n} \{\theta_i y_i - b(\theta_i) + \log f_0(y_i)\},$$

where $b(\theta) = \log \int f_0(y) \exp(\theta y) dy$ satisfies $b'(\theta_i) = \mu_i = \mu(\mathbf{x}_i)$ and $g(\mu_i) = \mathbf{x}_i'\beta$. Rathouz and Gao (2009) first considered the SGLM under the IID setup and develop the maximum likelihood method when $f_0(y)$ has a finite support. Huang (2014) used an empirical likelihood approach to approximate $f_0(y)$ nonparametrically and proposed the maximum empirical likelihood estimator (MELE) of β. Once the parameters in (6.45) are estimated, parametric fractional imputation in Section 6.1 can be developed from the estimated SGLM.

6.4 Fractional Imputation Using Mixture Models

We now discuss fractional imputation under the mixture model in (5.37) with a vector $\mathbf{Y}$. Let $\mathbf{Y} = (Y_1, \ldots, Y_p)$ be a p-dimensional random vector with density

$$f(\mathbf{y}; \pi, \theta) = \sum_{k=1}^{K} \pi_k f(\mathbf{y} \mid z = k, \theta_k), \tag{6.46}$$

where $\pi_k = P(z = k)$ satisfies $0 < \pi_1 < \pi_2 < \cdots < \pi_K < 1$ and $\sum_{k=1}^{K} \pi_k = 1$.

Let $(\mathbf{y}_{i,\text{obs}}, \mathbf{y}_{i,\text{mis}})$ be the observed and missing part of $\mathbf{y}_i$, respectively. The prediction model, the conditional distribution of $\mathbf{y}_{i,mis}$ given $\mathbf{y}_{i,\text{obs}}$, is

$$f(\mathbf{y}_{i,\text{mis}} \mid \mathbf{y}_{i,\text{obs}}; \pi, \theta) = \sum_{k=1}^{K} P(z = k \mid \mathbf{y}_{i,\text{obs}}; \theta, \pi) f(\mathbf{y}_{i,\text{mis}} \mid \mathbf{y}_{i,\text{obs}}, z_i = k; \theta_k),$$

$$\tag{6.47}$$

where

$$P(z = k \mid \mathbf{y}_{i,\text{obs}}; \theta, \pi) = \frac{\pi_k f(\mathbf{y}_{i,\text{obs}} \mid z = k; \theta_k)}{\sum_{l=1}^{K} \pi_l f(\mathbf{y}_{i,\text{obs}} \mid z = l; \theta_k)}$$

and

$$f(\mathbf{y}_{i,\text{mis}} \mid \mathbf{y}_{i,\text{obs}}; \theta_k) = \frac{f(\mathbf{y}_i \mid z = k, \theta_k)}{f(\mathbf{y}_{i,\text{obs}} \mid z = k; \theta_k)}$$

with $f(\mathbf{y}_{i,\text{obs}} \mid z = k; \theta_k) = \int f(\mathbf{y}_i \mid z = k; \theta_k) d\mathbf{y}_{i,\text{mis}}$.

To implement fractional imputation, we first estimate the parameters from the marginal distribution of the observed data and then apply the fractional imputation from the estimated prediction model in (6.47). Under model (6.46) and MAR, the density of the marginal distribution of the observed data is

$$f(\mathbf{y}_{i,\text{obs}}; \pi, \theta) = \sum_{k=1}^{K} \pi_k f(\mathbf{y}_{i,\text{obs}} \mid z_i = k, \theta_k). \tag{6.48}$$

If $f(\mathbf{y} \mid z = k, \theta_k)$ is Gaussian, then $f(\mathbf{y}_{i,\text{obs}} \mid z = k; \theta_k) = \int f(\mathbf{y}_i \mid z = k; \theta_k) d\mathbf{y}_{i,\text{mis}}$ is also Gaussian. Assuming that model (6.48) is identifiable, the following EM algorithm can be used to estimate the parameters.

[E-step] Using the current parameter values $(\pi^{(t)}, \theta^{(t)})$, compute

$$p_{ik}^{(t)} = \frac{f(\mathbf{y}_{i,obs} \mid z_i = k, \theta_k^{(t)})\pi_k^{(t)}}{\sum_{l=1}^{K} f(\mathbf{y}_{i,obs} \mid z_i = l, \theta_l^{(t)})\pi_l^{(t)}}.$$

[M-step] Update the parameters by maximizing

$$Q(\pi, \theta \mid \pi^{(t)}, \theta^{(t)}) = \sum_{i=1}^{n}\sum_{k=1}^{K} p_{ik}^{(t)} \left\{ \log \pi_k + \log f(\mathbf{y}_{i,obs} \mid z_i = k, \theta_k) \right\}$$

with respect to π and θ.

Once the parameters are estimated, we use the fractional imputation method to impute the missing values. To generate M imputed values from (6.47), we first let $(M_{i1}, M_{i2}, \ldots, M_{iK}) \sim \text{Multinomial}(M; \hat{\mathbf{p}}_i)$, where $\hat{\mathbf{p}}_i = (\hat{p}_{i1}, \ldots, \hat{p}_{iK})$. For each $k = 1, 2, \ldots, K$, we generate M_{ik} samples of $\mathbf{y}_{i,\text{mis}}$, say $\left\{ \mathbf{y}_{i,\text{mis}}^{*(kj)} : j = 1, \ldots, M_{ik} \right\}$, from the conditional distribution $f(\mathbf{y}_{i,\text{mis}} \mid \mathbf{y}_{i,\text{obs}}, z_i = k; \hat{\theta}_k)$.

In practice, the size of the mixture components, K, is often unknown and we need to estimate it from the sample data. Thus, K is the model complexity parameter to be determined from the sample. If K is larger than necessary, the proposed mixture model may be subject to overfitting and increase its variance. If K is small, then the approximation of the true distribution cannot provide accurate prediction due to its bias. Hence, we can allow the model complexity parameter K to depend on the sample size n, say $K = K(n)$. One can use 10-fold cross-validation to estimate K, which may involve heavy computation. Alternatively, one may consider using the Bayesian information criterion (Schwarz, 1978) to select K. Under multivariate missingness, we use the observed log-likelihood function to serve the role of the complete log-likelihood function in computing the information criterion, in the sense that

$$\text{BIC}(K) = -2\sum_{i=1}^{n} \log\left\{ \sum_{k=1}^{K} \hat{\pi}_k f(\mathbf{y}_{i,obs} \mid z_{ik} = 1; \hat{\theta}_k) \right\} + (\log n)(K + d), \quad (6.49)$$

where $(\hat{\pi}, \hat{\theta})$ are the estimators obtained from the proposed method and $d = \dim(\theta)$. See Section 6.6 for a more discussion of BIC.

Then, the final estimator of ψ can be obtained by solving the fractionally imputed estimating equation, given by

$$\sum_{i=1}^{n}\sum_{k=1}^{\hat{K}}\sum_{j=1}^{M_{ik}} w_{ikj}^* U(\psi; \mathbf{y}_i^{*(kj)}) = 0, \quad (6.50)$$

where $w_{ikj}^* = \hat{p}_{ik} M_{ik}^{-1}$ are the final fractional weights assigned to $\mathbf{y}_i^{*(kj)} =$

$(\mathbf{y}_{i,\text{obs}}, \mathbf{y}_{i,\text{mis}}^{*(kj)})$. By construction, the fractionally imputed estimating equation in (6.50) approximates

$$\sum_{i=1}^{n} E\left\{U(\theta; \mathbf{y}_i) \mid \mathbf{y}_{i,\text{obs}}\right\} = 0.$$

Sang et al. (2020) presents a more rigorous theory for the solution to (6.50). The Gaussian mixture model is a flexible model assumption and can be used to approximate general distribution reasonably well for some K (Bacharoglou, 2010).

If some part of $\mathbf{y}$ is always observed, then we do not have to estimate all the parameters in (6.46). The following example illustrates the idea.

Example 6.8. *For simplicity, suppose we have a bivariate random variables* (x, y)*. Assume that x is always observed and y is subject to missingness. Instead of assuming a joint mixture distribution as in (6.46), we can use the following conditional mixture model*

$$f(y \mid x; \alpha, \theta) = \sum_{k=1}^{K} \pi_k(x; \alpha) f_k(y \mid x, z = k; \theta_k), \qquad (6.51)$$

where $\pi_k(x; \alpha) = P(z = k \mid x; \alpha)$ is a parametric model for the conditional probability for $z = k$ with parameter α. For example,

$$\pi_k(x; \alpha) = \frac{\exp(\alpha_{k0} + \alpha_{k1} x)}{\sum_{k=1}^{K} \exp(\alpha_{k0} + \alpha_{k1} x)}.$$

In this case, the EM algorithm for estimating α and θ under nonresponse can be described as follows.

[E-step] Using the current parameter values $(\alpha^{(t)}, \theta^{(t)})$, compute

$$p_{ik}^{(t)} = \frac{\{f(y_i \mid x_i, z_i = k, \theta_k^{(t)})\}^{\delta_i} \pi_k(x_i; \alpha^{(t)})}{\sum_{k=1}^{K} \{f(y_i \mid x_i, z_i = k, \theta_k^{(t)})\}^{\delta_i} \pi_k(x_i; \alpha^{(t)})}.$$

[M-step] Update the parameters by maximizing

$$Q(\alpha, \theta \mid \alpha^{(t)}, \theta^{(t)}) = \sum_{i=1}^{n} \sum_{k=1}^{K} p_{ik}^{(t)} \{\log \pi_k(x_i; \alpha) + \delta_i \log f(y_i \mid x_i, z_i = k, \theta_k)\}$$

with respect to α and θ.

The conditional mixture model approach is more flexible model in the sense that the assumption for the marginal distribution of x is not necessary. Furthermore, it can easily handle the high dimensional covariates problem using the penalized regression technique in the M-step of the EM algorithm. See Lee and Kim (2021) for more details.

6.5   Fractional Imputation for Multivariate Categorical Data

We now discuss fractional imputation for multivariate categorical data. Let $\mathbf{Y} = (Y_1,\ldots,Y_p)$ be a p-dimensional categorical random vector with support $\mathcal{C}$. Define $\pi_c = P(\mathbf{Y} = \mathbf{c})$ for $\mathbf{c} \in \mathcal{C}$ such that $\sum_{c\in\mathcal{C}} \pi_c = 1$. Let n_c be the number of sample elements with $\mathbf{y}_i = \mathbf{c}$. That is, $n_c = \sum_{i=1}^n \mathbb{I}(\mathbf{y}_i = \mathbf{c})$. The log-likelihood function is

$$l_{com}(\pi) = \sum_{c\in\mathcal{C}} n_c \log(\pi_c) = \sum_{i=1}^n \sum_{c\in\mathcal{C}} \mathbb{I}(\mathbf{y}_i = \mathbf{c}) \log(\pi_c)$$

over the simplex with $\pi_c \in (0,1)$ for all $c \in \mathcal{C}$ and $\sum_{c\in\mathcal{C}} \pi_c = 1$. To avoid unnecessary details, we assume that $n_c > 0$ for all $c \in \mathcal{C}$.

Under nonresponse, we assume that observations have been grouped according to their missingness patterns. Define $\boldsymbol{\delta} = (\delta_1,\ldots,\delta_p)$ be the vector of response indicator functions with support $\mathcal{R}$, where

$$\delta_j = \begin{cases} 1 & \text{if } Y_j \text{ responds} \\ 0 & \text{otherwise.} \end{cases}$$

Also, we can decompose $\mathbf{y} = (\mathbf{y}_{\text{obs}}^{(\boldsymbol{\delta})}, \mathbf{y}_{\text{mis}}^{(\boldsymbol{\delta})})$ based on the realized $\boldsymbol{\delta}$, where $\mathbf{y}_{\text{obs}}^{(\boldsymbol{\delta})}$ and $\mathbf{y}_{\text{mis}}^{(\boldsymbol{\delta})}$ are the observed and missing part of $\mathbf{y}$ under the missing pattern characterized by $\boldsymbol{\delta}$. Let $\mathcal{D}$ be the support of $\mathbf{y}_{\text{obs}}^{(\boldsymbol{\delta})}$. Using the above notation, the observed log-likelihood can be written as

$$l_{\text{obs}}(\pi) = \sum_{i=1}^n \sum_{d\in\mathcal{D}} \mathbb{I}(\mathbf{y}_{i,\text{obs}}^{(\delta_i)} = d) \log(\pi_d^{(\delta_i)}) \tag{6.52}$$

where $\pi_d^{(\boldsymbol{\delta})} = P(\mathbf{y}_{\text{obs}}^{(\boldsymbol{\delta})} = d) = \sum_{\{c\in\mathcal{C}:\ \mathbf{y}_{obs}=d\}} \pi_c \mathbb{I}(\mathbf{y} = \mathbf{c})$. We now wish to compute the MLE of π that maximizes the observed log-likelihood in (6.52). However, the parameters are not necessarily identifiable in some cases. If the model is not identifiable, then the maximum likelihood estimators of the parameters do not exist. Statistical inference makes sense only under identifiable models.

The fully nonparametric model is to assign a different probability for different random events $\{\omega : \mathbf{Y}(\omega) = \mathbf{c}\}$ for $\mathbf{c} \in \mathcal{C}$. Thus, the probability $\pi_c = P(\mathbf{Y} = \mathbf{c})$ has only one restriction $\sum_{c\in\mathcal{C}} \pi_c = 1$ in the fully nonparametric approach. However, as explained in the previous section, such a nonparametric model approach can be subject to non-identifiability problem when there are missing values in the sample.

To avoid non-identifiability, we can impose additional restrictions in the parameters to obtain reduced models. The log-linear model is a model about π_c under Poisson distribution assumption.

Example 6.9. *Suppose that* $\mathbf{Y} = (Y_1, Y_2, Y_3)$ *with binary Y_js, taking values in $\{0,1\}$. The support of $\mathbf{Y}$ is*

$$
\begin{aligned}
\mathcal{C} &= \{(0,0,0),(0,0,1),(0,1,0),(0,1,1),(1,0,0),(1,0,1),(1,1,0),(1,1,1)\} \\
&:= \{c_1,\ldots,c_8\}.
\end{aligned}
$$

If we define $\pi_{ijk} = P(Y_1 = i, Y_2 = j, Y_3 = k)$, we can consider the following reduced models:

1. *[M1]* $\log(\pi_{ijk}) = \lambda + \alpha_i + \beta_j + \gamma_k$
2. *[M2]* $\log(\pi_{ijk}) = \lambda + \alpha_i + \beta_j + \gamma_k + (\alpha\beta)_{ij}$
3. *[M3]* $\log(\pi_{ijk}) = \lambda + \alpha_i + \beta_j + \gamma_k + (\beta\gamma)_{jk}$
4. *[M4]* $\log(\pi_{ijk}) = \lambda + \alpha_i + \beta_j + \gamma_k + (\alpha\gamma)_{ik}$
5. *[M5]* $\log(\pi_{ijk}) = \lambda + \alpha_i + \beta_j + \gamma_k + (\alpha\beta)_{ij} + (\beta\gamma)_{jk}$
6. *[M6]* $\log(\pi_{ijk}) = \lambda + \alpha_i + \beta_j + \gamma_k + (\alpha\beta)_{ij} + (\alpha\gamma)_{ik}$
7. *[M7]* $\log(\pi_{ijk}) = \lambda + \alpha_i + \beta_j + \gamma_k + (\alpha\gamma)_{ik} + (\beta\gamma)_{jk}$
8. *[M8]* $\log(\pi_{ijk}) = \lambda + \alpha_i + \beta_j + \gamma_k + (\alpha\beta)_{ij} + (\alpha\gamma)_{ik} + (\beta\gamma)_{jk}$
9. *[M9]* $\log(\pi_{ijk}) = \lambda + \alpha_i + \beta_j + \gamma_k + (\alpha\beta)_{ij} + (\alpha\gamma)_{ik} + (\beta\gamma)_{jk} + (\alpha\beta\gamma)_{ijk}$

with $\sum_i \alpha_i = \sum_j \beta_j = \sum_k \gamma_k = 0$, $\sum_i (\alpha\beta)_{ij} = \sum_j (\alpha\beta)_{ij} = 0$, $\sum_j (\beta\gamma)_{jk} = \sum_k (\beta\gamma)_{jk} = 0$, $\sum_i (\alpha\gamma)_{ik} = \sum_k (\alpha\gamma)_{ik} = 0$, $\sum_i (\alpha\beta\gamma)_{ijk} = \sum_j (\alpha\beta\gamma)_{ijk} = \sum_k (\alpha\beta\gamma)_{ijk} = 0$.

Generally speaking, the joint probability under model M can be written as

$$
\pi_c = \pi_c(\theta_M),
$$

where θ_M is the model parameter for model M. The dimension of θ_M determines the level of sparsity in the model. The observed log-likelihood function in (6.52) is now changed to

$$
l_{\text{obs}}(\theta_M) = \sum_{i=1}^{n} \sum_{d \in \mathcal{D}(\delta_i)} \mathbb{I}(\mathbf{y}_{i,\text{obs}} = d) \log\{\pi_d^{(\delta_i)}(\theta_M)\} \tag{6.53}
$$

where $\pi_d^{(\delta)}(\theta_M) = P(\mathbf{y}_{\text{obs}} = d; \theta_M) = \sum_{\{c \in \mathcal{C}:\ \mathbf{y}_{\text{obs}} = d\}} \pi_c(\theta_M) \mathbb{I}(\mathbf{y} = \mathbf{c})$. Here, we write $(\mathbf{y}_{i,\text{obs}}, \mathbf{y}_{i,\text{mis}}) = (\mathbf{y}_{i,\text{obs}}^{(\delta_i)}, \mathbf{y}_{i,\text{mis}}^{(\delta_i)})$ for simplicity of notation.

To find the maximizer of $l_{\text{obs}}(\theta_M)$ in (6.53), we use the EM algorithm of Dempster et al. (1977). In the E-step, we compute

$$
Q\left(\theta_M \mid \theta_M^{(t)}\right) \equiv E\{l_{\text{com}}(\theta_M) \mid \mathbf{Y}_{\text{obs}}; \theta_M^{(t)}\},
$$

where

$$
l_{\text{com}}(\theta_M) = \sum_{i=1}^{n} \sum_{\mathbf{c} \in \mathcal{C}} \mathbb{I}(\mathbf{y}_i = \mathbf{c}) \log\{\pi_c(\theta_M)\}.
$$

Note that the E-step in the categorical data is simply computing

$$p_i^{(t)}(\mathbf{c}) \equiv E\{\mathbb{I}(\mathbf{y}_i = \mathbf{c}) \mid \mathbf{y}_{i,\text{obs}}, \boldsymbol{\delta}_i; \theta_M^{(t)}\} \tag{6.54}$$

$$= E\{\mathbb{I}(\mathbf{y}_{i,\text{obs}} = \mathbf{c}_{\text{obs}(i)}, \mathbf{y}_{i,\text{mis}} = \mathbf{c}_{\text{mis}(i)}) \mid \mathbf{y}_{i,\text{obs}}, \boldsymbol{\delta}_i; \theta_M^{(t)}\}$$

$$= \mathbb{I}(\mathbf{y}_{i,\text{obs}} = \mathbf{c}_{\text{obs}(i)}) \cdot P(\mathbf{y}_{i,\text{mis}} = \mathbf{c}_{\text{mis}(i)} \mid \mathbf{y}_{i,\text{obs}}, \boldsymbol{\delta}_i; \theta_M^{(t)}),$$

where $\mathbf{c} = (\mathbf{c}_{\text{obs}(i)}, \mathbf{c}_{\text{mis}(i)})$ based on the response pattern of $\mathbf{y}_i$,

$$P(\mathbf{y}_{i,\text{mis}} = \mathbf{c}_{\text{mis}(i)} \mid \mathbf{y}_{i,\text{obs}}, \boldsymbol{\delta}_i; \theta_M^{(t)}) = \begin{cases} \dfrac{\pi_c(\theta_M^{(t)})}{\sum_{c \in \mathcal{C}_i} \pi_c(\theta_M^{(t)})} & \text{if } \mathbf{y}_{i,\text{obs}} = \mathbf{c}_{\text{obs}(i)} \\ 0 & \text{otherwise.} \end{cases}$$

and $\mathcal{C}_i = \{\mathbf{c} \in \mathcal{C} : \mathbf{y}_{i,obs} = \mathbf{c}_{\text{obs}(i)}\}$ is the subset of $\mathcal{C}$ whose $\mathbf{c}_{obs(i)}$ components are equal to $\mathbf{y}_{i,obs}$.

Using $p_i^{(t)}(\mathbf{c})$ in (6.54), we obtain

$$Q\left(\theta_M \mid \theta_M^{(t)}\right) = \sum_{\mathbf{c} \in \mathcal{C}} \sum_{i=1}^{n} p_i^{(t)}(\mathbf{c}) \log\{\pi_{\mathbf{c}}(\theta_M)\} = \sum_{\mathbf{c} \in \mathcal{C}} n_{\mathbf{c}}^{(t)} \log\{\pi_c(\theta_M)\},$$

where $n_c^{(t)} = \sum_{i=1}^{n} p_i^{(t)}(\mathbf{c})$. In the M-step, we update parameters by

$$\theta_M^{(t+1)} \leftarrow \underset{\theta_M}{\arg\max}\, Q\left(\theta_M \mid \theta_M^{(t)}\right).$$

Thus, in the M-step, we have only to replace n_c by $n_c^{(t)}$ in the formula for MLE of θ_M. See Table 6.3 for the illustration of the M-step for the setup in Example 6.9.

TABLE 6.3
M-Step Formula for Loglinear Models in Example 6.9

Model	Probability Formula	M-step Formula
M1	$\pi_{ijk} = \pi_{i++}\pi_{+j+}\pi_{++k}$	$\pi_{ijk}^{(t+1)} = (n_{i++}^{(t)} n_{+j+}^{(t)} n_{++k}^{(t)})/n^3$
M2	$\pi_{ijk} = \pi_{ij+}\pi_{++k}$	$\pi_{ijk}^{(t+1)} = (n_{ij+}^{(t)} n_{++k}^{(t)})/n^2$
M3	$\pi_{ijk} = \pi_{i++}\pi_{+jk}$	$\pi_{ijk}^{(t+1)} = (n_{i++}^{(t)} n_{+jk}^{(t)})/n^2$
M4	$\pi_{ijk} = \pi_{i+k}\pi_{+j+}$	$\pi_{ijk}^{(t+1)} = (n_{i+k}^{(t)} n_{+j+}^{(t)})/n^2$
M5	$\pi_{ijk} = \pi_{ij+}\pi_{+jk}/\pi_{+j+}$	$\pi_{ijk}^{(t+1)} = n_{ij+}^{(t)} n_{+jk}^{(t)}/(n_{+j+}^{(t)} n)$
M6	$\pi_{ijk} = \pi_{ij+}\pi_{i+k}/\pi_{i++}$	$\pi_{ijk}^{(t+1)} = n_{ij+}^{(t)} n_{i+k}^{(t)}/(n_{i++}^{(t)} n)$
M7	$\pi_{ijk} = \pi_{i+k}\pi_{+jk}/\pi_{++k}$	$\pi_{ijk}^{(t+1)} = n_{i+k}^{(t)} n_{+jk}^{(t)}/(n_{++k}^{(t)} n)$
M8	$\pi_{ijk} = \psi_{ij}\phi_{jk}\omega_{ik}$	Iterative Proportional Fitting

Once $\hat{\theta}_M$ is obtained from the above EM algorithm, we can use

$$AIC(M) = -2l_{obs}(\hat{\theta}_M) + 2p_M$$

to choose the best model, where p_M is the dimension of θ_M. The model with the smallest value of $AIC(M)$ will be chosen. The final joint probability under the chosen model $\hat{\pi}_c = \pi_c(\hat{\theta}_M)$ will be used for implementing fractional imputation.

The final joint probability under the chosen model $\hat{\pi}_c = \pi_c(\hat{\theta}_M)$ will be used for implementing fractional imputation. For each i, we observe $\mathbf{y}_{i,obs}$ and there are $M_i = |\mathcal{C}_i|$ possible values of $\mathbf{y}_{i,mis}$. Thus, the fractionally imputed values of $\mathbf{y}_i$ are given by

$$\mathbf{y}_i^{*(j)} = \mathbf{c}_j,$$

where $\mathbf{c}_j$ is the j-th element of $\mathcal{C}_i$. The fractional weight associated with $\mathbf{y}_i^{*(j)}$ is the estimated conditional probability of $\mathbf{y} = \mathbf{c}_j$ given $\mathbf{y}_{i,obs}$. That is,

$$w_{ij}^* = \frac{\hat{\pi}(\mathbf{c}_j)}{\sum_{c \in \mathcal{C}_i} \hat{\pi}(\mathbf{c})}, \tag{6.55}$$

where $\hat{\pi}(\mathbf{c}) = P(\mathbf{Y} = \mathbf{c}; \hat{\theta}_M)$ is the estimated joint probability using the MLE $\hat{\theta}_M$ under the final model M. The fractionally imputed data can be written

$$\text{FI-Data} = \{(w_{ij}^*, \mathbf{y}_i^{*(j)}) : i = 1, \ldots, n, j = 1, \ldots, M_i\}.$$

The "FI-Data" does not contain any missingness and users can use the "FI Data" to do their own analysis.

For variance estimation, we use the grouped jackknife method. We first partition the sample into L groups, $S = S_1 \cup \cdots \cup S_L$. Let $w_i^{(k)}$ be the jackknife weight of unit i when group k is removed. That is,

$$w_i^{(k)} = \begin{cases} 0 & \text{if } i \in S_k \\ (L-1)^{-1} & \text{other wise.} \end{cases}$$

Using $w_i^{(k)}$, the jackknife observed likelihood function is

$$l_{obs}^{(k)}(\theta_M) = \sum_{i=1}^{n} w_i^{(k)} \sum_{d \in \mathcal{C}_1(\delta_i)} \mathbb{I}(\mathbf{y}_{i,obs} = d) \log\{\pi_d^{(\delta_i)}(\theta_M)\} \tag{6.56}$$

and we can apply the same EM algorithm to obtain $\hat{\theta}_M^{(k)}$. Once $\hat{\theta}_M^{(k)}$ are obtained, the jackknife version of the fractional weights are computed as

$$w_{ij}^{*(k)} = \frac{\pi(\mathbf{c}_j; \hat{\theta}_M^{(k)})}{\sum_{c \in \mathcal{C}_i} \pi(\mathbf{c}; \hat{\theta}_M^{(k)})}.$$

Thus, we have a set of L replication weights, $w_i^{(k)} w_{ij}^{*(k)}$ are assigned to $\mathbf{y}_{ij}^*$ in the final FI data.

6.6   Model Selection

In this section, we introduce some model selection methods and discuss their application to missing data problem. We introduce Akaike Information Criterion (AIC) first and then introduce Bayesian Information Criterion (BIC).

Assume that we have a random sample $y_1, \ldots, y_n$ from a distribution with density $g(y)$ which is unknown. Suppose that we have K possible parametric models for y, denoted by $f_k(y; \theta_k)$ for $k = 1, \ldots, K$. Let $\hat{\theta}_k$ be the MLE of θ based on the current sample. Thus, we have K models, $\hat{f}_k(y) = f_k(y; \hat{\theta}_k)$, to choose from.

Among the K candidate models, we wish to find the best model in the sense of minimizing the Kullback-Leibler distance

$$
\begin{aligned}
KL(g, \hat{f}_k) &= \int g(y) \log \left\{ \frac{g(y)}{\hat{f}_k(y)} \right\} dy \\
&= \int g(y) \log\{g(y)\} dy - \int \log\{\hat{f}_k(y)\} g(y) dy \\
&= \text{Const.} - R(\hat{f}_k).
\end{aligned}
$$

Thus, we have only to find the model with the maximum value of $R(\hat{f}_k)$, where

$$
R(\hat{f}_k) = \int \log\{\hat{f}_k(y)\} g(y) dy = E_g\{\log \hat{f}_k(Y)\}.
$$

Using the second order Taylor expansion of $R(\hat{f}_k)$ around $\hat{\theta}_k = \theta_k$, we can obtain

$$
R(\hat{f}_k) \cong R(f_k) - 0.5 \text{tr}\{I_k(\theta_k) V(\hat{\theta}_k)\}, \tag{6.57}
$$

where

$$
I_k(\theta_k) = -E_g \left[\frac{\partial^2}{\partial \theta_k \partial \theta_k'} \log\{f_k(Y; \theta_k)\} \right]
$$

and

$$
V(\hat{\theta}_k) = n^{-1}\{I_k(\theta_k)\}^{-1} V_g\{S_k(\theta_k; Y)\}\{I_k(\theta_k)\}^{-1}
$$

with $S_k(\theta_k; y) = \partial \log f_k(y; \theta_k)/\partial \theta_k$.

Also, to estimate the first term of (6.57), we apply the same Taylor expansion on

$$
n^{-1} l_k(\theta_k) = \frac{1}{n} \sum_{i=1}^{n} \log f_k(y_i; \theta_k)
$$

around $\theta_k = \hat{\theta}_k$ to get

$$
R(f_k) \cong n^{-1} l_k(\hat{\theta}_k) - 0.5 \text{tr}\{I_k(\theta_k) V(\hat{\theta}_k)\}. \tag{6.58}
$$

Combining (6.57) with (6.58), we may use

$$AIC_k = 2nR(\hat{f}_k) = 2l_k(\hat{\theta}_k) - 2d_k^*, \tag{6.59}$$

where

$$d_k^* = n \cdot \text{tr}\{I_k(\theta_k)V(\hat{\theta}_k)\} = \text{tr}\left[\{I_k(\theta_k)\}^{-1}V_g\{S(\theta;Y)\}\right],$$

as a criterion for model selection. This is essentially the idea for AIC proposed by Akaike (1974) with a simple formula using $d_k^* = \dim(\theta_k)$. Takeuchi (1976) proposed a model-robust information criterion using

$$d_k^* = \text{tr}\left(\hat{I}_k^{-1}\hat{V}_k\right),$$

where $\hat{I}_k = -n^{-1}\sum_{i=1}^n \dot{S}_k(\hat{\theta}_k; y_i)$ and $\hat{V} = n^{-1}\sum_{i=1}^n S_k(\hat{\theta}_k; y_i)^{\otimes 2}$.

We now introduce Bayesian Information Criterion (BIC) of Schwarz (1978). Suppose that we have K different models: $(\mathcal{M}_1, \ldots, \mathcal{M}_K)$. In the Bayesian framework, we may wish to select the best model with the highest posterior probability of $P(\mathcal{M}_k \mid \text{data})$, where

$$P(\mathcal{M}_k \mid \text{data}) = \frac{P(\mathbf{y} \mid \mathcal{M}_k)P(\mathcal{M}_k)}{\sum_{k=1}^K P(\mathbf{y} \mid \mathcal{M}_k)P(\mathcal{M}_k)}.$$

If we use uniform prior on the model $P(\mathcal{M}_k) = 1/K$, the best model can be chosen with the largest value of the marginal density, which is given by

$$P(\mathbf{y} \mid \mathcal{M}_k) = \int L_k(\theta_k)p(\theta_k \mid \mathcal{M}_k)d\theta_k, \tag{6.60}$$

where $L_k(\theta_k) = \prod_{i=1}^n f_k(y_i; \theta_k)$. In the marginal density, the parameter θ_k is a nuisance parameter, and we integrate it out as in (6.60). The BIC is defined as

$$BIC_k = 2\log P(\mathbf{y} \mid \mathcal{M}_k).$$

To compute the BIC term approximately, we can apply the second order Taylor expansion (i.e. Laplace approximation) to get

$$
\begin{aligned}
P(\mathbf{y} \mid \mathcal{M}_k) &\cong L_k(\hat{\theta}_k) \int \exp\left\{-\frac{1}{2}(\theta_k - \hat{\theta}_k)'\{n\mathcal{I}_k(\hat{\theta}_k)\}(\theta_k - \hat{\theta}_k)\right\}p(\theta_k \mid \mathcal{M}_k)d\theta_k \\
&\cong L_k(\hat{\theta}_k)(2\pi)^{p_k/2}|n\mathcal{I}_k(\hat{\theta}_k)|^{-1/2} \\
&= L_k(\hat{\theta}_k)(2\pi)^{p_k/2}(n)^{-p_k/2}|\mathcal{I}_k(\hat{\theta}_k)|^{-1/2},
\end{aligned}
$$

where $p_k = \dim(\theta_k)$.

After ignoring the terms (which are free from n) and multiplying the remaining part by 2, we have

$$BIC_k \equiv 2\log P(\mathbf{y} \mid \mathcal{M}_k) \approx 2l_k(\hat{\theta}_k) - p_k \log(n).$$

Comparing the BIC formula with the AIC in (6.59), only the second terms (penality term) are different. Generally speaking, the BIC's penalty is larger than that of AIC. Thus, BIC is is more in favor of sparse models than the AIC.

Instead of selecting a single model, model averaging method can be more attractive especially in the context of imputation. Model selection provides less efficient estimation than the model averaging method. Post model-selection inference is a notoriously difficult problem. Bayesian model averaging method (Madigan and Raftery, 1994; Hoeting et al., 1999) can capture the uncertainly into the imputation procedure.

7

Propensity Scoring Approach

We now introduce a totally different approach to handling missing data. Instead of using imputation, which is a prediction approach, we use weighting for handling missing data. The weights, which is called the propensity score weights, are applied to the observed part of the data to obtain unbiased estimators of the parameters of interest. The propensity score weighting idea can be understood as the weighting approach in computing the Horvitz-Thompson estimator in survey sampling. The main difference is that we do not know the true response probability and use a statistical model to estimate the parameters to compute the propensity scores. The estimated parameters in the propensity score model make the resulting inference more complicated. We first start with a simple setup of known propensity score in Section 7.1. For unknown propensity score case, either maximum likelihood method or maximum entropy method can be used. Nonparametric method is introduced at the end of this chapter.

7.1 Introduction

Assume that the parameter of interest θ_0 is defined implicitly by $E\{U(\theta; Z)\} = 0$, and $U(\theta; z)$ is the estimating function for θ_0. Under complete response, a consistent estimator of θ can be obtained by solving

$$\hat{U}_n(\theta) \equiv n^{-1} \sum_{i=1}^{n} U(\theta; z_i) = 0 \tag{7.1}$$

for θ. Note that, by Lemma 4.1, the solution $\hat{\theta}$ is asymptotically unbiased for θ_0 if $\hat{U}(\theta_0)$ is asymptotically unbiased for zero. We assume that the solution to (7.1) is unique. Under some regularity conditions, $\hat{\theta}$ converges to θ_0 in probability, and the limiting distribution of $\hat{\theta}$ is normal.

Suppose that we can write $z_i = (x_i, y_i)'$, where x_i are always observed, and y_i are subject to missingness. Let δ_i be the response indicator function for y_i that takes the value one if and only if y_i is observed. The complete sample estimating equation (7.1) cannot be directly computed under the existence of missing data. To resolve this problem, we can consider an expected

DOI: 10.1201/9780429321740-7

estimating equation approach that computes the conditional expectation of $U(\theta; Z)$ given the observations, which requires correct specification of the conditional distribution of y_i on x_i. That is, we can consider

$$n^{-1} \sum_{i=1}^{n} [\delta_i U(\theta; x_i, y_i) + (1 - \delta_i) E\{U(\theta; x_i, y_i) \mid x_i, \delta_i = 0\}] = 0$$

and apply the imputation approach discussed in Chapter 4. Such an approach, called the prediction model approach, requires correct specification of the prediction model, the model for the conditional distribution of y on x and δ. While the prediction model approach provides efficient estimates when the prediction model is true, correct specification of the prediction model and parameter estimation can be difficult, especially when y is vector-valued. In this chapter, we consider an alternative approach based on modeling δ_i using all available information. Note that, even when y_i is a vector, we may have a scalar δ_i for unit nonresponse, so the modeling of δ_i may be easier than the modeling of y_i. This alternative approach, based on the model for δ_i, is called response probability model approach.

Under the existence of missing data, the complete case (CC) method, defined by the solution to

$$\sum_{i=1}^{n} \delta_i U(\theta; \mathbf{z}_i) = 0,$$

can lead to a biased estimator of θ, unless $Cov(\delta_i, U_i) = 0$, where $U_i = U(\theta_0; \mathbf{z}_i)$. Thus, unless the missing mechanism is missing completely at random (MCAR), the CC method results in biased estimation. Furthermore, the CC method does not make use of the observed information of x_i for $\delta_i = 0$. Thus, it is not fully efficient.

To correct the bias, a weighted complete case (WCC) estimator can be obtained by solving

$$\hat{U}_W(\theta) \equiv \frac{1}{n} \sum_{i=1}^{n} \delta_i \frac{1}{\pi_i} U(\theta; \mathbf{z}_i) = 0, \tag{7.2}$$

where $\pi_i = P(\delta_i = 1 \mid \mathbf{z}_i)$. The weight $1/\pi_i$ is often called the propensity score, termed by Rosenbaum and Rubin (1983). When the true propensity score is known, as in survey sampling, asymptotic properties of the WCC estimator can be obtained using the sandwich formula. Let $\hat{\theta}_W$ be the (unique) solution to (7.2) with π_i known. The solution is also called the Horvitz-Thompson estimator in survey sampling. Using Lemma 4.1, we have

$$\hat{\theta}_W - \theta_0 \cong -\left[E\{\dot{U}(\theta_0; Z)\}\right]^{-1} \hat{U}_W(\theta_0),$$

where $\dot{U}(\theta; \mathbf{z}) = \partial U(\theta; \mathbf{z})/\partial \theta'$. Thus,

$$E\{\hat{U}_W(\theta_0)\} = E\left[E\{\hat{U}_W(\theta_0) \mid \mathbf{z}\}\right] = E\{\hat{U}_n(\theta_0)\} = 0, \tag{7.3}$$

and the WCC estimator is asymptotically unbiased. Also, the asymptotic variance of $\hat{\theta}_W$ is computed by the sandwich formula

$$V\left(\hat{\theta}_W\right) \cong \tau^{-1} V\left\{\hat{U}_W(\theta_0)\right\}(\tau^{-1})',$$

where $\tau = E\left\{\dot{U}(\theta_0; Z)\right\}$ and, assuming that $Cov(\delta_i, \delta_j) = 0$ for $i \neq j$,

$$V\left\{\hat{U}_W(\theta_0)\right\} = V\left\{\hat{U}_n(\theta_0)\right\} + E\left\{n^{-2}\sum_{i=1}^{n}(\pi_i^{-1}-1)\,U(\theta_0; \mathbf{z}_i)^{\otimes 2}\right\}. \quad (7.4)$$

If $\mathbf{z}_1, \ldots, \mathbf{z}_n$ are independently and identically distributed, then

$$V\left\{\hat{U}_n(\theta_0)\right\} \cong E\left\{\frac{1}{n^2}\sum_{i=1}^{n}U(\theta_0; \mathbf{z}_i)^{\otimes 2} - \frac{1}{n}\bar{U}_n(\theta_0)^{\otimes 2}\right\},$$

where $\bar{U}_n(\theta_0) = n^{-1}\sum_{i=1}^{n}U(\theta_0; \mathbf{z}_i)$. Thus,

$$
\begin{aligned}
V\left\{\hat{U}_W(\theta_0)\right\} &= n^{-1}E\left\{n^{-1}\sum_{i=1}^{n}\pi_i^{-1}U(\theta_0; \mathbf{z}_i)^{\otimes 2} - \bar{U}_n(\theta_0)^{\otimes 2}\right\} \\
&\cong E\left\{n^{-2}\sum_{i=1}^{n}\pi_i^{-1}U(\theta_0; \mathbf{z}_i)^{\otimes 2}\right\},
\end{aligned}
\quad (7.5)
$$

and a consistent estimator for the variance of $\hat{\theta}_W$ is computed by

$$\hat{V}(\hat{\theta}_W) = \hat{\tau}^{-1}\hat{V}_u(\hat{\tau}^{-1})',$$

where $\hat{\tau} = n^{-1}\sum_{i=1}^{n}\delta_i\pi_i^{-1}\dot{U}(\hat{\theta}_W; \mathbf{z}_i)$ and $\hat{V}_u = n^{-2}\sum_{i=1}^{n}\delta_i\pi_i^{-2}U(\hat{\theta}_W; \mathbf{z}_i)^{\otimes 2}$.

Example 7.1. *Let the parameter of interest be $\theta = E(Y)$, and we use $U(\theta; \mathbf{z}) = (y - \theta)$ to compute θ. The WCC estimator of θ can be written as*

$$\hat{\theta}_W = \frac{\sum_{i=1}^{n}\delta_i y_i/\pi_i}{\sum_{i=1}^{n}\delta_i/\pi_i}. \quad (7.6)$$

The asymptotic variance of $\hat{\theta}_W$ in (7.6) is equal to, by (7.5),

$$E\left\{n^{-2}\sum_{i=1}^{n}\pi_i^{-1}(y_i - \theta)^2\right\}. \quad (7.7)$$

In the context of survey sampling, where the population size is equal to N and δ_i corresponds to the sampling indicator function, the estimator (7.6) is called the Hájek estimator. The asymptotic variance in (7.7) represents the asymptotic variance of the Hájek estimator under Poisson sampling when the finite population is a random sample from an infinite population, called superpopulation, and the parameter θ is the superpopulation parameter. See Godambe and Thompson (1986).

We now consider the case when x is decomposed into $x = (x_1, x_2)$ and MAR holds under x_1. That is,

$$P(\delta = 1 \mid x_1, y) = P(\delta = 1 \mid x_1). \tag{7.8}$$

Writing $\tilde{\pi}(x_1) = P(\delta = 1 \mid x_1)$, we note that $\tilde{\pi}(x_1)$ is the conditional expectation of $\pi(x) = P(\delta = 1 \mid x)$ given x_1. If the parameter of interest is $\theta = E(Y)$ and both $\pi(x)$ and $\tilde{\pi}(x_1)$ are known, there are two choices for deriving the WCC estimator of θ. The first one is obtained by solving

$$\hat{U}_{W1}(\theta) \equiv \sum_{i=1}^{n} \delta_i \frac{1}{\pi(x_i)} (y_i - \theta) = 0, \tag{7.9}$$

and the other is obtained by solving

$$\hat{U}_{W2}(\theta) \equiv \sum_{i=1}^{n} \delta_i \frac{1}{\tilde{\pi}(x_{1i})} (y_i - \theta) = 0. \tag{7.10}$$

Then, a question naturally arises: which one of the two is better? Kim et al. (2019) showed that the solution to (7.10) is more efficient than the solution to (7.9). The result is summarized in the following theorem.

Theorem 7.1. *Let $\hat{\theta}_1$ and $\hat{\theta}_2$ be the solutions to (7.9) and (7.10), respectively. Under (7.8), both $\hat{\theta}_1$ and $\hat{\theta}_2$ are asymptotically unbiased for $\theta_0 = E(Y)$ and*

$$V(\hat{\theta}_1) \geq V(\hat{\theta}_2) \tag{7.11}$$

asymptotically.

Proof. Asymptotic unbiasedness of $\hat{\theta}_1$ is already established in (7.3). To show the asymptotic unbiasedness of $\hat{\theta}_2$, we show that

$$E\left\{\hat{U}_{W2}(\theta_0)\right\} = E\left[\sum_{i=1}^{n} \frac{E(\delta_i \mid x_{1i}, y_i)}{\tilde{\pi}(x_{1i})} (y_i - \theta_0)\right] = E\left\{\sum_{i=1}^{n} (y_i - \theta_0)\right\} = 0.$$

To show (7.11), since

$$E\left\{\frac{\partial}{\partial \theta} \hat{U}_{W1}(\theta)\right\} = E\left\{\frac{\partial}{\partial \theta} \hat{U}_{W2}(\theta)\right\},$$

we have only to show that

$$V\{\hat{U}_{W1}(\theta_0)\} \geq V\{\hat{U}_{W2}(\theta_0)\}. \tag{7.12}$$

Using (7.5), we have

$$V\{\hat{U}_{W1}(\theta_0)\} \cong E\left\{\sum_{i=1}^{n} \{\pi(x_i)\}^{-1} (y_i - \theta_0)^2\right\}$$

and

$$V\{\hat{U}_{W2}(\theta_0)\} \cong E\left\{\sum_{i=1}^{n}\{\tilde{\pi}(x_{1i})\}^{-1}(y_i - \theta_0)^2\right\}.$$

Note that $f(x) = 1/x$ is a convex function at $x \in (0,1)$. Thus, using Jensen's inequality, we can show that

$$E\left\{\frac{1}{\pi(x_i)} \mid x_{1i}\right\} \geq \frac{1}{E\{\pi(x_i) \mid x_{1i}\}}.$$

Therefore, (7.12) is proved. □

We now consider an alternative expression of the response probability model. Using Bayes formula, we can first express

$$\frac{P(\delta = 0 \mid z)}{P(\delta = 1 \mid z)} = \frac{1-p}{p} \times \frac{f(z \mid \delta = 0)}{f(z \mid \delta = 1)},$$

where $z = (x, y)$ and $p = P(\delta = 1)$. Thus, we can express

$$\frac{1}{\pi(z)} = 1 + \left(\frac{1}{p} - 1\right) r(z), \tag{7.13}$$

where

$$r(z) = \frac{f(z \mid \delta = 0)}{f(z \mid \delta = 1)}$$

is a density ratio function. If $f(z \mid \delta = k)$ is $N(\mu_k, \sigma^2)$, then $\log\{r(z)\}$ is a linear function of z, which is equivalent to the logistic regression model for π (z).

Now, assume that $x = (x_1, x_2)$, we can obtain, similiarly to (7.13),

$$\frac{1}{\tilde{\pi}(x_1, y)} = 1 + \left(\frac{1}{p} - 1\right) \tilde{r}(x_1, y).$$

The new density ratio function $\tilde{r}(x_1, y)$ is based on the marginal model $f(x_1, y \mid \delta) = \int f(x, y \mid \delta) dx_2$. The following lemma presents the relationship between two density ratio functions.

Lemma 7.1. *The two density ratio functions satisfy*

$$E\left\{\frac{f(x, y \mid \delta = 0)}{f(x, y \mid \delta = 1)} \mid x_1, y, \delta = 1\right\} = \frac{f(x_1, y \mid \delta = 0)}{f(x_1, y \mid \delta = 1)}. \tag{7.14}$$

The proof of (7.14) can be established easily by the following result:

$$\begin{aligned} E\{r(x, y) \mid x_1, y, \delta = 1\} &= \int r(x, y) f(x_2 \mid x_1, y, \delta = 1) dx_2 \\ &= \frac{\int r(x, y) f(x, y \mid \delta = 1) dx_2}{f(x_1, y \mid \delta = 1)} \\ &= \frac{f(x_1, y \mid \delta = 0)}{f(x_1, y \mid \delta = 1)} = \tilde{r}(x_1, y). \end{aligned}$$

Now, recall that there are two WCC estimators of $\theta = E(Y)$:

$$\hat{\theta}_1 = n^{-1} \sum_{i=1}^{n} \delta_i \{1 + (p^{-1} - 1)r(x_i, y_i)\} y_i.$$

$$\hat{\theta}_2 = n^{-1} \sum_{i=1}^{n} \delta_i \{1 + (p^{-1} - 1)\tilde{r}(x_{1i}, y_i)\} y_i.$$

By Lemma 7.1, we have

$$E(\hat{\theta}_1 \mid x_1, y, \delta) = \hat{\theta}_2,$$

which implies that

$$E(\hat{\theta}_1) = E(\hat{\theta}_2)$$

and

$$V(\hat{\theta}_1) \geq V(\hat{\theta}_2).$$

Under MAR, we have $\pi(x, y) = \pi(x)$ and

$$\hat{\theta}_1 = n^{-1} \sum_{i=1}^{n} \delta_i \{1 + (p^{-1} - 1)r(x_i)\} y_i.$$

Under the reduced model MAR condition (7.8), we have $\tilde{\pi}(x_1, y) = \tilde{\pi}(x_1)$. Thus, the density ratio approach gives an alternative proof for Theorem 7.1.

Lemma 7.2. *If MAR condition given X holds (i.e. $Y \perp \delta \mid X$) and the reduced model for y holds such that*

$$f(y \mid x) = f(y \mid x_1) \tag{7.15}$$

for $x = (x_1, x_2)$, then we can obtain MAR given X_1. That is,

$$Y \perp \delta \mid X_1.$$

Proof. We have only to prove that

$$f(y \mid x_1, \delta) = f(y \mid x_1).$$

Now, using Bayes formula,

$$
\begin{aligned}
f(y \mid x_1, \delta) &= \frac{\int f(y \mid x, \delta) P(\delta \mid x) f(x_2 \mid x_1) f(x_1) dx_2}{\int \int f(y \mid x, \delta) P(\delta \mid x) f(x_2 \mid x_1) f(x_1) dx_2 dy} \\
&= \frac{\int f(y \mid x_1) P(\delta \mid x) f(x_2 \mid x_1) f(x_1) dx_2}{\int \int f(y \mid x_1) P(\delta \mid x) f(x_2 \mid x_1) f(x_1) dx_2 dy} \\
&= \frac{f(y \mid x_1) \int P(\delta \mid x) f(x_2 \mid x_1) dx_2}{\int f(y \mid x_1) \int P(\delta \mid x) f(x_2 \mid x_1) dx_2 dy} \\
&= \frac{f(y \mid x_1) P(\delta \mid x_1)}{\int f(y \mid x_1) P(\delta \mid x_1) dy} = f(y \mid x_1),
\end{aligned}
$$

where the second equality follows by MAR assumption and the reduced model assumption. $\qquad\square$

By Lemma 7.2, we can use the standard variable selection techniques in the regression of y on x among the set of respondents to choose the covariates that satisfying (7.15). Shortreed and Ertefaie (2017) also proposed variable selection technique for outcome model in the context of causal inference.

7.2 Regression Weighting Method

In practice, the propensity score is usually unknown, and the theory in Section 7.1 cannot be directly applied. In this section, we introduce a useful technique for obtaining a weighted estimator when the propensity score is unknown. Assume that auxiliary variables $\mathbf{x}_i$ are observed throughout the sample, and the response probability satisfies

$$\frac{1}{\pi_i} = \mathbf{x}_i'\lambda \tag{7.16}$$

for all unit i in the sample, where λ is unknown. We assume that an intercept is included in $\mathbf{x}_i$.

Under the response mechanism satisfying (7.16), according to Fuller et al. (1994), the regression estimator defined by

$$\hat{\theta}_{reg} = \sum_{i=1}^{n} \delta_i w_i y_i, \tag{7.17}$$

where

$$w_i = \left(\frac{1}{n}\sum_{i=1}^{n}\mathbf{x}_i\right)' \left(\sum_{i=1}^{n}\delta_i \mathbf{x}_i \mathbf{x}_i'\right)^{-1}\mathbf{x}_i$$

is asymptotically unbiased for $\theta = E(Y)$. Note that the regression weight (7.17) satisfies

$$\sum_{i=1}^{n}\delta_i w_i \mathbf{x}_i = \bar{\mathbf{x}}_n, \tag{7.18}$$

where $\bar{\mathbf{x}}_n = n^{-1}\sum_{i=1}^{n}\mathbf{x}_i$. Condition (7.18) is often called calibration equation.

To show this, note that we can write

$$\hat{\theta}_{reg} = \bar{\mathbf{x}}_n'\hat{\beta}_r,$$

where

$$\hat{\beta}_r = \left(\sum_{i=1}^{n}\delta_i \mathbf{x}_i \mathbf{x}_i'\right)^{-1}\sum_{i=1}^{n}\delta_i \mathbf{x}_i y_i.$$

Because an intercept term is included in $\mathbf{x}_i$, we can also express

$$\hat{\theta}_n \equiv \bar{y}_n = \bar{\mathbf{x}}_n'\hat{\beta}_n, \tag{7.19}$$

where

$$\hat{\beta}_n = \left(\sum_{i=1}^{n} \mathbf{x}_i \mathbf{x}_i' \right)^{-1} \sum_{i=1}^{n} \mathbf{x}_i y_i.$$

Thus, we can write

$$\hat{\theta}_{reg} - \hat{\theta}_n = \bar{\mathbf{x}}_n' \left(\sum_{i=1}^{n} \delta_i \mathbf{x}_i \mathbf{x}_i' \right)^{-1} \sum_{i=1}^{n} \delta_i \mathbf{x}_i \left(y_i - \mathbf{x}_i' \hat{\beta}_n \right),$$

so that

$$E\left(\hat{\theta}_{reg} - \hat{\theta}_n \mid \mathbf{X}, \mathbf{Y} \right) \cong \bar{\mathbf{x}}_n' \left(\sum_{i=1}^{n} \pi_i \mathbf{x}_i \mathbf{x}_i' \right)^{-1} \sum_{i=1}^{n} \pi_i \mathbf{x}_i \left(y_i - \mathbf{x}_i' \hat{\beta}_n \right),$$

where the expectation is taken with respect to the response mechanism. Thus, to show that $\hat{\theta}_{reg}$ is asymptotically unbiased, we have only to show that

$$\sum_{i=1}^{n} \pi_i \mathbf{x}_i \left(y_i - \mathbf{x}_i' \hat{\beta}_n \right) = \mathbf{0} \tag{7.20}$$

holds. By (7.16) and (7.19), we have

$$0 = \sum_{i=1}^{n} \left(y_i - \mathbf{x}_i \hat{\beta}_n \right) = \sum_{i=1}^{n} \pi_i \left(\lambda' \mathbf{x}_i \right) \left(y_i - \mathbf{x}_i' \hat{\beta}_n \right) = \lambda' \sum_{i=1}^{n} \pi_i \mathbf{x}_i \left(y_i - \mathbf{x}_i' \hat{\beta}_n \right),$$

which implies that (7.20) holds.

Example 7.2. *Assume that the sample is partitioned into G exhaustive and mutually exclusive groups, denoted by $A_1, \ldots, A_G$, where $|A_g| = n_g$. Assume a uniform response mechanism for each group. Thus, we assume that $\pi_i = p_g$ for some $p_g \in (0, 1]$ if $i \in A_g$. Let*

$$x_{ig} = \begin{cases} 1 & \text{if } i \in A_g \\ 0 & \text{otherwise.} \end{cases}$$

Then, $\mathbf{x}_i = (x_{i1}, \ldots, x_{iG})$ satisfies (7.16). The regression estimator (7.17) of $\theta = E(Y)$ can be written as

$$\hat{\theta}_{reg} = \frac{1}{n} \sum_{g=1}^{G} \frac{n_g}{r_g} \sum_{i \in A_g} \delta_i y_i = \frac{1}{n} \sum_{g=1}^{G} n_g \bar{y}_{Rg},$$

where $r_g = \sum_{i \in A_g} \delta_i$ is the realized size of respondents in group g and $\bar{y}_{Rg} = \left(\sum_{i \in A_g} \delta_i \right)^{-1} \sum_{i \in A_g} \delta_i y_i$. Because the covariate satisfies (7.16), the *regression estimator is asymptotically unbiased. The asymptotic variance of* $\hat{\theta}_{reg}$ *is*

$$V\left(\hat{\theta}_{reg}\right) = V\left(\hat{\theta}_n\right) + E\left\{n^{-2}\sum_{g=1}^{G}\left(\frac{n_g}{r_g}-1\right)\sum_{i\in A_g}(y_i-\bar{y}_{ng})^2\right\},$$

where $\bar{y}_{ng} = n_g^{-1}\sum_{i\in A_g} y_i$. *If we write*

$$V\left(\hat{\theta}_n\right) = E\left\{\frac{1}{n}S_n^2\right\} \doteq E\left\{\frac{1}{n^2}\sum_{i=1}^{n}(y_i-\bar{y}_n)^2\right\},$$

we can use

$$\sum_{i=1}^{n}(y_i-\bar{y}_n)^2 = \sum_{g=1}^{G}\sum_{i\in A_g}(y_i-\bar{y}_{ng})^2 + \sum_{g=1}^{G}n_g\left(\bar{y}_{ng}-\bar{y}_n\right)^2$$

so that

$$V\left(\hat{\theta}_{reg}\right) \doteq E\left[\frac{1}{n^2}\sum_{g=1}^{G}\sum_{i\in A_g}\left\{\frac{n_g}{r_g}(y_i-\bar{y}_{ng})^2 + (\bar{y}_{ng}-\bar{y}_n)^2\right\}\right].$$

This form shows that, compared with the complete sample case, the between-group variations are unchanged, but the within-group variances are increased by a factor of n_g/r_g.

To discuss variance estimation of the regression estimator in (7.17) with the covariates $\mathbf{x}_i$ satisfying (7.16), write

$$\bar{\mathbf{x}}_n'\hat{\beta}_r = \bar{\mathbf{x}}_n'\beta + \bar{\mathbf{x}}_n'\left(\hat{\beta}_r - \beta\right)$$

$$= \bar{\mathbf{x}}_n'\beta + \bar{\mathbf{x}}_n'\left(\sum_{i=1}^{n}\delta_i\mathbf{x}_i\mathbf{x}_i'\right)^{-1}\sum_{i=1}^{n}\delta_i\mathbf{x}_i\left(y_i-\mathbf{x}_i'\beta\right)$$

$$\cong \bar{\mathbf{x}}_n'\beta + \bar{\mathbf{x}}_n'\left(\sum_{i=1}^{n}\pi_i\mathbf{x}_i\mathbf{x}_i'\right)^{-1}\sum_{i=1}^{n}\delta_i\mathbf{x}_i\left(y_i-\mathbf{x}_i'\beta\right),$$

where β is the probability limit of $\hat{\beta}_r$. By the fact that 1 is included in $\mathbf{x}_i$ (indicating the existence of an intercept) and by (7.16), it can be shown that

$$\bar{\mathbf{x}}_n'\left(\sum_{i=1}^{n}\pi_i\mathbf{x}_i\mathbf{x}_i'\right)^{-1}\sum_{i=1}^{n}\delta_i\mathbf{x}_i\left(y_i-\mathbf{x}_i'\beta\right) = \frac{1}{n}\sum_{i=1}^{n}\frac{\delta_i}{\pi_i}\left(y_i-\mathbf{x}_i'\beta\right) \qquad (7.21)$$

and

$$V\left(\hat{\theta}_{reg}\right) \doteq V\left(\frac{1}{n}\sum_{i=1}^{n}d_i\right), \qquad (7.22)$$

where $d_i = \mathbf{x}_i'\beta + \delta_i \pi_i^{-1}(y_i - \mathbf{x}_i'\beta)$. Variance estimation can be implemented by using a standard variance estimation formula applied to $\hat{d}_i = \mathbf{x}_i'\hat{\beta}_r + \delta_i n w_i(y_i - \mathbf{x}_i'\hat{\beta}_r)$.

7.3 Propensity Score Method

We now consider a more realistic situation of the response probability being unknown. Suppose the true response probability is parametrically modeled by

$$\pi_i = \pi(z_i; \phi_0)$$

for some $\phi_0 \in \Omega$. If the maximum likelihood estimator of ϕ_0, denoted by $\hat{\phi}$, is available, then the propensity score (PS) (adjusted) estimator of θ, denoted by $\hat{\theta}_{PS}$, can be computed by solving

$$\hat{U}_{PS}(\theta) \equiv \frac{1}{n} \sum_{i=1}^{n} \delta_i \frac{1}{\hat{\pi}_i} U(\theta; z_i) = 0, \tag{7.23}$$

where $\hat{\pi}_i = \pi(z_i; \hat{\phi})$. Strictly speaking, the PS estimator in (7.23) is also a function of $\hat{\phi}$. Thus, we can write $(\hat{\theta}_{PS}, \hat{\phi})$ as the solution to $\hat{U}_1(\theta, \phi) = 0$ and $S(\phi) = 0$, where $\hat{U}_1(\theta, \hat{\phi}) = \hat{U}_{PS}(\theta)$, with $\hat{U}_{PS}(\theta)$ defined as in (7.23), and $S(\phi)$ is the score function for ϕ.

The following lemma presents some results on the relationship between the PS estimating function $\hat{U}_{PS}(\theta)$ and the score function $S(\phi)$.

Lemma 7.3. *Let*

$$U_1(\theta, \phi) = \sum_{i=1}^{n} u_{i1}(\theta, \phi),$$

where $u_{i1}(\theta, \phi) = u_{i1}(\theta, \phi; z_i, \delta_i)$, be an estimating equation satisfying

$$E\{U_1(\theta_0, \phi_0)\} = 0.$$

Let $\pi_i = \pi_i(\phi)$ be the probability of response. Then,

$$E\{-\partial U_1/\partial \phi'\} = Cov(U_1, S), \tag{7.24}$$

where S is the score function of ϕ.

Proof. Since $E\{U_1(\theta_0, \phi_0)\} = 0$, we have

$$
\begin{aligned}
0 &= \partial E\{U_1(\theta_0, \phi_0)\}/\partial\phi' \\
&= \sum_{i=1}^{n} \frac{\partial}{\partial\phi'} \int u_{i1}(\theta_0, \phi_0) f(\delta_i \mid z_i, \phi_0) f(z_i) d\delta_i dz_i \\
&= \sum_{i=1}^{n} \int \left[\frac{\partial}{\partial\phi'} u_{i1}(\theta_0, \phi_0) \right] f(\delta_i \mid z_i, \phi_0) f(z_i) d\delta_i dz_i \\
&\quad + \sum_{i=1}^{n} \int u_{i1}(\theta_0, \phi_0) \frac{\partial}{\partial\phi'} [f(\delta_i \mid z_i, \phi_0)] f(z_i) d\delta_i dz_i \\
&= E\{\partial U/\partial\phi'\} + E\{U(\theta_0, \phi_0) S(\phi_0)'\},
\end{aligned}
$$

which proves (7.24). $\qquad\square$

If we set $U_1(\theta, \phi) = S(\phi)$, then (7.24) reduces to $E\{-\partial S(\phi)/\partial\phi'\} = E\{S(\phi)^{\otimes 2}\}$, which is already presented in (2.5).

We now have the following theorem on the asymptotic property of the PS estimator obtained from (7.23).

Theorem 7.2. *Let $\hat{\theta}_{PS}$ be the solution to (7.23). Under some regularity conditions, we have*

$$
\sqrt{n}\left(\hat{\theta}_{PS} - \theta_0\right) \xrightarrow{\mathcal{L}} N(0, \sigma^2), \tag{7.25}
$$

where

$$
\sigma^2 = \tau^{-1} V \left\{ \frac{\delta}{\pi} U(\theta_0) + \kappa S(\phi_0) \right\} (\tau^{-1})'
$$

with $\tau = E\{\dot{U}(\theta_0; Z)\}$ and

$$
\kappa = -E \left\{ \frac{\delta}{\pi} U(\theta_0) S(\phi_0)' \right\} \mathcal{I}(\phi_0)^{-1}.
$$

Proof. The solution $(\hat{\theta}_{PS}, \hat{\phi})$ to

$$
\begin{aligned}
\hat{U}_{PS}(\theta, \phi) &= 0 \\
S(\phi) &= \mathbf{0}
\end{aligned}
$$

is asymptotically normal with mean $(\theta_0, \phi_0)'$ and variance $A^{-1}B(A^{-1})'$, where

$$
A = \begin{bmatrix} E\{-\partial\hat{U}_{PS}/\partial\theta'\} & E\{-\partial\hat{U}_{PS}/\partial\phi'\} \\ E\{-\partial S/\partial\theta'\} & E\{-\partial S/\partial\phi'\} \end{bmatrix} = \begin{bmatrix} A_{11} & A_{12} \\ 0 & A_{22} \end{bmatrix}
$$

$$
B = \begin{bmatrix} V\left(\hat{U}_{PS}\right) & C\left(\hat{U}_{PS}, S\right) \\ C\left(S, \hat{U}_{PS}\right) & V(S) \end{bmatrix} = \begin{bmatrix} B_{11} & B_{12} \\ B_{21} & B_{22} \end{bmatrix}.
$$

Then, using

$$A^{-1} = \begin{bmatrix} A_{11}^{-1} & -A_{11}^{-1}A_{12}A_{22}^{-1} \\ 0 & A_{22}^{-1} \end{bmatrix},$$

we have

$$Var\left(\hat{\theta}_{PS}\right) \cong A_{11}^{-1}\left[B_{11} - A_{12}A_{22}^{-1}B_{21} - B_{12}A_{22}^{-1}A_{12}' + A_{12}A_{22}^{-1}B_{22}A_{22}^{-1}A_{12}'\right]A_{11}^{'-1}.$$
$$\text{(7.26)}$$

By Lemma 7.3, $B_{22} = A_{22}$ and $B_{12} = A_{12}$. Thus,

$$V\left(\hat{\theta}_{PS}\right) \cong A_{11}^{-1}\left[B_{11} - B_{12}B_{22}^{-1}B_{21}\right]A_{11}^{'-1}. \qquad (7.27)$$

$$\square$$

Note that $\hat{\theta}_W = \hat{\theta}_W(\phi_0)$ computed from (7.2) with known π_i satisfies

$$V\left(\hat{\theta}_W\right) \cong A_{11}^{-1}B_{11}A_{11}^{-1'}.$$

Ignoring the smaller order terms, we have

$$V\left(\hat{\theta}_W\right) \geq V\left(\hat{\theta}_{PS}\right). \qquad (7.28)$$

The result of (7.28) means that the PS estimator with estimated π_i is more efficient than the PS estimator with known π_i. Such contradictory phenomenon has been discussed in Rosenbaum (1987), Robins et al. (1994), and Kim and Kim (2007). See also Henmi and Eguchi (2004).

Another way of understanding (7.27) is the projection in the Hilbert space. Given two random variables, say X and Y, with finite second moments with zero means, the projection of Y on the linear space generated by X, denoted by $\Pi(Y \mid X)$ satisfies the following two conditions:

1. $\Pi(Y \mid X) \in \mathcal{L}(X)$, where $\mathcal{L}(X) = \{a'X; a \in \mathbb{R}^p\}$ is the linear subspace generated by X and $p = \dim(X)$.

2. $Y - \Pi(Y \mid X)$ is orthogonal to all elements in $\mathcal{L}(X)$.

In the linear space of random variables, the norm of X and Y is the covariance and

$$\Pi(Y \mid X) = Cov(Y, X)\{Var(X)\}^{-1}X.$$

Also, we can define the projection of Y to the orthogonal complement of $\mathcal{L}(X)$ as

$$\Pi(Y \mid X^\perp) = Y - \Pi(Y \mid X).$$

Note that, as $E(X) = 0$,

$$E\{\Pi(Y \mid X^\perp)\} = E(Y)$$

and

$$V\left\{\Pi(Y\mid X^\perp)\right\} = V(Y) - Cov(Y,X)\left\{Var(X)\right\}^{-1}Cov(X,Y) \le V(Y).$$
(7.29)

Thus, the projection of Y onto $\mathcal{L}(X^\perp)$ with $E(X) = 0$ will always improve the efficiency.

Now, we can write $\hat\theta_{PS} = \hat\theta_W(\hat\phi)$. A standard Taylor linearization of $\hat\theta_{PS} \equiv \hat\theta_W(\hat\phi)$ leads to

$$\hat\theta_{PS} \cong \hat\theta_W(\phi_0) - E\left\{\frac{\partial}{\partial\phi'}\hat\theta_W(\phi_0)\right\}\left[E\left(\frac{\partial}{\partial\phi'}S(\phi_0)\right)\right]^{-1}S(\phi_0),$$
(7.30)

when $\hat\phi$ is the MLE. Using (7.24) and

$$V(S) = \mathcal{I}(\phi_0) = \{V(\hat\phi)\}^{-1},$$

we can write (7.30) as

$$\hat\theta_{PS} \cong \hat\theta_W - C(\hat\theta_W, S)\{V(S)\}^{-1}S(\phi_0),$$

which can be understood as a projection of $\hat\theta_W$ to orthogonal complement of the nuisance tangent space, the linear subspace generated by $S(\phi_0)$ with $E\{S(\phi_0)\} = 0$. That is, we have

$$\hat\theta_{PS} \cong \Pi(\hat\theta_W \mid S^\perp)$$
(7.31)

and, by (7.29), result (7.28) holds.

The variance formula is also useful for deriving a variance estimator for the PS estimators. If the response mechanism is ignorable (or MAR), we can further write

$$\pi_i = \pi(x_i; \phi_0)$$
(7.32)

for some $\phi_0 \in \Omega$, where x_i is completely observed in the sample. In this case, the propensity score can be estimated by the maximum likelihood method that solves

$$S(\phi) \equiv \sum_{i=1}^{n}\left\{\delta_i - \pi(x_i; \phi)\right\}\frac{1}{\pi(x_i; \phi)\{1 - \pi(x_i; \phi)\}}\dot\pi(x_i; \phi) = 0,$$
(7.33)

where $\dot\pi(x_i; \phi) = \partial\pi(x_i; \phi)/\partial\phi$. Under the logistic regression model

$$\pi(x_i; \phi) = \frac{\exp(\phi_0 + x_i\phi_1)}{1 + \exp(\phi_0 + x_i\phi_1)},$$

we have $\dot\pi(x_i; \phi) = \pi(x_i; \phi)\{1 - \pi(x_i; \phi)\}(1, x_i)'$.

Using (7.27), a plug-in variance estimator of the PS estimator is computed by

$$\hat V\left(\hat\theta_{PS}\right) = \hat A_{11}^{-1}\left(\hat B_{11} - \hat B_{12}\hat B_{22}^{-1}\hat B_{21}\right)\hat A_{11}'^{-1},$$

where $\hat{A}_{11} = n^{-1}\sum_{i=1}^{n} \delta_i \pi_i^{-1} \dot{U}(\hat{\theta}; z_i)$ and

$$
\begin{aligned}
\hat{B}_{11} &= n^{-2}\sum_{i=1}^{n} \delta_i \hat{\pi}_i^{-2} U(\hat{\theta}; z_i)^{\otimes 2}, \\
\hat{B}_{12} &= n^{-2}\sum_{i=1}^{n} \delta_i \hat{\pi}_i^{-1}(\hat{\pi}_i^{-1} - 1) U(\hat{\theta}; z_i)\mathbf{h}_i, \\
\hat{B}_{22} &= n^{-2}\sum_{i=1}^{n} \delta_i \hat{\pi}_i^{-1}(\hat{\pi}_i^{-1} - 1)\mathbf{h}_i \mathbf{h}_i',
\end{aligned}
$$

with $\hat{\theta} = \hat{\theta}_{PS}$, $\mathbf{h}_i = \dot{\pi}_i/(1 - \pi_i)$.

Example 7.3. *We consider the case of $\theta = E(Y)$ in Example 7.1, when the true response probability is unknown and is assumed to follow a parametric model (7.32). There are two types of PS estimators for θ. The first one is*

$$
\hat{\theta}_{PS1} = \frac{1}{n}\sum_{i=1}^{n} \frac{\delta_i}{\hat{\pi}_i} y_i,
$$

and the second one is a Hájek-type estimator of the form

$$
\hat{\theta}_{PS2} = \frac{\sum_{i=1}^{n} \delta_i y_i / \hat{\pi}_i}{\sum_{i=1}^{n} \delta_i / \hat{\pi}_i}.
$$

For the variance of $\hat{\theta}_{PS1}$, use (7.27) with $\hat{U}_1 = n^{-1}\sum_{i=1}^{n}(\delta_i y_i/\hat{\pi}_i - \theta)$, then

$$
\begin{aligned}
V\left(\hat{\theta}_{PS1}\right) &\cong V(\hat{\theta}_{W1}) - C(\hat{\theta}_{W1}, S)\left\{V(S)\right\}^{-1} C(S, \hat{\theta}_{W1}) \\
&= V\left\{\hat{\theta}_{W1} - B^*(\phi)' S(\phi)\right\},
\end{aligned}
$$

where $\hat{\theta}_{W1} = n^{-1}\sum_{i=1}^{n} \delta_i y_i/\pi_i$ and $B^*(\phi) = [V\{S(\phi)\}]^{-1} C\{S(\phi), \hat{\theta}_{W1}\}$. If the score function is of the form

$$
S(\phi) = \sum_{i=1}^{n}\left\{\frac{\delta_i}{\pi_i(\phi)} - 1\right\}\mathbf{h}_i(\phi),
$$

then

$$
B^*(\phi) = \left\{\sum_{i=1}^{n}(\pi_i^{-1} - 1)\mathbf{h}_i \mathbf{h}_i'\right\}^{-1}\sum_{i=1}^{n}(\pi_i^{-1} - 1)\mathbf{h}_i y_i. \tag{7.34}
$$

Thus, the asymptotic variance of $\hat{\theta}_{PS1}$ is equal to the variance of

$$
\hat{\theta}_{PS1,l} = \frac{1}{n}\sum_{i=1}^{n} B^*(\phi)' \mathbf{h}_i(\phi) + \frac{1}{n}\sum_{i=1}^{n}\frac{\delta_i}{\pi_i}\left\{y_i - B^*(\phi)' \mathbf{h}_i(\phi)\right\}, \tag{7.35}
$$

with the subscript l denoting linearization. That is, the asymptotic variance is equal to

$$V\left(\hat{\theta}_{PS1,l}\right) = V\left(\frac{1}{n}\sum_{i=1}^{n} y_i\right) + E\left\{\frac{1}{n^2}\sum_{i=1}^{n}\frac{1-\pi_i}{\pi_i}\left\{y_i - B^*(\phi_0)'\mathbf{h}_i(\phi_0)\right\}^2\right\},$$

and $B^*(\phi_0)$ in (7.34) minimizes the second term of the variance. For variance estimation, we can write

$$\hat{\theta}_{PS1,l} = \frac{1}{n}\sum_{i=1}^{n} d_i(\phi_0),$$

where

$$d_i(\phi) = B^*(\phi)'\mathbf{h}_i(\phi) + \frac{\delta_i}{\pi_i}\left\{y_i - B^*(\phi)'\mathbf{h}_i(\phi)\right\}.$$

With the linearized pseudo values $d_i(\phi)$ directly used for variance estimation, the variance of $\hat{\theta}_{PS1,l}$ is consistently estimated by

$$\frac{1}{n}\frac{1}{n-1}\sum_{i=1}^{n}\left(\hat{d}_i - \frac{1}{n}\sum_{j=1}^{n}\hat{d}_j\right)^2,$$

with $\hat{d}_i = d_i(\hat{\phi})$.

For the Hájek-type estimator $\hat{\theta}_{PS2}$, we can apply the same argument to show that $\hat{\theta}_{PS2}$ is asymptotically equivalent to

$$\hat{\theta}_{PS2,l} = \frac{1}{n}\sum_{i=1}^{n} B_2^*(\phi)'\mathbf{h}_i(\phi) + \frac{1}{n}\sum_{i=1}^{n}\frac{\delta_i}{\pi_i}\left\{y_i - \hat{\theta}_W - B_2^*(\phi)'\mathbf{h}_i(\phi)\right\}, \qquad (7.36)$$

where

$$B_2^* = \left\{\sum_{i=1}^{n}(\pi_i^{-1} - 1)\mathbf{h}_i\mathbf{h}_i'\right\}^{-1}\sum_{i=1}^{n}(\pi_i^{-1} - 1)\mathbf{h}_i(y_i - \hat{\theta}_W).$$

The pseudo value for variance estimation of $\hat{\theta}_{PS2}$ is

$$d_{2i}(\phi) = B_2^*(\phi)'\mathbf{h}_i(\phi) + \frac{\delta_i}{\pi_i}\left\{y_i - \hat{\theta}_W(\phi) - B_2^*(\phi)'\mathbf{h}_i(\phi)\right\}.$$

To improve the efficiency of the PS estimator in (7.23), one can consider a class of estimating equations of the form

$$\sum_{i=1}^{n}\delta_i\frac{1}{\hat{\pi}_i}\left\{U(\theta; x_i, y_i) - b(\theta; x_i)\right\} + \sum_{i=1}^{n} b(\theta; x_i) = 0, \qquad (7.37)$$

where $b(\theta; x_i)$ is to be determined. Assume that the estimated response probability is computed by $\hat{\pi}_i = \pi(x_i; \hat{\phi})$, where $\hat{\phi}$ is computed by

$$U_2(\phi) = \frac{1}{n} \sum_{i=1}^{n} \left(\frac{\delta_i}{\pi(x_i; \phi)} - 1 \right) \mathbf{h}_i(\phi) = \mathbf{0}, \qquad (7.38)$$

for some $\mathbf{h}_i(\phi) = \mathbf{h}(x_i; \phi)$. The score equation in (7.33) uses $\mathbf{h}_i(\phi) = \hat{\pi}_i(\phi)/\{1 - \pi_i(\phi)\}$. The following theorem, which was originally proved by Robins et al. (1994), presents asymptotic properties of the solution to the augmented estimating equation (7.37).

Theorem 7.3. *Assume that the response probability $P(\delta = 1 \mid x, y) = \pi(x)$ does not depend on the value of y. Let $\hat{\theta}_b$ be the solution to (7.37) for given $b(\theta; x_i)$. Under some regularity conditions, $\hat{\theta}_b$ is consistent and its asymptotic variance satisfies*

$$V(\hat{\theta}_b) \geq n^{-1} \tau^{-1} \left[V\{E(U \mid X)\} + E\{\pi^{-1} V(U \mid X)\} \right] (\tau^{-1})', \qquad (7.39)$$

where $\tau = E(\partial U/\partial \theta')$, and the equality in (7.39) holds when $b^*(\theta; x_i) = E\{U(\theta; x_i, y_i) \mid x_i\}$.

Proof. We first consider the case when the true response probability $\pi(x) = Pr(\delta = 1 \mid x)$ is known. Let

$$U_b(\theta) = \frac{1}{n} \sum_{i=1}^{n} b(\theta; x_i) + \frac{1}{n} \sum_{i=1}^{n} \frac{\delta_i}{\pi_i} \{U(\theta; x_i, y_i) - b(\theta; x_i)\},$$

where $\pi_i = \pi(x_i)$ and $b(\theta; x_i)$ is to be determined. Since $E\{U_b(\theta_0)\} = E\{U(\theta_0; x, y)\} = 0$, by Lemma 4.1, the solution $\hat{\theta}_b$ to $U_b(\theta) = 0$ is asymptotically unbiased for θ_0 and

$$V(\hat{\theta}_b) \cong \tau^{-1} V\{U_b(\theta_0)\}(\tau^{-1})',$$

where $\tau = E\{\partial U_b(\theta_0)/\partial \theta'\} = E\{\partial U(\theta_0; x, y)/\partial \theta'\}$.

Now, writing

$$U_b(\theta) = U_{b^*}(\theta) + D_b(\theta), \qquad (7.40)$$

where

$$U_{b^*}(\theta) = \frac{1}{n} \sum_{i=1}^{n} b^*(\theta; x_i) + \frac{1}{n} \sum_{i=1}^{n} \frac{\delta_i}{\pi_i} \{U(\theta; x_i, y_i) - b^*(\theta; x_i)\},$$

$$D_b(\theta) = \frac{1}{n} \sum_{i=1}^{n} \left(\frac{\delta_i}{\pi_i} - 1 \right) \{b^*(\theta; x_i) - b(\theta; x_i)\},$$

and $b^*(\theta; x_i) = E\{U(\theta; x_i, y_i) \mid x_i\}$, we have

$$V\{U_b(\theta)\} = V\{U_{b^*}(\theta)\} + V\{D_b(\theta)\} + 2Cov\{U_{b^*}(\theta), D_b(\theta)\}.$$

Note that

$$Cov\{U_{b^*}(\theta), D_b(\theta)\} = E\left\{\frac{1}{n^2}\sum_{i=1}^{n}\left(\frac{1}{\pi_i}-1\right)(U_i - b_i^*)(b_i^* - b_i)\right\},$$

where $U_i = U(\theta; x_i, y_i)$, $b_i = b(\theta; x_i)$, and $b_i^* = b^*(\theta; x_i)$. Because $E(U_i \mid x_i) = b_i^*$, the above covariance term is equal to zero and

$$V(\hat{\theta}_b) \cong \tau^{-1}[V\{U_{b^*}(\theta_0)\} + V\{D_b(\theta_0)\}](\tau^{-1})' \geq \tau^{-1}V\{U_{b^*}(\theta_0)\}(\tau^{-1})'.$$

Since

$$
\begin{aligned}
V\{U_{b^*}(\theta)\} &= V\left\{\frac{1}{n}\sum_{i=1}^{n}U(\theta; x_i, y_i)\right\} + E\left\{\frac{1}{n^2}\sum_{i=1}^{n}\left(\frac{1}{\pi_i}-1\right)(U_i - b_i^*)^2\right\} \\
&= n^{-1}V(U) + n^{-1}E\{(\pi^{-1}-1)V(U \mid X)\} \\
&= n^{-1}V\{E(U \mid X)\} + n^{-1}E\{\pi^{-1}V(U \mid X)\},
\end{aligned}
$$

result (7.39) holds when the true response probability is known.

Now, to discuss the case with the response probability unknown, let $\hat{\pi}_i = \pi(x_i; \hat{\phi})$, where $\hat{\phi}$ is estimated by solving (7.38). Writing

$$U_{1b}(\theta, \phi) = \frac{1}{n}\sum_{i=1}^{n}b(\theta; x_i) + \frac{1}{n}\sum_{i=1}^{n}\frac{\delta_i}{\pi(x_i; \phi)}\{U(\theta; x_i, y_i) - b(\theta; x_i)\},$$

the solution $(\hat{\theta}_b, \hat{\phi})$ to $U_{1b}(\theta, \phi) = 0$ and $U_2(\phi) = 0$ is consistent for (θ_0, ϕ_0). Using the same argument for deriving (7.26), we can show

$$V(\hat{\theta}_b) \cong \tau^{-1}V\{U_{1b}(\theta_0, \phi_0) - CU_2(\phi_0)\}(\tau^{-1})', \qquad (7.41)$$

where $C = (\partial U_{1b}/\partial \phi')(\partial U_2/\partial \phi')^{-1}$. Now, similar to (7.40), we can write

$$U_{1b}(\theta, \phi) - CU_2(\phi) = U_{b^*}(\theta) + D_{2b}(\theta, \phi),$$

where $D_{2b}(\theta, \phi) = D_b(\theta) - CU_2(\phi)$. Because $U_2(\phi)$ takes the form in (7.38), we can show that

$$Cov(U_{b^*}, D_{2b}) = E\left\{n^{-2}\sum_{i=1}^{n}(\pi_i^{-1}-1)(U_i - b_i^*)(b_i^* - b_i - C\mathbf{h}_i)\right\} = 0$$

and result (7.39) follows. $\square$

Note that, for the choice of $b^*(\theta; x_i) = E\{U(\theta; x_i, y_i) \mid x_i\}$, the resulting estimator achieves the lower bound (semi-parametric variance lower bound) in (7.39) regardless of whether $\pi_i = P(\delta_i = 1 \mid x_i)$ is known or estimated. For the case of known π_i, the lower bound in (7.39) was also discussed by Godambe and Joshi (1965) and Isaki and Fuller (1982) in the context of survey sampling.

Example 7.4. *Consider the sample from a linear regression model*

$$y_i = \mathbf{x}_i'\beta + e_i, \tag{7.42}$$

where $e_i's$ are independent with $E(e_i \mid \mathbf{x}_i) = 0$. Assume that $\mathbf{x}_i$ is available from the full sample and y_i is observed only when $\delta_i = 1$. The response propensity model follows from the logistic regression model with $\text{logit}(\pi_i) = \mathbf{x}_i'\phi$. We are interested in estimating $\theta = E(Y)$ from the partially observed data.

To construct the optimal estimator that achieves the minimum variance in (7.39), we can use $U_i(\theta) = y_i - \theta$ and $b_i^*(\theta) = \mathbf{x}_i'\beta - \theta$. Thus, the optimal estimator using $\hat{b}_i^*(\theta) = \mathbf{x}_i'\hat{\beta} - \theta$ in (7.37) is given by

$$\hat{\theta}_{opt}(\hat{\beta}) = \frac{1}{n}\sum_{i=1}^{n}\frac{\delta_i}{\hat{\pi}_i}y_i + \frac{1}{n}\left(\sum_{i=1}^{n}\mathbf{x}_i - \sum_{i=1}^{n}\frac{\delta_i}{\hat{\pi}_i}\mathbf{x}_i\right)'\hat{\beta}, \tag{7.43}$$

where $\hat{\beta}$ is any estimator of β satisfying the $\sqrt{n}$-consistency, which means $\sqrt{n}(\hat{\beta} - \beta) = O_p(1)$, with $X_n = O_p(1)$ denoting that X_n is bounded in probability. Note that the choice of $\hat{\beta}$ does not play any leading role in the asymptotic variance of $\hat{\theta}_{opt}(\hat{\beta})$. This is because

$$\hat{\theta}_{opt}(\hat{\beta}) \cong \hat{\theta}_{opt}(\beta_0) + E\left\{\frac{\partial}{\partial\beta'}\hat{\theta}_{opt}(\beta_0)\right\}\left(\hat{\beta} - \beta_0\right), \tag{7.44}$$

and, under the correct response model,

$$E\left\{\frac{\partial}{\partial\beta'}\hat{\theta}_{opt}(\beta_0)\right\} = E\left\{\frac{1}{n}\left(\sum_{i=1}^{n}\mathbf{x}_i - \sum_{i=1}^{n}\frac{\delta_i}{\hat{\pi}_i}\mathbf{x}_i\right)\right\} \cong \mathbf{0},$$

so the second term of (7.44) becomes negligible. Furthermore, it can be shown that the choice of $\hat{\phi}$ in $\hat{\pi}_i = \pi_i(\hat{\phi})$ does not matter asymptotically as long as the regression model holds.

7.4   Optimal PSEstimation

In Example 7.4, the optimal PS estimator is discussed under the assumption that the outcome model, which is the regression model (7.42) in this example, is correctly specified. If the regression model is not true, then optimality is not achieved. We now discuss optimal estimation with the propensity score method without introducing the outcome model. We assume that the propensity score is computed by (7.33). In general, the PS estimator $\hat{\theta}$ applied to $\theta = E(X)$ is not equal to the complete sample estimator $\hat{\theta}_n = n^{-1}\sum_{i=1}^{n}x_i$. Thus, the complete sample estimator $\bar{x}_n$ can be used to improve the efficiency of the PS estimator.

We now wish to develop efficient estimation incorporating the knowledge of $n^{-1}\sum_{i=1}^{n} x_i$. To achieve this goal, we consider a more general problem of minimizing the objective function

$$
Q = \begin{pmatrix} \hat{X}_1 - \mu_x \\ \hat{X}_2 - \mu_x \\ \hat{Y} - \mu_y \end{pmatrix}' \begin{pmatrix} V(\hat{X}_1) & C(\hat{X}_1,\hat{X}_2) & C(\hat{X}_1,\hat{Y}) \\ C(\hat{X}_1,\hat{X}_2) & V(\hat{X}_2) & C(\hat{X}_2,\hat{Y}) \\ C(\hat{X}_1,\hat{Y}) & C(\hat{X}_2,\hat{Y}) & V(\hat{Y}) \end{pmatrix}^{-1} \begin{pmatrix} \hat{X}_1 - \mu_x \\ \hat{X}_2 - \mu_x \\ \hat{Y} - \mu_y \end{pmatrix},
$$
(7.45)

where $\hat{X}_1$ and $\hat{X}_2$ are two unbiased estimators of μ_x and $\hat{Y}$ is an unbiased estimator of μ_y. The solution to the minimization is called General Least Squares (GLS) estimator or simply optimal estimator. The following lemma presents the optimal estimator in the sense that it has minimal asymptotic variance among linear unbiased estimators.

Lemma 7.4. *The optimal (GLS) estimator of (μ_x, μ_y) that minimizes Q in (7.45) is given by*

$$
\hat{\mu}_x^* = \alpha^* \hat{X}_1 + (1 - \alpha^*) \hat{X}_2
$$
(7.46)

and

$$
\hat{\mu}_y^* = \hat{Y} + B_1 \left(\hat{\mu}_x^* - \hat{X}_1 \right) + B_2 \left(\hat{\mu}_x^* - \hat{X}_2 \right),
$$
(7.47)

where

$$
\alpha^* = \frac{V(\hat{X}_2) - C(\hat{X}_1,\hat{X}_2)}{V(\hat{X}_1) + V(\hat{X}_2) - 2C(\hat{X}_1,\hat{X}_2)}
$$

and

$$
\begin{pmatrix} B_1 \\ B_2 \end{pmatrix} = \begin{pmatrix} V(\hat{X}_1) & C(\hat{X}_1,\hat{X}_2) \\ C(\hat{X}_1,\hat{X}_2) & V(\hat{X}_2) \end{pmatrix}^{-1} \begin{pmatrix} C(\hat{X}_1,\hat{Y}) \\ C(\hat{X}_2,\hat{Y}) \end{pmatrix}.
$$
(7.48)

Proof. First, do a mental partition of the matrix component of Q located in the middle of (7.45). Using the inverse of the partitioned matrix, we can write

$$
Q = Q_1 + Q_2,
$$

where

$$
Q_1 = \begin{pmatrix} \hat{X}_1 - \mu_x \\ \hat{X}_2 - \mu_x \end{pmatrix}' \begin{pmatrix} V(\hat{X}_1) & C(\hat{X}_1,\hat{X}_2) \\ C(\hat{X}_1,\hat{X}_2) & V(\hat{X}_2) \end{pmatrix}^{-1} \begin{pmatrix} \hat{X}_1 - \mu_x \\ \hat{X}_2 - \mu_x \end{pmatrix},
$$

$$
Q_2 = \left\{ \hat{Y} - \mu_y - B_1(\hat{X}_1 - \mu_x) - B_2(\hat{X}_2 - \mu_x) \right\}' V_{ee}^{-1} \left\{ \hat{Y} - \mu_y - B_1(\hat{X}_1 - \mu_x) \right.
$$
$$
\left. - B_2(\hat{X}_2 - \mu_x) \right\},
$$

and $V_{ee} = V(\hat{Y}) - (B_1, B_2)\{V(\hat{X}_1,\hat{X}_2)\}^{-1}(B_1,B_2)'$. Minimizing Q_1 with respect to μ_x yields $\hat{\mu}_x^*$ in (7.46), and minimizing Q_2 with respect to μ_y given $\hat{\mu}_x^*$ yields $\hat{\mu}_y^*$ in (7.47). $\qquad\square$

The optimal estimators in (7.46) and (7.47) are unbiased with minimum variance in the class of linear estimators of $\hat{X}_1$, $\hat{X}_2$, and $\hat{Y}$. The optimal estimator of μ_y takes the form of a regression estimator with $\hat{\mu}_x^*$ as the control. That is, the optimal estimator is the predicted value of $\hat{y} = b_0 + b_1 x$ evaluated at $x = \hat{\mu}_x^*$. Since $\hat{\mu}_x^* - \hat{X}_1 = (1 - \alpha^*)(\hat{X}_2 - \hat{X}_1)$ and $\hat{\mu}_x^* - \hat{X}_2 = -\alpha^*(\hat{X}_2 - \hat{X}_1)$, we can express

$$
\begin{aligned}
\hat{\mu}_y^* &= \hat{Y} + \{B_1(1 - \alpha^*) - B_2\alpha^*\}(\hat{X}_2 - \hat{X}_1) \\
&= \hat{Y} - C\left(\hat{Y}, \hat{X}_2 - \hat{X}_1\right)\left\{V\left(\hat{X}_2 - \hat{X}_1\right)\right\}^{-1}(\hat{X}_2 - \hat{X}_1).
\end{aligned}
$$

Thus, the optimal estimator $\hat{\mu}_y^*$ is the projection of $\hat{Y}$ onto the orthogonal complement of the subspace $\mathcal{L}(\hat{X}_1 - \hat{X}_2)$.

Under the missing data setup where x_i is always observed and y_i is subject to missingness, if we know π_i, then we can use $\hat{X}_1 = n^{-1}\sum_{i=1}^n x_i = \hat{X}_n$, $\hat{X}_2 = n^{-1}\sum_{i=1}^n \delta_i x_i / \pi_i = \hat{X}_W$, and $\hat{Y} = n^{-1}\sum_{i=1}^n \delta_i y_i / \pi_i = \hat{Y}_W$. In this case, $C(\hat{X}_2, \hat{X}_1) = V(\hat{X}_1)$, so $\alpha^* = 1$, which leads to $\hat{\mu}_x^* = \bar{X}_1$. In this case, the optimal estimator of μ_y reduces to

$$
\begin{aligned}
\hat{\mu}_y^* &= \hat{Y} + C\left(\hat{Y}, \hat{X}_2 - \hat{X}_1\right)\left\{V\left(\hat{X}_2 - \hat{X}_1\right)\right\}^{-1}(\hat{X}_1 - \hat{X}_2) \\
&= \hat{Y}_W + (\hat{X}_n - \hat{X}_W)' B^*,
\end{aligned}
$$

where

$$
B^* = E\left(\sum_{i=1}^n \frac{1 - \pi_i}{\pi_i} x_i x_i'\right)^{-1} E\left(\sum_{i=1}^n \frac{1 - \pi_i}{\pi_i} x_i y_i\right).
$$

In practice, we cannot estimate B^* and have to resort to a plug-in estimator

$$
\hat{\theta}_{opt} = \hat{Y}_W + (\hat{X}_n - \hat{X}_W)' \hat{B}^*, \tag{7.49}
$$

where

$$
\hat{B}^* = \left(\sum_{i=1}^n \delta_i \frac{1 - \pi_i}{\pi_i^2} x_i x_i'\right)^{-1} \left(\sum_{i=1}^n \delta_i \frac{1 - \pi_i}{\pi_i^2} x_i y_i\right).
$$

The estimator (7.49) is the asymptotically optimal estimator among the class of linear unbiased estimators. Furthermore, it satisfies the calibration constraint that establishes the identity between the optimal estimator and the full sample estimator under the special case of $y_i = x_i'\alpha$ for some α. To see this, note that $y_i = x_i'\alpha$ for all i implies $\hat{B}^* = \alpha$ and

$$
\begin{aligned}
\hat{\theta}_{opt} &= n^{-1}\sum_{i=1}^n \pi_i^{-1}\delta_i x_i'\alpha + \left\{n^{-1}\sum_{i=1}^n x_i - n^{-1}\sum_{i=1}^n \pi_i^{-1}\delta_i x_i\right\}'\alpha \\
&= n^{-1}\sum_{i=1}^n x_i'\alpha = n^{-1}\sum_{i=1}^n y_i.
\end{aligned}
$$

Note that the optimal estimator in (7.49) satisfies the calibration condition in (7.18). The calibration condition gives an extra protection against model misspecification. The calibration condition is also called the covariate balancing property (Imai and Ratkovic, 2014).

We now discuss the case when π_i is unknown and is estimated by $\hat{\pi}_i = \pi_i(\hat{\phi})$, where $\hat{\phi}$ is the maximum likelihood estimator of ϕ in the response probability model. In this case, the optimal estimator of μ_x is still equal to $\bar{x}_n$, but the optimal estimator of $\theta = E(Y)$ in (7.49) using $\hat{\pi}_i$ instead of π_i is not optimal because the covariance between $\hat{Y}_{PS}$ and $(\hat{X}_{PS}, \hat{X}_n)$ is different from the covariance between $\hat{Y}_W$ and $(\hat{X}_W, \hat{X}_n)$. To find the optimal estimator similar to (7.49), we can consider an estimator of the form

$$\hat{\theta}_B = \hat{Y}_{PS} + (\hat{X}_n - \hat{X}_{PS})B,$$

indexed by B, and find the optimal coefficient B^* that minimizes the variance of $\hat{\theta}_B$. By Lemma 7.4, the variance is minimized at

$$B^* = \left\{ V(\hat{X}_{PS} - \hat{X}_n) \right\}^{-1} C\left(\hat{X}_{PS} - \hat{X}_n, \hat{Y}_{PS} \right). \tag{7.50}$$

Using the argument for (7.31), we can write

$$\hat{X}_{PS} - \hat{X}_n = \Pi(\hat{X}_W \mid S^{\perp}) - \hat{X}_n = \Pi(\hat{X}_W - \bar{X}_n \mid S^{\perp})$$

and

$$V(\hat{X}_{PS} - \hat{X}_n) = V(\hat{X}_W - \hat{X}_n \mid S^{\perp}),$$

where $V(Y \mid X^{\perp}) = V(Y) - C(Y, X)\{V(X)\}^{-1}C(X, Y)$. Similarly, we can write

$$C\left(\hat{X}_{PS} - \hat{X}_n, \hat{Y}_{PS} \right) = C\left(\hat{X}_W - \hat{X}_n, \hat{Y}_W \mid S^{\perp} \right).$$

Thus, (7.50) can be written as

$$B^* = \left\{ V(\hat{X}_W - \hat{X}_n \mid S^{\perp}) \right\}^{-1} C\left(\hat{X}_W - \hat{X}_n, \hat{Y}_W \mid S^{\perp} \right). \tag{7.51}$$

Note that the optimal estimator from minimizing Q in (7.45) with $\hat{X}_1 = \hat{X}_n$, $\hat{X}_2 = \hat{X}_{PS}$, and $\hat{Y} = \hat{Y}_{PS}$ can be obtained by minimizing

$$Q(\mu_z) = \left(\hat{\mathbf{Z}} - \mu_z \right)' \left\{ V\left(\hat{\mathbf{Z}}_0 \mid S^{\perp} \right) \right\}^{-1} \left(\hat{\mathbf{Z}} - \mu_z \right), \tag{7.52}$$

where $\hat{\mathbf{Z}} = (\hat{X}_n, \hat{X}_{PS}, \hat{Y}_{PS})'$, $\hat{\mathbf{Z}}_0 = (\hat{X}_n, \hat{X}_W, \hat{Y}_W)'$ and $\mu_z = (\mu_x, \mu_x, \mu_y)'$. The optimal estimator minimizing Q in (7.52) can also be obtained by minimizing the augmented Q given by

$$Q^*(\mu_z, \phi) = \left(\begin{array}{c} \hat{\mathbf{Z}}_0 - \mu_z \\ S(\phi) \end{array} \right)' \left\{ \begin{array}{cc} V\left(\hat{\mathbf{Z}}_0 \right) & C\left\{ \hat{\mathbf{Z}}_0, S(\phi) \right\} \\ C\left\{ S(\phi), \hat{\mathbf{Z}}_0 \right\} & V\left\{ S(\phi) \right\} \end{array} \right\}^{-1} \left(\begin{array}{c} \hat{\mathbf{Z}}_0 - \mu_z \\ S(\phi) \end{array} \right). \tag{7.53}$$

To see the equivalence, note that we can establish

$$Q^*(\mu_z, \phi) = Q_1(\mu_z \mid \phi) + Q_2(\phi), \tag{7.54}$$

where

$$Q_1(\mu_z \mid \phi) = \left\{ \hat{\mathbf{Z}}_0 - \mu_z - BS(\phi) \right\}' \left\{ V\left(\hat{\mathbf{Z}}_0 \mid S^{\perp} \right) \right\}^{-1} \left\{ \hat{\mathbf{Z}}_0 - \mu_z - BS(\phi) \right\},$$

with $B = C\{\hat{\mathbf{Z}}_0, S(\phi)\} [V\{S(\phi)\}]^{-1}$, and

$$Q_2(\phi) = S(\phi)' \{V(S)\}^{-1} S(\phi).$$

The decomposition in (7.54) can be best understood by considering the kernel function of the Gaussian density model. That is, we note that (7.53) is the core part of the log-density of the multivariate normal distribution. The decomposition in (7.54) naturally follows by the property of normal distribution.

If $\hat{\phi}$ satisfies $S(\hat{\phi}) = 0$, then $Q_2(\hat{\phi}) = 0$ and $Q_1(\mu_z \mid \hat{\phi}) = Q(\mu_z)$ in (7.52). Thus, the effect of using an estimated propensity score can be easily taken into account by simply adding the score function for ϕ into the Q term. The optimization method, based on the augmented Q^* in (7.53), is especially useful for handling longitudinal missing data, where the missing pattern is more complicated. See Zhou and Kim (2012).

Example 7.5. *Under the response model where the score function for ϕ is*

$$S(\phi) = \frac{1}{n} \sum_{i=1}^{n} \left\{ \frac{\delta_i}{\pi_i(\phi)} - 1 \right\} \mathbf{h}_i(\phi),$$

the optimal coefficient in (7.51) can be written as

$$B_1^* = \left\{ V_{xx} - V_{xs} V_{ss}^{-1} V_{sx} \right\}^{-1} \left\{ V_{xy} - V_{xs} V_{ss}^{-1} V_{sy} \right\},$$

where

$$\begin{pmatrix} V_{xx} & V_{xy} & V_{xs} \\ V_{yx} & V_{yy} & V_{ys} \\ V_{sx} & V_{sy} & V_{ss} \end{pmatrix} = \begin{pmatrix} V(\hat{X}_d) & C(\hat{X}_d, \hat{Y}_W) & C(\hat{X}_d, S) \\ C(\hat{Y}_W, \hat{X}_d) & V(\hat{Y}_W) & C(\hat{Y}_W, S) \\ C(S, \hat{X}_d) & C(S, \hat{Y}_W) & V(S) \end{pmatrix},$$

and $\hat{X}_d = \hat{X}_W - \hat{X}_n$. Thus, a consistent estimator of B_1^* is

$$\hat{B}_1^* = (I_p, O_q) \left\{ \sum_{i=1}^{n} \delta_i b_i \begin{pmatrix} \mathbf{x}_i \\ \hat{\mathbf{h}}_i \end{pmatrix} \begin{pmatrix} \mathbf{x}_i \\ \hat{\mathbf{h}}_i \end{pmatrix}' \right\}^{-1} \sum_{i=1}^{n} \delta_i b_i \begin{pmatrix} \mathbf{x}_i \\ \hat{\mathbf{h}}_i \end{pmatrix} y_i, \tag{7.55}$$

where $b_i = \hat{\pi}_i^{-2}(1 - \hat{\pi}_i)$, $\hat{\mathbf{h}}_i = \mathbf{h}_i(\hat{\phi})$, p is the dimension of $\mathbf{x}_i$, and q is the dimension of ϕ.

The optimal PS estimator of $\theta = E(Y)$ is then given by

$$\hat{\theta}_{opt} = \hat{Y}_{PS} + \left(\hat{X}_n - \hat{X}_{PS} \right)' \hat{B}_1^*, \tag{7.56}$$

which was originally proposed in Cao et al. (2009). For variance estimation, we can use a linearization method. Since $S(\hat{\phi}) = 0$, we can write (7.56) as

$$\hat{\theta}_{opt} = \hat{Y}_{PS} + \left(\hat{X}_n - \hat{X}_{PS} \right)' \hat{B}_1^* + S(\hat{\phi})\hat{B}_2^*, \tag{7.57}$$

where

$$\begin{pmatrix} \hat{B}_1^* \\ \hat{B}_2^* \end{pmatrix} = \left\{ \sum_{i=1}^n \delta_i b_i \begin{pmatrix} \mathbf{x}_i \\ \hat{\mathbf{h}}_i \end{pmatrix} \begin{pmatrix} \mathbf{x}_i \\ \hat{\mathbf{h}}_i \end{pmatrix}' \right\}^{-1} \sum_{i=1}^n \delta_i b_i \begin{pmatrix} \mathbf{x}_i \\ \hat{\mathbf{h}}_i \end{pmatrix} y_i.$$

The estimated influence function for variance estimation can take the form

$$d_i = \mathbf{x}_i' \hat{B}_1^* + \hat{\mathbf{h}}_i' \hat{B}_2^* + \frac{\delta_i}{\hat{\pi}_i} \left\{ y_i - \mathbf{x}_i' \hat{B}_1^* - \hat{\mathbf{h}}_i' \hat{B}_2^* \right\},$$

and the final variance estimator of $\hat{\theta}_{opt}$ is

$$\frac{1}{n(n-1)} \sum_{i=1}^n \left(d_i - \bar{d}_n \right)^2.$$

7.5 Maximum Entropy Method

In this section, we introduce a different approach to PS estimation. We now treat δ as fixed and consider the conditional distribution of $(\mathbf{x}, y)$ given δ. Note that we can express

$$\begin{aligned}
E(Y \mid \delta = 0) &= E\{E(Y \mid \mathbf{x}, \delta = 0) \mid \delta = 0\} \\
&= \int E(Y \mid \mathbf{x}, \delta = 0) f(\mathbf{x} \mid \delta = 0) d\mu(\mathbf{x}) \\
&= \int E(Y \mid \mathbf{x}, \delta = 0) r(\mathbf{x}) f(\mathbf{x} \mid \delta = 1) d\mu(\mathbf{x}) \\
&= E\{E(Y \mid \mathbf{x}, \delta = 0) r(\mathbf{x}) \mid \delta = 1\},
\end{aligned}$$

where

$$r(\mathbf{x}) = \frac{f(\mathbf{x} \mid \delta = 0)}{f(\mathbf{x} \mid \delta = 1)} := \frac{f_0(\mathbf{x})}{f_1(\mathbf{x})}$$

is the density ratio function. Thus, writing $m_0(\mathbf{x}) = E(Y \mid \mathbf{x}, \delta = 0)$, we obtain

$$\frac{1}{n_1} \sum_{i=1}^n \delta_i m_0(\mathbf{x}_i) r(\mathbf{x}_i) = \frac{1}{n_0} \sum_{i=1}^n (1 - \delta_i) m_0(\mathbf{x}_i), \tag{7.58}$$

where $n_1 = \sum_{i=1}^{n} \delta_i$ and $n_0 = n - n_1$, as an integral equation for $r(\mathbf{x})$. If we further assume that $m_0(\mathbf{x}) = \mathbf{x}'\beta$ for some β, (7.58) reduces to

$$\frac{1}{n_1} \sum_{i=1}^{n} \delta_i r(\mathbf{x}_i)\mathbf{x}_i = \frac{1}{n_0} \sum_{i=1}^{n} (1 - \delta_i)\mathbf{x}_i. \tag{7.59}$$

Also, by $\int r f_1 d\mu = 1$, we impose

$$\frac{1}{n_1} \sum_{i=1}^{n} \delta_i r(\mathbf{x}_i) = 1 \tag{7.60}$$

as an additional constraint. Combining (7.59) and (7.60), we can express

$$\frac{1}{n_1} \sum_{i=1}^{n} \delta_i r(\mathbf{x}_i)(1, \mathbf{x}'_i) = \frac{1}{n_0} \sum_{i=1}^{n} (1 - \delta_i)(1, \mathbf{x}'_i) \tag{7.61}$$

as a finite-sample constraint for finding the density ratio function.

To estimate density ratio function, we consider the Kullback-Leibler divergence between f_0 and f_1, which is defined by

$$D_{\mathrm{KL}}(f_0, f_1) = \int \log\left(\frac{f_0}{f_1}\right) f_0 d\mu.$$

Note that we can express

$$
\begin{aligned}
D_{KL}(f_0, f_1) &= \int \log\left(\frac{f_0}{f_1}\right) f_0 d\mu - \int \frac{f_0}{f_1} f_1 d\mu + 1 \\
&= \int \log\left(\frac{f_0}{f_1}\right) \frac{f_0}{f_1} f_1 d\mu - \int \frac{f_0}{f_1} f_1 d\mu + 1 \\
&= \sup_{r>0}\left\{\int \{\log(r)\} r f_1 d\mu - \int r f_1 d\mu + 1\right\},
\end{aligned}
$$

so that the density ratio function can be understood as the maximizer of

$$Q(r) = \int r\{\log(r) - 1\} f_1 d\mu. \tag{7.62}$$

The sample version is

$$\hat{Q}(r) = \frac{1}{n_1} \sum_{i=1}^{n} \delta_i r(\mathbf{x}_i) \left[\log\{r(\mathbf{x}_i)\} - 1\right]. \tag{7.63}$$

Thus, we maximize $\hat{Q}(r)$ in (7.63) subject to the linear constraint (7.61). Using Lagrange multipler method, the primal optimization problem can be solved by finding the dual problem using

$$\log\{r(\mathbf{x}; \phi)\} = \phi_0 + \mathbf{x}'\phi_1 \tag{7.64}$$

for some parameter $\phi_0 \in \mathbb{R}$ and $\phi_1 \in \mathbb{R}^d$.

To see this, define

$$Q_\lambda(r) = \sum_{i=1}^{n} \delta_i r_i \{\log(r_i) - 1\} + \left\{ \sum_{i=1}^{n} \delta_i r_i(1, \mathbf{x}_i') - \frac{n_1}{n_0} \sum_{i=1}^{n} (1 - \delta_i)(1, \mathbf{x}_i') \right\} \lambda$$

where λ is a $d+1$ dimensional vector of Lagrange multipliers. Now,

$$\frac{\partial}{\partial r_i} Q_\lambda(r) = \delta_i \log(r_i) + \delta_i(1, \mathbf{x}_i')\lambda = 0$$

leads to

$$\log(r_i) = -(1, \mathbf{x}_i')\lambda,$$

which justifies (7.64).

Now, since

$$Q(r) = \int r\{\log(r) - 1\}f_1 d\mu = \int \log(r) f_0 d\mu - \int r f_1 d\mu,$$

the objective function reduces to

$$\hat{Q}(\phi) = \frac{1}{n_0} \sum_{i=1}^{n} (1 - \delta_i) \log \{r(\mathbf{x}_i; \phi)\} - \frac{1}{n_1} \sum_{i=1}^{n} \delta_i r(\mathbf{x}_i; \phi).$$

The maximizer of $\hat{Q}(\phi)$ is called the maximum entropy estimator of ϕ. The log-linear form in (7.64) is a parametric representation of the infinite dimensional density ratio function and it is derived from the outcome model $E(Y \mid \mathbf{x})$.

The maximizer of $\hat{Q}(\phi)$ can be obtained by solving

$$U(\phi) \equiv \frac{1}{n_1} \sum_{i=1}^{n} \delta_i \exp(\phi_0 + \mathbf{x}_i'\phi_1)(1, \mathbf{x}_i') - \frac{1}{n_0} \sum_{i=1}^{n} (1 - \delta_i)(1, \mathbf{x}_i') = \mathbf{0}, \quad (7.65)$$

as the estimating equation for ϕ. To obtain the solution, we first express the second part of estimating equation (7.65) as

$$\sum_{i=1}^{n} \delta_i w_i(\phi_1) (\mathbf{x}_i - \bar{\mathbf{x}}_0) = \mathbf{0}, \quad (7.66)$$

where

$$w_i(\phi_1) = \frac{\exp(\mathbf{x}_i'\phi_1)}{\sum_{i=1}^{n} \delta_i \exp(\mathbf{x}_i'\phi_1)}$$

and $\bar{\mathbf{x}}_0 = n_0^{-1} \sum_{i=1}^{n} (1 - \delta_i)\mathbf{x}_i$. Since the equality in (7.66) can be written as

$$\bar{\mathbf{x}}_1(\phi_1) = \bar{\mathbf{x}}_0,$$

where $\bar{\mathbf{x}}_1(\phi_1) = \sum_{i=1}^n \delta_i w_i(\phi_1)\mathbf{x}_i$, Newton's method for solving equation (7.66) is

$$\hat{\phi}_1^{(t+1)} = \hat{\phi}_1^{(t)} + \left\{\hat{\Sigma}_{xx}(\hat{\phi}_1^{(t)})\right\}^{-1} \left\{\bar{\mathbf{x}}_0 - \bar{\mathbf{x}}_1(\hat{\phi}_1^{(t)})\right\}, \tag{7.67}$$

where

$$\hat{\Sigma}_{xx}(\phi_1) = \sum_{i=1}^n \delta_i w_i(\phi_1)\{\mathbf{x}_i - \bar{\mathbf{x}}_1(\phi_1)\}^{\otimes 2}.$$

Once $\hat{\phi}_1$ is obtained, we can estimate ϕ_0 by solving the first part of (7.65).

The PS estimator using maximum entropy method satisfies the covariate-balancing property in the sense that

$$\frac{1}{n}\sum_{i=1}^n \frac{\delta_i}{\hat{\pi}_i}\left(1,\mathbf{x}_i'\right) = \frac{1}{n}\sum_{i=1}^n \left(1,\mathbf{x}_i'\right), \tag{7.68}$$

where $\hat{\pi}_i^{-1} = 1 + (n_0/n_1)r(\mathbf{x}_i;\hat{\phi})$. To see this, note that

$$\left\{\frac{1}{n_1}\sum_{i=1}^n \delta_i r(\mathbf{x}_i;\phi) - \frac{1}{n_0}\sum_{i=1}^n(1-\delta_i)\right\}\mathbf{x}_i = \frac{1}{n_0}\sum_{i=1}^n (\delta_i - 1)\mathbf{x}_i + \frac{1}{n_1}\sum_{i=1}^n \delta_i r(\mathbf{x}_i;\phi)\mathbf{x}_i$$

$$= \frac{1}{n_0}\sum_{i=1}^n \left\{\frac{\delta_i}{\pi(\mathbf{x}_i;\phi)} - 1\right\}\mathbf{x}_i,$$

where the second equality follows from the definition of $r(\mathbf{x}_i;\phi)$.

To discuss some asymptotic properties of the PS estimator using maximum entropy method, write

$$\hat{\theta}_{PS}(\hat{\phi}) = \frac{1}{n}\sum_{i=1}^n \delta_i \left\{1 + \frac{n_0}{n_1}r(\mathbf{x}_i;\hat{\phi})\right\}y_i.$$

Using Taylor expansion, we can obtain

$$\hat{\theta}_{PS}(\hat{\phi}) \cong \hat{\theta}_{PS}(\phi^*) + E\left\{\frac{\partial}{\partial\phi'}\hat{\theta}_{PS}(\phi^*)\right\}\left(\hat{\phi} - \phi^*\right) \tag{7.69}$$

and

$$\hat{\phi} - \phi^* \cong -\left[E\left\{\frac{\partial}{\partial\phi'}\hat{U}(\phi^*)\right\}\right]^{-1}\hat{U}(\phi^*),$$

where

$$\hat{U}(\phi) = \frac{1}{n}\sum_{i=1}^n \left(\frac{\delta_i}{\pi(\mathbf{x}_i;\phi)} - 1\right)(1,\mathbf{x}_i')' \tag{7.70}$$

and ϕ^* is the true parameter in the density ratio model (7.64).

The estimating equation in (7.70) is justified by the equality in (7.68), which is derived from (7.59), the empirical equation for density ratio function. Note that

$$E\{\hat{U}(\phi^*)\mid\delta\} = 0$$

under the assumption on the density ratio model.

If we define $\tilde{\mathbf{x}}_i = (1, \mathbf{x}_i')'$, after some algebra, we can express (7.69) as

$$
\begin{aligned}
\hat{\theta}_{PS}(\hat{\phi}) &\cong \frac{1}{n} \sum_{i=1}^{n} \left\{ \tilde{\mathbf{x}}_i' \beta^* + \frac{\delta_i}{\pi_i} \left(y_i - \tilde{\mathbf{x}}_i' \beta^* \right) \right\} \\
&:= \frac{1}{n} \sum_{i=1}^{n} d(\mathbf{x}_i, y_i, \delta_i; \phi),
\end{aligned}
\tag{7.71}
$$

where β^* satisfies $E(Y \mid \mathbf{x}) = \tilde{\mathbf{x}}' \beta^*$ and $\pi_i^{-1} = 1 + (n_0/n_1) r(\mathbf{x}_i; \phi^*)$. The linearization form in (7.71) is derived under the model (7.64) for the density ratio function. This linearization form clearly suggests that the PS estimator is approximately unbiased under the linear regression model $E(Y \mid \mathbf{x}) = \tilde{\mathbf{x}}' \beta^*$ and MAR. The PS model approach gives an extra protection against the failure of the assumed regression model. See Section 7.6 for details.

Thus, a consistent variance estimator is

$$
\hat{V}(\hat{\theta}_{PS}) \cong \hat{V}(\bar{d}_n) = \frac{1}{n} S_{\hat{d}}^2,
$$

where

$$
S_{\hat{d}}^2 = \frac{1}{n-1} \sum_{i=1}^{n} \left(\hat{d}_i - \bar{\hat{d}}_n \right)^2,
$$

$\hat{d}_i = \tilde{\mathbf{x}}_i' \hat{\beta} + \delta_i \hat{\pi}_i^{-1} (y_i - \tilde{\mathbf{x}}_i' \hat{\beta})$, and $\hat{\pi}_i^{-1} = 1 + (n_0/n_1) r(\mathbf{x}_i; \hat{\phi})$.

7.6 Doubly Robust Estimation

In this section, we consider some means of protection against the failure of the assumed model. The optimal PS estimator in (7.56) is asymptotically unbiased and optimal under the assumed response model. If the response model does not hold, then the validity of the optimal PS estimator is no longer guaranteed. An estimator is called doubly robust (DR) if it remains consistent if either model (outcome regression model or response model) is correctly specified. DR procedure offers some protection against misspecification of one model or the other. In this sense, it can be called doubly protected procedure, as termed by Kim and Park (2006).

To discuss DR estimators, consider the following outcome regression (OR) model

$$
E(y_i \mid \mathbf{x}_i) = m(\mathbf{x}_i; \beta_0)
$$

for some $m(\mathbf{x}_i; \beta_0)$ known up to β_0 and assume the MAR. For the response probability (RP) model, we can assume (7.32). Under these models, we can consider the following class of doubly robust estimators, with the subscript "DR" denoting "doubly robust":

$$\hat{\theta}_{DR} = \frac{1}{n} \sum_{i=1}^{n} \left\{ \hat{y}_i + \frac{\delta_i}{\hat{\pi}_i} (y_i - \hat{y}_i) \right\}, \tag{7.72}$$

where $\hat{y}_i = m(\mathbf{x}_i; \hat{\beta})$ for some function $m(\mathbf{x}_i; \beta_0)$ known up to β_0 and $\hat{\beta}$ is an estimator of β_0. The predicted value $\hat{y}_i$ is derived under the OR model while $\hat{\pi}_i$ is obtained from the RP model. Writing $\hat{\theta}_n = n^{-1} \sum_{i=1}^{n} y_i$, we have

$$\hat{\theta}_{DR} - \hat{\theta}_n = n^{-1} \sum_{i=1}^{n} \left(\frac{\delta_i}{\hat{\pi}_i} - 1 \right) (y_i - \hat{y}_i). \tag{7.73}$$

Taking expectation of the above, we note that the first term has approximate zero expectation if the response probability model is true. The second term has approximate zero expectation if the OR model is true. DR estimation has been considered by Robins et al. (1994), Bang and Robins (2005), Tan (2006), Kang and Schafer (2007), Cao et al. (2009), and Kim and Haziza (2014). In particular, the optimal PS estimator in (7.56) is doubly robust with $E(y_i \mid \mathbf{x}_i) = \mathbf{x}_i' B_0$ and $E(\delta_i \mid \mathbf{x}_i) = \pi_i(\phi_0)$ as long as $\hat{B}^*$ in (7.55) is consistent for B_0 under the OR model or $\hat{\phi}$ is consistent for ϕ_0 under the response probability model. Because the optimal PS estimator in (7.56) is obtained by minimizing the variance of the PS estimator under the response probability model, it is optimal when the assumed response probability model holds.

We now consider optimal estimation in the context of doubly robust estimation with a general form of the conditional expectation $m(\mathbf{x}_i; \beta_0)$ in the OR model. The optimality criteria for doubly robust estimation is somewhat unclear since there are two models involved. We consider the approach used in Cao et al. (2009) with the objective of minimizing the variance of DR estimator under the response probability model while maintaining the consistency of the point estimator under the OR model. In the class of DR estimators of the form

$$\hat{\theta}_{DR}(\hat{\beta}, \hat{\phi}) = \frac{1}{n} \sum_{i=1}^{n} \left[m(\mathbf{x}_i; \hat{\beta}) + \frac{\delta_i}{\pi_i(\hat{\phi})} \left\{ y_i - m(\mathbf{x}_i; \hat{\beta}) \right\} \right], \tag{7.74}$$

where $\hat{\phi}$ is the MLE, and $\hat{\beta}$ is to be determined. Rubin and van der Laan (2008) considered obtaining $\hat{\beta}$ that minimizes

$$\sum_{i=1}^{n} \frac{\delta_i}{\pi_i(\hat{\phi})} \left\{ \frac{1}{\pi_i(\hat{\phi})} - 1 \right\} \{y_i - m(\mathbf{x}_i; \beta)\}^2,$$

which can be justified when the response probability π_i is known, rather than estimated. To correctly account for the effect of estimating ϕ_0, one can use the linearization technique in Section 4.3 to obtain the optimal estimator. To be specific, we can account for the effect of π_i being estimated by writing (7.74) as

$$\hat{\theta}_{DR}(\hat{\beta},\hat{\phi},\mathbf{k}) = n^{-1} \sum_{i=1}^{n} \left[\mathbf{k}'\mathbf{h}_i(\hat{\phi}) + m(\mathbf{x}_i;\hat{\beta}) + \frac{\delta_i}{\pi_i(\hat{\phi})} \left\{ y_i - m(\mathbf{x}_i;\hat{\beta}) - \mathbf{k}'\mathbf{h}_i(\hat{\phi}) \right\} \right],$$
(7.75)

where $\mathbf{h}_i = (\partial \pi_i / \partial \phi)/(1 - \pi_i)$, and $(\hat{\beta}, \mathbf{k})$ is to be determined. Under the response probability model, the variance of the DR estimator is minimized by finding $(\hat{\beta}, \mathbf{k})$ that minimizes

$$\sum_{i=1}^{n} \frac{\delta_i}{\pi_i(\hat{\phi})} \left\{ \frac{1}{\pi_i(\hat{\phi})} - 1 \right\} \left\{ y_i - m(\mathbf{x}_i;\beta) - \mathbf{k}'\mathbf{h}_i(\hat{\phi}) \right\}^2.$$
(7.76)

Note that in this case there is no guarantee that the resulting estimator is efficient under the OR model. Also, the computation for minimizing (7.76) can be cumbersome.

Tan (2006) slightly changed the class of estimators to be

$$\hat{\theta}_{DR}(\mathbf{k}) = n^{-1} \sum_{i=1}^{n} \left\{ \mathbf{k}_1'\hat{\mathbf{m}}_i + \frac{\delta_i}{\pi_i(\hat{\phi})} (y_i - \mathbf{k}_1'\hat{\mathbf{m}}_i) \right\},$$
(7.77)

where $\mathbf{k}_1 = (k_0, k_1)'$, $\hat{\mathbf{m}}_i = (1, \hat{m}_i)$, $\hat{m}_i = m(\mathbf{x}_i; \hat{\beta})$, and $\hat{\beta}$ is optimally computed in advance under the OR model. Therefore, if we have some extra information about $V(y_i \mid \mathbf{x}_i)$, then we can incorporate it to obtain $\hat{\beta}$ in (7.77) while the DR estimator of Cao et al. (2009) does not use the information. The optimal estimator among the class in (7.77) can be obtained by minimizing

$$\sum_{i=1}^{n} \frac{\delta_i}{\pi_i(\hat{\phi})} \left\{ \frac{1}{\pi_i(\hat{\phi})} - 1 \right\} \left\{ y_i - \mathbf{k}_1'\hat{\mathbf{m}}_i - \mathbf{k}_2'\mathbf{h}_i(\hat{\phi}) \right\}^2.$$
(7.78)

The solution can be written as

$$\begin{pmatrix} \hat{\mathbf{k}}_1^* \\ \hat{\mathbf{k}}_2^* \end{pmatrix} = \left\{ \sum_{i=1}^{n} \delta_i b_i \begin{pmatrix} \hat{\mathbf{m}}_i \\ \hat{\mathbf{h}}_i \end{pmatrix} \begin{pmatrix} \hat{\mathbf{m}}_i \\ \hat{\mathbf{h}}_i \end{pmatrix}' \right\}^{-1} \sum_{i=1}^{n} \delta_i b_i \begin{pmatrix} \hat{\mathbf{m}}_i \\ \hat{\mathbf{h}}_i \end{pmatrix} y_i,$$
(7.79)

where $b_i = \hat{\pi}_i^{-1}(\hat{\pi}_i^{-1} - 1)$ and $\hat{\mathbf{h}}_i = \mathbf{h}_i(\hat{\phi})$. Note that the expected value of $\hat{\mathbf{k}}_1^*$ is approximately equal to $(0, 1)'$ under the OR model. Thus, under the OR model, the optimal DR estimator of Tan (2006) is asymptotically equivalent to (7.74). Furthermore, if $m(x_i; \beta) = \beta_0 + \beta_1 x_i$, then the optimal estimator is equal to the optimal PS estimator in (7.56). In fact, if both the OR model and the response probability model are correct, then the choice of $\hat{\beta}$ is not critical. This phenomenon, called the local efficiency of the DR estimator, was first discussed by Robins et al. (1994).

Remark 7.1. *We can construct a fractional imputation method that is doubly robust. In fractional imputation, several imputed values are used for each missing value, and fractional weights are assigned to the imputed values. Let* $y_{ij}^* = m(\mathbf{x}_j; \hat{\beta}) + \hat{e}_i$ *be the imputed value for unit j using donor i, where* $\hat{e}_i = y_i - m(\mathbf{x}_i; \hat{\beta})$. *The FI estimator can be written as*

$$\hat{\theta}_{FI} = n^{-1} \sum_{j=1}^{n} \left\{ \delta_j y_j + (1 - \delta_j) \sum_{i=1}^{n} w_{ij}^* \delta_i y_{ij}^* \right\}, \tag{7.80}$$

where w_{ij}^* are the fractional weights attached to unit j such that $\sum_{i=1}^{n} w_{ij}^* \delta_i = 1$. Note that (7.80) can be alternatively written as

$$\hat{\theta}_{FI} = n^{-1} \sum_{i=1}^{n} m(\mathbf{x}_i; \hat{\beta}) + n^{-1} \sum_{i=1}^{n} \delta_i \left\{ 1 + \sum_{j=1}^{n} (1 - \delta_j) w_{ij}^* \right\} \hat{e}_i. \tag{7.81}$$

Comparing (7.81) with (7.72), it follows that the FI estimator (7.80) is doubly robust if

$$1 + \sum_{j=1}^{n} (1 - \delta_j) w_{ij}^* = \frac{1}{\pi_i(\hat{\phi})}. \tag{7.82}$$

If the estimated propensity score satisfies

$$n^{-1} \sum_{i=1}^{n} \frac{\delta_i}{\pi_i(\hat{\phi})} = 1,$$

then the choice

$$w_{ij}^* = \frac{1/\pi_i(\hat{\phi}) - 1}{\sum_{k=1}^{n} \delta_k \left\{ 1/\pi_k(\hat{\phi}) - 1 \right\}} \tag{7.83}$$

satisfies (7.82), and the FI estimator is doubly robust.

Kim and Haziza (2014) considered a different method of parameter estimation for DR estimator. Note that (7.73) is a function of ϕ and β. Thus, one can consider minimizing $\{\hat{\theta}_{DR}(\beta, \phi) - \hat{\theta}_n)\}^2$ with respect to (β, ϕ) to obtain the estimating equation for (β, ϕ). Writing $\hat{\pi}_i = \pi(\mathbf{x}_i'\hat{\phi})$ and $\hat{y}_i = m(\mathbf{x}_i'\hat{\beta})$, the solution to this optimization problem is obtained by solving

$$\sum_{i=1}^{n} \left(\frac{\delta_i}{\pi(\mathbf{x}_i'\phi)} - 1 \right) \dot{m}(\mathbf{x}_i'\beta) \mathbf{x}_i = \mathbf{0} \tag{7.84}$$

and

$$\sum_{i=1}^{n} \frac{\delta_i}{\{\pi(\mathbf{x}_i'\phi)\}^2} \{y_i - m(\mathbf{x}_i'\beta)\} \dot{\pi}(\mathbf{x}_i'\phi) \mathbf{x}_i = \mathbf{0} \tag{7.85}$$

simultaneously, where $\dot{m}(x) = dm(x)/dx$ and $\dot{\pi}(x) = d\pi(x)/dx$. If $m(\mathbf{x}'\beta) = \mathbf{x}'\beta$ and $\text{logit}\{\pi(\mathbf{x}'\phi)\} = \mathbf{x}'\phi$, then (7.84) and (7.85) are changed to

$$\sum_{i=1}^{n} \left(\frac{\delta_i}{\pi(\mathbf{x}_i'\phi)} - 1 \right) \mathbf{x}_i = 0 \tag{7.86}$$

and

$$\sum_{i=1}^{n} \delta_i \left\{ \frac{1}{\pi(\mathbf{x}_i'\phi)} - 1 \right\} \mathbf{x}_i (y_i - \mathbf{x}_i'\beta) = \mathbf{0}, \tag{7.87}$$

respectively. One advantage of using the method of Kim and Haziza (2014) is that the effect of estimating the model parameters can be safely ignored, as $\hat{\theta}_{DR}(\hat{\beta}, \hat{\phi})$ satisfies $E\{\partial \hat{\theta}_{DR}/\partial \beta\} = 0$ and $E\{\partial \hat{\theta}_{DR}/\partial \phi\} = 0$ by construction. Thus, linearization variance estimation can be easily implemented by using the pseudo value of $d_i = \hat{y}_i + (\delta_i/\hat{\pi}_i)(y_i - \hat{y}_i)$. Yang et al. (2020) extended the idea to cover model selection with high dimensional covariates.

Doubly robust estimation is now extended to cover multiple OR and multiple RP models. See Han and Wang (2013) and Chen and Haziza (2017) for more details.

7.7 Empirical Likelihood Method

The empirical likelihood (EL) method, proposed by Owen (1988), has become a very powerful tool for nonparametric inference in statistics. It uses a likelihood-based approach without having to make a parametric distributional assumption about the data observed, often resulting in efficient estimation.

To discuss the empirical likelihood method under missing data, consider a multivariate random variable (X, Y) with distribution function $F(x, y)$ which is completely unspecified except that $E\{U(\theta_0; X, Y)\} = 0$ for some $\theta_0 \in \Omega \subset \mathbb{R}^p$, where $U(\theta_0; x, y)$ is m-dimensional vector that is linearly independent. If (x_i, y_i), $i = 1, 2, \ldots, n$, are n independent realizations of the random variable (X, Y), a consistent estimator of θ_0 can be obtained by solving

$$\sum_{i=1}^{n} U(\theta; x_i, y_i) = 0 \tag{7.88}$$

if $m = p$. If $m > p$, then the model is called over-identified, and the solution to (7.88) may not exist. The optimal estimation discussed in Section 7.4 covers an example of optimal estimation under over-identified models. The optimal estimation in Section 7.4 is an example of the generalized method of moments (GMM) technique which is very popular in econometrics. The GMM estimator is a class of estimator of the form

$$\hat{\theta}_W = \arg\min_{\theta \in \Omega} \frac{1}{n} \sum_{i=1}^{n} \{U(\theta; x_i, y_i)\}' W U(\theta; x_i, y_i), \qquad (7.89)$$

for some $m \times m$ symmetric matrix W. Note that the solution $\hat{\theta}_W$ is obtained by solving

$$U_W(\theta) \equiv n^{-1} \sum_{i=1}^{n} \{\dot{U}(\theta; x_i, y_i)\}' W \{U(\theta; x_i, y_i)\} = 0, \qquad (7.90)$$

where $\dot{U}(\theta; x, y) = \partial U(\theta; x, y)/\partial \theta'$. From (7.90), we can show that $\hat{\theta}_W$ is asymptotically unbiased for θ_0 with the asymptotic variance

$$V\left(\hat{\theta}_W\right) \cong \left\{E\left(\frac{\partial}{\partial \theta'} U_W(\theta)\right)\right\}^{-1} V\{U_W(\theta)\} \left\{E\left(\frac{\partial}{\partial \theta'} U_W(\theta)\right)'\right\}^{-1}$$

$$= n^{-1} \{\tau'W\tau\}^{-1} \tau'WV(U)W\tau \{\tau'W\tau\}^{-1},$$

where $\tau = E(\partial U/\partial \theta)$. The asymptotic variance is minimized at

$$W^* = \{V(U)\}^{-1}.$$

Thus, the optimal GMM estimator is defined as

$$\hat{\theta}^* = \arg\min \frac{1}{n} \sum_{i=1}^{n} \{U(\theta; x_i, y_i)\}' [V\{U(\theta)\}]^{-1} U(\theta; x_i, y_i).$$

The asymptotic variance of the optimal GMM estimator is

$$V(\hat{\theta}^*) = n^{-1} \left[\tau' \{V(U)\}^{-1} \tau\right]^{-1}. \qquad (7.91)$$

Instead of using GMM, empirical likelihood method can be also used to handle over-identified estimating equation problem. The EL approach is to find θ that maximizes the profile empirical likelihood function of θ:

$$L(\theta) = \max \left\{\prod_{i=1}^{n} p_i : p_i > 0, \sum_{i=1}^{n} p_i = 1, \text{ and } \sum_{i=1}^{n} p_i U(\theta; x_i, y_i) = 0\right\}. \qquad (7.92)$$

Using the Lagrange multiplier method, the empirical likelihood estimator can be obtained by maximizing

$$l_e(\theta) = \sum_{i=1}^{n} \log\{\hat{p}_i(\theta)\},$$

where $\hat{p}_i(\theta)$ is of the form

$$\hat{p}_i(\theta) = \frac{1}{n} \frac{1}{1 + \hat{\lambda}_\theta' U(\theta; x_i, y_i)},$$

and $\hat{\lambda}_\theta$ satisfies

$$\sum_{i=1}^{n} \hat{p}_i(\theta) U(\theta; x_i, y_i) = 0.$$

The following theorem, proved by Qin and Lawless (1994), presents an important asymptotic property of the maximum empirical likelihood estimator.

Theorem 7.4. *Under some regularity conditions, the maximizer $\hat{\theta}_{mel}$ of $L(\theta)$ in (7.92) satisfies*

$$\sqrt{n}\left(\hat{\theta}_{mel} - \theta_0\right) \overset{\mathcal{L}}{\longrightarrow} N(0, V),$$

where $V = [\tau'\{V(U)\}^{-1}\tau]^{-1}$ and $\tau = E(\partial U/\partial \theta)$.

Note that the asymptotic variance of $\hat{\theta}_{mel}$ is equal to the asymptotic variance (7.91) of the optimal GMM estimator. Therefore, the EL approach is an alternative way of implementing the optimal GMM estimation without specifying the variance-covariance matrices.

We now apply the EL method to the missing data setup. Assume that x_i is always observed, and y_i is subject to missingness. Let $\delta_i = 1$ if y_i is observed and $\delta_i = 0$ otherwise. Let $\pi(x; \phi) = P(\delta = 1 \mid x)$ be the response probability model. Let $S(\phi; \delta, x)$ be the score function of ϕ such that $E\{S(\phi; \delta, x)\} = 0$ holds. Thus, we have two moment conditions in applying the EL-based optimal PS estimation. One is

$$E\left\{\frac{\delta}{\pi(X; \phi)} U(\theta; X, Y)\right\} = 0,$$

and the other is the score equation for ϕ. Therefore, the maximum EL method is based on maximizing

$$L(\theta, \phi) = \max\left\{\prod_{i=1}^{n} p_i; \mathbf{p} \in B(\theta, \phi)\right\}, \tag{7.93}$$

where

$$B(\theta, \phi) = \left\{\mathbf{p}: p_i > 0, \sum_{i=1}^{n} p_i = 1, \sum_{i=1}^{n} p_i \frac{\delta_i}{\pi(x_i; \phi)} U(\theta; x_i, y_i) = 0, \sum_{i=1}^{n} p_i S(\phi; \delta_i, x_i) = 0\right\}.$$

Because of the equivalence between the optimal GMM estimator and the maximum EL estimator, we can establish that the maximum EL estimator obtained from (7.93) is asymptotically equivalent to the optimal PS estimator discussed in Section 7.4. See Qin et al. (2009) and Zhao et al. (2020) for more details.

7.8   Nonparametric Method

Instead of using parametric models for propensity scores, nonparametric approaches can also be used. For simplicity, we assume a bivariate data structure (x_i, y_i) with x_i fully observed and assume that the response mechanism is missing at random. Unlike the parametric model approach in Section 7.3, we assume that $\pi(x) = P(\delta = 1 \mid x)$ is completely unspecified, except that it is a smooth function of x with bounded partial derivatives of order 2.

Using the argument discussed in Section 6.2, a nonparametric regression estimator of $\pi(x) = E(\delta \mid x)$ can be obtained by

$$\hat{\pi}_h(x) = \frac{\sum_{i=1}^n \delta_i K_h(x_i, x)}{\sum_{i=1}^n K_h(x_i, x)}, \tag{7.94}$$

where K_h is the kernel function which satisfies certain regularity conditions, and h is the bandwidth. Once a nonparametric estimator of $\pi(x)$ is obtained, the nonparametric PS estimator $\hat{\theta}_{NPS}$ of $\theta_0 = E(Y)$ is given by

$$\hat{\theta}_{NPS} = \frac{1}{n} \sum_{i=1}^n \frac{\delta_i}{\hat{\pi}_h(x_i)} y_i. \tag{7.95}$$

The following theorem, proved by Chen et al. (2013), establishes the $\sqrt{n}$-consistency of the PS estimator of $\theta = E(Y)$.

Theorem 7.5. *Under some regularity conditions stated in Section 6.2, we have*

$$\hat{\theta}_{NPS} = \frac{1}{n} \sum_{i=1}^n \left[m(x_i) + \frac{\delta_i}{\pi(x_i)} \{y_i - m(x_i)\} \right] + o_p(n^{-1/2}), \tag{7.96}$$

where $m(x) = E(Y \mid x)$ and $\pi(X) = P(\delta = 1 \mid X)$. Furthermore, we have

$$\sqrt{n} \left(\hat{\theta}_{NPS} - \theta \right) \to N \left(0, \sigma_1^2 \right), \tag{7.97}$$

where $\sigma_1^2 = V\{m(X)\} + E\left[\{\pi(X)\}^{-1} V(Y \mid X)\right]$.

Note that the asymptotic variance of the nonparametric PS estimator is the same as that of the nonparametric fractional imputation estimator discussed in Theorem 6.2. The asymptotic variance is equal to the lower bound in (7.39).

For other nonparametric approaches, Hirano et al. (2003) used the sieve logistic regression for nonparametric PS estimation. The basic idea is to construct a sequence of basis functions, say $B_m(\mathbf{x}) = (b_{1m}(\mathbf{x}), b_{2m}(\mathbf{x}), \ldots, b_{mm}(\mathbf{x}))'$ of dimension m, and fit a logistic regression based on

$$\text{logit}\{\pi_m(\mathbf{x})\} = \sum_{j=1}^m \phi_j b_{mj}(\mathbf{x}).$$

Here, m is allowed to increase with n such that $m = o(n)$. Under some choice of m, the nonparametric PS estimator using $\hat{\pi}_m(\mathbf{x})$ achieves the efficiency lower bound in (7.97). Imai and Ratkovic (2014) noted that the sieve logistic regression method satisfies a covariate balancing property in that the corresponding score equation satisfies

$$\sum_{i=1}^{n} \left\{ \frac{\delta_i}{\pi_m(\mathbf{x}_i)} - 1 \right\} \mathbf{f}(\mathbf{x}_i) = \mathbf{0}, \tag{7.98}$$

where $\mathbf{f}(\mathbf{x}_i) = \pi_m(\mathbf{x}_i) B_m(\mathbf{x}_i)$. They further claim that any other choice of $\mathbf{f}(\mathbf{x})$ can make the resulting PS estimator robust. Graham et al. (2012) suggested including the basis functions of $E(Y \mid \mathbf{x})$ into the control variable $\mathbf{f}(\mathbf{x})$ in (7.98) to improve the efficiency of the resulting PS estimator.

Hainmueller (2012) proposed the so-called entropy balancing to find the propensity score weights using the Kullback-Leibler information criterion. The propensity weight is obtained by minimizing $\sum_{i=1}^{n} \delta_i \omega_i \log(\omega_i/q_i)$ such that

$$\sum_{i=1}^{n} \delta_i \omega_i \left(1, \mathbf{f}(\mathbf{x}_i)\right) = \frac{1}{n} \sum_{i=1}^{n} \left(1, \mathbf{f}(\mathbf{x}_i)\right),$$

where q_i are the base weights such as $q_i = 1/n_R$. Using the Lagrange multiplier method, the solution takes the form of

$$\omega_i^* = \frac{q_i \exp(\hat{\lambda}' \mathbf{f}(\mathbf{x}_i))}{\sum_{j=1}^{n} \delta_j q_i \exp(\hat{\lambda}' \mathbf{f}(\mathbf{x}_i))},$$

where $\hat{\lambda}$ is chosen to satisfy the calibration condition. Chan et al. (2016) generalize this idea further to develop a general calibration weighting method that satisfies the covariance balancing property with increasing dimensions of the control variables $\mathbf{f}(\mathbf{x})$. They further showed the global efficiency of the proposed calibration weighting PS estimator in the sense of achieving the efficiency lower bound in (7.97). Zhao (2019) generalized the idea further and developed a unified approach of covariate balancing PS method using tailored loss functions. The regularization techniques using penalty terms into the loss function can be naturally incorporated into the framework.

The covariate balancing condition, or calibration condition, in (7.98), can be relaxed. Zubizarreta (2015) relaxed the exact balancing constraints to some tolerance level. Wong and Chan (2018) used the theory of reproducing Kernel Hilbert space to develop an uniform approximate balance for covariate functions. Although the problem itself is an infinite dimensional optimization problem, they showed that it has a finite-dimensional representation and can be solved by eigenvalue optimization.

Exercises

1. Show (7.21).

2. Under the setup of Example 7.4, answer the following questions.

   (a) Write the optimal estimator $\hat{\theta}_{opt}(\hat{\beta})$ as $\hat{\theta}_{opt}(\hat{\beta}, \hat{\phi})$, where $\hat{\phi}$ is the estimator for computing $\hat{\pi}_i = \pi_i(\hat{\phi})$ in (7.43), and show that

$$E\left\{ \frac{\partial}{\partial \phi} \hat{\theta}_{opt}(\beta, \phi) \right\} = 0$$

   (b) Prove that, under the regression model in (7.42), $\hat{\theta}_{opt}(\hat{\beta}, \hat{\phi})$ is asymptotically equivalent to $\hat{\theta}_{opt}(\hat{\beta}, \phi_0)$ as long as $(\hat{\beta}, \hat{\phi})$ is consistent for (β, ϕ).

   (c) Prove that if the propensity scores are constructed to satisfy

$$\sum_{i=1}^{n} \frac{\delta_i}{\hat{\pi}_i} \mathbf{x}_i = \sum_{i=1}^{n} \mathbf{x}_i,$$

   then the PS estimator $\hat{\theta}_{PS}$ is optimal in the sense that it achieves the lower bound in (7.39).

3. Under the setup of Example 7.4 again, suppose that we are going to use

$$\hat{\theta}_p = \frac{1}{n} \sum_{i=1}^{n} \mathbf{x}_i' \hat{\beta}_c, \tag{7.99}$$

   where

$$\hat{\beta}_c = \left\{ \sum_{i=1}^{n} \delta_i c_i \mathbf{x}_i \mathbf{x}_i' \right\}^{-1} \sum_{i=1}^{n} \delta_i c_i \mathbf{x}_i y_i,$$

   and c_i is to be determined. Answer the following questions:

   (a) Show that $\hat{\theta}_p$ is asymptotically unbiased for $\theta = E(Y)$ regardless of the choice of c_i.

   (b) If $\mathbf{x}_i$ contains an intercept term, then the choice of $c_i = 1/\hat{\pi}_i$ makes the resulting estimator optimal in the sense that it achieves the lower bound in (7.39).

   (c) Instead of $\hat{\theta}_p$ in (7.99), suppose that an alternative estimator

$$\hat{\theta}_{p2} = \frac{1}{n} \sum_{i=1}^{n} \{ \delta_i y_i + (1 - \delta_i) \mathbf{x}_i' \hat{\beta}_c \}$$

   is used, where $\hat{\beta}_c$ is as defined in (7.99). Find a set of conditions for $\hat{\theta}_{p2}$ to be optimal.

4. Prove (7.54).

5. Suppose that the response probability is parametrically modeled by

$$\pi_i = \Phi(\phi_0 + \phi_1 x_i)$$

   for some (ϕ_0, ϕ_1), where $\Phi(\cdot)$ is the cumulative distribution function of the standard normal distribution. Assume that x_i is completely observed, and y_i is observed only when $\delta_i = 1$, where δ_i follows from Bernoulli distribution with parameter π_i.

   (a) Find the score equation for (ϕ_0, ϕ_1).
   (b) Discuss asymptotic variance of the PS estimator of $\theta = E(Y)$ using the MLE of (ϕ_0, ϕ_1).

6. Prove that minimizing (7.45) is algebraically equivalent to minimizing

$$Q_2 = \begin{pmatrix} \hat{X}_1 - \hat{X}_2 \\ \hat{Y} - \mu_y \end{pmatrix}' \begin{pmatrix} V(\hat{X}_1 - \hat{X}_2) & C(\hat{X}_1 - \hat{X}_2, \hat{Y}) \\ C(\hat{X}_1 - \hat{X}_2, \hat{Y}) & V(\hat{Y}) \end{pmatrix}^{-1} \begin{pmatrix} \hat{X}_1 - \hat{X}_2 \\ \hat{Y} - \mu_y \end{pmatrix},$$

   and the resulting optimal estimator minimizing Q_2 is given by

$$\hat{\mu}_y^* = \hat{Y} - \frac{C(\hat{X}_1 - \hat{X}_2, \hat{Y})}{V(\hat{X}_1 - \hat{X}_2)}\left(\hat{X}_1 - \hat{X}_2\right).$$

   Show that it is equal to the solution in (7.47).

7. Devise a linearization variance estimator for the doubly robust fractional imputation estimator in (7.80).

8. Let $\hat{\pi}_i = \pi(x_i; \hat{\phi})$ be the estimated response probability. Consider a regression estimator of the form

$$\hat{\theta}_{reg} = \sum_{i=1}^{n} w_i y_i,$$

   where

$$w_i = \left(n^{-1}\sum_{i=1}^{n} \mathbf{z}_i\right)' \left(\sum_{i=1}^{n} \mathbf{z}_i \mathbf{z}_i'\right)^{-1} \mathbf{z}_i$$

   and $\mathbf{z}_i' = (\hat{\pi}_i^{-1}, \mathbf{x}_i')$.

   (a) Show that $\hat{\theta}_{reg}$ is asymptotically unbiased under the response model $Pr(\delta = 1 \mid x) = \pi(x_i; \phi)$, where $\hat{\phi}$ is a consistent estimator of ϕ.
   (b) Construct a consistent variance estimator of $\hat{\theta}_{reg}$.

8

Nonignorable Missing Data

So far, we have assumed that the response mechanism is MAR. In this case, the response mechanism can be ignored and the resulting inference can be simple. If the response mechanism is not believed to be MAR, then we cannot simply ignore the response mechanism and a model for the response model needs to be specified. In this chapter, we introduce several approaches for handling nonignorable missing data. Model identification is an important issue under Non-MAR. We first start investigating the model identification problem in Section 8.1. Conditional likelihood approach in Section 8.2 is an interesting approach of handling nonignorable missing data when the response mechanism is known. The idea is extended further to pseudo likelihood approach in Section 8.3. When the explicit response model is used, as long as the model is identified, we can either use GMM in Section 8.4 or maximum likelihood method in Section 8.5.

8.1 Model Identification

We now consider the case of nonignorable missing data. This occurs when the probability of response depends on the variable that is not always observed. Let $\mathbf{x}_i$ be the variables that are always observed and y_i be the variable that is subject to missingness. Let δ_i be the response indicator function of y_i. In this case, the observed likelihood, conditional on $\mathbf{x}_i$'s, is

$$L_{\text{obs}}(\theta,\phi) = \prod_{\delta_i=1} f(y_i \mid \mathbf{x}_i;\theta) g(\delta_i \mid \mathbf{x}_i,y_i;\phi) \prod_{\delta_i=0} \int f(y_i \mid \mathbf{x}_i;\theta) g(\delta_i \mid \mathbf{x}_i,y_i;\phi) dy_i,$$

(8.1)

where $g(\delta_i \mid \mathbf{x}_i,y_i;\phi)$ is the conditional distribution of δ_i given $(y_i,\mathbf{x}_i)$, and ϕ is an unknown parameter. If $\mathbf{x}$ is null, then (8.1) becomes (2.13). If the response mechanism is ignorable in the sense that

$$g(\delta_i \mid \mathbf{x}_i,y_i;\phi) = g(\delta_i \mid \mathbf{x}_i;\phi),$$

then the observed likelihood in (8.1) can be written as

$$L_{\text{obs}}(\theta,\phi) = \prod_{\delta_i=1} f(y_i \mid \mathbf{x}_i;\theta) \times \prod_{i=1}^{n} g(\delta_i \mid \mathbf{x}_i;\phi) = L_1(\theta) \times L_2(\phi)$$

DOI: 10.1201/9780429321740-8

and the maximum likelihood estimator of θ can be obtained by maximizing $L_1(\theta)$. Otherwise, one needs to maximize the full likelihood (8.1) directly.

There are several problems in maximizing the full likelihood (8.1). First, the parameters in the full likelihood are not always identifiable. Second, the integrals in (8.1) are not easy to handle. Finally, inference with nonignorable missing data is sensitive to the failure of the assumed parametric model.

To discuss the identifiability of the models, we first introduce the following Lemma.

Lemma 8.1. *Define*

$$O(X,Y) = \frac{P(\delta = 0 \mid X,Y)}{P(\delta = 1 \mid X,Y)}$$

and

$$\tilde{O}(X) = \frac{P(\delta = 0 \mid X)}{P(\delta = 1 \mid X)}.$$

Then, we can obtain

$$\tilde{O}(x) = E\{O(X,Y) \mid x, \delta = 1\}. \tag{8.2}$$

Proof. Equality (8.2) can be easily proved by noting that

$$
\begin{aligned}
E\{O(X,Y) \mid x, \delta = 1\} &= \int O(x,y) f(y \mid x, \delta = 1) dy \\
&= \int \tilde{O}(x) f(y \mid x, \delta = 0) dy \\
&= \tilde{O}(x),
\end{aligned}
$$

where the second equality follows from

$$O(x,y) = \frac{P(\delta = 0 \mid x,y)}{P(\delta = 1 \mid x,y)} = \frac{P(\delta = 0 \mid x)}{P(\delta = 1 \mid x)} \times \frac{f(y \mid x, \delta = 0)}{f(y \mid x, \delta = 1)} = \tilde{O}(x) \times \frac{f(y \mid x, \delta = 0)}{f(y \mid x, \delta = 1)}$$

by Bayes formula. □

Lemma 8.1 is essentially the same as Lemma 7.1. If $P(\delta = 1 \mid x,y) = \pi(x,y;\phi)$ for some ϕ, by Lemma 8.1, we can consider the following log-likelihood function based on the conditional response probability

$$\tilde{l}(\phi) = \sum_{i=1}^{n} [\delta_i \log \tilde{\pi}(x_i;\phi) + (1-\delta_i) \log\{1 - \tilde{\pi}(x_i,\phi)\}],$$

where

$$\tilde{\pi}(x;\phi) = \frac{1}{1 + \tilde{O}(x;\phi)}$$

and $\tilde{O}(x;\phi) = E\{O(X,Y;\phi) \mid x, \delta = 1\}$. Note that the log-likelihood $\tilde{l}(\phi)$ can have a unique maximum only if the response model using $\tilde{O}(x;\phi)$ is identifiable. Thus, the identifiability of the original response model is closely related with whether the mapping $\phi \mapsto \tilde{\pi}(x;\phi)$ is one-to-one function of ϕ almost everywhether. In fact, Morikawa and Kim (2021) showed that it is the necessary and sufficient condition for the model identifiability of the response model. The following example, considered by Miao et al. (2016), gives an illustration of the identifiability issues.

Example 8.1. *Suppose that*

$$y_i \mid (x_i, \delta_i = 1) \sim N(\tau(x_i), \sigma^2).$$

Assume that x_i is always observed, but we observe y_i only when $\delta_i = 1$, where $\delta_i \sim Bernoulli\,[\pi_i(\phi)]$ and

$$\pi_i(\phi) = \frac{\exp(\phi_0 + \phi_1 x_i + \phi_2 y_i)}{1 + \exp(\phi_0 + \phi_1 x_i + \phi_2 y_i)}. \tag{8.3}$$

Then,

$$
\begin{aligned}
E\{O(x, Y; \phi) \mid x, \delta = 1\} &= E\{\exp(-\phi_0 - \phi_1 x - \phi_2 Y) \mid x, \delta = 1\} \\
&= \exp\{-\phi_0 - \phi_1 x - \phi_2 \tau(x) - \phi_2^2 \sigma^2 / 2\}.
\end{aligned}
$$

Therefore, the model is not identifiable if $\tau(x)$ is constant or linear.

In Example 8.1, if we can decompose $x = (x_1, x_2)$ and the response model in (8.3) is changed to

$$\pi_i(\phi) = \frac{\exp(\phi_0 + \phi_1 x_{1i} + \phi_2 y_i)}{1 + \exp(\phi_0 + \phi_1 x_{1i} + \phi_2 y_i)},$$

then

$$
\begin{aligned}
E_1\{O(x_1, Y; \phi) \mid x\} &= E_1\{\exp(-\phi_0 - \phi_1 x_1 - \phi_2 Y) \mid x\} \\
&= \exp\{-\phi_0 - \phi_1 x_1 - \phi_2 \tau(x) - \phi_2^2 \sigma^2 / 2\}.
\end{aligned}
$$

and the model is identifiable as long as $\tau(x)$ is a non-constant function of x_2. That is, if there is a covariate x_2 in $x = (x_1, x_2)$ such that the conditional distribution of y given x depends on the value of x_2, and $g(\delta|y, x)$ does not depend on x_2, then all parameters are identifiable. This is a special case of the following result, which was discussed in Wang et al. (2014).

Lemma 8.2. *Suppose that we can decompose the covariate vector $\mathbf{x}$ into two parts, $\mathbf{x}_1$ and $\mathbf{x}_2$, such that*

$$g(\delta|y, \mathbf{x}) = g(\delta|y, \mathbf{x}_1) \tag{8.4}$$

and, for any given $\mathbf{x}_1$, there exist $\mathbf{x}_2^{(a)} \neq \mathbf{x}_2^{(b)}$ such that

$$f(y \mid \mathbf{x}_1, \mathbf{x}_2 = \mathbf{x}_2^{(a)}) \neq f(y \mid \mathbf{x}_1, \mathbf{x}_2 = \mathbf{x}_2^{(b)}), \tag{8.5}$$

then, under some other minor conditions, all the parameters in f and g are identifiable.

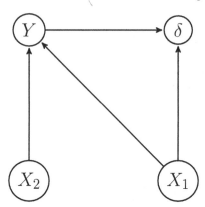

FIGURE 8.1
A DAG for understanding nonresponse instrumental variable X_2.

In the literature of measurement error model, where a covariate $\mathbf{x}^*$ associated with a study variable y^* is measured with error, valid estimators of regression parameters can be obtained by utilizing an instrument $\mathbf{z}$, that is correlated with $\mathbf{x}^*$ but independent of y^* conditioned on $\mathbf{x}^*$. In (8.4)–(8.5), we decompose the covariate vector $\mathbf{x}$ into two parts, $\mathbf{x}_1$ and $\mathbf{x}_2$, such that $\mathbf{x}_2$ plays the same role as an instrument, i.e., $\mathbf{z} = \mathbf{x}_2$ is correlated with $\mathbf{x}^* = (y, \mathbf{x}_1)$, and $\mathbf{z} = \mathbf{x}_2$ is independent of $y^* = \delta$ conditioned on $\mathbf{x}^* = (y, \mathbf{x}_1)$. Unconditionally, $\mathbf{x}_2$ may still be related to δ. Since y is subject to nonresponse, not measurement error, we name $\mathbf{x}_2$ as a *nonresponse instrument*. The nonresponse instrument $\mathbf{x}_2$ helps to identify the unknown quantities. Condition (8.4) can also be written as

$$Cov\left(\delta, \mathbf{x}_2 \mid y, \mathbf{x}_1\right) = 0.$$

That is, given $(y, \mathbf{x}_1)$, auxiliary variable $\mathbf{x}_2$ does not help in explaining δ. Figure 8.1 uses a directed acyclic graph (DAG) to explain this concept. In Figure 8.1, X_2 is used in explaining Y, but, it is related to δ only through Y.

The nonresponse instrument $\mathbf{x}_2$ defined by (8.4)–(8.5) may not be given in a particular application. Thus, we may have to search for an instrument from an available set of covariates that do not have missing values. Instrument search can be based on experience or a data-driven method; see, for example, Chen et al. (2020a) and Zhao et al. (2021).

Once the model identifiability is guaranteed, the observed likelihood has a unique maximum, and one can obtain the MLE that maximizes the observed likelihood (8.1). This is called the full likelihood or the fully parametric likelihood approach. To deal with the integral in the observed likelihood, numerical methods such as the EM algorithm can be used to compute the MLE; see Example 3.15. Baker and Laird (1988) discussed the EM method for a categorical y. Ibrahim et al. (1999) considered continuous y variable using a Monte Carlo

EM method of Wei and Tanner (1990). Chen and Ibrahim (2006) extended the method to generalized additive models under a parametric assumption on the response model.

Such a fully parametric approach in the nonignorable missing data case is known to be sensitive to the failure of the assumed parametric model (Kenward, 1998). Park and Brown (1994) used a Bayesian method to avoid the instability of the maximum likelihood estimators in the analysis of categorial missing data. Sensitivity analysis for nonignorable missingness can be a useful tool for addressing the issue associated with the nonignorable missingness. See Little (1995), Copas and Li (1997), Scharfstein et al. (1999), Copas and Eguchi (2001), Copas and Eguchi (2005), and Lin et al. (2012) for some examples of the sensitivity analysis of missing data inference under nonignorable nonresponse. Some theoretical limitation of the sensitivity analysis is discussed by Molenberghs et al. (2008).

8.2 Conditional Likelihood Approach

To avoid complicated computation, we now consider a likelihood-based approach of estimating parameters using only a part of the observed sample data. Recall that, if the parameter of interest is θ in $f(y \mid \mathbf{x}; \theta)$ and the response mechanism is ignorable, then the maximum likelihood method that maximizes

$$l_c(\theta) = \sum_{i=1}^{n} \delta_i \log\{f(y_i \mid \mathbf{x}_i, \delta_i = 1; \theta)\} \tag{8.6}$$

is consistent. The likelihood function in (8.6) is called the conditional likelihood because it is based on the conditional distribution given $\delta = 1$.

Following the decomposition below

$$f(y_i \mid \mathbf{x}_i) g(\delta_i \mid \mathbf{x}_i, y_i) = f_1(y_i \mid \mathbf{x}_i, \delta_i) g_1(\delta_i \mid \mathbf{x}_i),$$

the observed likelihood can be expressed as

$$L_{\text{obs}}(\theta, \phi) = \prod_{\delta_i=1} f_1(y_i \mid \mathbf{x}_i, \delta_i = 1) \times \prod_{i=1}^{n} g_1(\delta_i \mid \mathbf{x}_i). \tag{8.7}$$

The first component on the right hand side of (8.7) is the conditional likelihood

$$L_c(\theta) = \prod_{\delta_i=1} f_1(y_i \mid \mathbf{x}_i, \delta_i = 1) = \prod_{\delta_i=1} \left\{ \frac{f(y_i \mid \mathbf{x}_i; \theta) \pi(\mathbf{x}_i, y_i)}{\int f(y \mid \mathbf{x}_i; \theta) \pi(\mathbf{x}_i, y) dy} \right\}, \tag{8.8}$$

where

$$\pi(\mathbf{x}_i, y_i) = \pi_i = P(\delta_i = 1 \mid \mathbf{x}_i, y_i). \tag{8.9}$$

Unlike the observed likelihood (8.1), the conditional likelihood can be used even when $\mathbf{x}_i$'s associated with the missing y-values are not observed.

The score function derived from the conditional likelihood is

$$
\begin{aligned}
S_c(\theta) &= n^{-1} \frac{\partial}{\partial \theta} \ln L_c(\theta) \\
&= n^{-1} \sum_{i=1}^{n} \delta_i \left[S_i(\theta) - E\left\{ S_i(\theta) \mid \mathbf{x}_i, \delta_i = 1; \theta \right\} \right] \\
&= n^{-1} \sum_{i=1}^{n} \delta_i \left[S_i(\theta) - \frac{E\left\{ S_i(\theta)\pi_i \mid \mathbf{x}_i; \theta \right\}}{E\left(\pi_i \mid \mathbf{x}_i; \theta \right)} \right],
\end{aligned}
$$

where $S_i(\theta) = \partial \ln f(y_i \mid \mathbf{x}_i; \theta) / \partial \theta$. The second term $E\left\{ S_i(\theta) \mid \mathbf{x}_i, \delta_i = 1; \theta \right\}$ can be understood as a bias term of the complete-sample score function. If the response mechanism is MAR such that $\pi_i = \pi(\mathbf{x}_i)$, then the bias adjustment term is zero because

$$E\left\{ S_i(\theta) \mid \mathbf{x}_i, \delta = 1; \theta \right\} = E\left\{ S_i(\theta) \mid \mathbf{x}_i; \theta \right\} = 0.$$

Assume that π_i is known. Then maximizing the conditional likelihood (8.8) is to solve $S_c(\theta) = 0$, and we can apply the Fisher-scoring method. Note that

$$\frac{\partial}{\partial \theta'} S_c(\theta) = \frac{1}{n} \sum_{i=1}^{n} \delta_i \left[\frac{\partial}{\partial \theta'} S_i(\theta) - \frac{\partial}{\partial \theta'} \left\{ \frac{E\left\{ S_i(\theta)\pi_i \mid \mathbf{x}_i; \theta \right\}}{E\left(\pi_i \mid \mathbf{x}_i; \theta \right)} \right\} \right].$$

Writing $\dot{S}_i(\theta) = \partial S_i(\theta) / \partial \theta'$ and using

$$\frac{\partial}{\partial \theta'} E\left\{ S_i(\theta)\pi_i \mid \mathbf{x}_i; \theta \right\} = E\left\{ \dot{S}_i \pi_i \mid \mathbf{x}_i; \theta \right\} + E\left\{ S_i S_i' \pi_i \mid \mathbf{x}_i; \theta \right\}$$

and

$$\frac{\partial}{\partial \theta'} E\left\{ \pi_i \mid \mathbf{x}_i; \theta \right\} = E\left\{ \pi_i S_i' \mid \mathbf{x}_i; \theta \right\},$$

we have

$$
\begin{aligned}
\frac{\partial}{\partial \theta'} S_c(\theta) &= \frac{1}{n} \sum_{i=1}^{n} \delta_i \dot{S}_i(\theta) - \sum_{i=1}^{n} \delta_i \frac{E\left\{ \dot{S}_i \pi_i \mid \mathbf{x}_i; \theta \right\} + E\left\{ S_i S_i' \pi_i \mid \mathbf{x}_i; \theta \right\}}{E\left(\pi_i \mid \mathbf{x}_i; \theta \right)} \\
&\quad + \frac{1}{n} \sum_{i=1}^{n} \delta_i \frac{\left\{ E\left(S_i \pi_i \mid \mathbf{x}_i; \theta \right) \right\}^{\otimes 2}}{\left\{ E\left(\pi_i \mid \mathbf{x}_i; \theta \right) \right\}^2}.
\end{aligned}
$$

Hence,

$$
\begin{aligned}
\mathcal{I}_c(\theta) &= -E\left\{ \frac{\partial}{\partial \theta'} S_c(\theta) \right\} \\
&= E n^{-1} \sum_{i=1}^{n} \left[E\left\{ S_i S_i' \pi_i \mid \mathbf{x}_i; \theta \right\} - \frac{\left\{ E\left(S_i \pi_i \mid \mathbf{x}_i; \theta \right) \right\}^{\otimes 2}}{E\left(\pi_i \mid \mathbf{x}_i; \theta \right)} \right]. \tag{8.10}
\end{aligned}
$$

The Fisher-scoring method for obtaining the MLE from the conditional likelihood is then given by

$$\hat{\theta}^{(t+1)} = \hat{\theta}^{(t)} + \left\{ \mathcal{I}_c(\hat{\theta}^{(t)}) \right\}^{-1} S_c(\hat{\theta}^{(t)}).$$

To discuss some asymptotic properties, under some regularity conditions, it can be shown that the solution $\hat{\theta}_c$ to $S_c(\theta) = 0$ satisfies

$$\sqrt{n}(\hat{\theta}_c - \theta_0) \to_d N\left(0, \mathcal{I}_c^{-1}\right),$$

where $\mathcal{I}_c = \mathcal{I}_c(\theta_0)$ in (8.10).

We consider a class of estimating equations of the form

$$\sum_{i=1}^{n} \delta_i U(\theta; \mathbf{x}_i, y_i) = 0, \tag{8.11}$$

where the function U satisfies $E\{\delta U(\theta; \mathbf{x}, Y)\} = 0$. The choice of $S_c(\theta; \mathbf{x}, y) = S(\theta; \mathbf{x}, y) - E\{S(\theta; \mathbf{x}, Y) \mid \mathbf{x}, \delta = 1; \theta\}$ belong to the class in (8.11). The following theorem, also presented in Wang and Kim (2021b), establishes the optimality of the maximum conditional likelihood estimator.

Theorem 8.1. *Let $\hat{\theta}_u$ be the estimator obtained through solving (8.11), and assume that the regularity conditions for the following standard asymptotic expansion holds:*

$$\hat{\theta}_u = \theta - n^{-1} M_u^{-1} \sum_{i=1}^{n} \delta_i U(\theta_t; \mathbf{x}_i, y_i) + o_p(n^{-1/2}), \tag{8.12}$$

where $M_u = E\{\delta \dot{U}(\theta; \mathbf{X}, Y)\}$ and $\dot{U}(\theta; \mathbf{x}, y) = \partial U(\theta; \mathbf{x}, y)/\partial \theta'$. We have, ignoring the smaller order terms,

$$V(\hat{\theta}_u) \geq n^{-1} \mathcal{I}_c^{-1} = V(\hat{\theta}_c), \tag{8.13}$$

which suggests that the conditional MLE $\hat{\theta}_c$ achieves the lower bound in (8.13).

Proof. From $E\{\delta U(\theta; \mathbf{X}, Y)\} = 0$, we obtain

$$
\begin{aligned}
\frac{\partial}{\partial \theta'} E\{\delta U(\theta; \mathbf{X}, Y)\} &= E\{\delta \dot{U}(\theta; \mathbf{X}, Y)\} \\
&\quad + \int \delta U(\theta; \mathbf{x}, y) \left\{ \frac{\partial}{\partial \theta'} f(y \mid \mathbf{x}, \delta = 1; \theta) \right\} g(\mathbf{x} \mid \delta = 1) d(x, y) \\
&= E\{\delta \dot{U}(\theta; \mathbf{X}, Y)\} + E\{\delta U(\theta; \mathbf{X}, Y) S_c(\theta; \mathbf{X}, Y)'\} = 0,
\end{aligned}
$$

where $S_c(\theta; \mathbf{x}, \delta = 1) = \partial \log f(y \mid \mathbf{x}, \delta = 1)/\partial \theta$. Thus, we obtain

$$M_u = -E\{\delta U(\theta; \mathbf{X}, Y) S_c(\theta; \mathbf{X}, Y)'\} = -Cov\{\delta U(\theta; \mathbf{x}, Y), \delta S_c(\theta; \mathbf{x}, Y)\}. \tag{8.14}$$

If we use $U(\theta; \mathbf{x}, y) = S_c(\theta; \mathbf{x}, y)$ in (8.14), we obtain

$$\mathcal{I}_c \equiv E\{\delta \dot{S}_c(\theta; \mathbf{X}, Y)\} = -V\{\delta S_c(\theta; \mathbf{X}, Y)\}. \tag{8.15}$$

Now, note that

$$
\begin{aligned}
& V\left\{\delta M_u^{-1} U(\theta; \mathbf{X}, Y) + \delta \mathcal{I}_c^{-1} S_c(\theta; \mathbf{X}, Y)\right\} \\
= \ & V\left\{\delta M_u^{-1} U(\theta; \mathbf{X}, Y)\right\} + V\left\{\delta \mathcal{I}_c^{-1} S_c(\theta; \mathbf{X}, Y)\right\} \\
& + 2 Cov\left\{\delta M_u^{-1} U(\theta; \mathbf{X}, Y), \delta \mathcal{I}_c^{-1} S_c(\theta; \mathbf{X}, Y)\right\} \\
= \ & M_u^{-1} V\left\{\delta U(\theta; \mathbf{X}, Y)\right\} (M_u^{-1})' + \mathcal{I}_c^{-1} V\left\{\delta S_c(\theta; \mathbf{X}, Y)\right\} (\mathcal{I}_c^{-1})' \\
& + 2 M_u^{-1} Cov\left\{\delta U(\theta; \mathbf{X}, Y), \delta S_c(\theta; \mathbf{X}, Y)\right\} (\mathcal{I}_c^{-1})'.
\end{aligned}
$$

Using (8.14) and (8.15), we can establish

$$
\begin{aligned}
& V\left\{\delta M_u^{-1} U(\theta; \mathbf{X}, Y) + \delta \mathcal{I}_c^{-1} S_c(\theta; \mathbf{X}, Y)\right\} \\
= \ & M_u^{-1} V\left\{\delta U(\theta; \mathbf{X}, Y)\right\} (M_u^{-1})' - \mathcal{I}_c^{-1} \geq 0,
\end{aligned}
$$

which proves (8.13).   □

In practice, the response probability $\pi_i = \pi(x_i, y_i)$ is generally unknown. To apply the conditional likelihood method, π_i can be replaced by a consistent estimator $\hat{\pi}_i$. Such a consistent estimator cannot be obtained from $(y_i, \mathbf{x}_i)$ with $\delta_i = 1$ only. This will be further studied in §8.3 and §8.5.

The following example, originally presented in Chambers et al. (2012), indicates how to obtain estimators from both the full and conditional likelihood when π_i is a known function of y_i.

Example 8.2. *Assume that the original sample is a random sample from an exponential distribution with mean $\mu = 1/\theta$. That is, the probability density function of y is $f(y; \theta) = \theta \exp(-\theta y) I(y > 0)$. Suppose that we observe y_i only when $y_i > K$ for a known $K > 0$. Thus, the response indicator function is defined by $\delta_i = 1$ if $y_i > K$ and $\delta_i = 0$ otherwise. To compute the maximum likelihood estimator from the observed likelihood, note that*

$$S_{\text{obs}}(\theta) = \sum_{\delta_i=1} \left(\frac{1}{\theta} - y_i\right) + \sum_{\delta_i=0} \left\{\frac{1}{\theta} - E(y_i \mid \delta_i = 0)\right\}.$$

Since

$$E(Y \mid y \leq K) = \frac{1}{\theta} - \frac{K \exp(-\theta K)}{1 - \exp(-\theta K)},$$

the maximum likelihood estimator of θ can be obtained by the following iteration equation:

$$\left\{\hat{\theta}^{(t+1)}\right\}^{-1} = \bar{y}_r - \frac{n-r}{r}\left\{\frac{K \exp(-K\hat{\theta}^{(t)})}{1 - \exp(-K\hat{\theta}^{(t)})}\right\}, \tag{8.16}$$

where $r = \sum_{i=1}^{n} \delta_i$ and $\bar{y}_r = r^{-1} \sum_{i=1}^{n} \delta_i y_i$. *Similarly, we can derive the maximum conditional likelihood estimator. Note that* $\pi_i = Pr(\delta_i = 1 \mid y_i) = I(y_i > K)$ *and* $E(\pi_i) = E\{I(y_i > K)\} = \exp(-K\theta)$. *Thus, the conditional likelihood in (8.8) reduces to*

$$\prod_{\delta_i=1} \theta \exp\{-\theta(y_i - K)\}.$$

The maximum conditional likelihood estimator of θ is

$$\hat{\theta}_c = \frac{1}{\bar{y}_r - K}.$$

Since $E(y \mid y > K) = \mu + K$, *the maximum conditional likelihood estimator of μ, which is $\hat{\mu}_c = 1/\hat{\theta}_c$, is unbiased for μ.*

8.3 Pseudo Likelihood Approach

Now, we switch to another type of conditional likelihood for parameter estimation with nonignorable missing data. Assume that the joint density of $\mathbf{x}$ and y is $f(y \mid \mathbf{x}; \theta)p(\mathbf{x})$, where $f(y \mid \mathbf{x}; \theta)$ follows a parametric model and $p(\mathbf{x})$ is a density for the marginal distribution of $\mathbf{x}$. Also, $P(\delta = 1 \mid \mathbf{x}, y)$ is completely unspecified. As discussed in §8.1, we still need a nonresponse instrument for the identifiability of unknown quantities. Tang et al. (2003) first studied this problem by assuming that the entire covariate vector $\mathbf{x}$ is a nonresponse instrument. In this case, the response mechanism depends only on y. Thus, we have

$$P(\delta = 1 \mid \mathbf{x}, y) = P(\delta = 1 \mid y). \tag{8.17}$$

Under (8.17), we can construct the conditional likelihood as

$$L_c(\theta) = \prod_{\delta_i=1} f(\mathbf{x}_i \mid y_i, \delta_i = 1; \theta), \tag{8.18}$$

where

$$
\begin{aligned}
f(\mathbf{x} \mid y, \delta = 1; \theta) &= \frac{f(y \mid \mathbf{x}; \theta)p(\mathbf{x})P(\delta = 1 \mid \mathbf{x}, y)}{\int f(y \mid \mathbf{x}; \theta)p(\mathbf{x})P(\delta = 1 \mid \mathbf{x}, y)d\mathbf{x}} \\
&= \frac{f(y \mid \mathbf{x}; \theta)p(\mathbf{x})P(\delta = 1 \mid y)}{\int f(y \mid \mathbf{x}; \theta)p(\mathbf{x})P(\delta = 1 \mid y)d\mathbf{x}} \\
&= \frac{f(y \mid \mathbf{x}; \theta)p(\mathbf{x})}{\int f(y \mid \mathbf{x}; \theta)p(\mathbf{x})d\mathbf{x}}.
\end{aligned}
$$

Here, the second equality follows from (8.17). Thus, as long as (8.17) holds, we do not necessarily specify the nonresponse model. Instead, we need to specify the marginal distribution of $\mathbf{x}$, $p(\mathbf{x}) = p(\mathbf{x}; \alpha)$ for some unknown parameter α. Thus, the pseudo maximum likelihood estimator of θ is obtained by maximizing

$$L_p(\theta) = \prod_{\delta_i=1} \left\{ \frac{f(y_i \mid \mathbf{x}_i; \theta)\hat{p}(\mathbf{x}_i)}{\int f(y_i \mid \mathbf{x}; \theta)\hat{p}(\mathbf{x})d\mathbf{x}} \right\} \tag{8.19}$$

for given $\hat{p}(\mathbf{x}) = p(\mathbf{x}; \hat{\alpha})$. This is the basic idea of the pseudo maximum likelihood estimator of Tang et al. (2003). In particular, we can estimate $p(\mathbf{x})$ by the empirical distribution putting mass n^{-1} on each observed $\mathbf{x}_i$, $i = 1, ..., n$. The pseudo likelihood becomes

$$\prod_{\delta_i=1} \frac{f(y_i \mid \mathbf{x}_i; \theta)}{\sum_{l=1}^{n} f(y_i \mid \mathbf{x}_l; \theta)}. \tag{8.20}$$

Zhao and Ma (2018) later proved that the PMLE obtained from (8.20) is more efficient than the PMLE obtained from (8.19) using $\hat{p}(\mathbf{x}) = p(\mathbf{x}; \hat{\alpha})$.

The idea can be generalized further by making a weaker assumption, in which $\mathbf{x} = (\mathbf{x}_1, \mathbf{x}_2)$ and $\mathbf{x}_2$ is a nonresponse instrument, i.e., (8.4)–(8.5) hold. The main idea of this approach can be described as follows. Since δ and $\mathbf{x}_2$ are conditionally independent given $(y, \mathbf{x}_1)$, the conditional probability density of $\mathbf{x}_2$ given $(\delta, y, \mathbf{x}_1)$ satisfies

$$p(\mathbf{x}_2 \mid y, \mathbf{x}_1, \delta) = p(\mathbf{x}_2 \mid y, \mathbf{x}_1).$$

Then, we can use the observed y_i's and $\mathbf{x}_i$'s to estimate $p(\mathbf{x}_2|y, \mathbf{x}_1)$. If the parameter θ in $f(y|\mathbf{x}; \theta)$ is of interest, then we can estimate it by maximizing

$$\prod_{\delta_i=1} p(\mathbf{x}_{2i} \mid y_i, \mathbf{x}_{1i}) = \prod_{\delta_i=1} \frac{f(y_i \mid \mathbf{x}_i; \theta)p(\mathbf{x}_{2i} \mid \mathbf{x}_{1i})}{\int f(y_i \mid \mathbf{x}_{i1}, \mathbf{x}_2; \theta)p(\mathbf{x}_2|\mathbf{x}_{1i})d\mathbf{x}_2}, \tag{8.21}$$

where the equality follows from the well-known Bayes formula. Since $\mathbf{x}_i$, $i = 1, ..., n$, are fully observed, the conditional probability density $p(\mathbf{x}_2|\mathbf{x}_1)$ can be estimated using many well-established methods. Let $\hat{p}(\mathbf{x}_2|\mathbf{x}_1)$ be an estimated conditional probability density of $\mathbf{x}_2$ given $\mathbf{x}_1$. Substituting this estimate into the likelihood in (8.21), we can obtain the following pseudo likelihood:

$$L_p(\theta) = \prod_{\delta_i=1} \frac{f(y_i \mid \mathbf{x}_i; \theta)\hat{p}(\mathbf{x}_{2i} \mid \mathbf{x}_{1i})}{\int f(y_i \mid \mathbf{x}_{i1}, \mathbf{x}_2; \theta)\hat{p}(\mathbf{x}_2|\mathbf{x}_{1i})d\mathbf{x}_2}. \tag{8.22}$$

If $\hat{p}(\mathbf{x}_2|\mathbf{x}_1)$ is obtained from a nonparametric method such as Kernel regression, the pseudo likelihood in (8.22) becomes semi-parametric. For a parametric model $\hat{p}(\mathbf{x}_2|\mathbf{x}_1) = p(\mathbf{x}_2|\mathbf{x}_1; \hat{\alpha})$, the pseudo likelihood in (8.22) becomes

$$\prod_{\delta_i=1} \frac{f(y_i \mid \mathbf{x}_i; \theta)p(\mathbf{x}_{2i} \mid \mathbf{x}_{1i}; \hat{\alpha})}{\int f(y_i \mid \mathbf{x}_{i1}, \mathbf{x}_2; \theta)p(\mathbf{x}_2|\mathbf{x}_{1i}; \hat{\alpha})d\mathbf{x}_2}. \tag{8.23}$$

The pseudo maximum likelihood estimator (PMLE) of θ, denoted by $\hat{\theta}_p$, maximizing the pseudo likelihood in (8.23) can be obtained by solving

$$S_p(\theta;\hat{\alpha}) \equiv \sum_{\delta_i=1} [S(\theta;\mathbf{x}_i,y_i) - E\{S(\theta;\mathbf{x}_i,y_i) \mid y_i,\mathbf{x}_{1i};\theta,\hat{\alpha}\}] = 0$$

for θ, where $S(\theta;\mathbf{x},y) = \partial \log f(y \mid \mathbf{x};\theta)/\partial\theta$.

Under some regularity conditions, consistency and asymptotic normality of this estimator $\hat{\theta}$ is established in Tang et al. (2003), Jiang and Shao (2012), and Zhao and Shao (2015). That is,

$$\sqrt{n}(\hat{\theta}_p - \theta_0) \to_d N(0, \Sigma),$$

where θ_0 is the true value of the parameter θ and Σ is a covariance matrix. Due to the use of pseudo likelihood (the substitution of $p(\mathbf{x}_2|\mathbf{x}_1)$ by its estimator), the form of Σ is very complicated. Shao and Zhao (2013) proposed an approach for estimating Σ. Alternatively, we can apply bootstrapping to estimate Σ; see Shao and Zhao (2013).

We can apply sensitivity analysis or model selection to address the issue of parametric model assumption $f(y|\mathbf{x};\theta)$, which is crucial for the pseudo likelihood approach introduced in this section. See Fang and Shao (2016).

8.4 Generalized method of moments (GMM) Approach

In the previous section, $\pi(\mathbf{x},y)$ given by (8.9) is usually unknown and has to be estimated. To estimate $\pi(\mathbf{x},y)$, we assume (8.4)–(8.5) and the following parametric response model

$$\pi(\mathbf{x},y) = \pi(\mathbf{x}_1,y;\phi), \tag{8.24}$$

where ϕ is an unknown parameter vector, and π is a known strictly monotone and twice differentiable function of ϕ.

To estimate parameter ϕ, we consider the generalized method of moments (GMM) approach introduced in Section 7.7. The key idea of the GMM is to construct a set of L estimating functions

$$g_l(\phi,\mathbf{d}), \qquad l=1,...,L, \quad \phi \in \Psi,$$

where $\mathbf{d}$ is a vector of observations, Ψ is the parameter space containing the true parameter value ϕ, L is less than the dimension of Ψ, g_l's are non-constant functions with $E[g_l(\phi,\mathbf{d})] = 0$ for all l, and g_l's are not linearly dependent, i.e., the $L \times L$ matrix whose (l,l')th element being $E[g_l(\phi,\mathbf{d})g_{l'}(\phi,\mathbf{d})]$ is positive definite, which can usually be achieved by eliminating redundant functions when g_l's are linearly dependent. Let $\mathbf{d}_1, ..., \mathbf{d}_n$ be n independent vectors distributed as $\mathbf{d}$ and

$$G(\phi) = \left(\frac{1}{n} \sum_i g_1(\phi, \mathbf{d}_i), ..., \frac{1}{n} \sum_i g_L(\phi, \mathbf{d}_i) \right)', \quad \phi \in \Psi. \tag{8.25}$$

If L has the same dimension as ϕ, then we may be able to find a $\hat{\phi}$ such that $G(\hat{\phi}) = 0$. If L is larger than the dimension of ϕ, however, a solution to $G(\varphi) = 0$ may not exist. In any case, a GMM estimator of ϕ can be obtained using the following two-step algorithm:

1. Obtain $\hat{\phi}^{(1)}$ by minimizing $G'(\phi)G(\phi)$ over $\phi \in \Psi$.

2. Let $\hat{W}$ be the inverse of the $L \times L$ matrix whose (l, l')-th element is equal to $n^{-1} \sum_i g_l(\hat{\phi}^{(1)}, \mathbf{d}_i) g_{l'}(\hat{\phi}^{(1)}, \mathbf{d}_i)$. The GMM estimator $\hat{\phi}$ is obtained by minimizing $G'(\phi)\hat{W}G(\phi)$ over $\phi \in \Psi$.

To explain the use of GMM for parameter estimation with nonignorable missing data, we assume the following response model

$$P(\delta = 1 | \mathbf{x}, y) = \pi(\phi_0 + \mathbf{x}_1' \phi_1 + y\phi_2), \tag{8.26}$$

where π is defined in (8.24), and $\phi = (\phi_0, \phi_1, \phi_2)$ is a $(k+2)$-dimensional unknown parameter vector not depending on values of $\mathbf{x}$. A similar assumption to (8.26) was made in Qin et al. (2002) and Kott and Chang (2010).

We assume that $\mathbf{z} = \mathbf{x}_2$ has both continuous and discrete components and $\mathbf{x}_1$ is a continuous k-dimensional covariate. Let $\mathbf{z} = (z_d, \mathbf{z}_c)$, where $\mathbf{z}_c$ is m-dimensional continuous vector and z_d is discrete taking values $1, ..., J$. To estimate ϕ, the GMM can be applied with the following $L = k + m + J$ estimating functions:

$$\mathbf{g}(\phi, y, \mathbf{x}, \delta) = \boldsymbol{\xi}[\delta\omega(\phi) - 1], \tag{8.27}$$

where $\boldsymbol{\xi} = (\boldsymbol{\zeta}', \mathbf{z}_c', \mathbf{x}_1')'$, $\boldsymbol{\zeta}$ is a J-dimensional row vector whose lth component is $I(z_d = l)$, $I(A)$ is the indicator function of A, and $\omega(\phi) = \{\pi(\phi_0 + \mathbf{x}_1' \phi_1 + y\phi_2)\}^{-1}$.

The estimating function $\mathbf{g}$ is motivated by the fact that, when ϕ is the true parameter value,

$$
\begin{aligned}
E[\mathbf{g}(\phi, y, \mathbf{x}, \delta)] &= E\{\boldsymbol{\xi}[\delta w(\phi) - 1]\} \\
&= E(E\{\boldsymbol{\xi}[\delta w(\phi) - 1] \mid y, \mathbf{x}\}) \\
&= E\left\{\boldsymbol{\xi}\left[\frac{E(\delta \mid y, \mathbf{x})}{P(\delta = 1 \mid y, \mathbf{x}_1)} - 1\right]\right\} = 0.
\end{aligned}
$$

Let G be defined by (8.25) with g_l being the lth function of $\mathbf{g}$ and $\mathbf{d} = (y, \mathbf{x}', \delta)$, $\hat{W}$ be the $\hat{W}$ in the previously described two-step algorithm and $\hat{\phi}$ be the two-step GMM estimator of ϕ in (8.26). Under some regularity conditions, as discussed in Wang et al. (2014), we can establish that

$$\sqrt{n}(\hat{\phi} - \phi) \to_d N\left(0, (\Gamma' \Sigma^{-1} \Gamma)^{-1}\right), \tag{8.28}$$

where

$$\Gamma = \begin{bmatrix} E[\delta\boldsymbol{\zeta}^T\omega'(\phi)] & E[\delta\mathbf{x}_1\boldsymbol{\zeta}^T\omega'(\phi)] & E[\delta\boldsymbol{\zeta}^T y\omega'(\phi)] \\ E[\delta\mathbf{z}_c^T\omega'(\phi)] & E[\delta\mathbf{x}_1\mathbf{z}_c^T\omega'(\phi)] & E[\delta\mathbf{z}_c^T y\omega'(\phi)] \\ E[\delta\mathbf{x}_1^T\omega'(\phi)] & E[\delta\mathbf{x}_1\mathbf{x}_1^T\omega'(\phi)] & E[\delta\mathbf{x}_1^T y\omega'(\phi)] \end{bmatrix}, \tag{8.29}$$

and Σ is the positive definite $(k+m+J) \times (k+m+J)$ matrix with $E[g_l(\phi,y,\mathbf{x},\delta)g_{l'}(\phi,y,\mathbf{x},\delta)]$ as its (l,l')th element. Thus, the asymptotic result in (8.28) requires that Γ in (8.29) is of full rank. Also, the asymptotic covariance matrix $(\Gamma'\Sigma^{-1}\Gamma)^{-1}$ can be estimated by $(\hat{\Gamma}'\hat{\Sigma}^{-1}\hat{\Gamma})^{-1}$, where $\hat{\Gamma}$ is the $(k+m+J) \times (k+2)$ matrix whose lth row is

$$n^{-1}\sum_i \frac{\partial g_l(\hat{\phi},y_i,\mathbf{x}_i,\delta_i)}{\partial\phi}$$

and $\hat{\Sigma}$ is the $L \times L$ matrix whose (l,l')th element is $n^{-1}\sum_i g_l(\hat{\phi},y_i,\mathbf{x}_i,\delta_i)g_{l'}(\hat{\phi},y_i,\mathbf{x}_i,\delta_i)$.

Example 8.3. *Suppose that we are interested in estimating the parameters in the regression model*

$$y_i = \beta_0 + \beta_1 x_{1i} + \beta_2 x_{2i} + e_i, \tag{8.30}$$

where $E(e_i \mid \mathbf{x}_i) = 0$. Assume that y_i is subject to missingness and that

$$P(\delta_i = 1 \mid x_{1i}, x_{2i}, y_i) = \frac{\exp(\phi_0 + \phi_1 x_{1i} + \phi_2 y_i)}{1 + \exp(\phi_0 + \phi_1 x_{1i} + \phi_2 y_i)}.$$

Thus, x_{2i} is the nonresponse instrument variable in this setup. A consistent estimator of ϕ can be obtained by solving

$$\hat{U}_2(\phi) \equiv \sum_{i=1}^n \left\{ \frac{\delta_i}{\pi(x_{1i},y_i;\phi)} - 1 \right\} (1, x_{1i}, x_{2i}) = (0,0,0). \tag{8.31}$$

Roughly speaking, the solution to (8.31) exists almost surely if $E\{\partial\hat{U}_2(\phi)/\partial\phi\}$ is of full rank in the neighborhood of the true value of ϕ. If x_2 is a vector, then (8.31) is overidentified and the solution to (8.31) does not exist. In that case, the GMM algorithm can be used.

If x_{2i} is a categorical variable with category $\{1,\ldots,J\}$, then (8.31) can be written as

$$\hat{U}_2(\phi) \equiv \sum_{i=1}^n \left\{ \frac{\delta_i}{\pi(x_{1i},y_i;\phi)} - 1 \right\} (1, x_{1i}, \boldsymbol{\zeta}_i) = (0,0,\mathbf{0}'), \tag{8.32}$$

where $\boldsymbol{\zeta}_i$ is the J-dimensional row vector whose jth component is $I(x_{2i} = j)$.

Once the solution $\hat{\phi}$ to (8.31) or (8.32) is obtained, then a consistent estimator of $\beta = (\beta_0, \beta_1, \beta_2)$ can be obtained by solving

$$\hat{U}_1(\beta, \hat{\phi}) \equiv \sum_{i=1}^{n} \frac{\delta_i}{\hat{\pi}_i} \{y_i - \beta_0 - \beta_1 x_{1i} - \beta_2 x_{2i}\} (1, x_{1i}, x_{2i}) = (0, 0, 0) \qquad (8.33)$$

for β. The asymptotic variance of $\hat{\beta} = \hat{\beta}(\hat{\phi})$, computed from (8.33), can be obtained by

$$V(\hat{\theta}) \cong \left(\Gamma_a' \Sigma_a^{-1} \Gamma_a\right)^{-1},$$

where

$$\begin{array}{rcl}
\Gamma_a & = & E\{\partial \hat{U}(\theta)/\partial\theta\}, \\
\Sigma_a & = & V(\hat{U}), \\
\hat{U} & = & (\hat{U}_1', \hat{U}_2')',
\end{array}$$

and $\theta = (\beta, \phi)$. The nonresponse instrument variable approach does not use a fully parametric model for $f(y \mid x)$ and is less sensitive to the misspecification of the outcome model.

We now consider the following class of estimating equations for ϕ:

$$\hat{U}_h(\phi) \equiv \frac{1}{n} \sum_{i=1}^{n} \left\{ \frac{\delta_i}{\pi(\mathbf{x}_{1i}, y_i; \phi)} - 1 \right\} \mathbf{h}(\mathbf{x}_i; \phi) = 0 \qquad (8.34)$$

such that the solution exists uniquely. Note that, since $E\{\hat{U}_h(\phi)\} = 0$, the solution $\hat{\phi}_h$ to (8.34) is asymptotically unbiased regardless of choice of $\mathbf{h}(X; \phi)$. A natural question is this: what is the optimal choice of h in the sense of minimizing the asymptotic variance of $\hat{\phi}_h$? To answer the question, we introduce the following theorem.

Theorem 8.2. *Let $\hat{\phi}_h$ be the solution to (8.34). Under some regularity conditions, we obtain*

$$\sqrt{n}\left(\hat{\phi}_h - \phi\right) \xrightarrow{\mathcal{L}} N(0, A_h^{-1} B_h (A_h^{-1})'), \qquad (8.35)$$

where

$$\begin{array}{rcl}
A_h & = & E\{\mathbf{h} E(O \cdot S_0' \mid X)\}, \\
B_h & = & E\{E(O \mid X) \mathbf{h} \mathbf{h}'\},
\end{array}$$

$O(x, y) = \{1 - \pi(x, y)\}/\pi(x, y)$, $S_0 = S_0(\phi; x, y)$ *with*

$$\begin{array}{rcl}
S_\delta(\phi; x, y) & = & \dfrac{\partial}{\partial \phi} \{\delta \ln \pi(x, y; \phi) + (1 - \delta) \ln(1 - \pi(x, y; \phi))\} \\[2mm]
& = & \dfrac{\{\delta - \pi(x, y; \phi)\}}{\pi(x, y; \phi)\{1 - \pi(x, y; \phi)\}} \dfrac{\partial \pi(x, y; \phi)}{\partial \phi}.
\end{array} \qquad (8.36)$$

Proof. Since $\hat{\phi}_h$ is the solution to $\hat{U}_h(\phi) = 0$, we can apply the theory of estimating equation and use the sandwich formula to compute the asymptotic variance. Thus,

$$V(\hat{\phi}_h) \cong n^{-1} A_h^{-1} B_h (A_h^{-1})',$$

where

$$
\begin{aligned}
A_h &= E\{\partial \hat{U}_h / \partial \phi'\} \\
&= -E\left\{ \frac{1}{\pi(X_1,Y)} \mathbf{h}(X;\phi) \frac{\partial \pi(X_1,Y;\phi)}{\partial \phi'} \right\} \\
&= E\{\mathbf{h}(X;\phi) O(X_1,Y) S_0(X_1,Y)'\} \\
&= E\{\mathbf{h} E(O \cdot S_0' \mid X)\}
\end{aligned}
$$

and

$$
\begin{aligned}
B_h &= n \cdot V\{\hat{U}_h(\phi)\} \\
&= V\left\{ \frac{\pi(1-\pi)}{\pi^2} \mathbf{h}\mathbf{h}' \right\} \\
&= E\{\mathbf{h}\mathbf{h}' E(O \mid X)\}.
\end{aligned}
$$

$\square$

The optimal choice of $\mathbf{h}(x)$ in the class of the estimating equation in (8.34) can be obtained by finding the minimizer of the asymptotic variance in (8.35). We have the following corollary for the optimal estimation.

Corollary 8.1. *The asymptotic variance in (8.35) is minimized at*

$$
\begin{aligned}
\mathbf{h}^*(X;\phi) &= \frac{E(O \cdot S_0 \mid X)}{E(O \mid X)} & (8.37) \\
&= \frac{E_1\{\pi^{-1} O \cdot S_0 \mid X\}}{E_1\{\pi^{-1} O \mid X\}},
\end{aligned}
$$

where $E_1(\cdot \mid X)$ is the expectation with respect to $f_1(y \mid x) = f(y \mid x, \delta = 1)$.

Proof. First define $c = E(O \mid X)$, $\bar{s}_0 = E(O \cdot S_0 \mid X)$ and

$$\Sigma = E\left\{ c^{-1} \bar{s}_0 \bar{s}_0' \right\}.$$

Note that under the choice of $\mathbf{h}^*$ in (8.37), we have

$$A_{h^*} = B_{h^*} = \Sigma,$$

so

$$A_{h^*}^{-1} B_{h^*} (A_{h^*}^{-1})' = \Sigma^{-1}.$$

We now wish to show that

$$A_h^{-1} B_h (A_h^{-1})' - \Sigma^{-1} \geq 0, \qquad (8.38)$$

where $M \geq 0$ means that M is nonnegative definite.

Now, define

$$\mathbf{t}_1 = A_h^{-1}\left(\frac{\delta}{\pi} - 1\right)\mathbf{h}$$

and

$$\mathbf{t}_2 = \Sigma^{-1}\left(\frac{\delta}{\pi} - 1\right)\mathbf{h}^*.$$

Thus,

$$V(\mathbf{t}_1) = A_h^{-1}B_h(A_h^{-1})'$$

and

$$V(\mathbf{t}_2) = \Sigma^{-1}$$

hold. Note that

$$Cov(\mathbf{t}_1, \mathbf{t}_2) = A_h^{-1}E\{ch(\mathbf{h}^*)'\}\Sigma^{-1} = \Sigma^{-1} = V(\mathbf{t}_2),$$

and so

$$V(\mathbf{t}_1 - \mathbf{t}_2) = V(\mathbf{t}_1) - V(\mathbf{t}_2) \geq 0,$$

which proves (8.38).   □

Thus, by Corollary 8.1, we can use

$$\mathbf{h}^*(X;\phi) = \frac{\hat{E}_1\{\pi^{-1}O \cdot S_0 \mid X\}}{\hat{E}_1\{\pi^{-1}O \mid X\}} \tag{8.39}$$

in (8.34) to obtain the optimal GMM estimator of ϕ, where $\hat{E}_1(\cdot)$ is a consistent estimator of $E_1(\cdot)$. However, to compute the $\mathbf{h}^*(X;\phi)$ in (8.39), one need to correctly specify the model for $f_1(y \mid x) = f(y \mid x, \delta = 1)$. If the model is correct, then the solution to (8.34) using $\mathbf{h}^*(X;\phi)$ in (8.39) is optimal. If the model is incorrect, the solution is still consistent as it still belongs to the GMM class in (8.34). The idea of optimal estimation using (8.39) was first proposed by Rotnitzky and Robins (1997) and later generalized by Morikawa and Kim (2021).

To solve a nonlinear equation $U(\phi) = 0$ such as (8.34), one might use the Newton method

$$\hat{\phi}^{(t+1)} = \hat{\phi}^{(t)} - \left\{\dot{U}(\hat{\phi}^{(t)})\right\}^{-1}U(\hat{\phi}^{(t)}), \tag{8.40}$$

where $\dot{U}(\phi) = \partial U(\phi)/\partial\phi'$. However, the partial derivative $\dot{U}(\phi)$ is not symmetric, and the iterative computation in (8.40) can have numerical problems. To deal with the problem, we can use

$$\hat{\phi}^{(t+1)} = \hat{\phi}^{(t)} - \left\{\dot{U}(\hat{\phi}^{(t)})'\dot{U}(\hat{\phi}^{(t)})\right\}^{-1}\dot{U}(\hat{\phi}^{(t)})'U(\hat{\phi}^{(t)}), \tag{8.41}$$

which is essentially equivalent to finding $\hat{\phi}$ that minimizes $Q(\phi) = U(\phi)'U(\phi)$.

To end this section we discuss the parametric model assumption (8.24). First, sensitivity analysis discussed in Section 8.1 can be applied to address the modeling issue. Second, model and/or variable selection can also be a useful tool to find a reasonable parametric model; see, for example, Wang et al. (2021) and Chen et al. (2020a). Finally, we may also replace (8.24) with a semi-parametric model that is more general flexible than any fully parametric model; see, for example, Shao and Wang (2016) and Shao (2018).

8.5   Exponential Tilting Model

Under the response model in (8.24), we can assume that δ_i are generated from a Bernoulli distribution with probability $\pi_i(\phi) = \pi(\mathbf{x}_i, y_i; \phi)$ for some ϕ. To estimate ϕ, we can consider maximizing the observed likelihood function

$$L_{obs}(\phi) = \prod_{i=1}^{n} \{\pi_i(\phi)\}^{\delta_i} \left[\int \{1 - \pi_i(\phi)\} f(y \mid \mathbf{x}_i) dy\right]^{1-\delta_i},$$

where $f(y \mid \mathbf{x})$ is the true conditional distribution of y given $\mathbf{x}$. The MLE of ϕ can be obtained by solving the observed score function $S_{obs}(\phi) = \partial \log L_{obs}(\phi)/\partial \phi = 0$. Finding the solution to the observed score equation can be computationally challenging because it involves integration with unknown parameters. An alternative way of finding the MLE of ϕ is to solve the mean score equation $\bar{S}(\phi) = 0$, where

$$\bar{S}(\phi) = \sum_{i=1}^{n} [\delta_i S_1(\phi; \mathbf{x}_i, y_i) + (1 - \delta_i) E\{S_0(\phi; \mathbf{x}_i, Y) \mid \mathbf{x}_i, \delta_i = 0\}], \quad (8.42)$$

and $S_\delta(\phi; \mathbf{x}, y)$ is defined in (8.36).

To compute the conditional expectation in (8.42), we use the following relationship:

$$P(y_i \in B \mid \mathbf{x}_i, \delta_i = 0)$$
$$= P(y_i \in B \mid \mathbf{x}_i, \delta_i = 1) \times \frac{P(\delta_i = 0 \mid \mathbf{x}_i, y_i \in B)/P(\delta_i = 1 \mid \mathbf{x}_i, y_i \in B)}{P(\delta_i = 0 \mid \mathbf{x}_i)/P(\delta_i = 1 \mid \mathbf{x}_i)}.$$

Thus, we can write the conditional distribution of the missing data given x as

$$f_0(y_i \mid \mathbf{x}_i) = f_1(y_i \mid \mathbf{x}_i) \times \frac{O(\mathbf{x}_i, y_i)}{E\{O(\mathbf{x}_i, Y_i) \mid \mathbf{x}_i, \delta_i = 1\}}, \quad (8.43)$$

where $f_\delta(y_i \mid \mathbf{x}_i) = f(y_i \mid \mathbf{x}_i, \delta_i = \delta)$, and

$$O(\mathbf{x}_i, y_i) = \frac{P(\delta_i = 0 \mid \mathbf{x}_i, y_i)}{P(\delta_i = 1 \mid \mathbf{x}_i, y_i)} \quad (8.44)$$

is the conditional odds of nonresponse. If the response probability in (8.24) follows from a logistic regression model

$$\pi(\mathbf{x}_i, y_i) \equiv P\left(\delta_i = 1 \mid \mathbf{x}_i, y_i\right) = \frac{\exp\left(\phi_0 + \phi_1\mathbf{x}_i + \phi_2 y_i\right)}{1 + \exp\left(\phi_0 + \phi_1\mathbf{x}_i + \phi_2 y_i\right)}, \qquad (8.45)$$

the odds function (8.44) can be written as $O\left(x_i, y_i\right) = \exp\left\{-\phi_0 - \phi_1\mathbf{x}_i - \phi_2 y_i\right\}$, and the expression (8.43) can be simplified to

$$f_0\left(y_i \mid \mathbf{x}_i\right) = f_1\left(y_i \mid \mathbf{x}_i\right) \times \frac{\exp\left(-\phi_2 y_i\right)}{E\left\{\exp\left(-\phi_2 Y\right) \mid \mathbf{x}_i, \delta_i = 1\right\}}. \qquad (8.46)$$

Model (8.46) states that the density for the nonrespondents is an exponential tilting of the density for the respondents. If $\phi_2 = 0$, the the response mechanism is ignorable and $f_0(y \mid \mathbf{x}) = f_1(y \mid \mathbf{x})$. Kim and Yu (2011) also used an exponential tilting model to compute the conditional expectation $E_0(y \mid \mathbf{x})$ nonparametrically from the respondents. Equation (8.46) implies that we only need the response model (8.45) and the conditional distribution of study variable given the auxiliary variables for respondents $f_1(y \mid \mathbf{x})$, which is relatively easy to verify from the observed part of the sample.

If $f_1(y \mid \mathbf{x})$ is known, we can obtain the maximum likelihood estimator of ϕ by solving

$$\sum_{i=1}^{n} \left[\delta_i S_1(\phi; \mathbf{x}_i, y_i) + (1 - \delta_i) \frac{E_1\{O(\mathbf{x}_i, Y; \phi)S_0(\phi; \mathbf{x}_i, Y) \mid \mathbf{x}_i\}}{E_1\{O(\mathbf{x}_i, Y; \phi) \mid \mathbf{x}_i\}} \right] = 0, \quad (8.47)$$

where $E_1(\cdot \mid \mathbf{x})$ is the expectation with respect to $f_1(y \mid \mathbf{x})$. To compute the MLE from (8.47) with unknown f_1, Riddles et al. (2016) considered a parametric approach by assuming that

$$f_1(y \mid \mathbf{x}) = f_1(y \mid \mathbf{x}; \gamma), \qquad (8.48)$$

for some γ. To estimate the response model parameter ϕ, we first need to obtain a consistent estimator of γ. For example, the MLE of γ can be computed by

$$\hat{\gamma} = \arg\max_{\gamma} \sum_{i=1}^{n} \delta_i \log f_1(y_i \mid \mathbf{x}_i; \gamma). \qquad (8.49)$$

Once $\hat{\gamma}$ is obtained from (8.49), the mean score equation can be written, by (8.42) and (8.55), as

$$\bar{S}\left(\phi \mid \hat{\gamma}\right) \equiv \sum_{i=1}^{n} \left[\delta_i S_1(\phi; \mathbf{x}_i, y_i) + (1 - \delta_i) \frac{\hat{E}_1\{O(\mathbf{x}_i, Y; \phi)S_0(\phi; \mathbf{x}_i, Y) \mid \mathbf{x}_i\}}{\hat{E}_1\{O(\mathbf{x}_i, Y; \phi) \mid \mathbf{x}_i\}} \right] = 0,$$

$$(8.50)$$

where $\hat{E}_1(\cdot \mid x)$ is the expectation with respect to $f_1(y \mid x; \hat{\gamma})$. To solve (8.50) for ϕ, either a Newton method or the EM algorithm can be used.

To discuss the EM algorithm to solve (8.50), we first consider the simple case when y is a categorical variable with M categories, taking values in $\{1, 2, \ldots, M\}$. Let $f_\delta(y \mid \mathbf{x}; \gamma)$ be the probability mass function of y conditional on $\mathbf{x}$ and δ. In this case, we can express (8.50) as

$$\bar{S}(\phi, \hat{\gamma}) = \sum_{\delta_i=1} S(\phi; \delta_i, \mathbf{x}_i, y_i) +$$

$$\sum_{\delta_i=0} \left\{ \frac{\sum_{y=1}^{M} S(\phi; \delta_i, \mathbf{x}_i, y) \exp(-\phi_2 y) f_1(y \mid \mathbf{x}_i; \hat{\gamma}, \phi)}{\sum_{y=1}^{M} \exp(-\phi_2 y) f_1(y \mid \mathbf{x}_i; \hat{\gamma}, \phi)} \right\} = 0. \qquad (8.51)$$

To solve (8.51) by the EM algorithm, we first compute the fractional weights

$$w_{iy}^{*(t)} = \frac{\exp(-\hat{\phi}_2^{(t)} y) f_1(y \mid \mathbf{x}_i; \hat{\gamma})}{\sum_{y=1}^{M} \exp(-\hat{\phi}_2^{(t)} y) f_1(y \mid \mathbf{x}_i; \hat{\gamma})}$$

using the current value $\hat{\phi}^{(t)}$ of ϕ. This is the E-step of the EM algorithm. In the M-step, the parameter estimate is updated by solving

$$\sum_{\delta_i=1} S(\phi; \delta_i, \mathbf{x}_i, y_i) + \sum_{\delta_i=0} \sum_{y=1}^{M} w_{iy}^{*(t)} S(\phi; \delta_i, \mathbf{x}_i, y) = 0.$$

If Y is continuous, an algorithm similar to the Monte Carlo EM algorithm can be implemented as follows:

(Step 1) Generate $y_{ij}^* \sim f(y \mid \mathbf{x}_i, \delta_i = 1, \hat{\gamma})$ for each nonrespondent i and $j = 1, 2, \ldots, m$, where $\hat{\gamma}$ is obtained from (8.49).

(Step 2) Using the current value of $\phi^{(t)}$ and the Monte Carlo sample from (Step 1), compute

$$\bar{S}_p(\phi \mid \phi^{(t)}) = \sum_{i=1}^{n} \left\{ \delta_i S(\phi; \delta_i, \mathbf{x}_i, y_i) + (1 - \delta_i) \sum_{j=1}^{m} w_{ij}^{*(t)} S(\phi; \delta_i, \mathbf{x}_i, y_{ij}^*) \right\},$$
$$(8.52)$$

where

$$w_{ij}^{*(t)} = \frac{O_{ij}^{*(t)}}{\sum_{k=1}^{m} O_{ik}^{*(t)}}, \qquad (8.53)$$

$O_{ij}^{*(t)} = 1/\pi_{ij}^*(\phi^{(t)}) - 1$, and $\pi_{ij}^*(\phi) = \pi(\mathbf{x}_i, y_{ij}^*; \phi)$.

(Step 3) Find the solution $\phi^{(t+1)}$ to $\bar{S}_p(\phi \mid \phi^{(t)}) = 0$, where $\bar{S}_p(\phi \mid \phi^{(t)})$ is computed from (Step 2).

(Step 4) Update $t = t + 1$ and repeat (Step 2)-(Step 3) until convergence.

In the above algorithm, (Step 1) and (Step 2) correspond to the E-step and (Step 3) corresponds to the M-step of the EM algorithm. Unlike the usual Monte Carlo EM algorithm, (Step1) is not repeated. The proposed method can be regarded as a special application of the parametric fractional imputation of Kim (2011) under nonignorable nonresponse. If the response model is of a logistic form in (8.54), the fractional weights in (8.53) can be simply expressed as

$$
w_{ij}^{*(t)} = \frac{\exp\left(-\phi_2^{(t)} y_{ij}^*\right)}{\sum_{k=1}^m \exp\left(-\phi_2^{(t)} y_{ik}^*\right)}.
$$

Riddles et al. (2016) provided some asymptotic properties of the PS estimator using $\hat{\phi}$ obtained from (8.50) with the parametric model assumption for $f_1(y \mid \mathbf{x})$.

Morikawa et al. (2017) relaxed the parametric model assumption for $f_1(y \mid \mathbf{x})$. Instead, they use a kernel-based nonparametric regression in computing $\hat{E}_1(\cdot \mid \mathbf{x})$ in (8.50) and developed a semiparametric MLE of ϕ.

We now consider an extension to semi-parametric regression model. The response probability follows from a logistic regression model

$$
\pi(\mathbf{x}_i, y_i) \equiv Pr\left(\delta_i = 1 \mid \mathbf{x}_i, y_i\right) = \frac{\exp\left\{g(\mathbf{x}_i) + \phi y_i\right\}}{1 + \exp\left\{g(\mathbf{x}_i) + \phi y_i\right\}}, \qquad (8.54)
$$

where $g(\mathbf{x})$ is completely unspecified. The expression (8.43) can be used to obtain

$$
f_0\left(y_i \mid \mathbf{x}_i\right) = f_1\left(y_i \mid \mathbf{x}_i\right) \times \frac{\exp\left(\gamma y_i\right)}{E\left\{\exp\left(\gamma Y\right) \mid \mathbf{x}_i, \delta_i = 1\right\}}, \qquad (8.55)
$$

where $\gamma = -\phi$ and $f_1(y \mid \mathbf{x})$ is the conditional density of y given $\mathbf{x}$ and $\delta = 1$.

Model (8.55) states that the density for the nonrespondents is an exponential tilting of the density for the respondents. The parameter γ is the tilting parameter that determines the amount of departure from the ignorability of the response mechanism. If $\gamma = 0$, the the response mechanism is ignorable and $f_0(y \mid \mathbf{x}) = f_1(y \mid \mathbf{x})$.

If γ is known, we can estimate $E(Y \mid \mathbf{x}, \delta = 0)$ by

$$
\hat{E}_0(Y \mid \mathbf{x}; \gamma) = \frac{\sum_{i=1}^n \delta_i \exp(\gamma y_i) K_h(x - x_i) y_i}{\sum_{i=1}^n \delta_i \exp(\gamma y_i) K_h(x - x_i)},
$$

where $K_h(x)$ is a kernel function with bandwidth h. Thus, a semiparametric imputation estimator for $\theta = E(Y)$ can be obtained by

$$
\hat{\theta}_I = \frac{1}{n} \sum_{i=1}^n \left\{\delta_i y_i + (1 - \delta_i) \hat{E}_0(Y \mid \mathbf{x}_i; \gamma)\right\}.
$$

Under this model, we can use

$$
E\left\{\frac{\delta}{\pi(\mathbf{x}, y)} - 1 \mid \mathbf{x}\right\} = 0
$$

to obtain

$$\exp\{g(\mathbf{x})\} = \frac{E\{\delta \exp(\gamma y) \mid \mathbf{x}\}}{E\{1 - \delta \mid \mathbf{x}\}}.$$

For known γ case, we can use kernel regression estimator

$$\exp\{\hat{g}_\gamma(x)\} = \frac{\sum_{i=1}^n \delta_i \exp(\gamma y_i) K_h(x - x_i)}{\sum_{i=1}^n (1 - \delta_i) K_h(x - x_i)}$$

to obtain the following profile response probability

$$\hat{\pi}_p(x_i, y_i; \gamma) = \frac{\exp\{\hat{g}_\gamma(x_i) - \gamma y_i\}}{1 + \exp\{\hat{g}_\gamma(x_i) - \gamma y_i\}}.$$

For estimation of γ, Shao and Wang (2016) suggested using the GMM method based on some moment conditions. Uehara and Kim (2020) also suggested using the profile maximum likelihood method. The idea is to use the profile log likelihood

$$l_{p,\text{com}}(\gamma) = \sum_{i=1}^n [\delta_i \log\{\hat{\pi}_p(x_i, y_i; \gamma)\} + (1 - \delta_i) \log\{1 - \hat{\pi}_p(x_i, y_i; \gamma)\}]$$

and apply a version of EM algorithm using (8.55) in the E-step.

8.6   Latent Variable Approach

Another approach of modeling nonignorable nonresponse is to assume a latent variable that is related to the survey variable and then assume that the study variable is observed if and only if the latent variable exceeds a threshold (say zero). The latent variable approach is very popular in econometrics in explaining self-selection bias (Heckman, 1979). O'Muircheartaigh and Moustaki (1999) also considered the latent variable approach to model item nonresponse in attitude scale. In the survey sampling context, Copas and Farewell (1998) introduced a variable called 'enthusiasm-to-respond' to the survey, which is expected to be related to probabilities of unit and item response. The use of continuous latent variables to model item nonresponse was considered in Moustaki and Knott (2000). Matei and Ranalli (2015) also considered latent modeling approach for handling non-ignorable nonresponse problems.

To explain the idea, consider the following example considered in Little and Rubin (2002, Example 15.7).

Example 8.4. *Suppose that the original study variable y follows a normal-theory linear model given by*

$$y_i = \mathbf{x}_i'\beta + e_i \tag{8.56}$$

and $e_i \sim N(0, \sigma^2)$. Let δ_i be the response indicator function for y_i. The response model is

$$\delta_i = \begin{cases} 1 & \text{if } z_i > 0 \\ 0 & \text{if } z_i \leq 0, \end{cases}$$

where z_i is the latent variable representing the level of survey participation and follows

$$z_i = \mathbf{x}_i'\gamma + u_i$$

and

$$\begin{pmatrix} e_i \\ u_i \end{pmatrix} \sim N\left[\begin{pmatrix} 0 \\ 0 \end{pmatrix}, \begin{pmatrix} \sigma^2 & \rho\sigma \\ \rho\sigma & 1 \end{pmatrix}\right]. \tag{8.57}$$

Here, the variance of u_i is set to 1 for the sake of model identifiability.

By the property of normal distribution, we can derive

$$\begin{aligned} P(\delta_i = 1 \mid \mathbf{x}_i, y_i) &= P(z_i \geq 0 \mid \mathbf{x}_i, y_i) \\ &= 1 - \Phi\left\{\frac{\mathbf{x}_i'\gamma + \rho\sigma^{-1}(y_i - \mathbf{x}_i'\beta)}{\sqrt{1-\rho^2}}\right\}. \end{aligned}$$

When $\rho \neq 0$, the response probability depends on y_i, which is subject to missingnenss, and the response mechanism becomes nonignorable. If $\rho = 0$, then the response mechanism is ignorable.

In the above example, z_i is always missing but is useful in representing the response probability as a function of parameters in the latent variable model. The latent model is then viewed as a *hurdle model* since crossing a hurdle or threshold leads to participation. The classic early application of the latent model to nonignorable missing (or selection bias) was to labor supply, where z is the unobserved desire or propensity to work, while y is the actual hours worked. In Example 8.4, the observed likelihood is

$$L_{obs} = \prod_{i=1}^{n} \{P(z_i \leq 0 \mid \mathbf{x}_i)\}^{1-\delta_i} \{f(y_i \mid \mathbf{x}_i, z_i > 0)P(z_i > 0 \mid \mathbf{x}_i)\}^{\delta_i}.$$

This likelihood function is applicable to general models, not just linear models with joint normal errors.

In the normal case,

$$E(y_i \mid \mathbf{x}_i, z_i > 0) = \mathbf{x}_i'\beta + \rho\sigma\lambda(\mathbf{x}_i'\gamma),$$

where $\lambda(z) = \phi(z)/\Phi(z)$, which is often called inverse Mills ratio (Amemiya, 1985). To estimate the parameters, two-step estimation procedure, proposed by Heckman (1979), can be easily implemented as follows:

[Step 1] Estimate γ by applying the probit regression of δ_i on $\mathbf{x}_i$ since $P(\delta = 1 \mid \mathbf{x}) = \Phi(\mathbf{x}'\gamma)$.

[Step 2] Using only the cases with $\delta_i = 1$, fit a linear regression model

$$y_i = \mathbf{x}_i'\beta + \sigma_{12}q_i + \nu_i,$$

where ν_i is an error term and $q_i = \lambda(\mathbf{x}_i'\hat{\gamma})$ with $\hat{\gamma}$ obtained from [Step 1].

In Step 2, the covariate q_i is used to correct for the selection bias in the sample. The computation for the two-step method is relatively easy. Instead of the above two-step method, EM algorithm can also be used with additional computational cost. Note that, as $\lambda(z) \cong a + bz$, we may write

$$E(y_i \mid \mathbf{x}_i, z_i > 0) \cong a + \mathbf{x}_i'\beta + b\mathbf{x}_i'\gamma,$$

which leads to obvious multicollinearity problems (Nawata and Nagase, 1996). To avoid this nonidentifiability problem, one regressor in estimating γ may be excluded from the model in estimating β, which might limit the applicability of the latent variable approach in practice. For more details of the latent variable approach to handle selection bias, see Chapter 16 of Cameron and Trivedi (2005).

8.7 Callbacks

We now assume a nonignorable nonresponse mechanism of the form

$$P(\delta_i = 1 \mid \mathbf{x}_i, y_i) = \pi(\phi; \mathbf{x}_i, y_i) = \frac{\exp(\phi_0 + \mathbf{x}_i, \phi_1 + y_i\phi_2)}{1 + \exp(\phi_0 + \mathbf{x}_i'\phi_1 + y_i\phi_2)} \qquad (8.58)$$

and discuss how to obtain a consistent estimator of the response probability under the existence of missing data. Clearly, the score equation

$$\sum_{i=1}^{n} \{\delta_i - \pi(\phi; \mathbf{x}_i, y_i)\} (\mathbf{x}_i', y_i) = 0$$

cannot be solved because y_i are not observed when $\delta_i = 0$.

To estimate the parameters in (8.58), we consider the special case when there are some callbacks among nonrespondents. That is, among the elements with $\delta_i = 0$, further efforts are made to obtain the observation of y_i. Let $\delta_{2i} = 1$ if the element i is selected for a callback or $\delta_i = 1$ and $\delta_{2i} = 0$ otherwise. We assume that the selection mechanism for the callback depends only on δ_i. That is,

$$P(\delta_2 = 1 \mid \mathbf{x}, y, \delta) = \begin{cases} 1 & \text{if } \delta = 1 \\ \nu & \text{if } \delta = 0 \end{cases} \qquad (8.59)$$

for some $\nu \in (0, 1]$. The following lemma shows that the response probability can be estimated from the original sample and the callback sample.

Lemma 8.3. *Assume that the response mechanism satisfies (8.58) and the followup sample is randomly selected among nonrespondents with probability ν. Then, the response probability among the set with $\delta_{i2} = 1$ can be expressed as*

$$P(\delta_i = 1 \mid \mathbf{x}_i, y_i, \delta_{2i} = 1) = \frac{\exp(\phi_0^* + \mathbf{x}_i'\phi_1^* + y_i\phi_2^*)}{1 + \exp(\phi_0^* + \mathbf{x}_i'\phi_1^* + y_i\phi_2^*)}, \qquad (8.60)$$

where $\phi_0^* = \phi_0 - \ln(\nu)$, $(\phi_1^*, \phi_2^*) = (\phi_1, \phi_2)$, and (ϕ_0, ϕ_1, ϕ_2) is defined in (8.58).

Proof. By Bayes formula,

$$\frac{P(\delta = 1 \mid \mathbf{x}, y, \delta_2 = 1)}{P(\delta = 0 \mid \mathbf{x}, y, \delta_2 = 1)} = \frac{P(\delta_2 = 1 \mid \mathbf{x}, y, \delta = 1)}{P(\delta_2 = 1 \mid \mathbf{x}, y, \delta = 0)} \times \frac{P(\delta = 1 \mid \mathbf{x}, y)}{P(\delta = 0 \mid \mathbf{x}, y)}.$$

By (8.59), the above formula reduces to

$$\frac{P(\delta = 1 \mid \mathbf{x}, y, \delta_2 = 1)}{P(\delta = 0 \mid \mathbf{x}, y, \delta_2 = 1)} = \frac{1}{\nu} \times \frac{P(\delta = 1 \mid \mathbf{x}, y)}{P(\delta = 0 \mid \mathbf{x}, y)}.$$

Taking the logarithm of the above equality, we have

$$\phi_0^* + \mathbf{x}'\phi_1^* + y\phi_2^* = \phi_0 - \ln(\nu) + \mathbf{x}'\phi_1 + y\phi_2.$$

Because the above relationship holds for all $\mathbf{x}$ and y, we have $\phi_0^* = \phi_0 - \ln(\nu)$ and $(\phi_1^*, \phi_2^*) = (\phi_1, \phi_2)$. $\qquad\qquad\square$

By Lemma 8.3, the MLE of ϕ^* can be obtained by maximizing the conditional likelihood. That is, we solve

$$\sum_{i=1}^{n} \delta_{2i} \left\{ \delta_i - \pi(\phi^*; \mathbf{x}_i, y_i) \right\} (\mathbf{x}_i, y_i) = 0 \qquad (8.61)$$

and then apply the transformation in Lemma 8.1. In particular, the MLE for the slope (ϕ_1, ϕ_2) in (8.58) can be directly computed by solving (8.61). Asymptotic variance of $(\hat{\phi}_1, \hat{\phi}_2)$ is directly obtained by the asymptotic variance of $(\hat{\phi}_1^*, \hat{\phi}_2^*)$. For more theoretical details, see Scott and Wild (1997).

In practice, the sample from the callback is also subject to missingness and, in this case, the score equation (8.61) is not directly applicable. Now, assume that there are several followups to increase the number of respondents. Let A_1 be the set of respondents who provided answers to the surveys at the initial contact. Suppose that there are $T - 1$ followups made to those who remain nonrespondents in the survey. Let $A_2(\subset A_1)$ be the set of respondents who provided answers to the surveys at the time of the second contact. By definition, A_2 contains those already provided answers in the first contact. Thus, $A_1 \subset A_2$. Similarly, we can define A_3 be the set of respondents who provided answers at the time of the third contact, or the second followup. Continuing the process, we can define $A_1, \ldots, A_T$ such that

$$A_1 \subset \cdots \subset A_T.$$

Followup can also be called call-back. Suppose that there are T attempts (or $T-1$ followups) to obtain the survey response y_i, and let δ_{it} be the response indicator function for y_i at the t-th attempt. If $\delta_{iT} = 0$, then the unit never responds and it is called hardcore nonrespondent (Drew and Fuller, 1980). Using the definition of A_t, we can write $\delta_{it} = 1$ if $i \in A_t$ and $\delta_{it} = 0$ otherwise.

When the study variable y is categorical with K categories, Drew and Fuller (1980) proposed using a multinomial distribution with $T \times K + 1$ cells, where the cell probabilities are defined by

$$
\begin{aligned}
\pi_{tk} &= \gamma(1-p_k)^{t-1}p_k f_k, \\
\pi_0 &= (1-\gamma) + \gamma \sum_{k=1}^{K}(1-p_k)^T f_k,
\end{aligned}
$$

p_k is the response probability for category k, f_k is the population proportion such that $\sum_{k=1}^{K} f_k = 1$, and $1-\gamma$ is a proportion of hard-core nonrespondents. Thus, π_{tk} means the response probability that an individual in category k will respond at time t, and π_0 is the probability that an individual will not have responded after T trials. Under simple random sampling, the maximum likelihood estimator of the parameter can be easily obtained by maximizing the log-likelihood

$$
\log L = \sum_{t=1}^{T}\sum_{k=1}^{K} n_{tk}\log \pi_{tk} + n_0 \log \pi_0,
$$

where n_{tk} is the number of elements in the k-th category responding on the t-th contact, and n_0 is the number of individual who did not respond up to the T-th contact.

Alho (1990) considered the same problem with continuous y variable under simple random sampling. Let p_{it} be the conditional probability of $\delta_{it} = 1$, conditional on y_i and $\delta_{i,t-1} = 0$, and assume the logistic regression model

$$
p_{it} = P(\delta_{it} = 1 \mid \delta_{i,t-1} = 0, y_i) = \frac{\exp(\alpha_t + x_i\phi_1 + y_i\phi_2)}{1 + \exp(\alpha_t + x_i\phi_1 + y_i\phi_2)}, \quad t = 1,2,\ldots,T,
$$

$$(8.62)$$

for the conditional response probability of δ_{it}. Here, we assume $\delta_{i0} \equiv 0$.

To estimate the parameters in (8.62), Alho (1990) also assumed that $(\delta_{i1}, \delta_{i2} - \delta_{i1}, \ldots, \delta_{iT} - \delta_{i,T-1}, 1 - \delta_{iT})$ follows a multinomial distribution with parameter vector $(\pi_{i1}, \pi_{i2}, \ldots, \pi_{iT}, 1 - \sum_{t=1}^{T}\pi_{it})$, where

$$
\pi_{it} = P(\delta_{i,t-1} = 0, \delta_{it} = 1 \mid y_i). \tag{8.63}
$$

Thus, we can write $\pi_{it} = p_{it}\prod_{k=1}^{t-1}(1-p_{ik})$. Under this setup, Alho (1990) considered maximizing the following conditional likelihood

$$L(\phi) \;=\; \prod_{\delta_{iT}=1} P(\delta_{i1}=1 \mid y_i, \delta_{iT}=1)^{\delta_{i1}} \prod_{t=2}^{T} \left\{ P\left(\delta_{it}=1 \mid y_i, \delta_{i,t-1}=0, \delta_{iT}=1\right) \right\}^{\delta_{it}}$$

$$= \; \prod_{R_i=1} \left(\frac{\pi_{i1}}{1-\pi_{i,T+1}} \right)^{\delta_{i1}} \prod_{t=2}^{T} \left(\frac{\pi_{it}}{1-\pi_{i,T+1}} \right)^{\delta_{it}-\delta_{i,t-1}}, \tag{8.64}$$

where $\pi_{i,T+1} = 1 - \sum_{t=1}^{T} \pi_{it}$. To avoid the non-identifiability problem, Alho (1990) imposed

$$\sum_{i \in A_{t-1}} \delta_{it} \exp\left(-\alpha_t - \phi y_i\right) = n - (n_1 + \cdots + n_t), \tag{8.65}$$

for $t = 1, 2, \ldots, T$. Note that (8.65) computes α_t given ϕ. To incorporate the observed auxiliary information outside A_t, one can add the following constraints

$$\sum_{i=1}^{n}(1-\delta_{i,t-1})\frac{\delta_{iT}}{\hat{p}_{it}} \;=\; \sum_{i=1}^{n}(1-\delta_{i,t-1}), \quad t = 1,2,\ldots,T \tag{8.66}$$

$$\sum_{i=1}^{n}\frac{\delta_{iT}}{(1-\hat{\pi}_{i,T+1})}x_i \;=\; \sum_{i=1}^{n}x_i. \tag{8.67}$$

A constrained optimization algorithm can be used to find the constrained maximum likelihood estimators.

Instead of the maximum likelihood method from the conditional likelihood, a calibration approach can also be used. Kim and Im (2012) proposed solving

$$\sum_{i=1}^{n}\delta_{i,t-1}(x_i,y_i) + \sum_{i=1}^{n}(1-\delta_{i,t-1})\frac{\delta_{it}}{p_{it}}(x_i,y_i)$$

$$= \sum_{i=1}^{n}\delta_{i,T-1}(x_i,y_i) + \sum_{i=1}^{n}(1-\delta_{i,T-1})\frac{\delta_{iT}}{p_{iT}}(x_i,y_i) \tag{8.68}$$

for $t = 1, 2, \ldots, T-1$, and

$$\sum_{i=1}^{n}(1-\delta_{i,t-1})\frac{\delta_{it}}{p_{it}} = \sum_{i=1}^{n}(1-\delta_{i,t-1}), \quad t = 1,2,\ldots,T. \tag{8.69}$$

Note that both terms in (8.68) estimates $\sum_{i=1}^{n} y_i$ unbiasedly under the conditional response model.

Now, under the conditional response model in (8.62), we have, by (8.69),

$$\sum_{i=1}^{n}(1-\delta_{i,t-1})\delta_{it}\{1+\exp(-\alpha_t - \phi_1 x_i - \phi_2 y_i)\} = \sum_{i=1}^{n}(1-\delta_{i,t-1}) \tag{8.70}$$

and the solution $\hat{\alpha}_t$ to (8.70) can be written

$$\exp(-\hat{\alpha}_t) = \frac{\sum_{i=1}^n (1 - \delta_{i,t-1})(1 - \delta_{it})}{\sum_{i=1}^n (1 - \delta_{i,t-1})\delta_{it}\exp(-\phi_1 x_i - \phi_2 y_i)}. \tag{8.71}$$

Inserting (8.71) into (8.68) and (8.69), we have

$$(\hat{X}_{R(t)}, \hat{Y}_{R(t)}) + \hat{N}_{M(t)} \left(\frac{\sum_{i=1}^n w_{i,t}(\phi)(x_i, y_i)}{\sum_{i=1}^n w_{i,t}(\phi)} \right)$$

$$= (\hat{X}_{R(T)}, \hat{Y}_{R(T)}) + \hat{N}_{M(T)} \left(\frac{\sum_{i=1}^n w_{i,T}(\phi)(x_i, y_i)}{\sum_{i=1}^n w_{i,T}(\phi)} \right)$$

for $t = 1, 2, \ldots, T-1$, where $(\hat{X}_{R(t)}, \hat{Y}_{R(t)}) = \sum_{i=1}^n \delta_{it}(x_i, y_i)$, $\hat{N}_{M(t)} = \sum_{i=1}^n (1 - \delta_{it})$, and $w_{it}(\phi) = (1 - \delta_{i,t-1})\delta_{it}\exp(-\phi_1 x_i - \phi_2 y_i)$. Thus, we have $p + q$ parameters ($p = \dim(x)$ and $q = \dim(y)$) with $(p+q)(T-1)$ equations. When $T > 2$, we have more equations than parameters, and so we can apply the GMM technique to compute the estimates.

8.8   Capture-Recapture (CR) Experiment

In this section, we consider the case of making two independent attempts to obtain response for (x, y), where y is subject to missingness, and x is always observed. The classical capture-recapture (CR) sampling setup can be applied to estimate the response probability. Capture-recapture (CR) sampling is very popular in estimating the population size of wildlife animals. Amstrup et al. (2005) provided a comprehensive summary of the existing methods for CR analysis. Huggins and Hwang (2011) reviewed the conditional likelihood approach in CR experiments.

To apply the conditional likelihood approach, we assume that the two response indicators, δ_{1i} and δ_{2i}, are assumed to be independently generated from Bernoulli distributions with probabilities

$$\pi_{1i}(\phi) = P(\delta_{1i} = 1 \mid x_i, y_i) = \frac{\exp(\phi_0 + \phi_1 x_i + \phi_2 y_i)}{1 + \exp(\phi_0 + \phi_1 x_i + \phi_2 y_i)}$$

and

$$\pi_{2i}(\phi^*) = P(\delta_{2i} = 1 \mid x_i, y_i) = \frac{\exp(\phi_0^* + \phi_1^* x_i + \phi_2^* y_i)}{1 + \exp(\phi_0^* + \phi_1^* x_i + \phi_2^* y_i)},$$

respectively, where $\phi = (\phi_0, \phi_1, \phi_2)$ and $\phi^* = (\phi_0^*, \phi_1^*, \phi_2^*)$. Write $\Phi = (\phi, \phi^*)$. An efficient estimator of Φ can be obtained by maximizing the conditional likelihood

$$
\begin{aligned}
L_C(\Phi) \;=\; &\prod_{i \in A_1/A_2} \frac{\pi_{1i}(\phi)\{1 - \pi_{2i}(\phi^*)\}}{p_i(\phi, \phi^*)} \prod_{i \in A_1 \cap A_2} \frac{\pi_{1i}(\phi)\pi_{2i}(\phi^*)}{p_i(\phi, \phi^*)} \\
&\prod_{i \in A_2/A_1} \frac{\{1 - \pi_{1i}(\phi)\}\pi_{2i}(\phi^*)}{p_i(\phi, \phi^*)},
\end{aligned}
$$

where $A_2/A_1 = A_2 \cap A_1^c$ and A_1 is the set of sample elements with $\delta_{1i} = 1$, A_2 is the set of sample elements with $\delta_{2i} = 1$, and $p_i(\phi, \phi^*) = 1 - \{1 - \pi_{1i}(\phi)\}\{1 - \pi_{2i}(\phi^*)\}$. The conditional likelihood is obtained by considering the conditional distribution of $(\delta_{1i}, \delta_{i2})$ given that unit i is selected in either of the two samples. The log-likelihood of the conditional distribution is

$$
\begin{aligned}
l_C(\Phi) \;=\; &\sum_{i \in A_1} \log(\pi_{1i}) + \sum_{i \in A_2} \log(\pi_{2i}) + \sum_{i \in A_1/A_2} \log(1 - \pi_{2i}) \\
&+ \sum_{i \in A_2/A_1} \log(1 - \pi_{1i}) - \sum_{i \in A_1 \cup A_2} \log(p_i).
\end{aligned}
$$

The conditional maximum likelihood estimator (CMLE) that maximizes the conditional likelihood can be obtained by solving $S_C(\Phi) = 0$, where $S_C(\Phi) = \partial l_C(\Phi)/\partial \Phi = (S'_{C1}(\Phi), S'_{C2}(\Phi))'$ with

$$
S_{C1}(\Phi) \equiv \sum_{i \in A_1} (1, \mathbf{x}'_i, y_i)' - \sum_{i \in A_1 \cup A_2} \frac{\pi_{1i}(\phi)}{p_i(\phi, \phi^*)} (1, \mathbf{x}'_i, y_i)'
$$

and

$$
S_{C2}(\Phi) \equiv \sum_{i \in A_2} (1, \mathbf{x}'_i, y_i)' - \sum_{i \in A_1 \cup A_2} \frac{\pi_{2i}(\phi)}{p_i(\phi, \phi^*)} (1, \mathbf{x}'_i, y_i)'.
$$

Once the CMLE of Φ, denoted by $\hat{\Phi}$, is obtained, we can construct the following propensity score estimator of $\theta = E(Y)$ based on $A_1 \cup A_2$ by

$$
\hat{\theta} = \frac{\sum_{i \in A_1 \cup A_2} p_i^{-1}(\hat{\phi}, \hat{\phi}^*) y_i}{\sum_{i \in A_1 \cup A_2} p_i^{-1}(\hat{\phi}, \hat{\phi}^*)}.
$$

Asymptotic properties of the above PS estimator can be obtained by a standard linearization argument.

Exercises

1. Consider a bivariate categorical variable (x, y), where x takes values among $\{1, \ldots, K\}$ and y is either 0 or 1. We assume a fully

nonparametric model on (x,y) such that $P(X = i, Y = j) = p_{ij}$ with $\sum_{i=1}^{K} \sum_{j=0}^{1} p_{ij} = 1$. Assume that x_i is fully observed, and y_i is subject to missingness with probability

$$P(\delta_i = 1 \mid x_i, y_i) = \frac{\exp(\phi_0 + \phi_1 y_i)}{1 + \exp(\phi_0 + \phi_1 y_i)}$$

for some (ϕ_0, ϕ_1). Answer the following questions:

(a) Show that the model for observed data is identifiable if $K \geq 3$.

(b) Discuss how to construct a GMM estimator of (ϕ_0, ϕ_1).

(c) Discuss how to construct a pseudo likelihood estimator of (ϕ_0, ϕ_1).

2. Under the setup of Example 2.1, answer the following questions:

(a) Show that

$$E(y_i \mid x_i, y_i \leq 0; \theta) = x_i'\beta - \sigma\lambda\left(-\frac{x_i'\beta}{\sigma}\right),$$

where $\lambda(z) = \phi(z)/\Phi(z)$.

(b) Discuss how to implement EM algorithm for estimating β when σ is known.

(c) Discuss how to implement EM algorithm for estimating $\theta = (\beta, \sigma)$.

3. Under the setup of Example 8.4, discuss how to implement EM algorithm for estimating γ and β.

4. Suppose that $f(y \mid x, \delta = 1)$ is a normal distribution with mean $\beta_0 + \beta_1 x$ and variance σ^2. Assume that

$$P(\delta = 1 \mid x, y) = \frac{\exp(\phi_0 + \phi_1 x + \phi_2 y)}{1 + \exp(\phi_0 + \phi_1 x + \phi_2 y)}.$$

Using (8.55), show that the conditional distribution $f(y \mid x, \delta = 0)$ also follows from a normal distribution with mean $\beta_0 - \phi_2\sigma^2 + \beta_1 x$ and variance σ^2.

5. Assume that $y_i \mid \delta_i$ follows a normal distribution with mean $\mu_1\delta_i + \mu_0(1 - \delta_i)$ and variance σ^2. Assume $\pi = P(\delta = 1)$ is known. Prove that the response probability can be written as

$$P(\delta = 1 \mid y) = \frac{\exp(\phi_0 + \phi_1 y)}{1 + \exp(\phi_0 + \phi_1 y)} \tag{8.72}$$

for some (ϕ_0, ϕ_1). Express ϕ_0 and ϕ_1 in terms of μ_0, μ_1, and π.

6. Consider the problem of estimating $\theta = E(Y)$ under the existence of missing data in y. Assume that the response model satisfies (8.72) with known ϕ_1. Answer the following questions.

(a) Show that, using (8.55) or other formulas,

$$\hat{\mu}_0 = \frac{\sum_{i=1}^{n} \delta_i \exp(-\phi y_i) y_i}{\sum_{i=1}^{n} \delta_i \exp(-\phi y_i)} \tag{8.73}$$

is asymptotically unbiased for $\mu_0 = E(Y \mid \delta = 0)$.

(b) The prediction estimator

$$\hat{\theta}_p = \frac{1}{n} \sum_{i=1}^{n} \{\delta_i y_i + (1 - \delta_i)\hat{\mu}_0\},$$

where $\hat{\mu}_0$ is defined in (8.73), is algebraically equivalent to the PS estimator

$$\hat{\theta}_{PS} = \frac{1}{n} \sum_{i=1}^{n} \delta_i \frac{1}{\hat{\pi}_i} y_i,$$

where $\hat{\pi}_i$ is the estimated response probability computed by

$$\sum_{i=1}^{n} \left(\frac{\delta_i}{\hat{\pi}_i} - 1 \right) = 0.$$

7. Derive (8.64).

8. Suppose that we are interested in estimating the current employment status of a certain population, and we have four followups in the survey. Suppose that we have obtained the following results:

TABLE 8.1
Realized Responses in a Survey of Employment Status

Status	$T = 1$	$T = 2$	$T = 3$	$T = 4$	No Response
Employment	81,685	46,926	28,124	15,992	
Unemployment	1,509	948	597	352	32,350
Not in labor force	57,882	32,308	19,086	10,790	

(a) Compute the full sample likelihood function under the assumption that there is no nonresponse after four followups.

(b) Compute the conditional likelihood function and obtain the parameter values that maximize the conditional likelihood.

(c) Use the Drew and Fuller (1980) method to compute the maximum likelihood estimator.

(d) Discuss how to compute the standard errors of the MLE in (c).

9. Assume that two voluntary samples, A_1 and A_2, are obtained with a nested structure. That is, $A_2 \subset A_1$. Let δ_{1i} and δ_{2i} be the response indicator functions of the first sample A_1 and the second sample A_2, respectively. Assume that A_1 has the probability of survey participation given by

$$P(\delta_{1i} = 1 \mid y_{1i}) = \frac{\exp(\phi_0 + \phi_1 y_{1i})}{1 + \exp(\phi_0 + \phi_1 y_{1i})}$$

for some (ϕ_0, ϕ_1), where y_{1i} is the value of y of unit i at the time of realizing $\delta_{1i} = 1$. To estimate ϕ_1, we obtain a second voluntary sample A_2, by asking the same questions again, where the probability of the second survey participation

$$P(\delta_{2i} = 1 \mid y_{2i}, \delta_{1i} = 1) = \frac{\exp(\phi_0^* + \phi_1 y_{2i})}{1 + \exp(\phi_0^* + \phi_1 y_{2i})}$$

for some ϕ_0^*, where y_{2i} is the value of y of unit i at the time of realizing $\delta_{2i} = 1$. Answer the following questions:

(a) Show that a consistent estimator of (ϕ_0^*, ϕ_1) is obtained by solving

$$\sum_{i \in A_1} \delta_{2i} \{1 + \exp(-\phi_0^* - \phi_1 y_{2i})\} (1, y_{1i}) = \sum_{i \in A_1} (1, y_{1i}).$$

(b) Assuming that the marginal probability $\pi = P(\delta_{1i} = 1)$ is known, discuss how to construct a PS estimator of $\theta = E(Y)$.

(c) Derive the asymptotic variance of the PS estimator in (b).

9

Longitudinal and Clustered Data

In a longitudinal study, we collect data from every sampled subject (or unit) at multiple time points. Under cluster sampling, we obtain data from units within each sampled cluster. Longitudinal or clustered data are often encountered in medical studies, population health, social studies, and economics. Related statistical analyses typically estimate or make inference on the mean of the study response variable or the relationship between the response and some covariates. Longitudinal or clustered data look similar with multivariate data, but the main difference is the former studies one response variable measured at different time points or units, whereas the latter concerns several different variables. There are two major approaches for longitudinal or clustered data analysed under complete data. One is based on modeling the marginal distribution (or mean and variance) of the responses without requiring a correct specification of the correlation structure of longitudinal data; for example, the generalized estimation equation (GEE) approach. The linear model approach is a special case of GEE. The other approach is based on a mixed-effect model, which applies to the conditional distribution (or mean and variance) of the responses given some random effects.

Missing data in the study variable is a serious impediment to performing a valid statistical analysis, because the response probability usually (directly or indirectly) depends on the value of the response and missing mechanism is often nonignorable. In this chapter, we introduce some methods of handling longitudinal or cluster data with missing values. These methods are introduced in each section that makes a particular assumption about missing data mechanism.

9.1 Ignorable Missing Data

Let y_{it} be the response at time point t for subject i, $\mathbf{y}_i = (y_{i1}, ..., y_{iT})$, δ_{it} be the indicator of whether y_{it} is observed, $\boldsymbol{\delta}_i = (\delta_{i1}, ..., \delta_{iT})$, and $\mathbf{x}_{it}$ be a covariate vector whose values are always observed, $\mathbf{x}_i = (\mathbf{x}_{i1}, ..., \mathbf{x}_{iT})$. The covariate $\mathbf{x}_i$ may be cross-sectional or longitudinal.

DOI: 10.1201/9780429321740-9

We assume that $(\mathbf{y}_i, \boldsymbol{\delta}_i, \mathbf{x}_i)$, $i = 1, ..., n$, are independent. Then the general definition of ignorable missing data reduces to

$$q(\boldsymbol{\delta}_i \mid \mathbf{y}_i, \mathbf{x}_i) = q(\boldsymbol{\delta}_i \mid \mathbf{y}_{i,\text{obs}}, \mathbf{x}_i), \tag{9.1}$$

where $\mathbf{y}_{i,\text{obs}}$ contains observed components of $\mathbf{y}_i$. An assumption stronger than (9.1) is that the response mechanism is covariate-dependent, i.e.,

$$q(\boldsymbol{\delta}_i \mid \mathbf{y}_i, \mathbf{x}_i) = q(\boldsymbol{\delta}_i \mid \mathbf{x}_i). \tag{9.2}$$

If (9.2) does not hold, then (9.1) appears unnatural when $\mathbf{y}_i$ is a cluster of data; for example, if y_{it}'s are responses from a sampled household, then it is often not true that the probability of a unit not responding depends on the respondents in the same household. In the situation where components of $\mathbf{y}_i$ are sampled at T ordered time points, (9.1) is unnatural unless missing pattern is monotone in the sense that if y_{it} is missing, so is y_{is} for any $s > t$, as it is hard to imagine that the probability of observing y_{it} depends on an observed y_{is} at a future time point $s > t$.

Thus, in this section we focus on monotone missing data under assumption (9.1) or general type missing data under assumption (9.2). If we adopt a parametric approach, then methods for multivariate responses in previous chapters can be applied here. If we assume a covariate-dependent response mechanism, then we do not need a parametric assumption on $f(\mathbf{y}_i)$ or $f(\mathbf{y}_i \mid \mathbf{x}_i)$. In this case, PS methods can be applied.

In the rest of this section, we consider an imputation method that assumes monotone ignorable missingness but does not require a parametric model on $\mathbf{y}_i$ (Paik, 1997). To use this method, we need to assume y_{i1} observed or use one benchmarking covariate as y_{i1} (such as the baseline observations in a clinical study).

For a missing y_{it} with the last observed value at time point $r < t$, y_{it} is imputed by $\hat{\phi}_{t,r}(\mathbf{y}_{jr})$, where $\mathbf{y}_{ir} = (y_{i1}, ..., y_{ir})$ and $\hat{\phi}_{t,r}$ is an estimated conditional expectation

$$\phi_{t,r}(\mathbf{y}_{ir}) = E(Y_{it} \mid \mathbf{y}_{ir}, \mathbf{x}_i, \delta_{i(r+1)} = 0, \delta_{ir} = 1) = E(Y_{it} \mid \mathbf{y}_{ir}, \mathbf{x}_i, \delta_{i(r+1)} = 1), \tag{9.3}$$

using data from all units with $\delta_{i(r+1)} = 1$, $r = 1, ..., t-1$, $t = 2, ..., T$. When $\mathbf{y}_i$ is multivariate normal, the conditional expectation in (9.3) is linear in $\mathbf{y}_{ir}$. Thus, $\hat{\phi}_{t,r}$ is the linear regression function fitted using y_{it} as the response and $y_{i1}, ..., y_{ir}$ as predictors. Data for this regression fitting are from units with $\delta_{i(r+1)} = 1$; the observed $\mathbf{y}_{ir}$ is used as a predictor and either observed y_{it} or previously imputed values of missing y_{it} are used as responses.

Note that the previously imputed values can be used as responses, but not as predictors in the regression fitting. For each fixed t, imputation can be done sequentially for $r = t-1, t-2, ..., 2$, so that previously imputed responses can be used.

When $\mathbf{y}_i$ is not normal, however, the conditional expectation in (9.3) is not linear in $\mathbf{y}_{ir}$, even if $\phi_{t,r}(\mathbf{y}_{ir})$ is linear in the case of no missing data. Hence, some nonparametric or semiparametric regression methods need to be applied to obtain $\hat{\phi}_{t,r}$. Since nonparametric or semiparametric regression will be presented later in more complicated situations, we omit the discussion here.

After missing values are imputed, the mean of y_{it} for each t can be estimated by the sample mean at t by treating imputed values as observed. If a regression between y_{it} and $\mathbf{x}_i$ needs to be fitted, we can also use standard methods by treating imputed values as observed.

To assess the variances of point estimators, however, we cannot treat imputed values as observed data and apply standard variance estimation methods. Adjustments have to be made or a bootstrap method that consists of a re-imputation component can be applied.

9.2 Nonignorable Monotone Missing Data

In this section, we consider nonignorable missing data with a monotone missing pattern, i.e., if y_{it} is missing at a time point t, then y_{is} is also missing at any $s > t$. Monotone missingness is also referred to as *dropout*. In this section, we introduce methods under different parametric or nonparametric assumptions on the propensity $q(\boldsymbol{\delta} \mid \mathbf{x}, \mathbf{y})$ and the conditional probability density $p(\mathbf{y} \mid \mathbf{x})$.

9.2.1 Parametric Models

We first consider the fully parametric case, i.e., both $p(\mathbf{y} \mid \mathbf{x})$ and $q(\boldsymbol{\delta} \mid \mathbf{x}, \mathbf{y})$ are parametric, say $p(\mathbf{y} \mid \mathbf{x}) = f(\mathbf{y} \mid \mathbf{x}; \theta)$ and $q(\boldsymbol{\delta} \mid \mathbf{x}, \mathbf{y}) = g(\boldsymbol{\delta} \mid \mathbf{x}, \mathbf{y}; \phi)$, where f and g are known functions, and θ and ϕ are unknown parameter vectors. As we pointed out in Chapter 8, in general, the parameters θ and ϕ are not identifiable, i.e., two different sets of (θ, ϕ) may produce the same data.

The concept of the nonresponse instrument in handling nonignorable missing data has been introduced in Chapter 8. The same idea can be applied here. Let $\mathbf{x}_2$ be a nonresponse instrument, i.e., $\mathbf{x} = (\mathbf{x}_1, \mathbf{x}_2)$ and

$$q(\boldsymbol{\delta} \mid \mathbf{x}, \mathbf{y}) = q(\boldsymbol{\delta} \mid \mathbf{x}_1, \mathbf{y}) \quad \text{and} \quad p(\mathbf{y} \mid \mathbf{x}_1, \mathbf{x}_2) \neq p(\mathbf{y} \mid \mathbf{x}_1), \qquad (9.4)$$

so θ and ϕ are identifiable and can be estimated by maximizing the parametric likelihood

$$\prod_{\delta_i=1} f(\mathbf{y}_i \mid \mathbf{x}_i; \theta) g(\boldsymbol{\delta}_i \mid \mathbf{x}_{1i}, \mathbf{y}_i; \phi) \prod_{\delta_i=0} \int f(\mathbf{y} \mid \mathbf{x}_i; \theta) g(\boldsymbol{\delta}_i \mid \mathbf{x}_{1i}, \mathbf{y}; \phi) d\mathbf{y}.$$

The integral may not have an explicit form, and numerical methods are needed.

Parametric methods can be sensitive to model violations. We introduce two semiparametric methods next.

9.2.2   Nonparametric $p(\mathbf{y} \mid \mathbf{x})$

We consider nonparametric $p(\mathbf{y} \mid \mathbf{x})$ and parametric propensity $q(\boldsymbol{\delta} \mid \mathbf{x}, \mathbf{y}) = q(\boldsymbol{\delta} \mid \mathbf{x}_1, \mathbf{y}) = g(\boldsymbol{\delta} \mid \mathbf{x}_1, \mathbf{y}; \phi)$, where $\mathbf{x}' = (\mathbf{x}_1', \mathbf{x}_2')$ and $\mathbf{x}_2$ is a nonresponse instrument satisfying (9.4). In addition, for longitudinal data, it makes sense to assume that the dropout at time point t is statistically unrelated to future values $y_{t+1}, ..., y_T$. Thus,

$$P(\delta_t = 1 \mid \delta_{t-1} = 1, y_1, ..., y_T, \mathbf{x}) = P(\delta_t = 1 \mid \delta_{t-1} = 1, y_1, ..., y_t, \mathbf{x}_1). \quad (9.5)$$

Consider now the situation where $\mathbf{x}_1$ has a continuous component $\mathbf{x}_{1c}$ and a discrete component u_d taking values $1, ..., R$. Assume that

$$P(\delta_t = 1 \mid \delta_{t-1} = 1, y_1, ..., y_t, \mathbf{x}_1) = \psi(\alpha_{tu_d} + \beta_{tu_d} y_t + \mathbf{w}_t' \gamma_{tu_d}), \qquad t = 1, ..., T, \quad (9.6)$$

where $\mathbf{w}_t' = (y_1, ..., y_{t-1}, \mathbf{x}_{1c}')$, ψ is an increasing function on $(0,1]$, α_{tu_d}, β_{tu_d}, and components of γ_{tu_d} are unknown parameters possibly depending on u_d.

In applications, we may consider some special cases of (9.6). For example,

$$P(\delta_t = 1 \mid \delta_{t-1} = 1, y_1, ..., y_t, \mathbf{x}_1) = \psi(\alpha_t + \beta_t y_t), \qquad t = 1, ..., T,$$

or

$$P(\delta_t = 1 \mid \delta_{t-1} = 1, y_1, ..., y_t, \mathbf{x}_1) = \psi(\alpha_t + \beta_t y_t + \gamma_t y_{t-1}), \qquad t = 1, ..., T,$$

where γ_t is an unknown parameter.

We adopt the GMM method described in Section 8.4 to estimate the unknown parameters in the propensity. The key is to construct a set of L estimation functions under condition (9.6).

First, we consider the case where $\mathbf{x} = \mathbf{x}_2$ is a q-dimensional continuous covariate and $\mathbf{x}_1 = 0$ in dropout propensity model (9.6), which means $\mathbf{w}_t = (y_1, ..., y_{t-1})'$. For each t, there are $t+1$ parameters in model (9.6). Therefore, the total number of parameters in the propensity for all time points is $T(T+3)/2$. To apply the GMM, we need at least $T(T+3)/2$ functions. When T is not small, optimization over $T(T+3)/2$ parameters simultaneously has two crucial issues. First, there could be a large computational rounding error which results in large standard errors of the GMM estimators. Second, the computation speed is slow since the optimization is done in a $T(T+3)/2$ dimensional space. Therefore, we consider estimating the $(t+1)$-dimensional parameter $\phi_t = (\alpha_t, \beta_t, \gamma_t^\mathsf{T})$ for each separate t. For $t = 1, ..., T$, consider the following $L = t + q$ functions for the GMM:

$$\mathbf{g}_t(\vartheta, \mathbf{y}, \mathbf{x}_2, \boldsymbol{\delta}) = \begin{pmatrix} \delta_{t-1}[\delta_t \omega(\vartheta) - 1] \\ \mathbf{x}_2' \delta_{t-1}[\delta_t \omega(\vartheta) - 1] \\ \mathbf{w}_t' \delta_{t-1}[\delta_t \omega(\vartheta) - 1] \end{pmatrix}, \quad (9.7)$$

where $\vartheta = (\vartheta_1, \vartheta_2, \vartheta_3')$, ϑ_3 is a $(t-1)$-dimensional column vector, $\omega(\vartheta) =$

$[\psi(\vartheta_1 + \vartheta_2 y_t + \mathbf{w}_t'\vartheta_3)]^{-1}$, and $\mathbf{w}_t$ is defined in (9.6). Since ϕ_t is $(t+1)$-dimensional, the minimum requirement for q is $q = 1$. The GMM estimator $\hat{\phi}_t = (\hat{\alpha}_t, \hat{\beta}_t, \hat{\gamma}_t^{\mathsf{T}})$ can be obtained using the two-step algorithm described in Section 8.4 with g_l being the l-th component function of $\mathbf{g}_t$ in (9.7).

The following theorem establishes the consistency and asymptotic normality of the GMM estimator of ϕ_t for every t. It also derives a consistent estimator of the asymptotic covariance matrix of the GMM estimator.

Theorem 9.1. *Suppose that the parameter space Θ_t containing the true value ϕ_t is an open subset of $\mathcal{R}^{t+1}$ and model (9.6) holds. Assume further the following conditions.*

(C1) $E(\|\mathbf{x}_2\|^2) < \infty$ and there exists a neighborhood $\mathbf{N}$ of ϕ_t such that

$$E\left[\delta_t \sup_{\vartheta \in \mathbf{N}} \left\{(1+\|\mathbf{x}_2\|^2)\omega^2(\vartheta) + |\boldsymbol{\xi}_t||\boldsymbol{\xi}_{t+1}||\omega'(\vartheta)| + \|\boldsymbol{\xi}_{t+1}\|^2|\omega''(\vartheta)|\right\}\right] < \infty,$$

where $\boldsymbol{\xi}_t = (1, \mathbf{x}_2, y_1, ..., y_{t-1})'$, $\omega'(\vartheta) = \omega'(\vartheta_1 + \vartheta_2 y_t + \mathbf{w}_t'\vartheta_3)$, $\omega''(\vartheta) = \omega''(\vartheta_1 + \vartheta_2 y_t + \mathbf{w}_t'\vartheta_3)$, $\omega(s) = [\psi(s)]^{-1}$, $|\mathbf{a}|$ is the L_1 norm of a vector $\mathbf{a}$ and $\|\mathbf{a}\| = \sqrt{\text{trace}(\mathbf{a}^{\mathsf{T}}\mathbf{a})}$ is the L_2 norm for a vector or matrix $\mathbf{a}$.

(C2) For $t = 1, ..., T$, the $(t+q) \times (t+1)$ matrix

$$\Gamma_t = \left[E\{\boldsymbol{\xi}_t \delta_t \omega'(\phi_t)\}, E\{\boldsymbol{\xi}_t y_t \delta_t \omega'(\phi_t)\}, E\{\boldsymbol{\xi}_t \mathbf{w}_t' \delta_t \omega'(\phi_t)\}\right]$$

is of full rank.

Then, we have the following conclusions as $n \to \infty$.

(i) There exists $\{\hat{\phi}_t\}$ such that $P(\mathbf{s}(\hat{\phi}_t) = 0) \to 1$, $\mathbf{G}(\hat{\phi}_t) \to_p 0$, and $\hat{\phi}_t \to_p \phi_t$, where $\mathbf{s}(\vartheta) = -\partial\mathbf{G}'(\vartheta)\hat{\mathbf{W}}\mathbf{G}(\vartheta)/\partial\vartheta$ and $\to_p$ denotes convergence in probability.

(ii) For any sequence $\{\tilde{\phi}\}$ satisfying $\mathbf{s}(\tilde{\phi}) = 0$ and $\tilde{\phi} \to_p \phi_t$,

$$\sqrt{n}(\tilde{\phi} - \phi_t) \to_d N\left(0, (\boldsymbol{\Gamma}'\boldsymbol{\Sigma}^{-1}\boldsymbol{\Gamma})^{-1}\right),$$

where $\to_d$ is convergence in distribution and $\boldsymbol{\Sigma}$ is the positive definite $(t+q) \times (t+q)$ matrix whose (l, l')th element is $E[g_l(\phi_t, \mathbf{y}, \mathbf{x}_2, \boldsymbol{\delta})g_{l'}(\phi_t, \mathbf{y}, \mathbf{x}_2, \boldsymbol{\delta})]$.

(iii) Let $\hat{\boldsymbol{\Gamma}}$ be the $(t+q) \times (t+1)$ matrix whose l-th row is

$$\frac{1}{n}\sum \frac{\partial g_l(\vartheta, \mathbf{y}_i, \mathbf{x}_{2i}, \boldsymbol{\delta}_i)}{\partial\vartheta}\bigg|_{\vartheta=\hat{\phi}_t}$$

and $\hat{\boldsymbol{\Sigma}}$ be the $(t+q) \times (t+q)$ matrix whose (l, l')-th element is

$$\frac{1}{n}\sum g_l(\hat{\phi}_t, \mathbf{y}_i, \mathbf{x}_{2i}, \boldsymbol{\delta}_i)g_{l'}(\hat{\phi}_t, \mathbf{y}_i, \mathbf{x}_{2i}, \boldsymbol{\delta}_i).$$

Then $\hat{\boldsymbol{\Gamma}}'\hat{\boldsymbol{\Sigma}}^{-1}\hat{\boldsymbol{\Gamma}} \to_p \boldsymbol{\Gamma}'\boldsymbol{\Sigma}^{-1}\boldsymbol{\Gamma}$.

Consider now the general case where $\mathbf{x}' = (\mathbf{x}'_1, \mathbf{x}'_2)$, $\mathbf{x}'_1 = (x_{1d}, \mathbf{x}'_{1c})$, and $\mathbf{x}'_2 = (x_{2d}, \mathbf{x}'_{2c})$, where $\mathbf{x}_{1c}$ and $\mathbf{x}_{2c}$ are continuous r- and q-dimensional covariate vectors, respectively, and x_{1d} and x_{2d} are discrete covariates taking values $1, ..., K$ and $1, ..., M$, respectively. Assume that the dropout propensity follows model (9.6). To apply GMM to estimate the parameter vector $\phi_{tk} = (\alpha_{tk}, \beta_{tk}, \gamma'_{tk})$ in the category of $u_d = k$, we construct the following $L = r + t + q + M$ functions:

$$\mathbf{g}_t(\vartheta, \mathbf{y}, \mathbf{x}, \boldsymbol{\delta}) = I(u_d = k) \begin{pmatrix} \boldsymbol{\zeta}' \delta_{t-1}[\delta_t \omega(\vartheta) - 1] \\ \mathbf{x}'_{2c} \delta_{t-1}[\delta_t \omega(\vartheta) - 1] \\ \mathbf{w}'_t \delta_{t-1}[\delta_t \omega(\vartheta) - 1] \end{pmatrix}, \qquad (9.8)$$

where $\boldsymbol{\zeta}$ is the M-dimensional row vector whose lth component is $I(z_d = l)$, $I(A)$ is the indicator function of A, $\mathbf{w}_t = (y_1, ..., y_{t-1}, \mathbf{x}'_{1c})$, $\omega(\vartheta) = [\psi(\vartheta_1 + \vartheta_2 y_t + \mathbf{w}'_t \vartheta_3)]^{-1}$, $\vartheta = (\vartheta_1, \vartheta_2, \vartheta'_3)$, and ϑ_3 is a $(r + t - 1)$-dimensional column vector. Because ϕ_{tk} is $(r + t + 1)$-dimensional, we require that $M + q \geq 2$.

The consistency and asymptotic normality of the GMM estimators can be established under similar conditions in Theorem 9.1 by replacing $\boldsymbol{\xi}_t$ with $(\boldsymbol{\zeta}', \mathbf{x}'_{2c}, y_1, ..., y_{t-1}, \mathbf{x}'_{1c})'$ and the general $\omega(\vartheta)$ with the specific $\omega(\vartheta)$ in (9.8).

Once parameters in the dropout propensity are estimated, we can obtain nonparametric estimators of some parameters in the marginal distribution of $\mathbf{y}$ or the joint distribution of $\mathbf{y}$ and $\mathbf{x}$. For example, the estimation of the marginal mean of $\mathbf{y}$ is often the main focus in areas such as clinical studies and sample surveys. We consider the general case where $\mathbf{x}' = (\mathbf{x}'_1, \mathbf{x}'_2)$ and both $\mathbf{x}_1$ and $\mathbf{x}_2$ may have continuous and discrete components.

First, consider the situation where $\mathbf{x}_1$ is continuous. For any $t = 1, ..., T$, let $\hat{\phi}_t = (\hat{\alpha}_t, \hat{\beta}_t, \hat{\gamma}'_t)$ be the GMM estimator of the unknown parameter under model (9.6), and y_{ti}'s, $\mathbf{w}_{ti}$'s, and δ_{ti}'s be the realized values of y_t, $\mathbf{w}'_t = (y_1, ..., y_{t-1}, \mathbf{x}'_1)$, and δ_t from the sampled unit $i = 1, ..., n$, respectively. Since

$$P(\delta_t = 1 \mid \mathbf{x}, \mathbf{y}) = \prod_{s=1}^{t} P(\delta_s = 1 \mid \mathbf{x}, \mathbf{y}, \delta_{s-1} = 1) = \prod_{s=1}^{t} \psi(\alpha_s + \beta_s y_s + \mathbf{w}_s \gamma_s),$$

which can be estimated by

$$\hat{\pi}_{ti} = \prod_{s=1}^{t} \psi(\hat{\alpha}_s + \hat{\beta}_s y_{si} + \mathbf{w}_{si} \hat{\gamma}_s), \qquad (9.9)$$

the marginal distribution of y_t can be estimated by the empirical distribution putting mass p_{ti} to each observed y_{ti}, where p_{ti} is proportional to $\delta_{ti}/\hat{\pi}_{ti}$ for a fixed t. The marginal mean of y_t, $\mu_t = E(y_t)$, can be estimated by a PS estimator

$$\tilde{\mu}_t = \frac{1}{n} \sum_{i=1}^{n} \frac{\delta_{ti} y_{ti}}{\hat{\pi}_{ti}} \quad \text{or} \quad \hat{\mu}_t = \sum_{i=1}^{n} \frac{\delta_{ti} y_{ti}}{\hat{\pi}_{ti}} \bigg/ \sum_{i=1}^{n} \frac{\delta_{ti}}{\hat{\pi}_{ti}}. \qquad (9.10)$$

Similarly, we can estimate $E(y_t y_s)$ for any $s \leq t$ (and hence the covariance or the correlation between y_t and y_s) by using (9.10) with y_{ti} replaced by $y_{ti} y_{si}$. Replacing y_s by $\mathbf{x}$, we can also obtain estimators of covariances between y_t and any covariate.

For every t, we establish the following general theorem that can be applied to show the asymptotic normality of various estimators and derive asymptotic covariance estimators, which allows us to carry out a large sample inference such as setting confidence intervals.

Theorem 9.2. *Assume the conditions in Theorem 9.1 with* $\boldsymbol{\xi}_t = (\boldsymbol{\zeta}', \mathbf{x}_{2c}, \mathbf{w}_t')'$ *for* $t = 1, ..., T$. *Let* $\boldsymbol{f}(\vartheta, \mathbf{d})$ *be an m-dimensional function with* $E[\boldsymbol{f}(\phi, \mathbf{d})] = \varphi$, *where* $\phi = (\phi_1, ..., \phi_t)$ *is the parameter vector in the dropout propensity and let* $\mathbf{g}(\vartheta) = (\mathbf{g}_1(\vartheta)', ..., \mathbf{g}_t(\vartheta)')'$, *where* $\vartheta = (\vartheta_1, ..., \vartheta_t)$ *and* $\mathbf{g}_s(\vartheta) = \mathbf{g}_s(\vartheta_s)$ *is the estimation functions for estimating* ϕ_s. *Let*

$$\hat{\varphi} = \frac{1}{n} \sum_{i=1}^{n} \boldsymbol{f}(\hat{\phi}, \mathbf{d}_i), \tag{9.11}$$

where $\hat{\phi} = (\hat{\phi}_1, ..., \hat{\phi}_t)$ *with* $\hat{\phi}_s$ *being the GMM estimator of* ϕ_s *in Theorem 9.1. Assume further the following condition:*

(C3) $\|E[\boldsymbol{f}(\phi, \mathbf{d})\boldsymbol{f}'(\phi, \mathbf{d})]\| < \infty$, $\|E[\nabla \boldsymbol{f}(\phi)]\| < \infty$, *where* $\nabla \boldsymbol{\eta}(\phi) = \partial \boldsymbol{\eta}(\vartheta)/\partial \vartheta|_{\vartheta = \phi}$ *for a function* $\boldsymbol{\eta}(\vartheta)$, *and there exists a neighborhood* $\boldsymbol{N}$ *of* ϕ *such that*

$$E \left[\sup_{\vartheta \in \boldsymbol{N}} \left\| \frac{\partial^2 \boldsymbol{f}(\vartheta, \mathbf{d})}{\partial \vartheta \partial \vartheta'} \right\| \right] < \infty.$$

Then we have the following conclusions as $n \to \infty$.

(i) $\sqrt{n}(\hat{\varphi} - \varphi) \to_d N(0, \boldsymbol{\Omega})$. *Here* $\boldsymbol{\Omega} = \mathbf{K}' \boldsymbol{\Lambda} \mathbf{K}$, $\boldsymbol{\Lambda} = E[\mathbf{h}(\phi, \varphi, \mathbf{d})\mathbf{h}'(\phi, \varphi, \mathbf{d})]$, $\mathbf{H} = (\boldsymbol{\Gamma}_1' \mathbf{W} \boldsymbol{\Gamma}_1)^{-1} \boldsymbol{\Gamma}_1' \mathbf{W}$, $\mathbf{K} = [-\boldsymbol{\Gamma}_2 \mathbf{H}, \mathbf{I}_{m \times m}]'$, $\boldsymbol{\Gamma}_2 = E[\nabla \boldsymbol{f}(\phi, \mathbf{d})]$, $\mathbf{h}(\vartheta, \boldsymbol{\psi}, \mathbf{d}) = [\mathbf{g}'(\vartheta, \mathbf{d}), (\boldsymbol{f}(\vartheta, , \mathbf{d}) - \boldsymbol{\psi})']'$ *and*

$$\mathbf{W} = \begin{pmatrix} \boldsymbol{\Sigma}_1^{-1} & & \\ & \ddots & \\ & & \boldsymbol{\Sigma}_t^{-1} \end{pmatrix}$$

with $\boldsymbol{\Sigma}_s = [E\{\mathbf{g}_s(\phi)\mathbf{g}_s(\phi)'\}]^{-1}$.

(ii) Let $\hat{\boldsymbol{\Omega}} = \hat{\mathbf{K}}' \hat{\boldsymbol{\Lambda}} \hat{\mathbf{K}}$, *where*

$$\hat{\boldsymbol{\Lambda}} = \frac{1}{n} \sum_{i=1}^{n} \mathbf{h}(\hat{\phi}, \hat{\varphi}, \mathbf{d}_i)\mathbf{h}'(\hat{\phi}, \hat{\varphi}, \mathbf{d}_i), \quad \hat{\mathbf{H}} = (\hat{\boldsymbol{\Gamma}}_1' \hat{\mathbf{W}} \hat{\boldsymbol{\Gamma}}_1)^{-1} \hat{\boldsymbol{\Gamma}}_1' \hat{\mathbf{W}},$$

$$\hat{\boldsymbol{\Gamma}}_1 = \frac{1}{n} \sum_{i=1}^{n} \nabla \mathbf{g}(\hat{\phi}, \mathbf{d}_i), \quad \hat{\mathbf{K}} = \left[-\hat{\boldsymbol{\Gamma}}_2 \hat{\mathbf{H}}, \mathbf{I}_{m \times m} \right]', \quad \hat{\boldsymbol{\Gamma}}_2 = \frac{1}{n} \sum_{i=1}^{n} [\nabla \boldsymbol{f}(\hat{\phi}, \mathbf{d}_i)]$$

$$\text{and} \quad \hat{\boldsymbol{W}} = \begin{pmatrix} \hat{\boldsymbol{\Sigma}}_1^{-1} & & \\ & \ddots & \\ & & \hat{\boldsymbol{\Sigma}}_t^{-1} \end{pmatrix},$$

with $\hat{\boldsymbol{\Sigma}}_s = 1/n \sum_i \mathbf{g}_s(\hat{\phi}, \mathbf{d}_i) \mathbf{g}_s(\hat{\phi}, \mathbf{d}_i)'$. Then $\hat{\boldsymbol{\Omega}} \to_p \boldsymbol{\Omega}$.

If we take $\boldsymbol{f}(\vartheta, \mathbf{d}_i) = y_{ti}\delta_{ti}/\pi_{ti}(\vartheta)$, then $\hat{\varphi}$ in (9.11) is $\tilde{\mu}_t$ in (9.10). If

$$\boldsymbol{f}(\vartheta, \mathbf{d}_i) = (y_{ti}\delta_{ti}/\pi_{ti}(\vartheta), \ \delta_{ti}/\pi_{ti}(\vartheta))',$$

then Theorem 9.2 and the Delta method imply that

$$\sqrt{n}(\hat{\mu}_t - \mu_t) \to_d N(0, \mathbf{a}'\tilde{\boldsymbol{\Sigma}}\mathbf{a}),$$

where $\tilde{\boldsymbol{\Sigma}} = \mathbf{K}'\boldsymbol{\Lambda}\mathbf{K}$ with $\mathbf{K}$ and $\boldsymbol{\Lambda}$ given in Theorem 9.2 and $\mathbf{a} = (1, -\mu_t)'$. Furthermore, $\hat{\mathbf{a}}'\hat{\mathbf{K}}'\hat{\boldsymbol{\Lambda}}\hat{\mathbf{K}}\hat{\mathbf{a}}$ is a consistent estimator of $\mathbf{a}'\tilde{\boldsymbol{\Sigma}}\mathbf{a}$, where $\hat{\mathbf{K}}$ and $\hat{\boldsymbol{\Lambda}}$ are given in Theorem 9.2 and $\hat{\mathbf{a}} = (1, -\hat{\mu}_t)'$.

For estimating $E(y_t y_s)$, $s \le t$, or $E(y_t \mathbf{x})$, we can obtain the asymptotic results by replacing y_{ti} in the previous $\boldsymbol{f}(\vartheta, \mathbf{d}_i)$ by $y_{ti}y_{si}$ or $y_{ti}\mathbf{x}_i$. The details are omitted.

Next, consider the case where $\mathbf{x}_1$ has a discrete component u_d taking values $k = 1, ..., K$, and (9.6) holds. For every k, we can apply the previous results using data with $x_{1d} = k$ to obtain estimators of parameters in the conditional distribution of $\mathbf{y}$ or $(\mathbf{y}, \mathbf{x})$ given $x_{1d} = k$. Then, the parameters in the unconditional distribution of $\mathbf{y}$ or $(\mathbf{y}, \mathbf{x})$ can be obtained by taking averages. For example, an estimator of $\mu_{t,k} = E(y_t \mid x_{1d} = k)$ is

$$\hat{\mu}_{t,k} = \sum_{i=1}^n I(x_{1di} = k) \frac{\delta_{ti} y_{ti}}{\hat{\pi}_{ti}} \Big/ \sum_{i=1}^n I(x_{1di} = k) \frac{\delta_{ti}}{\hat{\pi}_{ti}}.$$

Asymptotic results for these estimators similar to those in Theorem 9.2 can be established.

9.2.3 Nonparametric Propensity

Now, we consider nonparametric propensity and the parametric model

$$p(\mathbf{y} \mid \mathbf{x}) = \prod_{t=1}^T f_t(y_t \mid \mathbf{v}_{t-1}, \theta_t), \tag{9.12}$$

where $f_t(y_t \mid \mathbf{v}_{t-1}, \theta_t)$ is the probability density of y_t given $\mathbf{v}_{t-1} = (y_1, ..., y_{t-1}, \mathbf{x})$, f_t's are known functions, and θ_t's are distinct unknown parameter vectors. The parameter of interest is $\theta = (\theta_1, ..., \theta_T)$. We still assume that there is a nonresponse instrument $\mathbf{x}_2$ satisfying (9.4).

Consider first the case of $\mathbf{x} = \mathbf{x}_2$ ($\mathbf{x}_1 = 0$). When $t = 1$, under the assumed conditions,

$$p(\mathbf{x} \mid y_1, \delta_1 = 1) = \frac{p(y_1 \mid \mathbf{x})p(\mathbf{x})}{\int p(y_1 \mid \mathbf{x})p(\mathbf{x})d\mathbf{x}}.$$

Hence, using the theory in Section 8.3, we consider the likelihood

$$\prod_{\delta_{i1}=1} \frac{f_1(y_{i1} \mid \mathbf{x}_i; \theta_1)p(\mathbf{x}_i)}{\int f_1(y_{i1} \mid \mathbf{x}; \theta_1)p(\mathbf{x})d\mathbf{x}}.$$

Substituting $p(\mathbf{x})$ by the nonparametric empirical distribution of $\mathbf{x}$ putting mass n^{-1} to each $\mathbf{x}_i$, we obtain an estimator $\hat{\theta}_1$ by maximizing the pseudo likelihood

$$\prod_{\delta_{i1}=1} \frac{f_1(y_{i1} \mid \mathbf{x}_i; \theta_1)}{\sum_{j=1}^{n} f_1(y_{i1} \mid \mathbf{x}_j; \theta_1)}.$$

For $t = 2, ..., T$, suppose that $\hat{\theta}_1, ..., \hat{\theta}_{t-1}$ have been obtained. Consider the likelihood

$$\prod_{\delta_{it}=1} p(\mathbf{x}_i \mid y_{i1}, ..., y_{it}, \delta_{it} = 1) = \prod_{\delta_{it}=1} \frac{p(y_{i1}, ..., y_{it} \mid \mathbf{x}_i)p(\mathbf{x}_i)}{\int p(y_{i1}, ..., y_{it} \mid \mathbf{x})p(\mathbf{x})d\mathbf{x}}.$$

Under (9.12),

$$p(y_{i1}, ..., y_{it} \mid \mathbf{x}_i) = f_t(y_{it} \mid \mathbf{v}_{i(t-1)}, \theta_t) \prod_{s=1}^{t-1} f_s(y_{is} \mid \mathbf{v}_{i(s-1)}, \theta_s),$$

where $\mathbf{v}_{is} = (y_{i1}, ..., y_{is}, \mathbf{x}_i)$. Replacing each θ_s by the previously obtained $\hat{\theta}_s$ and $p(\mathbf{x}_i)$ by the nonparametric empirical distribution of $\mathbf{x}$, we estimate θ_t by maximizing the pseudo likelihood

$$\prod_{\delta_{it}=1} \frac{f_t(y_{it} \mid \mathbf{v}_{i(t-1)}, \theta_t) \prod_{s=1}^{t-1} f_s(y_{is} \mid \mathbf{v}_{i(s-1)}, \hat{\theta}_s)}{\sum_{j=1}^{n} \left\{ f_t(y_{it} \mid \mathbf{x}_j, y_{i1}, ..., y_{i(t-1)}, \theta_t) \prod_{s=1}^{t-1} f_s(y_{is} \mid \mathbf{x}_j, y_{i1}, ..., y_{i(s-1)}, \hat{\theta}_s) \right\}}.$$

$$(9.13)$$

Note that all observed values up to time t are included in this likelihood.

If we do not substitute $\theta_1, ..., \theta_{t-1}$ by their estimates, in theory, we can estimate $(\theta_1, ..., \theta_t)$ by maximizing (9.13) with $\hat{\theta}_s$ replaced by θ_s, $s = 1, ..., t - 1$. However, the computation may not be feasible because the dimension of $(\theta_1, ..., \theta_t)$ is much higher than the dimension of θ_t.

Consider now the general case where $\mathbf{x}' = (\mathbf{x}_1', \mathbf{x}_2')$. Note that

$$
\begin{aligned}
p(\mathbf{x}_2 \mid y_1, ..., y_t, \mathbf{x}_1, \delta_t = 1) &= p(\mathbf{x}_2 \mid y_1, ..., y_t, \mathbf{x}_1) \\
&= \frac{p(y_1, ..., y_t \mid \mathbf{x}_1, \mathbf{x}_2)p(\mathbf{x}_2 \mid \mathbf{x}_1)}{\int p(y_1, ..., y_t \mid \mathbf{x}_1, \mathbf{x}_2)p(\mathbf{x}_2 \mid \mathbf{x}_1)d\mathbf{x}_2}.
\end{aligned}
$$

If a parametric model on $p(\mathbf{x}_2|\mathbf{x}_1) = g(\mathbf{x}_2|\mathbf{x}_1;\xi)$ is assumed, where ξ is an unknown parameter vector, ξ can be estimated by $\hat{\xi}$ using the likelihood based on $\mathbf{x}_1,...,\mathbf{x}_n$. Thus, we can use the following likelihood for the estimation of θ_t:

$$\prod_{\delta_{it}=1} \frac{f_t(y_{it}|\mathbf{v}_{i(t-1)},\theta_t) \prod_{s=1}^{t-1} f_s(y_{is}|\mathbf{v}_{i(s-1)},\hat{\theta}_s)g(\mathbf{x}_{2i}|\mathbf{x}_{1i};\hat{\xi})}{\int f_t(y_{it}|\mathbf{x}_{1i},\mathbf{x}_2,y_{i1},...,y_{i(t-1)},\theta_t) \prod_{s=1}^{t-1} f_s(y_{is}|\mathbf{x}_{1i},\mathbf{x}_2,y_{i1},...,y_{i(s-1)},\hat{\theta}_s)g(\mathbf{x}_2|\mathbf{x}_{1i};\hat{\xi})d\mathbf{x}_2}.$$

To consider asymptotic properties, we focus on the situation where $\mathbf{x} = \mathbf{x}_2$. The following two additional conditions are needed:

$$\pi_t = P(\delta_t = 1) > 0, \quad t = 1,...,T, \tag{9.14}$$

and, for any θ_t in the parameter space that is not the same as the true parameter value θ_t^0 and any function ψ of $(y_1,...,y_t,\theta_t)$,

$$P\left\{(y_1,...,y_t): \frac{f_t(y_t \mid \mathbf{v}_{t-1},\theta_t)}{f_t(y_t \mid \mathbf{v}_{t-1},\theta_t^0)} = \psi(y_1,...,y_t,\theta_t) \text{ for any } \mathbf{x}\right\} < 1. \tag{9.15}$$

Consistency and asymptotic normality of $\hat{\theta}_t$ can be established using a standard argument.

We now derive an asymptotic representation of $\sqrt{n}(\hat{\theta}_t - \theta_t^0)$, which allows us to obtain an easy-to-compute consistent estimator of the asymptotic covariance matrix of $\sqrt{n}(\hat{\theta}_t - \theta_t^0)$ without knowing its actual form. The asymptotic covariance matrix of $\sqrt{n}(\hat{\theta}_t - \theta_t^0)$ is very complicated because of the fact that $\hat{\theta}_t$ is defined in terms of previous estimators $\hat{\theta}_1,...,\hat{\theta}_{t-1}$ and the empirical distribution of $\mathbf{x}$.

Theorem 9.3. *Assume (9.6), (9.12), (9.14), (9.15), and the following two conditions.*

1. *The functions f_t's in (9.12) are continuously twice differentiable with respect to θ_t, and $E\left[\frac{\partial^2 H_t(\varphi_t^0)}{\partial\theta_t\partial\theta_t'}\right]$ is positive definite, where $H_t(\varphi_t) = \delta_t \log G_t(\varphi_t)$ and*

$$G_t(\varphi_t) = \frac{f_t(y_t \mid \mathbf{v}_{t-1},\theta_t) \prod_{s=1}^{t-1} f_s(y_s \mid \mathbf{v}_{s-1},\theta_s)p(\mathbf{x})}{\int f_t(y_t \mid \mathbf{x},y_1,...,y_{t-1},\theta_t) \prod_{s=1}^{t-1} f_s(y_s \mid \mathbf{x},y_1,...,y_{s-1},\theta_s)p(\mathbf{x})d\mathbf{x}}.$$

2. *There exists an open subset Ω_t containing θ_t^0 such that*

$$\sup_{\theta_t\in\Omega_t} E\left\|\frac{\partial^2 H_t(\theta_t,\varphi_{t-1}^0)}{\partial\theta_t\partial\theta_j'}\right\| < M_{tj}, \quad j = 1,...,t,$$

where M_{tj} are integrable functions and $\|A\|^2 = \text{trace}(A'A)$ for a matrix A.

Then, as $n \to \infty$,

$$\sqrt{n}(\hat{\theta}_t - \theta_t^0) = \frac{1}{\sqrt{n}} \sum_{i=1}^{n} \psi_t(W_t^{(i)}, A_t, \varphi_t^0) + o_p(1) \to_d N(0, \Sigma_t), \qquad (9.16)$$

where $\to_d$ denotes convergence in distribution, $o_p(1)$ denotes a quantity converging to 0 in probability, Σ_t is the covariance matrix of $\psi_t(W_{it}, A_t, \varphi_t^0)$, $W_{it} = (\mathbf{v}_{it}, \delta_{it})$, $i = 1, ..., n$, $A_1 = A_{11}$, $A_t = (A_{t-1}, A_{t1}, ..., A_{tt}), t \geq 2$,

$$A_{tj} = E\left[\frac{\partial^2 H_t(\varphi_t^0)}{\partial \theta_t \partial \theta_j'}\right], \quad j = 1, ..., t,$$

$$\psi_t(W_{it}, A_t, \varphi_t^0) = -A_{tt}^{-1}\left\{\frac{\partial H_{it}(\varphi_t^0)}{\partial \theta_t} + 2h_{1t}(\mathbf{x}_i, \varphi_t^0) + \sum_{j=1}^{t-1} A_{tj}\psi_j(W_{ij}, A_j, \varphi_j^0)\right\},$$
$$(9.17)$$

$$\psi_1(W_{i1}, A_1, \varphi_1^0) = -A_{11}^{-1}\left\{\frac{\partial H_{i1}(\theta_1^0, F)}{\partial \theta_1} + 2h_{11}(\mathbf{x}_i, \varphi_1^0)\right\}, \qquad (9.18)$$

and F is the distribution function of $\mathbf{x}_i$.

The functions ψ_t, $t = 1, ..., T$, are defined iteratively according to (9.17)-(9.18) and, hence, their covariance matrices are very complicated. One may apply a bootstrap method to obtain estimators of Σ_t's, but in each bootstrap replication, maximizing a bootstrap analog of (9.13) is required, which results in a very large amount of computation. Instead, we propose the following estimator of Σ_t, utilizing the representation in (9.16). Let $D_{it} = \psi_t(W_{it}, A_t, \varphi_t^0)$. Since $\Sigma_t = \text{Var}(D_{it})$, the sample covariance matrix based on $D_{1t}, ..., D_{nt}$ is a consistent estimator of Σ_t. However, D_{it} contains the unknown φ_t^0 and A_t. Substituting D_{it} by $\hat{D}_{it} = \psi_t(W_{it}, \hat{A}_t, \hat{\varphi}_t)$, $i = 1, ..., n$, where $\hat{A}_t = (\hat{A}_{t-1}, \hat{A}_{t1}, ..., \hat{A}_{tt})$ and

$$\hat{A}_{tj} = \frac{1}{n} \sum_{i=1}^{n} \frac{\partial^2 H_t^{(i)}(\varphi_t)}{\partial \theta_t \partial \theta_j'}\bigg|_{\varphi_t = \hat{\varphi}_t}, \quad j = 1, ..., t,$$

we define the sample covariance matrix based on $\hat{D}_{1t}, ..., \hat{D}_{nt}$ as our estimator $\hat{\Sigma}_t$. This estimator is easy to compute, using (9.17)-(9.18). Under the conditions listed in Theorem 9.4, $\hat{\Sigma}_t$ is consistent.

Theorem 9.4. *Assume that the conditions in Theorem 9.3 hold and that*

1. $\sup_{\|w\| \leq c} \|\psi_t(w, \hat{A}_t, \hat{\varphi}_t) - \psi_t(w, A_t, \varphi_t^0)\| = o_p(1)$ for any $c > 0$.

2. There exist a constant $c_0 > 0$ and a function $h(w) \geq 0$ such that $E[h(W_t^{(1)})] < \infty$ and $P(\|\psi_t(w, \hat{A}_t, \hat{\varphi}_t)\|^2 \leq h(w)$ for all $\|w\| \geq c_0) \to 1$.

Then, as $n \to \infty$, $\|\hat{\Sigma}_t - \Sigma_t\| = o_p(1)$.

The proofs of Theorems 9.3 - 9.4 can be found in Shao and Zhao (2013).

9.3   Past-Value-Dependent Missing Data

Longitudinal data with nonmonotone missing responses are typically nonignorable and hard to handle. Some assumptions are needed. For example, if we assume that

$$q(\boldsymbol{\delta}_i \mid \mathbf{y}_i, \mathbf{x}_i) = q(\boldsymbol{\delta}_i \mid y_{it})$$

without necessarily assuming a parametric form for q, then the method described in §9.2.3 can be applied. We omit the details that can be found in Tang et al. (2003) and Jiang and Shao (2012).

In this section, we consider longitudinal data with nonmonotone missing responses under the following past-value-dependent missingness:

$$q(\delta_{it} \mid \mathbf{y}_i, \mathbf{x}_i, \delta_{is}, s \neq t) = q(\delta_{it} \mid \mathbf{y}_{i(t-1)}, \mathbf{x}_i), \quad t = 2, ..., T, \qquad (9.19)$$

where $\mathbf{y}_{i(t-1)} = (y_{i1}, ..., y_{i(t-1)})$. In other words, at time point t, the missingness propensity depends on covariates and all past y-values (whether or not they are observed), but does not depend on the current and future y-values. This propensity is ignorable if we add the monotone missingness condition, but it is nonignorable if missing is nonmonotone, since some of $y_{i1}, ..., y_{i(t-1)}$ may be missing. We still assumed that at $t = 1$, there is no missing value.

9.3.1   Three Different Approaches

The first approach is parametric modeling under (9.19). With parametric models assumed for $q(\delta_{it} \mid \mathbf{y}_{i(t-1)}, x_i)$ and $p(\mathbf{y}_i \mid \mathbf{x}_i)$, the parametric approach estimates model parameters using the maximum likelihood or some Bayesian methods. Zhou and Kim (2012) provided a unified approach of the PS estimation under parametric models for the propensity scores with monotone missing data.

The second approach is to artificially create a dataset with monotone missing data by using only observed y-values from a sampled subject up till its first missing y-value. After that, the missing and artificially discarded data are "missing" at random. We can then apply methods appropriate for monotone missing data under ignorable missing (e.g., Section 9.1) to the reduced dataset. We call this method *censoring at the first missing* or censoring for short. Although the censoring approach produces consistent estimators, it is not efficient when T is not small, since many observed data can be discarded.

The third approach uses the same idea in the imputation method for monotone missing data described in Section 9.1. However, because missing is nonignorable, the imputation procedure is much more complicated. Furthermore, in most cases we have to use nonparametric or at least semiparametric regression in the imputation process.

9.3.2 Imputation Models under Past-Value-Dependent Nonmonotone Missing

It can be shown that, under missing mechanism (9.19),

$$
\begin{aligned}
&E(Y_{it} \mid \mathbf{y}_{i(t-1)}, \mathbf{x}_i, \delta_{i1} = \cdots = \delta_{i(t-1)} = 1, \delta_{it} = 0) \\
=\ &E(Y_{it} \mid \mathbf{y}_{i(t-1)}, \mathbf{x}_i, \delta_{i1} = \cdots = \delta_{i(t-1)} = 1, \delta_{it} = 1)
\end{aligned}
\qquad t = 2, ..., T. \quad (9.20)
$$

This means that the conditional expectation of a missing y_{it}, given that y_{it} is the first missing value and given observed values $y_{i1}, ..., y_{i(t-1)}$ and $\mathbf{x}_i$, is the same as the conditional expectation of an observed y_{it} given that $y_{i1}, ..., y_{i(t-1)}$ are observed and given observed values $y_{i1}, ..., y_{i(t-1)}$ and $\mathbf{x}_i$. We can make use of this to carry out imputation.

Also, it can be shown that, for a missing y_{it} with $r+1$ as the first time point of having a missing value ($r = 1, ..., t-2$),

$$
\begin{aligned}
&E(Y_{it} \mid \mathbf{y}_{ir}, \mathbf{x}_i, \delta_{i1} = \cdots = \delta_{ir} = 1, \delta_{i(r+1)} = 0, \delta_{it} = 0) \\
=\ &E(Y_{it} \mid \mathbf{y}_{ir}, \mathbf{x}_i, \delta_{i1} = \cdots = \delta_{ir} = 1, \delta_{i(r+1)} = 1, \delta_{it} = 0) \\
&\qquad\qquad\qquad\qquad\qquad r = 1, ..., t-2, \quad t = 3, ..., T.
\end{aligned}
\qquad (9.21)
$$

This means that the conditional expectation of a missing y_{it}, given that $y_{i(r+1)}$ is the first missing value and given observed values $y_{i1}, ..., y_{ir}$ and $\mathbf{x}_i$, is the same as the conditional expectation of a missing y_{it}, given that $y_{i1}, ..., y_{i(r+1)}$ are observed and given observed values $y_{i1}, ..., y_{ir}$ and $\mathbf{x}_i$.

We use (9.20)–(9.21) as imputation models. Like the monotone missing case, the number of imputation models is $T(T-1)/2$. Note that models in (9.20) are the same as those in (9.3) with $r = t-1$, but models in (9.21) are different from those in (9.3).

We now explain how to use (9.20)–(9.21) for imputation. Let t be a fixed time point > 1 (it does not matter which t we start). First, consider subjects whose first missing occurs at time point t, i.e., $r = t-1$. Denote the first line of (9.20) by $\phi_{t,t-1}(\mathbf{y}_{i(t-1)}, \mathbf{x}_i)$. If the function $\phi_{t,t-1}$ is known, then a natural imputed value for a missing y_{it} is $\phi_{t,t-1}(\mathbf{y}_{i(t-1)}, \mathbf{x}_i)$. Since $\phi_{t,t-1}$ is usually unknown, we have to estimate it. Since $\phi_{t,t-1}$ cannot be estimated by regressing missing y_{it} on $(\mathbf{y}_{i(t-1)}, \mathbf{x}_i)$ based on data from subjects with missing y_{it} values, we need to use (9.20), i.e., the fact that $\phi_{t,t-1}$ is the same as the quantity on the second line of (9.20), which can be estimated by regressing y_{it} on $(\mathbf{y}_{i(t-1)}, \mathbf{x}_i)$, using data from all subjects having observed y_{it} and observed $\mathbf{y}_{i(t-1)}$. Denote the resulting estimate by $\hat{\phi}_{t,t-1}$. (The form of $\hat{\phi}_{t,t-1}$ will be given in Section 9.3.3.) Then, the missing y_{it} of subject i whose first missing is at time point t can be imputed by $\hat{\phi}_{t,t-1}(\mathbf{y}_{i(t-1)}, \mathbf{x}_i)$. Model (9.20) allows us to use data from subjects without any missing values in estimating the regression function $\phi_{t,t-1}$.

The case of $r < t-1$ is more complicated. For a subject whose first missing is at time point $r+1$ with $r < t-1$, a missing y_{it} can be imputed by $\phi_{t,r}(\mathbf{y}_{ir}, \mathbf{x}_i)$, which denotes the quantity on the first line of (9.21), if $\phi_{t,r}$ is known. Since $\phi_{t,r}$ is unknown, we need to estimate it. To estimate $\phi_{t,r}$ by regression we need some values of y_{it} as responses. Unlike the case of $r = t - 1$, the conditional expectation on the second line of (9.21) is also conditional on a missing y_{it} ($\delta_{it} = 0$), although $y_{i1}, ..., y_{ir}$ and $\mathbf{x}_i$ are observed. Suppose that imputation is carried out sequentially for $r = t - 1, t - 2, ..., 1$. Then, for a given $r < t - 1$, the missing y_{it} values from subjects whose first missing is at time point $r + 2$ have already been imputed. (For $r = t - 2$, imputed values are obtained in the previous discussion for subjects whose first missing is at $t = r + 2$.) We can then fit a regression between imputed y_{it} and observed $(\mathbf{y}_{ir}, \mathbf{x}_i)$, using data from all subjects with already imputed y_{it} (as responses) and observed $y_{i1}, ..., y_{ir}$ and $\mathbf{x}_i$ (as predictors) and $\delta_{i(r+1)} = 1$. Denote the resulting estimate by $\hat{\phi}_{t,r}$. (The form of $\hat{\phi}_{t,r}$ will be given in Section 9.3.3.) Then the missing y_{it} of subject i whose first missing is at time point $r + 1$ can be imputed by $\hat{\phi}_{t,r}(\mathbf{y}_{ir}, \mathbf{x}_i)$. Model (9.21) allows us to use previously imputed values of y_{it} in the regression estimation of $\phi_{t,r}$.

We illustrate the proposed imputation process in the case of $T = 4$ (Table 9.1). The horizontal direction in Table 9.1 corresponds to time points and the vertical direction corresponds to 8 different missing patterns, where each pattern is represented by a 4-dimensional vector of 0's and 1's with 0 indicating a missing value and 1 indicating an observed value. It does not matter at which $t = 2, ..., T$ the imputation starts, but within each t, imputation is sequential.

1. [Step A] Consider first the imputation at $t = 3$. There are two steps (the block in Table 9.1 under title $t = 3$). At step 1, we impute the missing data at $t = 3$ with the first missing at time 3 ($r = 2$), i.e., patterns 2 and 6. According to imputation model (9.20), we fit a regression using the data in patterns 3 and 8 indicated by $+$ (used as predictors) and $\times$ (used as responses). Then, imputed values (indicated by $\bigcirc$) are obtained from the fitted regression using the data indicated by $*$ as predictors. At step 2, we impute the missing data at $t = 3$ with the first missing at time 2 ($r = 1$), i.e., patterns 1 and 5. According to imputation model (9.21), we fit a regression using data in patterns 2 and 6 indicated by $+$ (as predictors) and $\otimes$ (previously imputed values used as responses). Then, imputed values (indicated by $\bigcirc$) are obtained from the fitted regression using the data indicated by $*$ as predictors.

2. [Step B] Consider next the imputation at $t = 2$ (the block in Table 9.1 under title $t = 2$). This is the simplest case: the missing data at $t = 2$ are in patterns 1, 4, 5, and 7; we fit a regression using the data in patterns 2, 3, 6, and 8 indicated by $+$ (as predictors) and $\times$ (as responses). Then, imputed values (indicated by $\bigcirc$) are obtained from the fitted regression using the data indicated by $*$ as predictors.

3. [Step C] Finally, consider the imputation at $t = 4$ (the block in Table 9.1 under title $t = 4$). At step 1, we impute the missing data at time

TABLE 9.1
Illustration of Imputation Process When $T = 4$

	$t=3$ Step 1: $r=2$				$t=3$ Step 2: $r=1$				$t=2$ $r=1$			
Pattern	Time 1	2	3	4	Time 1	2	3	4	Time 1	2	3	4
(1,0,0,0)					*		○		*	○		
(1,1,0,0)	*	*	○		+		⊗		+	×		
(1,1,1,0)	+	+	×						+	×		
(1,0,1,0)									*	○		
(1,0,0,1)					*		○		*	○		
(1,1,0,1)	*	*	○		+		⊗		+	×		
(1,0,1,1)									*	○		
(1,1,1,1)	+	+	×						+	×		

	$t=4$ Step 1: $r=3$				$t=4$ Step 2: $r=2$				$t=4$ Step 3: $r=1$			
Pattern	Time 1	2	3	4	Time 1	2	3	4	Time 1	2	3	4
(1,0,0,0)									*			○
(1,1,0,0)					*	*		○	+			⊗
(1,1,1,0)	*	*	*	○	+	+		⊗	+			⊗
(1,0,1,0)									*			○
(1,0,0,1)												
(1,1,0,1)												
(1,0,1,1)												
(1,1,1,1)	+	+	+	×								

+: observed data used in regression fitting as predictors
×: observed data used in regression fitting as responses
⊗: imputed data used in regression fitting as responses
*: observed data used as predictors in imputation
○: imputed values

4 with the first missing at time 4 (pattern 3). According to imputation model (9.20), we fit a regression using the data in pattern 8
indicated by $+$ (as predictors) and $\times$ (as responses). Then, imputed
values (indicated by $\bigcirc$) are obtained from the fitted regression using the data indicated by $*$ as predictors. At step 2, the missing
values at $t = 4$ with the first missing at time 3 are in pattern 2.
According to imputation model (9.21), we fit a regression using the
data in pattern 3 indicated by $+$ (as predictors) and $\bigotimes$ (previously
imputed values used as responses). Then, imputed values (indicated
by $\bigcirc$) are obtained from the fitted regression using the data indicated by $*$ as predictors. At step 3, the missing values at $t = 4$ with
the first missing at time 2 are in patterns 1 and 4. According to imputation model (9.21), we fit a regression using the data in patterns
2 and 3 indicated by $+$ (as predictors) and $\bigotimes$ (previously imputed
values used as responses). Then, imputed values (indicated by $\bigcirc$)
are obtained from the fitted regression using the data indicated by
$*$ as predictors.

One may wonder why we don't use observed y_{it} values in the estimation
of $\phi_{t,r}$ when $r < t-1$. For a subject with missing y_{it} and the first missing at
time point $r+1 < t$, in general,

$$E(y_{it} \mid \mathbf{y}_{ir}, \mathbf{x}_i, \delta_{i1} = \cdots = \delta_{ir} = 1, \delta_{it} = 0) \neq E(y_{it} \mid \mathbf{y}_{ir}, \mathbf{x}_i, \delta_{i1} = \cdots = \delta_{ir} = 1, \delta_{it} = 1)$$

unless the missing is ignorable. Therefore, we cannot use observed y_{it} values
in the estimation of $\phi_{t,r}$ when $r < t-1$.

After missing values are imputed, the mean of y_{it} for each t can be estimated by the sample mean at t by treating imputed values as observed. If
a regression model between y_{it} and $\mathbf{x}_i$ needs to be fitted, we can also use
standard methods by treating imputed values as observed.

9.3.3   Nonparametric Regression Imputation

The imputation procedure described in Section 9.3.2 requires that we
regress observed or already imputed y_{it} on $(\mathbf{y}_{ir}, \mathbf{x}_i)$, using subjects with some
particular missing patterns. In this section, we specify the regression method.
Because missing is nonignorable, conditional expectations in (9.20)–(9.21) depend not only on the distribution of $\mathbf{y}_i$, but also on the propensity. Thus,
parametric regression requires parametric models on both $p(\mathbf{y}_i \mid \mathbf{x}_i)$ and the
propensity. Furthermore, even if $p(\mathbf{y}_i \mid \mathbf{x}_i)$ is normal, conditional expectations
in (9.20)–(9.21) are not linear because of the nonignorable missing mechanism.

Nonparametric regression model is robust because it avoids specifying a
parametric model on the propensity that cannot be verified using data. We
now describe a kernel nonparametric regression method to estimate $\phi_{t,r}$.

Let $Z_{ir} = (\mathbf{y}_{ir}, \mathbf{x}_i)$ and

$$\phi_{t,t-1}(u) = E(y_{it} \mid Z_{ir} = u, \delta_{i1} = \cdots = \delta_{it} = 1).$$

Under the condition given in (9.20), the kernel regression estimator of $\phi_{t,t-1}(u)$ is

$$\hat{\phi}_{t,t-1}(u) = \sum_{i=1}^{n} K_{t,t-1}\left(\frac{u - Z_{i(t-1)}}{h}\right) I_{t,t-1,i} y_{it} \bigg/ \sum_{i=1}^{n} K_{t,t-1}\left(\frac{u - Z_{i(t-1)}}{h}\right) I_{t,t-1,i},$$

where $K_{t,t-1}$ is a probability density function, $h > 0$ is a bandwidth, and

$$I_{t,t-1,i} = \begin{cases} 1 & \delta_{i1} = \cdots = \delta_{it} = 1 \\ 0 & \text{otherwise.} \end{cases}$$

For $t = 2, ..., T$, a missing y_{it} with observed $Z_{i(t-1)}$ is imputed by

$$\tilde{y}_{it} = \hat{\phi}_{t,t-1}(Z_{i(t-1)}).$$

For $r = 1, ..., t-2$, let

$$\phi_{t,r}(u) = E(y_{it} | Z_{ir} = u, \delta_{i1} = \cdots = \delta_{i(r+1)} = 1, \delta_{it} = 0),$$

the conditional expectation in (9.21). Its kernel estimator is

$$\hat{\phi}_{t,r}(u) = \sum_{i=1}^{n} K_{t,r}\left(\frac{u - Z_{ir}}{h}\right) I_{t,r,i} \tilde{y}_{it} \bigg/ \sum_{i=1}^{n} K_{t,r}\left(\frac{u - Z_{ir}}{h}\right) I_{t,r,i},$$

where $\tilde{y}_{it}$ is a previously imputed value and

$$I_{t,r,i} = \begin{cases} 1 & \delta_{it} = 0, \delta_{i1} = \cdots = \delta_{i(r+1)} = 1 \\ 0 & \text{otherwise.} \end{cases}$$

A missing y_{it} with the first missing at $r+1$ is imputed by

$$\tilde{y}_{it} = \hat{\phi}_{t,r}(Z_{ir}).$$

9.3.4 Dimension Reduction

The dimension of the regressor Z_{ir} increases with T and the dimension of the covariate vector. As the dimension of regressor increases, the number of observations needed for kernel regression escalates exponentially. Unless we have a very large sample under each imputation model, kernel regression imputation in Section 9.3.3 may break down because of the sparseness of relevant data points. This is the so-called *curse of dimensionality* well known in nonparametric regression.

Nonparametric regression imposes no condition on $p(\mathbf{y}_i|\mathbf{x}_i)$, and no assumption on the propensity other than the past-data-dependent missing assumption (9.19). To deal with the curse of dimensionality, we consider the following semiparametric model:

$$q(\delta_{it} \mid \mathbf{y}_i, \mathbf{x}_i, \delta_{is}, s \neq t) = q(\delta_{it} \mid Z'_{i(t-1)}\beta_{t-1}, \Delta_{i(t-1)}), \quad t = 2, ..., T, \quad (9.22)$$

where $\Delta_{i(t-1)} = (\delta_{i1}, ..., \delta_{i(t-1)})$. That is, the dependence of the propensity on the high-dimensional $Z_{i(t-1)}$ is through a one-dimensional projection $Z'_{i(t-1)}\beta_{t-1}$. It follows from (9.22) that

$$
\begin{aligned}
& E(y_{it} \mid Z'_{i(t-1)}\beta_{t,t-1}, \delta_{i1} = \cdots = \delta_{i(t-1)} = 1, \delta_{it} = 0) \\
= \; & E(y_{it} \mid Z'_{i(t-1)}\beta_{t,t-1}, \delta_{i1} = \cdots = \delta_{i(t-1)} = 1, \delta_{it} = 1)
\end{aligned}
\qquad t = 2, ..., T,
$$
$$(9.23)$$

and, when $r+1$ is the first time point of having a missing value,

$$
\begin{aligned}
& E(Y_{it} \mid Z'_{ir}\beta_{t,r}, \delta_{i1} = \cdots = \delta_{ir} = 1, \delta_{i(r+1)} = 0, \delta_{it} = 0) \\
= \; & E(Y_{it} \mid Z'_{ir}\beta_{t,r}, \delta_{i1} = \cdots = \delta_{ir} = 1, \delta_{i(r+1)} = 1, \delta_{it} = 0) \\
& \hspace{6cm} r = 1, ..., t-2, \quad t = 3, ..., T, \qquad (9.24)
\end{aligned}
$$

where $\beta_{t,r}$ is $\beta_{t-1}(A_{ir})$ with all components of A_{ir} equal to 1, $r = 1, ..., t-1$. Under (9.22), (9.23) and (9.24) replace (9.20) and (9.21), respectively, as our imputation models.

If $\beta_{t,r}$'s are known, then we can apply the imputation method in Section 9.3.2 using a one-dimensional kernel regression with Z_{ir} replaced by $Z'_{ir}\beta_{t,r}$. In general, $\beta_{t,r}$'s are unknown. We first apply the sliced inverse regression (Li, 1991) under model (9.22) to obtain consistent estimators of $\beta_{t,r}$'s. We then apply the one-dimensional kernel regression based on (9.23)–(9.24) with $\beta_{t,r}$'s replaced by their estimators.

The Sliced Inverse Regression

For each t and r, we obtain an estimator of $\beta_{t,r}$ using the sliced inverse regression based on model (9.22) and the observed data on $\delta_{i(r+1)}$ and Z_{ir} from subjects with $\delta_{i1} = \cdots = \delta_{ir} = 1$. The detailed procedure is given as follows. Let $t = 2, ..., T$ be fixed and $r+1$ be the first missing time point.

1. Compute $D = [(\bar{Z}_{r1} - \bar{Z}_r)(\bar{Z}_{r1} - \bar{Z}_r)' + (\bar{Z}_{r0} - \bar{Z}_r)(\bar{Z}_{r0} - \bar{Z}_r)']/2$, where $\bar{Z}_{r1}$ is the sample mean of Z_{ir}'s from subjects with $\delta_{i1} = \cdots = \delta_{i(r+1)} = 1$, $\bar{Z}_{r0}$ is the sample mean of Z_{ir}'s from subjects with $\delta_{i1} = \cdots = \delta_{ir} = 1$ and $\delta_{i(r+1)} = 0$, and $\bar{Z}_r$ is the sample mean of Z_{ir}'s from subjects with $\delta_{i1} = \cdots = \delta_{ir} = 1$.

2. Compute S, the sample covariance matrix of Z_{ir}'s from subjects with $\delta_{i1} = \cdots = \delta_{ir} = 1$.

3. Compute $\hat{\beta}_{t,r}$, our estimator of $\beta_{t,r}$, which is the eigenvector corresponding to the largest eigenvalue of the matrix $D^{-1}S$.

One-Dimensional Kernel Regression Imputation

The regression functions

$$
\phi_{t,t-1}(u'\beta_{t,t-1}) = E(Y_{it} \mid Z'_{i(t-1)}\beta_{t,t-1} = u'\beta_{t,t-1}, \delta_{i1} = ... = \delta_{i(t-1)} = 1, \delta_{it} = 1)
$$

and

$$
\phi_{t,r}(u'\beta_{t,r}) = E(Y_{it} \mid Z'_{ir}\beta_{t,r} = u'\beta_{t,r}, \delta_{i1} = ... = \delta_{ir} = 1, \delta_{i(r+1)} = 1, \delta_{it} = 0)
$$

are both one-dimensional functions. Once we have estimators $\hat{\beta}_{t,r}$, they can be estimated by

$$\hat{\phi}_{t,r}(u'\hat{\beta}_{t,r}) = \sum_{i=1}^{n} K\left(\frac{u'\hat{\beta}_{t,r} - Z'_{ir}\hat{\beta}_{t,r}}{h}\right) I_{t,r,i}\tilde{y}_{it} \Big/ \sum_{i=1}^{n} K\left(\frac{u'\hat{\beta}_{t,r} - Z'_{ir}\hat{\beta}_{t,r}}{h}\right) I_{t,r,i},$$

$r = 1, ..., t-1$, $t = 2, ..., T$, where $\tilde{y}_{it}$ is y_{it} when $r = t-1$ and is imputed from $\hat{\phi}_{t,r}(Z'_{ir}\hat{\beta}_{t,r})$ if $r < t-1$.

One of the conditions for the sliced inverse regression is that $\mathbf{y}_i$ has an elliptically symmetric distribution such as multivariate normal. If this condition does not hold, then $\hat{\beta}_{t,r}$ may be inconsistent and the sample means based on imputed data may be biased.

Last-Value-Dependent Missing

Another dimension reduction method is based on the assumption of last-value-dependent missing mechanism

$$q(\delta_{it} \mid \mathbf{y}_i, \mathbf{x}_i, \delta_{is}, s \neq t) = q(\delta_{it}|y_{i(t-1)}), \quad t = 2, ..., T. \tag{9.25}$$

Under (9.25),

$$E(Y_{it} \mid y_{i(t-1)}, \delta_{it} = 0, \delta_{i(t-1)} = 1) = E(y_{it} \mid y_{i(t-1)}, \delta_{it} = 1, \delta_{i(t-1)} = 1), \quad t = 2, ..., T, \tag{9.26}$$

and

$$E(Y_{it} \mid y_{ir}, \delta_{it} = \cdots = \delta_{i(r+1)} = 0, \delta_{ir} = 1)$$
$$= E(y_{it} \mid y_{ir}, \delta_{it} = \cdots = \delta_{i(r+2)} = 0, \delta_{i(r+1)} = \delta_{ir} = 1) \tag{9.27}$$

for $r = 1, ..., t-2; t = 2, ..., T$, where, for each missing y_{it}, y_{ir} is the last observed component from the same unit. Equations (9.26) and (9.27) replace equations (9.20) and (9.21), respectively. The imputation procedure is similar to that in Section 9.3.2. Since the conditional expectations in (9.26)–(9.27) involve only one y-value (the last observed y_{ir}), the kernel regression in Section 9.3.3 can be applied with the one-dimensional "covariate" y_{ir}. Thus, we achieve the dimensional reduction using assumption (9.25).

9.3.5 Simulation Study

We present here results from a simulation study to evaluate the performance of the sample mean as an estimator of $E(y_{it})$ based on several methods, when the sample size $n = 1,000$, the total number of time points $T = 4$, and there are no covariates.

For comparison, we consider five estimators: the sample mean of the complete data, which is used as the gold standard; the sample mean of subjects without any missing value, which ignores missing data; the sample mean based on censoring and linear regression imputation (see Section 9.3.1), which first discards all observations of a subject after the first missing time point in order to create a dataset with "monotone missing" and then apply linear regression imputation to the created monotone missing dataset as described in Paik (1997); the sample mean based on the imputation method introduced in Sections 9.3.2 and 9.3.3, that we call nonparametric regression imputation; and the sample mean based on the imputation method introduced in §9.3.2 and the sliced inverse regression in §9.3.4, that we call *semiparametric regression imputation*. Censoring and linear regression imputation and semiparametric regression imputation produce consistent estimators if $E(y_{it} \mid Z_{ir})$ is linear for all t and r (e.g., y_i is multivariate normal), otherwise they produce biased estimators.

We simulated data in two situations, a normal case and a log-normal case. In the normal case, y_i's were independently generated from a multivariate normal distribution with mean vector $(1.33, 1.94, 2.73, 3.67)$ and the covariate matrix having an AR(1) structure with correlation coefficient 0.7; all data at $t = 1$ were observed; missing data at $t = 2, 3, 4$ were generated according to

$$P(\delta_{it} = 0 \mid Z_{i(t-1)}, A_{i(t-1)}) = \Phi(0.6 - 0.6 Z'_{i(t-1)} \beta_{t-1}), \qquad (9.28)$$

where Φ is the standard normal distribution function, β_{t-1} is a $(t-1)$-vector whose j-th component is

$$\frac{j + (1 - \delta_j)j}{\sum_{k=1}^{T} \{k + (1 - \delta_k)k\}}, \quad j = 1, ..., t-1.$$

The probabilities of missing patterns under model (9.28) are given in Table 9.2. In the log-normal case, the log of the components of y_i's were independently generated from the multivariate normal distribution with mean vector $(1.33, 1.77, 2.25, 2.76)$ and the same covariance matrix as in the normal case. The missing mechanism remains the same except that the right hand side of (9.28) is changed to $\Phi(2 - 0.5 Z'_{i(t-1)} \beta_{t-1})$. The probabilities of missing patterns are also given in Table 9.2.

Table 9.3 reports (based on 1,000 simulation runs) the relative bias and variance of mean estimators, the mean of bootstrap variance (BV) estimators (based on 200 bootstrap replications), the coverage probability of approximate 95% confidence intervals (CI) obtained using point estimator $\pm 1.96 \times \sqrt{\text{bootstrap variance}}$, and the length of CI. The following is a summary of the results in Table 9.3.

1. Bias. The nonparametric regression imputation method produces estimators with negligible biases at all time points in both normal and log-normal cases. The sample mean based on ignoring all missing data is clearly biased. Although in some cases the bias is small, the corresponding CI has very low coverage probability, because the variance of the sample mean is also very small. The sample means

TABLE 9.2
Probabilities of Missing Patterns in the Simulation Study $(T = 4)$

	Missing Pattern	Probability of Missing Pattern		
		Normal Case	Log-normal Case	
Monotone	$(1,0,0,0)$	0.051	0.111	
	$(1,1,0,0)$	0.054 total = 0.195	0.043 total = 0.187	
	$(1,1,1,0)$	0.090	0.033	
Intermittent	$(1,0,0,1)$	0.089	0.105	
	$(1,0,1,0)$	0.055 total = 0.432	0.031 total = 0.336	
	$(1,0,1,1)$	0.139	0.118	
	$(1,1,0,1)$	0.149	0.082	
Complete	$(1,1,1,1)$	0.373	0.477	

based on censoring and linear regression imputation are well estimated in the normal case with negligible biases, but significantly biased in the log-normal case because a wrong linear regression is used in imputation. The semiparametric regression imputation method produces negligible biases in the normal case but large biases in the log-normal case because the elliptically symmetric distribution condition does not hold.

2. Variance. In the normal case where all three imputation methods are correct, semiparametric regression imputation is the most efficient method, and nonparametric regression imputation is the least efficient. This is because semiparametric regression uses more information. The censoring and linear regression imputation method is more efficient than nonparametric regression imputation when linear regression is correct (in the normal case), but is less efficient than semiparametric regression because it discards data. In the log-normal case, however, nonparametric regression imputation is the only correct method and has a smaller variance than the other two methods, which shows the robustness of nonparametric regression.

3. Bootstrap and CI. The bootstrap variance estimator performs well in all cases, even when the mean estimator is biased. The related CI has a coverage probability close to the nominal level of 95% when the mean estimator has little bias.

9.3.6 Wisconsin Diabetes Registry Study

The following analysis of data from the Wisconsin Diabetes Registry Study (WDRS) is given in Xu (2007). The WDRS is a geographically defined population-based incident cohort study. All individuals with newly diagnosed

TABLE 9.3
Simulation Results for Mean Estimation for the Simulation Study in §9.3.5

Method	Quantity	Normal Case			Log-Normal Case		
		$t=2$	$t=3$	$t=4$	$t=2$	$t=3$	$t=4$
I	relative bias	0.0%	0.0%	0.0%	0.0%	0.0%	0.0%
	variance $\times 10^3$	0.981	0.954	0.917	0.154	0.430	1.193
	bootstrap variance $\times 10^3$	0.999	0.995	0.992	0.157	0.416	1.164
	CI coverage rate	94.9%	95.9%	95.6%	94.5%	94.5%	95.0%
	CI length	0.124	0.123	0.123	1.539	2.497	4.181
II	relative bias	10.2%	6.8%	3.5%	31.3%	28.4%	17.1%
	variance $\times 10^3$	1.334	1.448	1.212	0.322	0.868	1.789
	bootstrap variance $\times 10^3$	1.393	1.424	1.269	0.334	0.835	1.755
	CI coverage rate	0.0%	0.1%	4.3%	0.0%	0.0%	4.4%
	CI length	0.146	0.148	0.139	2.242	3.529	5.128
III	relative bias	0.0%	0.1%	0.1%	8.5%	14.6%	15.8%
	variance $\times 10^3$	1.281	1.984	2.859	0.261	1.176	3.726
	bootstrap variance $\times 10^3$	1.328	2.094	3.066	0.244	1.018	3.323
	CI coverage rate	95.3%	95.1%	95.1%	62.0%	33.9%	34.2%
	CI length	0.143	0.179	0.216	1.913	3.875	7.025
IV	relative bias	0.1%	0.1%	−0.3%	0.1%	0.9%	0.8%
	variance $\times 10^3$	1.349	2.986	3.884	0.182	0.848	2.365
	bootstrap variance $\times 10^3$	1.399	3.047	4.369	0.182	0.824	2.441
	CI coverage rate	95.6%	96.0%	96.5%	93.8%	94.2%	95.1%
	CI length	0.146	0.215	0.257	1.658	3.459	5.912
V	relative bias	0.1%	0.4%	0.2%	0.1%	5.9%	6.2%
	variance $\times 10^3$	1.349	1.643	1.764	0.182	1.167	2.212
	bootstrap variance $\times 10^3$	1.375	1.736	1.906	0.182	1.054	2.812
	CI coverage rate	95.6%	94.0%	94.8%	93.8%	88.8%	90.0%
	CI length	0.146	0.163	0.170	1.658	3.960	6.049

Method I: complete data
Method II: ignoring all missing data
Method III: censoring and linear regression imputation
Method IV: the nonparametric regression imputation in Sections 9.3.2 and 9.3.3
Method V: the semiparametric regression imputation in Sections 9.3.2 and 9.3.4

type I diabetes between May 1987 and April 1992 in southern and central Wisconsin were invited to enroll in the study. Subjects were asked to measure glycosylated haemoglobin (GHb) at each ordinary visit to a local physician, or every four months by submitting a blood specimen using prestamped mailing kits.

One of the main interests in this study is how GHb changes with duration defined as the number of years after diagnosis at year 0. The average GHb values within each year is the response from each subject. However, compliance was not constant, and some subjects could go one or two years without submitting blood samples. In some cases, they resumed mailing blood samples. At the beginning of the study, the number of subjects was 521. Table 9.4 shows the number of subjects with responses from the baseline year to the 5th year.

TABLE 9.4
Realized Missing Percentages in a WDRS Survey

Duration (years)	0	1	2	3	4	5
Number of observed data	521	450	417	441	357	321
Missing percentage	0%	14%	20%	15%	31%	38%

To estimate the mean of GHb from duration 0 (baseline) to duration 5, four methods are applied: the sample mean of subjects without any missing value (the naive method), the sample mean based on censoring and linear regression imputation, the sample mean based on nonparametric regression imputation, and the sample mean based on semiparametric regression imputation. The estimated means are plotted against durations in Figure 9.1. The estimated mean of GHb based on nonparametric and semiparametric regression imputation increases initially and levels off after duration 3 (3 years after diagnosis), whereas the estimated means of GHb based on the other two methods display continued increase after duration 3. They have a larger increase than the naive method of ignoring missing data.

Clinical experience indicates that GHb values do not increase several years after the diagnosis of diabetes, which supports the theory that nonparametric regression imputation provides the most plausible result from the medical or epidemiological point of view.

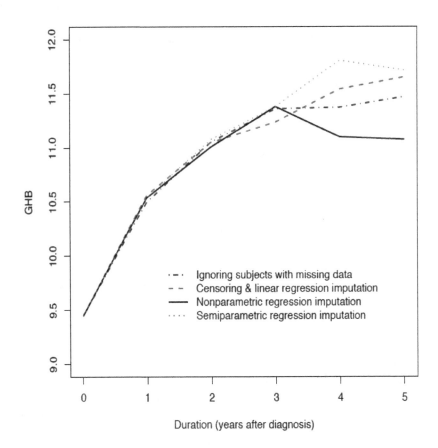

FIGURE 9.1
Estimated means of GHb by duration.

9.4 Random-Effect-Dependent Missing Data

The random-effect-dependent propensity model assumes that there exists a subject-level unobserved random effect $\mathbf{b}_i$ such that

$$q(\delta_i \mid \mathbf{y}_i, \mathbf{x}_i, \mathbf{b}_i) = q(\delta_i \mid \mathbf{x}_i, \mathbf{b}_i), \qquad (9.29)$$

which is often called the *shared parameter model* (Follmann and Wu, 1995). Since $\mathbf{b}_i$ is unobserved, this propensity is nonignorable, which is common when a mixed-effect model is assumed for complete data or when cluster sampling is applied in surveys. As a specific example for (9.29), imagine a cluster sampling case, where $\mathbf{y}_i$ is from a particular household (cluster) and a single person completes survey forms for all persons in the household. It is likely that the missing probability depends on a household-level variable (the person who completes survey forms), not on any within-household variable. We assume that, in addition to (9.29), $\mathbf{y}_i$ has at least one observed component.

When there are no missing data, we assume a linear mixed-effect model

$$\mathbf{y}_i = \mathbf{x}_i\beta + \mathbf{z}_i\mathbf{b}_i + \epsilon_i, \qquad (9.30)$$

where $\mathbf{b}_i$ is a subject-level random-effect vector, $E(\mathbf{b}_i) = 0$, $\mathbf{b}_i$ is independent of $\mathbf{x}_i$, $\mathbf{z}_i$ is a submatrix of $\mathbf{x}_i$, ϵ_i is a within-subject error vector, $\mathbf{b}_i$ and ϵ_i are independent and unobserved, $E(\epsilon_i) = 0$, and $V(\epsilon_i) = V(\mathbf{x}_i)$, a covariance matrix possibly depending on $\mathbf{x}_i$. Under (9.30), assumption (9.29) holds if and only if the missing probability of a component y_{it} depends on $(\mathbf{b}_i, \mathbf{x}_i)$ but not on ϵ_i.

9.4.1 Three Existing Approaches

Under missing mechanism (9.29), there are three existing approaches. The first one is the parametric likelihood approach. Under (9.29),

$$
\begin{aligned}
p(\mathbf{y}_i, \delta_i, \mathbf{b}_i \mid \mathbf{x}_i) &= p(\mathbf{y}_i \mid \mathbf{b}_i, \mathbf{x}_i)p(\delta_i \mid \mathbf{y}_i, \mathbf{b}_i, \mathbf{x}_i)p(\mathbf{b}_i \mid \mathbf{x}_i) \\
&= p(\mathbf{y}_i \mid \mathbf{b}_i, \mathbf{x}_i)p(\delta_i \mid \mathbf{b}_i, \mathbf{x}_i)p(\mathbf{b}_i \mid \mathbf{x}_i),
\end{aligned}
$$

where $p(\cdot \mid \cdot)$ is the generic notation for conditional likelihood. Let $\mathbf{y}_{i,\mathrm{mis}}$ be the missing components of $\mathbf{y}_i$. Assuming parametric models on $p(\mathbf{y}_i \mid \mathbf{b}_i, \mathbf{x}_i), p(\delta_i \mid \mathbf{b}_i, \mathbf{x}_i)$, and $p(\mathbf{b}_i \mid \mathbf{x}_i)$, we obtain the following parametric likelihood

$$\prod_i \int \left(\int p(\mathbf{y}_i \mid \mathbf{b}_i, \mathbf{x}_i) d\mathbf{y}_{i,\mathrm{mis}} \right) p(\delta_i \mid \mathbf{b}_i, \mathbf{x}_i)p(\mathbf{b}_i \mid \mathbf{x}_i) d\mathbf{b}_i. \qquad (9.31)$$

The integration is necessary since $\mathbf{y}_{i,\mathrm{mis}}$'s and $\mathbf{b}_i$'s are not observed. The parameters in $p(\mathbf{y}_i \mid \mathbf{b}_i, \mathbf{x}_i)$ can be estimated as long as they can be identified under some conditions. The likelihood in (9.31) involves intractable integrals except for some very special cases, hence, the difficulty of computing maximum likelihood. To avoid the difficulty, parametric fractional imputation of Kim (2011) can be applied. Yang et al. (2013) provides a detailed description of the FI method for the shared parameter model in (9.29). In general, shared parameter models require untestable assumptions, and the parametric approach is sensitive to parametric assumptions on various conditional probability densities.

The second approach is a semiparametric method specifying only the first- and second-order conditional moments. Assumptions (9.29)–(9.30) imply that

$$E(\mathbf{y}_i \mid \mathbf{x}_i, \mathbf{b}_i, \delta_i) = \mathbf{x}_i \beta + \mathbf{z}_i \mathbf{b}_i \qquad \text{and} \qquad V(\mathbf{y}_i \mid \mathbf{x}_i, \mathbf{b}_i, \delta_i) = V(\mathbf{x}_i).$$

Then, we have a conditional model

$$E(\mathbf{y}_i \mid \mathbf{x}_i, \delta_i) = E[E(\mathbf{y}_i \mid \mathbf{x}_i, \mathbf{b}_i, \delta_i) \mid \mathbf{x}_i, \delta_i] = \mathbf{x}_i \beta + \mathbf{z}_i E(\mathbf{b}_i \mid \mathbf{x}_i, \delta_i). \qquad (9.32)$$

If $E(\mathbf{b}_i \mid \mathbf{x}_i, \delta_i)$ can be approximated by a simple form, then β may be estimated using this approximate conditional model (ACM). This approach is known as the ACM approach. However, one must deal with the following two issues:

1. How to find a reasonable approximation $E(\mathbf{b}_i \mid \mathbf{x}_i, \delta_i)$? Follmann and Wu (1995) showed that, if $p(\delta_i \mid \mathbf{b}_i)$ is in an exponential family and if δ_{it}'s are conditionally i.i.d. given $\mathbf{b}_i$, then $E(\mathbf{b}_i \mid \delta_i)$ is a monotone function of a summary statistic S_i. As a result, they suggested that $E(\mathbf{b}_i \mid \delta_i)$ can be approximated by a linear or polynomial function of S_i. This approximation, however, may not be good enough. Furthermore, the exponential family assumption is somewhat restrictive.

2. The parameter β may not be identifiable after $E(\mathbf{b}_i \mid \mathbf{x}_i, \delta_i)$ is approximated through some simple function, i.e., β may be confounded with some parameters in the ACM (Albert and Follmann, 2000). For example, suppose that the covariate and $\mathbf{b}_i = b_i$ are both univariate so that $E(y_{it} \mid x_{it}, b_i) = \beta_0 + \beta_1 x_{it} + b_i x_{it}$. If S_i is univariate and $E(b_i \mid \mathbf{x}_i, \delta_i)$ is approximated by $\gamma_0 + \gamma_1 S_i$, then the ACM is $E(y_{it} \mid x_{it}, b_i) \approx \beta_0 + \beta_1 x_{it} + \gamma_0 x_{it} + \gamma_1 S_i x_{it}$ and β_1 is confounded with γ_0. Non-identifiability seriously limits the scope of the application of the ACM approach, although in some situations β_1 can be estimated with additional work.

The third approach is the method of grouping, which can be applied according to the following steps:

1. Find a summary statistic S_i such that

$$E(\mathbf{b}_i \mid \mathbf{x}_i, \delta_i) = E(\mathbf{b}_i \mid S_i) \qquad \text{and} \qquad E(\mathbf{b}_i \mathbf{b}_i' \mid \mathbf{x}_i, \delta_i) = E(\mathbf{b}_i \mathbf{b}_i' \mid S_i).$$
$$(9.33)$$

2. Assume that S_i is discrete and takes values $s_1, ..., s_L$. Then , we divide the sample into L groups according the value of S_i.

3. Obtain estimate $\hat{\beta}_l$ using data in the l-th group and the following model:

$$
\begin{aligned}
E(\mathbf{y}_{i,\text{obs}} \mid \mathbf{x}_{i,\text{obs}}, S_i = s_l) &= \mathbf{x}_{i,\text{obs}}\beta + \mathbf{z}_{i,\text{obs}}E(\mathbf{b}_i \mid S_i = s_l) \\
V(\mathbf{y}_{i,\text{obs}} \mid \mathbf{x}_{i,\text{obs}}, S_i = s_l) &= \mathbf{z}_{i,\text{obs}}V(\mathbf{b}_i \mid S_i = s_l)\mathbf{z}'_{i,\text{obs}} \quad (9.34) \\
&\quad + E(V(\mathbf{x}_i)|\mathbf{x}_{i,\text{obs}}, S_i = s_l),
\end{aligned}
$$

where $\mathbf{y}_{i,\text{obs}}$ is the vector of observed $\mathbf{y}_i$-values and $\mathbf{x}_{i,\text{obs}}$ and $\mathbf{z}_{i,\text{obs}}$ are submatrices of $\mathbf{x}_i$ and $\mathbf{z}_i$ corresponding to $\mathbf{y}_{i,\text{obs}}$. In (9.34), $E(\mathbf{b}_i|S_i = s_l)$ is viewed as an unobserved random effect. We can use the weighted least squares estimator

$$
\hat{\beta}_l = \left(\sum_{i \in \mathcal{G}_l} \mathbf{x}'_{i,\text{obs}} \hat{V}^{-1}_{i,\text{obs}} \mathbf{x}_{i,\text{obs}} \right)^{-1} \sum_{i \in \mathcal{G}_l} \mathbf{x}'_{i,\text{obs}} \hat{V}^{-1}_{i,\text{obs}} \mathbf{y}_{i,\text{obs}},
$$

where $\mathcal{G}_l$ is the l-th group, $\hat{V}_{i,\text{obs}}$ is an estimator of $V_{i,\text{obs}} = V(\mathbf{y}_{i,\text{obs}}|\mathbf{x}_{i,\text{obs}}, S_i = s_l)$ as specified in (9.34). Many statistical packages can be used to obtain $\hat{\beta}_l$. Since we view $E(\mathbf{b}_i|S_i = s_l)$ as an unobserved random effect, we do not need to estimate or approximate it.

4. Because $E(\mathbf{b}_i|S_i)$ is not necessarily equal to 0, $\hat{\beta}_l$ with a fixed l is not approximately unbiased for β. Let $p_l = P(S_i = s_l)$. Since

$$
\sum_l p_l E(\mathbf{b}_i|S_i = s_l) = E(\mathbf{b}_i) = 0,
$$

$\sum_l p_l \hat{\beta}_l$ is approximately unbiased for β. After replacing the unknown p_l by its estimator n_l/n, where n_l is the number of subjects in $\mathcal{G}_l$, we obtain the following approximately unbiased estimator of β:

$$
\hat{\beta} = \sum_l \frac{n_l}{n} \hat{\beta}_l.
$$

The key to this grouping method is that, although each $\hat{\beta}_l$ is biased for β ($E(\mathbf{b}_i|S_i = l) \neq 0$), the linear combination $\hat{\beta}$ is approximately unbiased for β because the average of the $E(\mathbf{b}_i|S_i = s_l)$ is 0. This method avoids the estimation of $E(\mathbf{b}_i|S_i)$ (which is what the original ACM does) and, hence, it does not have the problem of parameter confounding.

If the summary statistic S_i is continuous (or discrete but takes many values), then we need to replace S_i by a function of S_i taking discrete values. Details are given in Section 9.4.2.

9.4.2   Summary Statistics

As we discussed previously, the starting point for the ACM or the group method is to find a summary statistic. We define a summary statistic S_i to be a function of $(\mathbf{x}_i, \delta_i)$ such that (9.33) holds. The second condition in (9.33) is not needed if we use ordinary least squares instead of weighted least squares in model fitting. A trivial summary statistic is $(\mathbf{x}_i, \delta_i)$ itself, but a simple S_i with low dimension is desired.

The following lemma is useful for finding an S_i satisfying (9.33).

Lemma 9.1. *Assume (9.29)-(9.30). A sufficient condition for (9.33) is that there exists a measurable function g such that*

$$p(\delta_i \mid \mathbf{x}_i, \mathbf{b}_i) = g(\mathbf{b}_i, S_i).$$

We now derive S_i under some nonparametric or semiparametric models on the missing mechanism.

Conditionally i.i.d. Model

We start with the simplest case where $p(\delta_i \mid \mathbf{x}_i, \mathbf{b}_i) = p(\delta_i \mid \mathbf{b}_i)$ (i.e., the propensity does not depend on covariates) and components of δ_i are conditionally i.i.d., given $\mathbf{b}_i$. That is,

$$
\begin{aligned}
p(\delta_i \mid \mathbf{b}_i) &= \prod_{t=1}^{T} P(\delta_{it} = 1 \mid \mathbf{b}_i)^{\delta_{it}} P(\delta_{it} = 0 \mid \mathbf{b}_i)^{1-\delta_{it}} \\
&= P(\delta_{it} = 1 \mid \mathbf{b}_i)^{R_i} P(\delta_{it} = 0 \mid \mathbf{b}_i)^{T-R_i},
\end{aligned}
$$

which is a function of $\mathbf{b}_i$ and $R_i = \sum_{t=1}^{T} \delta_{it}$ is the number of observed components of $\mathbf{y}_i$. According to Lemma 9.1, $S_i = R_i$ is a discrete summary statistic.

Conditionally Independent Model with Time Trend

Wu and Follmann (1999) considered a propensity model with time trend where the components of δ_i are independent given $\mathbf{b}_i$ and satisfy

$$
P(\delta_{it} = 1 \mid \mathbf{x}_i, \mathbf{b}_i) = \frac{\exp\{\phi_1(\mathbf{b}_i) + \phi_2(\mathbf{b}_i)s_t\}}{1 + \exp\{\phi_1(\mathbf{b}_i) + \phi_2(\mathbf{b}_i)s_t\}}, \quad t = 1, ..., T. \tag{9.35}
$$

Here, $s_1, ..., s_T$ are time-related values (fixed and identical for all sampled subjects) and ϕ_1 and ϕ_2 are unknown nonparametric functions of $\mathbf{b}_i$. Then,

$$
\begin{aligned}
p(\delta_i \mid \mathbf{x}_i, \mathbf{b}_i) &= \prod_{t=1}^{T} \left[\frac{P(\delta_{it} = 1 \mid \mathbf{x}_i, \mathbf{b}_i)}{P(\delta_{it} = 0 \mid \mathbf{x}_i, \mathbf{b}_i)} \right]^{\delta_{it}} P(\delta_{it} = 0 \mid \mathbf{x}_i, \mathbf{b}_i) \\
&= \frac{\exp\{\phi_1(\mathbf{b}_i)R_i + \phi_2(\mathbf{b}_i)R_{si}\}}{\prod_{t=1}^{T} [1 + \exp\{\phi_1(\mathbf{b}_i) + \phi_2(\mathbf{b}_i)s_t\}]},
\end{aligned}
$$

where $R_i = \sum_{t=1}^{T} \delta_{it}$ and $R_{si} = \sum_{t=1}^{T} \delta_{it} s_t$. According to Lemma 9.1, the summary statistic is $S_i = (R_i, R_{si})$. The component R_{si} accounts for the time trend in the propensity model.

An extension to model (9.35) is to replace s_t by a covariate x_{it}. The resulting summary statistic is $S_i = (R_i, R_{xi})$ with $R_i = \sum_{t=1}^{T} \delta_{it}$ and $R_{xi} = \sum_{t=1}^{T} \delta_{it} x_{it}$.

Markov Dependency Model

In previous cases, the components of δ_i are assumed to be conditionally independent given $(\mathbf{x}_i, \mathbf{b}_i)$. To consider a conditionally dependent model, we assume $p(\delta_i \mid \mathbf{x}_i, \mathbf{b}_i) = p(\delta_i \mid \mathbf{b}_i)$ and that δ_{it}'s follow a stationary Markov chain model, given $\mathbf{b}_i$. Let $\phi_{u,v} = P(\delta_{it} = u \mid \delta_{i(t-1)} = v, \mathbf{b}_i)$. Then

$$
\begin{aligned}
p(\delta_i \mid \mathbf{b}_i) &= P(\delta_{i1} = 1 \mid \mathbf{b}_i)^{\delta_{i1}} P(\delta_{i1} = 0 \mid \mathbf{b}_i)^{1-\delta_{i1}} \\
&\quad \times \prod_{t=2}^{T} \left[\phi_{0,0}^{(1-\delta_{it})(1-\delta_{i(t-1)})} \phi_{1,0}^{\delta_{it}(1-\delta_{i(t-1)})} \phi_{0,1}^{(1-\delta_{it})\delta_{i(t-1)}} \phi_{1,1}^{\delta_{it}\delta_{i(t-1)}} \right] \\
&= P(\delta_{i1} = 1 \mid \mathbf{b}_i)^{\delta_{i1}} P(\delta_{i1} = 0 \mid \mathbf{b}_i)^{1-\delta_{i1}} \\
&\quad \times \phi_{0,0}^{T-\delta_{iT}+G_i} \phi_{1,0}^{R_i-\delta_{iT}-G_i} \phi_{0,1}^{R_i-\delta_{i1}-G_i} \phi_{1,1}^{G_i},
\end{aligned}
$$

where $R_i = \sum_{t=1}^{T} \delta_{it}$ and $G_i = \sum_{t=2}^{T} \delta_{it} \delta_{i(t-1)}$. It follows from Lemma 9.1 that a summary statistic is $S_i = (\delta_{i1}, \delta_{iT}, R_i, G_i)$. Compared with the summary statistic R_i in the i.i.d. case, we find that three statistics, δ_{i1}, δ_{iT}, and G_i, are needed to account for Markov dependency.

If $p(\delta_i \mid \mathbf{x}_i, \mathbf{b}_i)$ depends on $\mathbf{x}_i$, similar but more complicated summary statistics can be derived.

Monotone Missing Data

Since monotone missing has fewer missing patterns, simpler summary statistics can often be obtained. For example, when missing is monotone, $\sum_{t=1}^{R_i} \delta_{it} s_t = \sum_{t=1}^{R_i} s_t$; the summary statistic $(\delta_{i1}, \delta_{iT}, R_i, G_i)$ under the Markov dependency model reduces to R_i, because $(\delta_{i1}, \delta_{iT}, G_i)$ is a function of R_i.

Approximate Summary Statistics

Some summary statistics, such as R_i and R_{si}, are discrete. However, some are continuous or discrete with many categories and the grouping method cannot be directly applied.

Let $\tilde{S}_i$ be a function of S_i such that (1) $\tilde{S}_i$ is discrete with a reasonable number of categories and (2) in each group defined by a category of $\tilde{S}_i$, values of $E(b_i \mid S_i)$ are similar. We can then call $\tilde{S}_i$ an approximate summary statistic and apply the grouping method using $\tilde{S}_i$ to create groups. If $E(b_i \mid S_i)$ is observed, then any method in classification or clustering may be applied to find a good approximate summary statistic $\tilde{S}_i$. Some examples can be found in Xu and Shao (2009).

9.4.3   Simulation Study

Some simulations were conducted to study the performance of the grouping method. We used the following response model

$$y_{it} = \beta_0 + \beta_1 x_{it} + b_{i0} + b_{i1}x_{it} + e_{it}, \tag{9.36}$$

where $i = 1, ..., 1,000$ is the index for subjects, each subject has $t = 1, ..., 5$ repeated measurements, $\beta_0 = 1$ and $\beta_1 = 1$ are parameters, x_{it} is a one-dimensional covariate to be specified later, b_{i0} and b_{i1} are random effect variables following a bivariate normal distribution with mean 0, $V(b_{i0}) = 1$, $V(b_{i1}) = 2$, and correlation coefficient 0.3, and e_{it}'s are i.i.d. and follows a standard normal distribution and are independent of $b_i = (b_{i0}, b_{i1})$.

After data were generated according to (9.36), missing data were generated according to

$$\text{logit}\{P(\delta_{it} = 0 \mid b_i, x_{it})\} = \gamma_0 + \gamma_1 b_{i1} + \lambda_0 x_{it} + \lambda_1 b_{i1} x_{it}, \tag{9.37}$$

where δ_{it}'s are conditionally independent given b_i and x_i. The overall missingness proportions are between 35% and 50%.

The following methods for the estimation of β_1 were compared in the simulation: (1) the naive maximum likelihood estimator ignoring missing data; (2) ACM-R, the ACM method using R_i being the number of observed components in y_i as the summary statistic; (3) ACM-S, the ACM method using a correctly derived S_i from model (9.37) as the summary statistic; (4) GRP-R, the grouping method using R_i as the summary statistic; (5) GRP-S, the grouping method using a correctly derived S_i from model (9.37) as the summary statistic. Note that R_i is a correct summary statistic if $\lambda_0 = \lambda_1 = 0$ in (9.37).

For the grouping method, if S_i is continuous, then an approximate summary statistic is used. For the ACM, S_i is included as a covariate in the following linear model

$$y_{it} = \beta_0^* + \beta_1^* x_{it} + \beta_2^* S_i + \beta_3^* S_i x_{it} + b_{i0}^* + b_{i1}^* x_{it} + \epsilon_{ij}^*.$$

The estimator $\hat{\beta}_1^*$ is not an estimator of β_1. Instead, we use $\hat{\beta}_1^* + \hat{\beta}_3^* \bar{S}$ as an estimator of β_1, where $\bar{S}$ is the sample mean of S_i's.

From the 500 replications of the simulation, we computed the empirical bias and mean squared error (MSE) of the estimator of β_1 and the coverage probability (CP) of the related 95% confidence interval using bootstrap for variance estimation. The results are given in Table 9.5. Four different situations (Cases I-IV) with different missing mechanisms and covariate types were considered. The following is a description of these four cases and a summary of the results in Table 9.5.

Case I. The parameters in model (9.37) are $\gamma_0 = 0.5$, $\gamma_1 = 1$, and $\lambda_0 = \lambda_1 = 0$, which corresponds to the conditional i.i.d. model in Section 9.4.2 and a

TABLE 9.5

Simulation Results for the Simulation Study in §9.4.3 Based on 500 Runs

Method	Case I			Case II		
	Bias	MSE	CP(%)	Bias	MSE	CP(%)
Naive	0.0985	0.0120	16.4	0.3695	0.1405	0
ACM-R	−0.0026	0.0026	94.8	0.0657	0.0097	78.2
ACM-S				0.0455	0.0074	89.2
GRP-R	−0.0013	0.0026	95.4	0.1413	0.0286	46.4
GRP-S				0.0210	0.0093	94.6

Method	Case III			Case IV		
	Bias	MSE	CP(%)	Bias	MSE	CP(%)
Naive	−0.0091	0.0049	92.8	0.0467	0.0068	81.8
ACM-R	−0.0126	0.0052	91.8	0.0090	0.0049	91.0
ACM-S				0.0052	0.0049	91.0
GRP-R	−0.0058	0.0078	94.2	0.0085	0.0063	91.5
GRP-S	−0.0028	0.0051	95.0	0.0028	0.0063	94.0

Naive: maximum likelihood estimator ignoring missing data
ACM-R: ACM with summary statistic R
ACM-S: ACM with summary statistic S
GRP-R: grouping with summary statistic R
GRP-S: grouping with summary statistic S
R: number of observed components
S: derived summary statistic

correct summary statistic is R_i. Thus, ACM-S and GRP-S are the same as ACM-R and GRP-R, respectively.

The naive method has bias about 10% of the true value ($\beta_1 = 1$) and the CP of the 95% confidence interval is only 16.4%. The ACM and grouping methods are almost the same. They are much better than the naive method and have nearly 95% CP.

Case II. To account for time-dependent missingness, we used a propensity model that depends on both random effects and a time covariate $x_{it} = s_t$, where $s_t = t$. The parameters in model (9.37) are $\gamma_0 = \gamma_1 = 0$, $\lambda_0 = -0.2$, and $\lambda_1 = 0.3$. Under this time-dependent missing mechanism, the proportion of missing data increases as t increases. According to the discussion in Section 9.4.2 and by the fact that $\gamma_0 = \gamma_1 = 0$, a correct summary statistic is $S_i = \sum_{t=1}^{5} a_{it} s_t$ instead of R_i.

The naive method is heavily biased and has CP $= 0$. Since R_i is not the right summary statistic, both ACM-R and GRP-R are biased with low CP values. ACM-S and GRP-S use the correct summary statistic S_i. However, ACM-S still has a 4.6% relative bias, which results in a low CP of 89.2%. This may be caused by the fact that $E(\mathbf{b}_i \mid S_i)$ is not linear in S_i. GRP-S is much better than GRP-R, indicating the importance of having a correct summary statistic. It is also better than ACM-S, since no linear approximation to $E(\mathbf{b}_i \mid S_i)$ is required in the grouping method.

Case III. We considered a discrete covariate $x_{it} = x_i$ that does not depend on t and takes only two values, 0.5 and -1. The parameters in (9.37) are $\gamma_0 = 0.1$, $\gamma_1 = 0.2$, $\lambda_0 = 0.3$ and $\lambda_1 = 1$. Subjects with different x_i values have opposite signs of random slopes in the missing mechanism. The result in Section 9.4.2 indicates that a correct summary statistic is the 2-dimensional statistic $S_i = (R_i, x_i)$.

The naive method fares well in terms of bias and its CP is 92.8%. This may be because the dataset has two subsets according to the value of x_i and, within each subset, the naive method is biased (e.g., the results in Case I) but the biases are canceled in the overall estimation. The ACM-R does poorly compared to the naive method. We did not use ACM-S in this case, because how to apply ACM using a 2-dimensional summary statistic was not addressed previously. GRP-R uses an incorrect summary statistic, but its performance is acceptable. GRP-S uses a correct summary statistic and performs the best among all methods under consideration.

Case IV. We considered a continuous covariate $x_{it} \sim N(\log t, 0.5)$. The parameters in (9.37) are $\gamma_0 = \gamma_1 = 0$, $\lambda_0 = 0.1$ and $\lambda_1 = \sqrt{2}/10$. According to the result in Section 9.4.2, a correct summary statistic is $S_i = \sum_{t=1}^{5} a_{it} x_{it}$, which is a continuous statistic. Thus, for the grouping approach, we apply the GUIDE (Loh, 2002) to find an approximate summary statistic for grouping.

The naive method has a 4.7% bias and a low CP of 81.8%. ACM-R and ACM-S are about the same, although the former uses the wrong summary statistic R_i and the latter uses the correct summary statistic S_i and is less biased. GRP-R uses the wrong summary statistic R_i, and is slightly worse than ACM-S. GRP-S uses the correct summary statistic and has the smallest bias and a CP closest to 95% among all methods under consideration.

9.4.4   Modification of Diet in Renal Disease

We now present a real data example in Xu and Shao (2009). The modification of diet in renal disease (MDRD) study was a randomized clinical trial of patients with progressive renal disease. The intervention examined here is the two levels of dietary intake of protein and phosphorous. The primary interest of the trial is to compare two treatments in terms of the rate reduction of their renal disease. The primary outcome measure is the decline in glomerular filtration rate (GFR), a continuous measure of how rapidly the kidneys filter blood. GFR is measured serially every 4 months. The primary period of interest is from month 16 to month 36.

A total of 520 patients were randomized to one of the two treatment levels: low protein diet (Diet L) and very low protein diet (Diet VL). Relevant summaries of the two treatment groups are given in Table 9.6.

TABLE 9.6
MDRD Study with 2 Treatments

	Treatment	
	Diet VL	Diet L
Number of patients	259	261
Proportion of missing data	44%	45%
Median number of GFR per patient	3	3
Number (%) of patients with complete data	41 (15.8%)	39 (14.9%)
Number (%) of patients with monotone missingness	191 (73.7%)	197 (75.5%)
Number (%) of patients with intermittent missingness	27 (10.5%)	25 (9.6%)

We consider the following response model when there is no missing:

$$\text{GFR}_{it} = \beta_0 + \beta_1 s_t + \beta_2 x_i + \beta_3 x_i s_t + b_{i0} + b_{i1} s_t + \epsilon_{it},$$

where x_i is the treatment indicator ($x_i = 0$ for Diet VL and $x_i = 1$ for Diet L), $s_t = 12 + 4t$ is the duration of the study that does not depend on i, $t = 1, ..., 6$, β_j's are unknown parameters, and b_{i0} and b_{i1} are random subject effects. Note that β_1 is the slope over time for GFR with Diet VL treatment and $\beta_1 + \beta_3$ is the slope over time for GFR with Diet L treatment. Thus, β_3 is the difference in slope for GFR between two treatments and is the main focus of our analysis. A positive β_3 leads to the conclusion that Diet L is better whereas a negative β_3 leads to the conclusion that Diet VL is better.

The overall proportion of missing data is about 45%. One issue related to the missing data is that when a patient's GFR drops to some level the kidney function would be impaired and no outcome could be obtained as a result. We calculate the ordinary least squares estimates of GFR slopes for patients with at least two observations. The individual slopes from the patients with missing data are significantly more negative than those from the patients with complete data (p-value = 0.004922, Wilcoxon rank sum test). Another characteristic of the missing data in this example is the notable time trend of missingness proportions. We analyze this dataset under the following propensity model:

$$\text{logit}\{P(\delta_{it} = 0 \mid b_i, x_{it})\} = \phi_2(b_i)s_t.$$

Here, the summary statistic is $S_i = \sum_{t=1}^{6} \delta_{it} s_t$. Because the majority of subjects have monotone missingness (Table 9.6), in this example, S_i is close to R_i which is the number of observed components of $\mathbf{y}_i$.

We apply the five methods in the simulation study in Section 9.4.3 to the

MDRD data: (1) The naive method ignoring missing data; (2) ACM-R, the ACM using R_i as the summary statistic; (3) ACM-S, the ACM using S_i as the summary statistic; (4) GRP-R, the grouping method using R_i as the summary statistic; (5) GRP-S, the grouping method using S_i as the summary statistic. When we apply ACM-S, we fit the model

$$\begin{aligned}
\text{GFR}_{it} = \ & \beta_0^* + \beta_1^* s_t + \beta_2^* x_i + \beta_3^* S_i + \beta_4^* x_i s_t + \beta_5^* x_i S_i + \beta_6^* s_t S_i \\
& + \beta_7^* x_i s_t S_i + b_{i0}^* + b_{i1}^* s_t + \epsilon_{it}.
\end{aligned}$$

We consider $\hat{\beta}_4^* + \hat{\beta}_6^*(\bar{S}_1 - \bar{S}_0) + \hat{\beta}_7^* \bar{S}_1$, instead of $\hat{\beta}_4^*$, as an estimator of β_3, where $\bar{S}$ is the sample mean of S_i's, and $\bar{S}_k$ is the sample mean of S_i under treatment k for $k = 0, 1$. For ACM-R, we can simply substitute S_i by R_i.

TABLE 9.7
Estimates and 95% Confidence Upper and Lower Limits of β_3 in the MDRD Study

Method	Estimate	Lower Limit	Upper Limit
Naive	0.06	−0.04	0.17
ACM-R	0.06	−0.11	0.21
ACM-S	0.01	−0.16	0.18
GRP-R	−0.03	−0.19	0.13
GRP-S	−0.04	−0.16	0.08

Table 9.7 shows the estimates and 95% confidence upper and lower limits of β_3 using the five methods. The estimates from the naive method, ACM-R, and ACM-S are all positive, whereas estimates from GRP-R and GRP-S are negative. All confidence intervals include 0 so that we cannot reject the hypothesis of $\beta_3 = 0$ at the significance level of 5%, which may be due to low power: we only have about 260 subjects in each treatment and about 45% of them have missing values. However, the results indicate that the grouping method and the ACM approach may lead to different conclusions on which diet is better; the ACM approach agrees with the naive method and is in favor of Diet L, whereas the grouping method is in favor of Diet VL.

10

Application to Survey Sampling

In this chapter, we consider the problem of parameter estimation in the context of survey sampling. Estimation with data from a probability sample can be viewed as a special case of the missing data problem , where the sample is treated as the set of respondents and the sampling mechanism is known. The main difference is that the first order inclusion probability is not necessarily a known function of the observations in the sample. Missing data in the probability sample can be viewed as a two-phase sampling problem. Multiple imputation and fractional imputation can be developed but should incorporate the sampling design features into estimation. Synthetic imputation in Section 10.8 can be a useful tool for combining information from different surveys.

10.1 Introduction

To formally define the setup, let $U = \{1, 2 \ldots, N\}$ be the index set of a finite population and let I_i be the sample indicator function such that $I_i = 1$ indicates the selection of unit i for the sample, and $I_i = 0$ otherwise. The probability $\pi_i = P(I_i = 1 \mid i \in U)$ is often called the *first-order inclusion probability* and is known in probability sampling. Thus, estimation with data from a probability sample is a special case of the missing data problem where the sample is treated as the set of respondents and the sampling mechanism is known.

Let y_i be the realized value of a random variable Y for unit i. Assume that Y follows a distribution with density $f(y; \theta)$, for some unknown parameter θ. The model for generating the finite population is often called the *superpopulation model*. Let A be the set of indices in the sample. If we use

$$\sum_{i \in A} S(\theta; y_i) = 0,$$

where $S(\theta; y) = \partial \log f(y; \theta)/\partial \theta$, to estimate θ from the sample, the solution is consistent if I_i is independent of y_i. Such condition is very close to the condition of missing completely at random (MCAR). If the parameter of interest is for the conditional distribution of y given x, denoted by $f_1(y \mid x; \theta)$, then the

DOI: 10.1201/9780429321740-10

sample score equation

$$\sum_{i \in A} S_1(\theta; x_i, y_i) = 0, \tag{10.1}$$

where $S_1(\theta; x, y) = \partial \log f_1(y \mid x; \theta)/\partial \theta$, provides a consistent estimator of θ if

$$Cov \{I_i, S_1(\theta; x_i, Y) \mid x_i\} = 0.$$

This condition will hold if $P(I_i = 1 \mid x_i, y_i)$ is a function of x_i only. Thus, if the sampling design is such that

$$E(I_i \mid x_i, y_i) = E(I_i \mid x_i) \tag{10.2}$$

holds for all (x_i, y_i), then the sampling design is called noninformative in the sense that we can use the sample score equation (10.1) to estimate θ. Note that the definition of a noninformative sampling design is specific to the model considered. If the model is about $f(x \mid y)$, the conditional distribution of x given y, then the condition for a noninformative sampling design is changed to

$$E(I_i \mid x_i, y_i) = E(I_i \mid y_i).$$

The noninformative sampling condition is essentially the MAR condition in missing data.

If the sampling design is informative for estimating θ in $f_1(y \mid x; \theta)$ in the sense that (10.2) does not hold, then the sample score equation (10.1) leads to a biased estimate. To remove the bias, the pseudo maximum likelihood estimator, defined by solving

$$\sum_{i \in A} \frac{1}{\pi_i} S_1(\theta; x_i, y_i) = 0, \tag{10.3}$$

is often used. Because

$$Cov \{I_i \pi_i^{-1}, S_1(\theta; x_i, y_i) \mid x_i\} \quad = \quad 0,$$

the weighted score equation using the survey weight $d_i = 1/\pi_i$ leads to a consistent estimator of θ. Conditions ensuring consistency and asymptotic normality of the pseudo maximum likelihood estimator (PMLE) are established, for example, in Binder (1983), Rubin-Bleuer and Schiopu-Kratina (2005), and Han and Wellner (2021).

In fact, since

$$Cov \{I_i \pi_i^{-1} q(x_i), S_1(\theta; x_i, y_i) \mid x_i\} \quad = \quad 0$$

for any $q(x_i)$, the solution to

$$\sum_{i \in A} \frac{1}{\pi_i} S_1(\theta; x_i, y_i) q(x_i) = 0 \tag{10.4}$$

is consistent for θ, regardless of the choice of $q(x)$. See Magee (1998).

Example 10.1. *Suppose that we are interested in estimating* $\beta = (\beta_0, \beta_1)$ *for the linear regression model*

$$y_i = \beta_0 + \beta_1 x_i + e_i, \tag{10.5}$$

where $E(e_i \mid x_i) = 0$ *and* $Cov(e_i, e_j \mid x) = 0$ *for* $i \neq j$. *We consider the following class of estimators*

$$\hat{\beta} = \left(\sum_{i \in A} d_i \mathbf{x}_i \mathbf{x}_i' q_i \right)^{-1} \sum_{i \in A} d_i \mathbf{x}_i y_i q_i, \tag{10.6}$$

where $\mathbf{x}_i = (1, x_i)'$, $q_i = q(x_i)$ *and* $d_i = \pi_i^{-1}$. *Since*

$$\hat{\beta} - \beta = \left(\sum_{i \in A} d_i \mathbf{x}_i \mathbf{x}_i' q_i \right)^{-1} \sum_{i \in A} d_i \mathbf{x}_i e_i q_i,$$

where $e_i = y_i - \mathbf{x}_i' \beta$,

$$E \left(\hat{\beta} - \beta \right) \cong E \left\{ \left(\sum_{i=1}^{N} \mathbf{x}_i \mathbf{x}_i' q_i \right)^{-1} \sum_{i=1}^{N} \mathbf{x}_i e_i q_i \right\} \cong 0,$$

where the expectation is with respect to the joint distribution of the superpopulation model (10.5) and the sampling mechanism. The anticipated variance, which is the total variance with respect to the joint distribution, is

$$V \left(\hat{\beta} - \beta \right) \cong \left\{ E \left(\sum_{i=1}^{N} \mathbf{x}_i \mathbf{x}_i' q_i \right) \right\}^{-1} V \left\{ \sum_{i \in A} d_i \mathbf{x}_i e_i q_i \right\} E \left\{ \left(\sum_{i=1}^{N} \mathbf{x}_i \mathbf{x}_i' q_i \right) \right\}^{-1} \tag{10.7}$$

and

$$
\begin{aligned}
&V \left\{ \sum_{i \in A} d_i \mathbf{x}_i e_i q_i \right\} \\
&= E \left\{ V \left(\sum_{i \in A} d_i \mathbf{x}_i e_i q_i \mid \mathcal{F}_N \right) \right\} + V \left\{ E \left(\sum_{i \in A} d_i \mathbf{x}_i e_i q_i \mid \mathcal{F}_N \right) \right\} \\
&= E \left\{ \sum_{i=1}^{N} \sum_{j=1}^{N} (\pi_{ij} - \pi_i \pi_j) d_i d_j \mathbf{x}_i \mathbf{x}_j' e_i e_j q_i q_j \right\} + V \left\{ \sum_{i=1}^{N} \mathbf{x}_i e_i q_i \right\} \\
&= E \left\{ \sum_{i=1}^{N} E(d_i e_i^2 \mid \mathbf{x}_i) \mathbf{x}_i \mathbf{x}_i' q_i^2 \right\},
\end{aligned}
$$

where the conditional expectation on $\mathcal{F}_N = \{(\mathbf{x}_i, y_i); i \in U\}$ *denotes the expectation with respect to the sampling design treating the population values* of $(\mathbf{x}_i, y_i)$ as fixed. Thus, the optimal choice of q_i that minimizes the total variance in (10.7) is

$$q_i^* = \{E(d_i e_i^2 \mid \mathbf{x}_i)\}^{-1}. \tag{10.8}$$

To estimate q_i^*, the estimated GLS method or the variance function estimation technique (Davidian and Caroll, 1987) can be used. Fuller (2009, Chapter 6) discussed the optimal estimation of β using $\hat{q}_i^* = q^*(\mathbf{x}_i; \hat{\alpha})$, where $\hat{\alpha}$ is a consistent estimator of α in the model $q^*(x; \alpha) = E(d_i e_i^2 \mid \mathbf{x}_i; \alpha)$. The effect of replacing α with $\hat{\alpha}$ in $\hat{q}_i$ is negligible in variance estimation of $\hat{\beta}$.

10.2   Calibration Estimation

In survey sampling, we often have one or more auxiliary variables observed throughout the population. Let $\mathbf{x}_i$ be a p-dimensional vector of auxiliary variables whose population total $\mathbf{X} = \sum_{i=1}^{N} \mathbf{x}_i$ is known. In this case, it is often desirable to achieve consistency with $\mathbf{X}$ in the estimation. That is, for $\hat{Y} = \sum_{i \in A} w_i y_i$, weights are desired to satisfy

$$\sum_{i \in A} w_i \mathbf{x}_i = \mathbf{X}, \tag{10.9}$$

which is often called the *calibration condition* or *benchmarking condition*.

To uniquely determine the calibration weight, we can formulate the problem as minimizing

$$Q(w, d) = \sum_{i \in A} d_i \left(\frac{w_i}{d_i} - 1\right)^2 q_i \tag{10.10}$$

subject to (10.9) where $q_i = q(\mathbf{x}_i)$ is a known function of $\mathbf{x}_i$. In this case, the solution can be written as

$$w_i = d_i \left\{ 1 + \left(\mathbf{X} - \hat{\mathbf{X}}_{HT}\right)' \left(\sum_{i \in A} d_i q_i^{-1} \mathbf{x}_i \mathbf{x}_i'\right)^{-1} q_i^{-1} \mathbf{x}_i' \right\}, \tag{10.11}$$

where $\hat{\mathbf{X}}_{HT} = \sum_{i \in A} d_i \mathbf{x}_i$. Note that the calibration estimator using (10.11) can be written as

$$\hat{Y}_{reg} = \hat{Y}_{HT} + (\mathbf{X} - \hat{\mathbf{X}})' \hat{\beta}_q \tag{10.12}$$

where $\hat{\beta}_q = \left(\sum_{i \in A} d_i q_i^{-1} \mathbf{x}_i \mathbf{x}_i'\right)^{-1} \sum_{i \in A} d_i q_i^{-1} \mathbf{x}_i y_i$. Under the regression superpopulation model

$$y_i = \mathbf{x}_i' \beta + \epsilon_i,$$

with $\epsilon_i \sim (0, q_i^2 \sigma^2)$, Isaki and Fuller (1982) showed that the regression estimator in (10.12) achieves the Godambe-Joshi lower bound, which is the lower bound of the anticipated variance under the superpopulation model. Deville and Särndal (1992) generalize the calibration estimation using a more general objective function $Q(w,d)$. Often, we have $w_i = d_i g(\mathbf{x}_i; \hat{\lambda})$, where $d_i = \pi_i^{-1}$ and $\hat{\lambda}$ is determined from (10.9). The functional form of $g(\cdot)$ is determined from the objective function for calibration.

Assume that $\hat{\lambda}$ converges in probability to λ_0 where $g(\mathbf{x}_i; \lambda_0) = 1$. The choice of $g(\mathbf{x}_i; \hat{\lambda}) = 1 + q_i^{-1} \mathbf{x}_i' \hat{\lambda}$ leads to the regression estimator with weights in (10.11). To make the calibration weights always positive, the exponential tilting calibration estimator, discussed in Kim (2010), uses $g(\mathbf{x}_i; \hat{\lambda}) = \exp(q_i^{-1} \mathbf{x}_i' \hat{\lambda})$ and is asymptotically equivalent to the regression estimator.

The following theorem, originally proved by Kim and Park (2010), presents some asymptotic properties of the calibration estimator.

Theorem 10.1. *Let $\hat{Y}_{cal} = \sum_{i \in A} w_i y_i$, where $w_i = d_i g(\mathbf{x}_i; \hat{\lambda})$ and $\hat{\lambda}$ is the unique solution to (10.9). Assume that $\hat{\lambda}$ converges in probability to λ_0 where $g(\mathbf{x}_i; \lambda_0) = 1$. Under some regularity conditions, the calibration estimator is asymptotically equivalent to*

$$\hat{Y}_{IV} = \sum_{i \in A} d_i y_i + \left\{ \mathbf{X} - \sum_{i \in A} d_i \mathbf{x}_i \right\}' B_z, \qquad (10.13)$$

where the subscript "IV" stands for "instrumental variable" and

$$B_z = \left(\sum_{i=1}^{N} \mathbf{z}_i \mathbf{x}_i' \right)^{-1} \sum_{i=1}^{N} \mathbf{z}_i y_i \qquad (10.14)$$

with $\mathbf{z}_i = \partial g(\mathbf{x}_i; \lambda)/\partial \lambda$ evaluated at $\lambda = \lambda_0$.

Proof. Consider

$$\hat{Y}_B(\lambda) = \sum_{i \in A} d_i g(\mathbf{x}_i; \lambda) y_i + \left\{ \mathbf{X} - \sum_{i \in A} d_i g(\mathbf{x}_i; \lambda) \mathbf{x}_i \right\}' B,$$

where B is a p-dimensional vector. Note that we have $\hat{Y}_B(\hat{\lambda}) = \hat{Y}_{cal}$ for any choice of B. To find the particular choice B^* of B such that $\hat{Y}_{B^*}(\hat{\lambda})$ is asymptotically equal to $\hat{Y}_{B^*}(\lambda_0)$, by the theory of Randles (1982), we only have to find B that satisfies

$$E\left\{ \partial \hat{Y}_B(\lambda_0)/\partial \lambda \right\} = 0. \qquad (10.15)$$

Thus, the choice of $B = B_z$ in (10.14) satisfies (10.15) and the asymptotic equivalence holds as $g(\mathbf{x}_i; \lambda_0) = 1$. $\qquad \square$

By Theorem 10.1, the calibration estimator is consistent and has asymptotic variance

$$V\left(\hat{Y}_{cal} \mid \mathcal{F}_N \right) \cong V\left\{ \sum_{i \in A} d_i \left(y_i - \mathbf{x}_i' B_z \right) \mid \mathcal{F}_N \right\}, \qquad (10.16)$$

where the expectation conditional on $\mathcal{F}_N$ refers to the expectation with respect to the sampling mechanism. The consistency does not depend on the validity of the outcome regression model such as (10.5). However, the variance in (10.16) will be small if the regression model holds for the finite population at hand. Thus, the estimator is model-assisted, not model-dependent, in the sense that the regression model is used only to improve the efficiency. Model assisted estimation is very popular in sample surveys. If we have $\hat{B}_z = \left(\sum_{i \in A} d_i \mathbf{z}_i \mathbf{x}_i' \right)^{-1} \sum_{i \in A} d_i \mathbf{z}_i y_i$ instead of the B_z in (10.14), the resulting estimator in (10.13) is called the *instrumental-variable calibration estimator*, with $\mathbf{z}_i$ being the instrumental variable.

For variance estimation, writing (10.16) as

$$V\left(\hat{Y}_{cal} \mid \mathcal{F}_N\right) \cong V\left\{ \sum_{i \in A} d_i g(\mathbf{x}_i; \lambda_0) \left(y_i - \mathbf{x}_i' B_z \right) \mid \mathcal{F}_N \right\} \qquad (10.17)$$

and applying the standard variance formula to $\hat{\eta}_i = g(\mathbf{x}_i; \hat{\lambda}) \left(y_i - \mathbf{x}_i' \hat{B}_z \right)$ will provide a consistent variance estimator.

If the outcome regression model is nonlinear, e.g. $E(y_i \mid \mathbf{x}_i) = m(\mathbf{x}_i; \beta)$ for some nonlinear function m, we can directly apply the predicted value, $\hat{m}_i = m(\mathbf{x}_i; \hat{\beta})$, to obtain the prediction estimator

$$\hat{Y}_p = \sum_{i=1}^{N} \hat{m}_i,$$

which does not necessarily satisfy design consistency. Here, design consistency means convergence in probability to the target parameter under the sampling mechanism. To achieve design consistency, the following bias-corrected prediction estimator

$$\hat{Y}_{p,bc} = \sum_{i=1}^{N} \hat{m}_i + \sum_{i \in A} d_i \left(y_i - \hat{m}_i \right) \qquad (10.18)$$

can be used. The above bias-corrected prediction estimator is design consistent and has asymptotic variance

$$V\left(\hat{Y}_{p,bc} \mid \mathcal{F}_N\right) \cong V\left[\sum_{i \in A} d_i \left\{ y_i - m(\mathbf{x}_i; \beta) \right\} \mid \mathcal{F}_N \right].$$

That is, the effect of $\hat{\beta}$ in $\hat{m}_i = m(\mathbf{x}_i; \hat{\beta})$ can be ignored in variance estimation. The bias-corrected prediction estimator in (10.18) is asymptotically equivalent to the calibration estimator using

$$\sum_{i \in A} w_i \hat{m}_i = \sum_{i=1}^{N} \hat{m}_i$$

as a constraint for calibration equation. Wu and Sitter (2001) developed such a calibration procedure which is called model calibration. Breidt and Opsomer (2017) provided a comprehensive overview of calibration weighting using modern prediction techniques.

Example 10.2. *To extend the idea of model calibration to regression analysis with survey data, suppose that we are interested in estimating the parameter in the regression model*

$$E(Y \mid x_1, x_2) = m(\beta_0 + \beta_1 x_1 + \beta_2 x_2),$$

where $m(\cdot)$ is known and $\beta = (\beta_0, \beta_1, \beta_2)$ is unknown. The estimating equation for β using sample A can be written as

$$\hat{U}_a(\beta) \equiv \sum_{i \in A} d_i \{y_i - m(x_{1i}, x_{2i}; \beta)\} h(x_{1i}, x_{2i}; \beta) = 0 \qquad (10.19)$$

for some $h(x_{1i}, x_{2i}; \beta)$ such that $\hat{U}_a(\beta)$ is linearly independent.

   Now, suppose that we observe x_{1i} and y_i throughout the finite population. We wish to incorporate the partial information from the finite population. To do this, suppose that we have a "working" model for $E(Y \mid X_1)$:

$$E(Y \mid X_1) = m(\alpha_0 + \alpha_1 X_1) \qquad (10.20)$$

for some $\alpha = (\alpha_0, \alpha_1)$. Note that since (x_{1i}, y_i) are observed, we can use the population observations to estimate $\hat{\alpha}_N$. Let $\hat{\alpha}_N$ be obtained by solving $\sum_{i=1}^{N} U(\alpha; x_{1i}, y_i) = 0$, for some U which is unbiased to zero under model (10.20). Once $\hat{\alpha}_N$ is obtained, we can use this extra information to improve the efficiency of $\hat{\beta}$ in (10.19). One way is to use the optimal estimator of β given by

$$\hat{\beta}^* = \hat{\beta} + \widehat{Cov}(\hat{\beta}, \hat{\alpha})\{\hat{V}(\hat{\alpha})\}^{-1}(\hat{\alpha}_N - \hat{\alpha}), \qquad (10.21)$$

where $\hat{\alpha}$ is obtained from

$$\sum_{i \in A} d_i U(\alpha; x_{1i}, y_i) = 0.$$

The optimal estimator in (10.21) is first proposed by Chen and Chen (2000).

   To apply the model calibration, we can find the maximizer of $Q(w, d)$ in (10.10) subject to

$$\sum_{i \in A} w_i U(\hat{\alpha}_N; x_{1i}, y_i) = 0. \qquad (10.22)$$

Constraint (10.22) incorporates the extra information in the population level. Once the solution $\hat{w}_i$ is obtained, we can use

$$\sum_{i \in A} \hat{w}_i \{y_i - m(x_{1i}, x_{2i}; \beta)\} h(x_{1i}, x_{2i}; \beta) = 0$$

to estimate β. More discussion will be presented in Section 11.7.

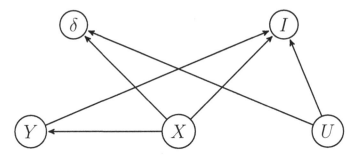

FIGURE 10.1
A DAG for a setup where PMAR holds but SMAR does not hold. Variable U is latent in the sense that it is never observed.

10.3   Propensity Score Weighting Method

We now consider the case of unit nonresponse in survey sampling. Assume that $\mathbf{x}_i$ is observed throughout the sample, and y_i is observed only if $\delta_i = 1$. We assume that the response mechanism does not depend on y. Thus, we assume that

$$P(\delta = 1|\mathbf{x}, y) = P(\delta = 1|\mathbf{x}) = p(\mathbf{x}; \phi_0) \qquad (10.23)$$

for some unknown vector ϕ_0. The first equality implies that the data are missing-at-random (MAR) in the population model, a model for the finite population. Or, one may assume that

$$P(\delta = 1|\mathbf{x}, y, I = 1) = P(\delta = 1|\mathbf{x}, I = 1), \qquad (10.24)$$

which can be called *sample MAR (SMAR)*, while condition (10.23) can be called *population MAR (PMAR)*. Unless the sampling design is non-informative, the two MAR conditions, (10.23) and (10.24), are different. In survey sampling, assumption (10.23) is more appropriate because an individual's decision on whether or not to respond to a survey depends on his or her own characteristics. See Berg et al. (2016) for further discussion on this topic.

For example, as in Figure 10.1, Y is conditionally independent of δ given X. However, Y is not conditionally independent of δ given X and I. Augmenting X by including the sampling weights does not solve the problem. The existence of latent variable U which is correlated with I and δ makes SMAR unachievable.

Given the response model (10.23), a consistent estimator of ϕ_0 can be obtained by solving

$$\hat{U}_h(\phi) \equiv \sum_{i \in A} d_i \left\{ \frac{\delta_i}{p(\mathbf{x}_i; \phi)} - 1 \right\} \mathbf{h}(\mathbf{x}_i; \phi) = \mathbf{0} \qquad (10.25)$$

for some $\mathbf{h}(\mathbf{x}; \phi)$ such that $\partial \hat{U}_h(\phi)/\partial \phi$ is of full rank. If we choose

$$\mathbf{h}(\mathbf{x}_i; \phi) = p(\mathbf{x}_i; \phi)\{\partial \mathrm{logit}\, p(\mathbf{x}_i; \phi)/\partial \phi\},$$

then (10.25) is equal to the design-weighted score equation for ϕ.

The estimating equation in (10.25) can be derived from the maximum entropy method for density ratio estimation. Let $f_k(\mathbf{x})$ be the density function of $\mathbf{x}$ given $\delta = k$, for $k = 0, 1$. Using Bayes formula, we can express

$$P(\delta = 1 \mid \mathbf{x}) = \frac{\pi_1 f_1(\mathbf{x})}{\pi_1 f_1(\mathbf{x}) + \pi_0 f_0(\mathbf{x})}, \qquad (10.26)$$

where $\pi_1 = P(\delta = 1)$ and $\pi_0 = P(\delta = 0)$. Thus, writing $r(\mathbf{x}) = f_0(\mathbf{x})/f_1(\mathbf{x})$, we can express

$$\frac{1}{p(\mathbf{x})} = 1 + \frac{\pi_0}{\pi_1} r(\mathbf{x}).$$

Since π_0/π_1 is a constant, there is an one-to-one correspondence between the model for $p(\mathbf{x})$ and the model for $r(\mathbf{x})$. The ratio $r(\mathbf{x})$ is called the density ratio function.

Suppose that we employ a parametric model such as

$$\log\{r(\mathbf{x})\} = \phi_0 + \sum_{k=1}^{L} \phi_k b_k(\mathbf{x}), \qquad (10.27)$$

for some basis functions $\{b_1(\mathbf{x}), \ldots, b_L(\mathbf{x})\}$. To estimate the model parameter in (10.27), we can apply the maximum entropy method described in Section 7.5.

Recall that the density ratio function can be understood as the maximizer of $Q(r)$ in (7.62). Under a parametric model for $r(\mathbf{x}) = r(\mathbf{x}; \phi)$ such as (10.27), the finite population version of the objective function for maximum entropy method is

$$Q(\phi) = \frac{1}{N_0} \sum_{i=1}^{N} (1 - \delta_i) \log\{r(\mathbf{x}_i; \phi)\} - \frac{1}{N_1} \sum_{i=1}^{N} \delta_i r(\mathbf{x}_i; \phi),$$

with constraint $N_1^{-1} \sum_{i=1}^{N} \delta_i r(\mathbf{x}_i; \phi) = 1$, where $N_k = \sum_{i=1}^{N} I(\delta_i = k)$.

Using Horvitz-Thompson estimation, we can use

$$\hat{Q}(\phi) = \frac{1}{\hat{N}_0} \sum_{i \in A} d_i (1 - \delta_i) \log\{r(\mathbf{x}_i; \phi)\} - \frac{1}{\hat{N}_1} \sum_{i \in A} d_i \delta_i r(\mathbf{x}_i; \phi)$$

as a sample-based objective function, where $\hat{N}_k = \sum_{i \in A} d_i I(\delta_i = k)$. It turns out that the maximizer of $\hat{Q}(\phi)$ can be obtained by solving

$$\sum_{i \in A} d_i \delta_i \left\{ 1 + \frac{\hat{N}_0}{\hat{N}_1} r(\mathbf{x}_i; \phi) \right\} b(\mathbf{x}_i; \phi) = \sum_{i \in A} d_i b(\mathbf{x}_i; \phi), \tag{10.28}$$

where $b(\mathbf{x}; \phi) = \partial \log r(\mathbf{x}; \phi)/\partial \phi$. Note that (10.28) belongs to the class of estimating equations in (10.25).

Once $\hat{\phi}_h$ is computed from (10.25), the propensity score (PS) estimator of $Y = \sum_{i=1}^{N} y_i$ is given by

$$\hat{Y}_{PS} = \sum_{i \in A_R} d_i g(\mathbf{x}_i; \hat{\phi}_h) y_i, \tag{10.29}$$

where $A_R = \{i \in A : \delta_i = 1\}$ is the set of respondents and $g(\mathbf{x}_i; \hat{\phi}_h) = \{p(\mathbf{x}_i; \hat{\phi}_h)\}^{-1}$. Afterward, we can apply the argument of Theorem 10.1 to show that $\hat{Y}_{PS}$ is asymptotically equivalent to

$$\begin{aligned}
\tilde{Y}_{PS} &= \sum_{i \in A_R} d_i g(\mathbf{x}_i; \phi_0) y_i + \left\{ \sum_{i \in A} d_i \mathbf{h}_i - \sum_{i \in A_R} d_i g(\mathbf{x}_i; \phi_0) \mathbf{h}_i \right\}' B_z \\
&= \sum_{i \in A_R} d_i \{p(\mathbf{x}_i; \phi_0)\}^{-1} y_i + \left\{ \sum_{i \in A} d_i \mathbf{h}_i - \sum_{i \in A_R} d_i \{p(\mathbf{x}_i; \phi_0)\}^{-1} \mathbf{h}_i \right\}' B_z,
\end{aligned}$$

$$\tag{10.30}$$

where

$$B_z = \left(\sum_{i=1}^{N} \delta_i \mathbf{z}_i \mathbf{h}_i' \right)^{-1} \sum_{i=1}^{N} \delta_i \mathbf{z}_i y_i$$

and $\mathbf{z}_i = \partial g(\mathbf{x}_i; \phi)/\partial \phi$ evaluated at $\phi = \phi_0$. Thus, the asymptotic variance is equal to

$$\begin{aligned}
V\left(\tilde{Y}_{PS} \mid \mathcal{F}_N\right) &= V\left(\hat{Y}_{HT} \mid \mathcal{F}_N\right) + V\left\{ \sum_{i \in A_R} d_i p_i^{-1} \left(y_i - \mathbf{h}_i' B_z\right) \mid \mathcal{F}_N \right\} \\
&= V\left(\hat{Y}_{HT} \mid \mathcal{F}_N\right) + E\left\{ \sum_{i \in A} d_i^2 (p_i^{-1} - 1) \left(y_i - \mathbf{h}_i' B_z\right)^2 \mid \mathcal{F}_N \right\},
\end{aligned}$$

where $p_i = p(\mathbf{x}_i; \phi_0)$ and the second equality follows from independence among

δ_i's. Note that

$$E\left\{\sum_{i \in A} d_i^2(p_i^{-1}-1)\left(y_i - \mathbf{h}_i'B_z\right)^2\right\}$$

$$= E\left[\sum_{i \in A} d_i^2(p_i^{-1}-1)\left\{y_i - E(y_i \mid \mathbf{x}_i) + E(y_i \mid \mathbf{x}_i) - \mathbf{h}_i'B_z\right\}^2\right]$$

$$= E\left[\sum_{i \in A} d_i^2(p_i^{-1}-1)\left\{y_i - E(y_i \mid \mathbf{x}_i)\right\}^2\right]$$

$$+ E\left[\sum_{i \in A} d_i^2(p_i^{-1}-1)\left\{E(y_i \mid \mathbf{x}_i) - \mathbf{h}_i'B_z\right\}^2\right]$$

and the cross product term is zero because $y_i - E(y_i \mid \mathbf{x}_i)$ is conditionally unbiased for zero, conditional on $\mathbf{x}_i$. Thus, we have

$$V\left(\tilde{Y}_{PS} \mid \mathcal{F}_N\right) \geq V_l \equiv V\left(\hat{Y}_{HT} \mid \mathcal{F}_N\right) + E\left[\sum_{i \in A} d_i^2(p_i^{-1}-1)\left\{y_i - E(y_i \mid \mathbf{x}_i)\right\}^2 \mid \mathcal{F}_N\right].$$
(10.31)

Kim and Riddles (2012) established (10.31) and showed that the equality in (10.31) holds if $\hat{\phi}_h$ satisfies

$$\sum_{i \in A} d_i\left\{\frac{\delta_i}{p(\mathbf{x}_i;\phi)}-1\right\}E(Y \mid \mathbf{x}_i) = 0.$$
(10.32)

Any PS estimator that has the asymptotic variance V_l in (10.31) is optimal in the sense that it achieves the lower bound of the asymptotic variance among the class of PS estimators with $\hat{\phi}_h$ satisfying (10.25). The PS estimator using the maximum likelihood estimator of ϕ_0 does not necessarily achieve the lower bound of the asymptotic variance.

Condition (10.32) provides a way of constructing an optimal PS estimator. First, we need an assumption for $E(Y \mid \mathbf{x})$, which is often called the *outcome regression model*. If the outcome regression model is a linear regression model of the form $E(Y \mid \mathbf{x}) = \beta_0 + \beta_1'\mathbf{x}$, an optimal PS estimator of θ can be obtained by solving

$$\sum_{i \in A} d_i\frac{\delta_i}{p_i(\phi)}(1,\mathbf{x}_i) = \sum_{i \in A} d_i(1,\mathbf{x}_i).$$
(10.33)

Condition (10.33) is appealing because it says that the PS estimator applied to $y = a + \mathbf{b}'\mathbf{x}$ leads to the original HT estimator. Condition (10.33) is essentially the covariate-balancing property and is also called the *calibration condition* in survey sampling. The calibration condition applied to $\mathbf{x}$ makes full use of the information contained in it if the study variable y is well approximated by a linear function of $\mathbf{x}$. Lesage et al. (2019) extended the calibration idea to handle nonignorable nonresponse using instrumental variables.

We now discuss variance estimation of PS estimators of the form (10.29) where $\hat{p}_i = p_i(\hat{\phi})$ is constructed to satisfy (10.25). By (10.30), we can write

$$\hat{Y}_{PS} = \sum_{i \in A} d_i \eta_i(\phi_0) + o_p\left(n^{-1/2}N\right), \tag{10.34}$$

where

$$\eta_i(\phi) = \mathbf{h}_i B_z + \frac{\delta_i}{p_i(\phi)}\left(y_i - \mathbf{h}_i' B_z\right). \tag{10.35}$$

To derive the variance estimator, we assume that the variance estimator $\hat{V} = \sum_{i \in A}\sum_{j \in A} \Omega_{ij} q_i q_j$ satisfies $\hat{V}/V(\hat{q}_{HT}|\mathcal{F}_N) = 1 + o_p(1)$ for some Ω_{ij} related to the joint inclusion probability, where $\hat{q}_{HT} = \sum_{i \in A} d_i q_i$ for any q with a finite fourth moment.

To obtain the total variance, the *reverse framework* of Fay (1992), Shao and Steel (1999), and Kim et al. (2006b) is considered. In this framework, the finite population is divided into two groups, a population of respondents and a population of nonrespondents, so the response indicator is extended to the entire population as $\mathcal{R}_N = \{\delta_1, \delta_2, \ldots, \delta_N\}$. Given the population, the sample A is selected according to a probability sampling design. Then, we have both respondents and nonrespondents in the sample A. The total variance of $\hat{\eta}_{HT} = \sum_{i \in A} d_i \eta_i$ can be written as

$$V(\hat{\eta}_{HT}|\mathcal{F}_N) = E\{V(\hat{\eta}_{HT}|\mathcal{F}_N, \mathcal{R}_N)|\mathcal{F}_N\} + V\{E(\hat{\eta}_{HT}|\mathcal{F}_N, \mathcal{R}_N)|\mathcal{F}_N\} := V_1 + V_2. \tag{10.36}$$

The conditional variance term $V(\hat{\eta}_{HT}|\mathcal{F}_N, \mathcal{R}_N)$ in (10.36) can be estimated by

$$\hat{V}_1 = \sum_{i \in A}\sum_{j \in A} \Omega_{ij} \hat{\eta}_i \hat{\eta}_j, \tag{10.37}$$

where $\hat{\eta}_i = \eta_i(\hat{\phi})$ is defined in (10.35) with B_z replaced by a consistent estimator such as

$$\hat{B}_z = \left(\sum_{i \in A_R} d_i \hat{\mathbf{z}}_i \mathbf{h}_i'\right)^{-1} \sum_{i \in A_R} d_i \hat{\mathbf{z}}_i y_i$$

and $\hat{\mathbf{z}}_i = \mathbf{z}(\mathbf{x}_i; \hat{\phi})$ is the value of $\mathbf{z}_i = \partial g(\mathbf{x}_i; \phi)/\partial\phi$ evaluated at $\phi = \hat{\phi}$. To show that $\hat{V}_1$ is also consistent for V_1 in (10.36), it suffices to show that $V\{nN^{-2} \cdot V(\hat{\eta}_{HT}|\mathcal{F}_N, \mathcal{R}_N)|\mathcal{F}_N\} = o(1)$, which follows by some regularity conditions on the first and the second-order inclusion probabilities and the existence of the fourth moment. See Kim et al. (2006b). The second term V_2 in (10.36) is

$$V\{E(\hat{\eta}_{HT}|\mathcal{F}_N, \mathcal{R}_N)|\mathcal{F}_N\} = V\left(\sum_{i=1}^{N} \eta_i \Big| \mathcal{F}_N\right)$$

$$= \sum_{i=1}^{N} \frac{1 - p_i}{p_i}\left(y_i - \mathbf{h}_i' B_z\right)^2.$$

A consistent estimator of V_2 can be derived as

$$\hat{V}_2 = \sum_{i \in A_R} d_i \frac{1 - \hat{p}_i}{\hat{p}_i^2} \left(y_i - \mathbf{h}_i' \hat{B}_z \right)^2 . \qquad (10.38)$$

Therefore,

$$\hat{V} \left(\hat{Y}_{PS} \right) = \hat{V}_1 + \hat{V}_2 \qquad (10.39)$$

is consistent for the variance of the PS estimator defined in (10.29) with $\hat{p}_i = p_i(\hat{\phi})$ satisfying (10.25), where $\hat{V}_1$ is in (10.37) and $\hat{V}_2$ is in (10.38).

Note that the first term of the total variance is $V_1 = O_p(n^{-1}N^2)$, but the second term is $V_2 = O_p(N)$. Thus, when the sampling fraction nN^{-1} is negligible, that is, $nN^{-1} = o(1)$, the second term V_2 can be ignored and $\hat{V}_1$ is a consistent estimator of the total variance. Otherwise, the second term V_2 should be taken into consideration so that a consistent variance estimator can be constructed as in (10.39).

10.4 Multiple Imputation

To discuss multiple imputation in complex sampling, we first discuss Bayesian inference under informative sampling. The main difficult in implementing Bayesian inference under complex sampling is that there is no intrinsic likelihood function under complex sampling. Consider the problem of making inference about θ in the statistical model $f(y; \theta)$. Instead of obtaining a random sample, suppose that we have the following two-phase sampling structure.

1. Obtain a finite population $\mathcal{F}_N = \{y_1, \ldots, y_N\}$ from the superpopulation model with density $f(y; \theta)$.

2. From the finite population $\mathcal{F}_N$, we obtain a probability sample with known first-order inclusion probability π_i.

We observe the realized values y_i of the study variable in the sample.

We are interested in estimating superpopulation model parameters under informative sampling. The pseudo maximum likelihood method is to maximize

$$l_w(\theta) = \sum_{i \in A} w_i \log f(y_i; \theta) \qquad (10.40)$$

with respect to $\theta \in \Theta$, where w_i is the sampling weight of unit i and Θ is the parameter space. The maximizer can be written as the solution to the following (pseudo) score equation

$$\hat{S}_w(\theta) \equiv N^{-1} \sum_{i \in A} w_i S(\theta; y_i) = 0, \tag{10.41}$$

where $S(\theta; y) = \partial \log f(y; \theta)/\partial \theta$.

Under some regularity conditions, we have

$$\sqrt{n}(\hat{\theta} - \theta) \mid \theta \longrightarrow N(0, \Sigma_\theta) \tag{10.42}$$

in distribution as $n \to \infty$, where

$$\Sigma_\theta = \mathcal{I}(\theta)^{-1} \Sigma_S(\theta_0) \{\mathcal{I}(\theta)^{-1}\}',$$

$\mathcal{I}(\theta) = E\{I(\theta; Y)\}$, $I(\theta; y) = -\partial^2 \log f(y; \theta)/\partial\theta\partial\theta^T$, $\Sigma_S(\theta) = \lim_{n \to \infty} [nV\{\hat{S}_w(\theta)\}]$, and $V\{\hat{S}_w(\theta)\}$ is the variance of $\hat{S}_w(\theta)$ in (10.41). The reference distribution in (10.42) is the joint distribution of the super-population model and the sampling mechanism. The CLT in (10.42) is established, for example, in Rubin-Bleuer and Schiopu-Kratina (2005).

For statistical inference about θ, instead of using the asymptotic normality of the PMLE of $\hat{\theta}$, we may use the asymptotic normality of $\hat{S}_w(\theta) = N^{-1} \sum_{i \in S} w_i S(\theta; y_i)$. That is,

$$\sqrt{n} S_w(\theta) \mid \theta \xrightarrow{\mathcal{L}} N(0, \Sigma_S(\theta)), \tag{10.43}$$

where $\Sigma_S(\theta) = nV\{\hat{S}_w(\theta) \mid \theta\}$. The approximate distribution in (10.43) can also be used to develop a Bayesian method using a prior for θ. By treating (10.43) as the likelihood part in the posterior distribution, approximate Bayesian inference for θ can be made even under informative sampling. That is, given a prior $\pi(\theta)$, we can use

$$p(\theta \mid \text{data}) = \frac{\phi(\hat{S}_w(\theta); 0, \hat{\Sigma}_s(\theta)/\sqrt{n})\pi(\theta)}{\int \phi(S_w(\theta); 0, \hat{\Sigma}_s(\theta)/\sqrt{n})\pi(\theta)d\theta}, \tag{10.44}$$

where $\phi(\mathbf{x}; \mu, \Sigma)$ is the density function of $N(\mu, \Sigma)$ distribution. The approximate posterior in (10.44) first appeared in Wang et al. (2018).

To introduce multiple imputation, we first partition the original sample into two parts, $Y_{\text{com}} = (Y_{\text{obs}}, Y_{\text{mis}})$, the observed part and the missing part, respectively. In classical multiple imputation, as discussed in Section 5.2, the imputed values are generated by the following data augmentation algorithm:

P-Step Given the current imputed data $Y^*_{\text{com}} = (Y_{\text{obs}}, Y^*_{\text{mis}})$, generate θ^* from the posterior density,

$$\theta^* \sim \frac{L_s(\theta \mid Y^*_{\text{com}})\pi(\theta)}{\int L_s(\theta \mid Y^*_{\text{com}})\pi(\theta)d\theta}, \tag{10.45}$$

where $L_s(\theta \mid Y_{\text{com}})$ is the complete-data likelihood of θ, and $\pi(\theta)$ is the prior density of θ.

I-Step For each unit, generate $y_{i,\mathrm{mis}}$ from the imputation model evaluated at θ^*, $y_{i,\mathrm{mis}}^* \sim f(y_{i,\mathrm{mis}} \mid y_{i,\mathrm{obs}}; \theta^*)$.

Under an non-informative sampling design, the sample-data likelihood can be based on the population model, i.e., $L_s(\theta \mid Y_{\mathrm{com}}) = \prod_{i \in A} f(y_i; \theta)$.

For informative sampling designs, the sample likelihood function is unknown. Instead, using the approximate posterior distribution in (10.44), Kim and Yang (2017) propose using the following P-step in the data augmentation algorithm:

New P-step. Given the imputed values y_i^*, generate

$$\theta^* \sim p(\theta \mid Y_{\mathrm{com}}^*) = \frac{g(\hat{\theta}^* \mid \theta)\pi(\theta)}{\int g(\hat{\theta}^* \mid \theta)\pi(\theta)d\theta}, \tag{10.46}$$

where $\hat{\theta}^* = \hat{\theta}(Y_n^*)$ is the solution to (10.41) using the imputed values generated from the I-step, and $g(\hat{\theta} \mid \theta)$ is the density of the sampling distribution of $\hat{\theta}$. Since the exact sampling distribution is unknown, we can use the normal distribution in (10.42) as an approximation. Instead of (10.46), one can also use the approximate posterior distribution in (10.44).

Roughly speaking, in the original P-step, the likelihood function of θ is based on the sample data Y_{com}, whereas in the new P-step, the likelihood function of θ is replaced by the sampling distribution of the pseudo maximum likelihood estimator $\hat{\theta}(Y_{\mathrm{com}})$. The new data augmentation algorithm implies that the posterior density based on the observed data is

$$p(\theta|Y_{\mathrm{obs}}) = \frac{\int g(\hat{\theta}|\theta)\pi(\theta)dY_{\mathrm{mis}}}{\int \int g(\hat{\theta}|\theta)\pi(\theta)dY_{\mathrm{mis}}d\theta}. \tag{10.47}$$

Once M imputed datasets are generated from the above data augmentation method, we can apply Rubin's formula to combine estimates from each dataset. Goh and Kim (2021) developed multiple imputation accounting for model selection uncertainty.

10.5 Fractional Imputation

We now consider item nonresponse in sample surveys. Imputation is a popular technique for handling item nonresponse. Filling in missing values would enable valid comparisons of different analyses as they all start from the same (complete) dataset. Imputation also makes full use of the information in the partial responses. For example, domain estimation after imputation provides more efficient estimates than direct estimation because imputation borrows strength from observations outside the domains.

We first assume that x_i is observed throughout the sample and y_i is subject to missingness. A natural approach is to obtain a model for the conditional distribution of y_i given x_i and generate imputed values from the conditional distribution. In survey sampling, two models need to be distinguished. The population model refers to the original distribution that generates the population, and the sample model refers to the conditional distribution of the sample data given that they are selected in the sample. That is, we have

$$f_s\left(y \mid x\right) = f_p\left(y \mid x\right) \frac{P\left(I = 1 \mid x, y\right)}{P\left(I = 1 \mid x\right)},$$

where $f_s(\cdot)$ is the density for the sample distribution, and $f_p(\cdot)$ is the density for the population distribution. Unless the sample design is noninformative in the sense that it satisfies (10.2), the two models are not the same.

Generally speaking, we are interested in the parameters of the population model. We assume the MAR in the population level:

$$f\left(y \mid x, \delta = 1\right) = f\left(y \mid x, \delta = 0\right). \tag{10.48}$$

Condition (10.48), called the population MAR (PMAR), is the classical MAR condition in survey sampling literature. When x_i is always observed and PMAR holds, then the imputed value of y_i can be generated from $f\left(y_i \mid x_i\right)$, the population model of y_i given x_i. When estimating the parameters in the population model, we need to use the sampling weights because the sampling design can be informative. That is, we use

$$\sum_{i \in A} w_i \delta_i S(\theta; y_i, x_i) = 0 \tag{10.49}$$

to estimate θ in $f(y \mid x; \theta)$, where w_i is the sampling weight of unit i such that $\sum_{i \in A} w_i y_i$ is a design-consistent estimator of Y. Let $y_{i1}^*, \ldots, y_{im}^*$ be m imputed values generated from a proposal distribution $f_0(y \mid x)$. The choice of the proposal distribution is somewhat arbitrary. If we do not have a good guess about θ, we may use

$$f_0(y \mid x) = \hat{f}(y \mid \delta = 1) = \frac{\sum_{i \in A} w_i \delta_i I(y_i = y)}{\sum_{i \in A} w_i \delta_i},$$

which estimates the marginal distribution of y_i using the set of respondents. If x is categorical, then we can use

$$f_0(y \mid x) = \frac{\sum_{i \in A} w_i \delta_i I(x_i = x, y_i = y)}{\sum_{i \in A} w_i \delta_i I(x_i = x)}.$$

For continuous x, we may use a kernel-type proposal distribution

$$f_0(y \mid x) = \frac{w_i \delta_i K_h(x_i, x) K_h(y_i, y)}{\sum_{i \in A} w_i \delta_i K_h(x_i, x)}.$$

To generate m imputed values from $f_0(y \mid x_i)$, one can use the following systematic sampling algorithm:

1. Generate $u_1 \sim U(0, 1/m)$.
2. Compute $u_k = u_1 + (k-1)/m$ for $k = 2, \ldots, m$.
3. For each j, choose

$$y_{ij}^* = F_0^{-1}(u_j \mid x_i), \qquad (10.50)$$

where $F_0(y \mid x) = \sum_{y_i < y} f_0(y_i \mid x)/\{\sum_i f_0(y_i \mid x)\}$ is the cumulative distribution function derived from $f_0(y \mid x)$.

In practice, to remove the discontinuity points of F_0, we use the interpolation technique when computing $F_0(y \mid x)$. That is, we can express the interpolated cumulative distribution function (CDF) $\tilde{F}_0(y \mid x)$ as

$$\tilde{F}_0(y \mid x) = F_0(y_{(i)} \mid x) + (y - y_{(i)}) \frac{F_0(y_{(i+1)} \mid x) - F_0(y_{(i)} \mid x)}{y_{(i+1)} - y_{(i)}} \quad \text{if } y_{(i)} \le y < y_{(i+1)}.$$

The interpolated CDF can be used in (10.50).

The fractional weight associated with y_{ij}^* is computed as

$$w_{ij0}^* = \frac{f(y_{ij}^* \mid x_i; \hat{\theta})/f_0(y_{ij}^* \mid x_i)}{\sum_{k=1}^{m} f(y_{ik}^* \mid x_i; \hat{\theta})/f_0(y_{ik}^* \mid x_i)}.$$

When m is small, the fractional weights can be further modified in the calibration step. The proposed calibration equation for improving the fractional weights in this case is

$$\sum_{i \in A} \sum_{j=1}^{m} w_i (1 - \delta_i) w_{ij}^* S(\hat{\theta}; x_i, y_{ij}^*) = 0 \qquad (10.51)$$

and $\sum_{j=1}^{m} w_{ij}^* = 1$ for each i with $\delta_i = 0$, where $\hat{\theta}$ is computed from (10.49). Using the idea of regression weighting, the final calibration fractional weights can be computed by

$$w_{ij}^* = w_{ij0}^* + w_{ij0}^* \Delta \left(S_{ij}^* - \bar{S}_{i\cdot}^* \right), \qquad (10.52)$$

where $S_{ij}^* = S(\hat{\theta}; x_i, \mathbf{y}_{ij}^*)$, $\bar{S}_{i\cdot}^* = \sum_{j=1}^{m} w_{ij0}^* S_{ij}^*$, and

$$\Delta = -\left\{ \sum_{i \in A} w_i (1 - \delta_i) \sum_{j=1}^{m} w_{ij0}^* S_{ij}^* \right\}' \left[\sum_{i \in A} w_i (1 - \delta_i) \sum_{j=1}^{m} w_{ij0}^* \left(S_{ij}^* - \bar{S}_{i\cdot}^* \right)^{\otimes 2} \right]^{-1}.$$

See also (6.10) and its related discussion.

The calibration condition (10.51) guarantees that the imputed score equation leads to the same $\hat{\theta}$, computed from (10.49). Once the FI data are created, the fractionally imputed estimator of $Y = \sum_{i=1}^{N} y_i$ is obtained by

$$\hat{Y}_{FI} = \sum_{i \in A} w_i \left\{ \delta_i y_i + (1 - \delta_i) \sum_{j=1}^{m} w_{ij}^* y_{ij}^* \right\}.$$

For variance estimation, a replication method can be used. Let $w_i^{(k)}$ be the k-th replication weights such that

$$\hat{V}_{rep} = \sum_{k=1}^{L} c_k (\hat{Y}^{(k)} - \hat{Y})^2$$

is consistent for the variance of $\hat{Y} = \sum_{i \in A} w_i y_i$, where L is the replication size, c_k is the k-th replication factor that depends on the replication method and the sampling mechanism, and $\hat{Y}^{(k)} = \sum_{i \in A} w_i^{(k)} y_i$ is the k-th replicate of $\hat{Y}$.

To apply the replication method to fractional imputation, we first apply the replication weights $w_i^{(k)}$ in (10.49) to compute $\hat{\theta}^{(k)}$. Once $\hat{\theta}^{(k)}$ is obtained, we use the same imputed values to compute the initial replication fractional weights

$$w_{ij0}^{*(k)} = \frac{f(y_{ij}^* \mid x_i; \hat{\theta}^{(k)}) / f_0(y_{ij}^* \mid x_i)}{\sum_{l=1}^{m} f(y_{il}^* \mid x_i; \hat{\theta}^{(k)}) / f_0(y_{il}^* \mid x_i)}$$

and then apply the same calibration

$$\sum_{i \in A} \sum_{j=1}^{m} w_i^{(k)} (1 - \delta_i) w_{ij}^{*(k)} S(\hat{\theta}^{(k)}; x_i, y_{ij}^*) = 0 \qquad (10.53)$$

and $\sum_{j=1}^{m} w_{ij}^{*(k)} = 1$, to obtain the final replicate fractional weights $w_{ij}^{*(k)}$. Via the regression weighting method, the final replicate fractional weights are computed similarly to (10.52) using $\hat{\theta}^{(k)}$ and the replicated weights. Afterwards,

$$\hat{Y}_{FI}^{(k)} = \sum_{i \in A} w_i^{(k)} \left\{ \delta_i y_i + (1 - \delta_i) \sum_{j=1}^{m} w_{ij}^{*(k)} y_{ij}^* \right\}$$

can be used to compute the replication variance estimator

$$\hat{V}_{rep}(\hat{Y}_{FI}) = \sum_{k=1}^{L} c_k (\hat{Y}_{FI}^{(k)} - \hat{Y}_{FI})^2.$$

The replication method is very useful for multipurpose estimation. For example, if another parameter of interest is $\Psi = P(Y < 3)$, then the FI estimator of Ψ is

$$\hat{\Psi}_{FI} = \sum_{i \in A} w_i \left\{ \delta_i I(y_i < 3) + (1 - \delta_i) \sum_{j=1}^{m} w_{ij}^* I(y_{ij}^* < 3) \right\},$$

and its replication variance estimator can be computed similarly using

$$\hat{\Psi}_{FI}^{(k)} = \sum_{i \in A} w_i^{(k)} \left\{ \delta_i I(y_i < 3) + (1 - \delta_i) \sum_{j=1}^{m} w_{ij}^{*(k)} I(y_{ij}^* < 3) \right\}.$$

Example 10.3. *Suppose that a sample of bivariate (x_i, y_i) are generated from*

$$y_i = \beta_0 + \beta_1 x_i + e_i, \tag{10.54}$$

with $e_i \sim N(0, \sigma^2)$. Assume that we observe (x_i, y_i) for $\delta_i = 1$ and observe x_i only if $\delta_i = 0$. Assume that the response mechanism is PMAR.

To implement a fractional imputation with $m = 10$ under this setup, we can use the following steps:

[Step 1] Obtain a fully efficient estimator of $\theta = (\beta_0, \beta_1, \sigma^2)$ by solving

$$\sum_{i \in A} w_i \delta_i S(\theta; x_i, y_i) = 0,$$

where $S(\theta; x, y)$ is the score function for the conditional distribution in (10.54). Under the normal distribution, we have

$$S(\theta; x, y) = \begin{bmatrix} (y - \beta_0 - \beta_1 x)/\sigma^2 \\ (y - \beta_0 - \beta_1 x)x/\sigma^2 \\ \{(y - \beta_0 - \beta_1 x)^2 - \sigma^2\}/(2\sigma^4) \end{bmatrix}.$$

[Step 2] For each missing unit i, first generate $m_1 \gg m = 10$, (say, $m_1 = 1,000$) imputed values of y_i from $f(y \mid x_i; \hat{\theta})$, where $\hat{\theta}$ is obtained from [Step 1].

[Step 3] Among the m_1 imputed values of y_i, denoted by $y_{i1}^*, \ldots, y_{i,m_1}^*$, select a subsample of size m using an efficient sampling design. One simple way to obtain an efficient sample is to do systematic sampling from the population of $\{y_{i1}^*, \ldots, y_{i,m_1}^*\}$, sorted by the ascending order (or half-ascending and half descending order).

[Step 4] Use the imputed values selected from [Step 3] to find the calibration weights that satisfy $\sum_{j=1}^{m} w_{ij}^{*(k)} = 1$ and (10.51).

For variance estimation, the replication method described above can be used. In this case, the imputed values are not changed for each replication. Only the fractional weights are changed. In the k-th replication, the replicate $\hat{\theta}^{(k)}$ is first computed by solving

$$\sum_{i \in A} w_i^{(k)} \delta_i S(\theta; x_i, y_i) = 0.$$

Using $\hat{\theta}^{(k)}$, the initial replication fractional weights are computed by

$$w_{ij0}^{*(k)} = \frac{f(y_{ij}^* \mid x_i; \hat{\theta}^{(k)}) / f(y_{ij}^* \mid x_i; \hat{\theta})}{\sum_{s=1}^m f(y_{is}^* \mid x_i; \hat{\theta}^{(k)}) / f(y_{is}^* \mid x_i; \hat{\theta})}.$$

The replication fractional weights are further modified to satisfy $\sum_{j=1}^m w_{ij}^* = 1$ and (10.53).

10.6 Fractional Hot Deck Imputation

Hot deck imputation creates imputed values from realized observations. That is, no artificial values are created. The hot deck imputation method is very popular in household surveys. Fractional hot deck imputation was proposed by Kalton and Kish (1984) as a way of achieving efficient hot deck imputation. Kim and Fuller (2004) and Fuller and Kim (2005) provided a rigorous treatment of fractional hot deck imputation and discussed variance estimation. However, their approach is not directly applicable to multivariate missing data. Hot deck imputation for multivariate missing data with an arbitrary missing pattern is challenging because it is difficult to preserve the covariance structure in the data after hot deck imputation.

Let $\mathbf{y}$ be a K-dimensional vector of study variables and let $(\mathbf{y}_{i,\text{obs}}, \mathbf{y}_{i,\text{mis}})$ be the (observed, missing) part of $\mathbf{y}_i$. To study multivariate fractional hot deck imputation, first consider the simple case of categorical $\mathbf{y}$. In this case, under MAR, the joint distribution of $\mathbf{y}$ can be computed by the EM algorithm with fractional imputation implemented, as discussed in Example 6.2. In this case, the fractional weights

$$w_{ij}^* = \frac{\pi(\mathbf{y}_{i,\text{obs}}, \mathbf{y}_{i,\text{mis}}^{*(j)})}{\sum_k \pi(\mathbf{y}_{i,\text{obs}}, \mathbf{y}_{i,\text{mis}}^{*(k)})} \tag{10.55}$$

are assigned for all possible values of $\mathbf{y}_{i,\text{mis}}$. If a fixed number m is used for imputation, then a fractional imputation of size m can be constructed by a PPS (Probability Proportional to Size) sampling of m values from all possible values of $\mathbf{y}_{i,\text{mis}}$, where the selection probability for the j-th enumeration of

$\mathbf{y}_{i,\text{mis}}$ is w_{ij}^* in (10.55). The resulting FI estimator will have equal fractional weights $w_{ij}^* = 1/m$.

For continuous $\mathbf{y}$ variables, we consider a discrete approximation. The basic step in the proposed imputation is temporary replacement of the original data by a discrete approximation that temporarily replaces the original data. Each continuous variable is transformed into a discrete variable by dividing the range into a small finite number of segments. Let $\tilde{Y}_{ki}$ denote the discrete version of Y_{ki}. One simple way of computing the discrete version is to divide the range into groups of equal length. Note that, if Y_k is observed then $\tilde{Y}_k$ is observed.

Once $\tilde{Y}_k$'s are constructed, we use the observed value of $\tilde{Y}_k$ to compute the joint probability of $(\tilde{Y}_1, \tilde{Y}_2, \ldots, \tilde{Y}_K)$. Let $\tilde{\pi}(\tilde{y}_1, \tilde{y}_2, \ldots, \tilde{y}_K)$ be the joint probability of obtaining $(\tilde{Y}_1, \tilde{Y}_2, \ldots, \tilde{Y}_K) = (\tilde{y}_1, \tilde{y}_2, \ldots, \tilde{y}_K)$. The joint probability can be obtained by the EM algorithm for partially classified categorical data using the partially observed value of $\tilde{y}_{ik}$'s.

In the EM algorithm, the E-step is essentially the same as applying the fully efficient fractional imputation (FEFI) method of Fuller and Kim (2005) using all possible combinations of imputed values. The imputed values for $\tilde{\mathbf{y}}_{i,\text{mis}(i)}$, where $\text{mis}(i)$ is the index of nonresponding items in unit i, are taken from the support of $\tilde{\mathbf{y}}_{\text{mis}(i)}$ matching the support of the respondents. Let $\tilde{\mathbf{y}}_{\text{mis}(i)}^{*(j)}, (j = 1, \ldots, M_i)$ be the set of possible values of $\tilde{\mathbf{y}}_{\text{mis}(i)}$ in the sample. Once the realized values of $\tilde{\mathbf{y}}_{\text{mis}(i)}$ are imputed, we can use the idea of EM by weighting (Ibrahim, 1990) to compute the fractional weights, as in Example 6.2. The fractional weight assigned to $\tilde{\mathbf{y}}_{i,\text{mis}(i)} = \tilde{\mathbf{y}}_{\text{mis}(i)}^{*(j)}$ at the t-th EM iteration is

$$\tilde{w}_{ij(t)}^* = \frac{\tilde{\pi}_t(\tilde{\mathbf{y}}_{i,\text{obs}}, \tilde{\mathbf{y}}_{\text{mis}(i)}^{*(j)})}{\sum_j \tilde{\pi}_t(\tilde{\mathbf{y}}_{i,\text{obs}}, \tilde{\mathbf{y}}_{\text{mis}(i)}^{*(j)})}, \qquad (10.56)$$

where $\tilde{\pi}_t(\tilde{y}_1, \ldots, \tilde{y}_K)$ is the current value of the joint probabilities $P(\tilde{Y}_1 = \tilde{y}_1, \ldots, \tilde{Y}_K = \tilde{y}_K)$ evaluated at the t-th EM algorithm. If unit i has no missing data, then $\tilde{w}_{ij(t)}^* = 1$. Computing fractional weights in (10.56) corresponds to the E-step of the EM algorithm. Once the FEFI is constructed as above, the M-step is used to update the joint probabilities via the weighted average of the FEFI data using fractional weights. That is,

$$\tilde{\pi}_{t+1}(\tilde{y}_1, \ldots, \tilde{y}_K) = \left(\sum_{i \in A} w_i\right)^{-1} \sum_{i \in A} \sum_{j=1}^{M_i} w_i \tilde{w}_{ij(t)}^* I\left\{\tilde{y}_{i1}^{*(j)} = \tilde{y}_1, \ldots, \tilde{y}_{iK}^{*(j)} = \tilde{y}_K\right\},$$
$$(10.57)$$

where w_i is the sampling weight for unit i and $\tilde{y}_{ijk}^*$ is the j-th imputed value for $\tilde{Y}_{ik}$. If $\tilde{Y}_{ik}$ is observed, then $\tilde{y}_{ijk}^*$ is the observed value.

Once the joint probabilities are computed, a set of imputed values of size m is constructed by doing PPS sampling with probability

$$w_{ij}^* = \frac{\tilde{\pi}(\tilde{\mathbf{y}}_{i,\text{obs}}, \tilde{\mathbf{y}}_{i,\text{mis}}^{*(j)})}{\sum_k \tilde{\pi}(\tilde{\mathbf{y}}_{i,\text{obs}}, \tilde{\mathbf{y}}_{i,\text{mis}}^{*(k)})}, \tag{10.58}$$

for each i in the sample. In PPS sampling, the size measures are proportional to the original fractional weights after accounting for certainty selection. That is, we first select the candidates with $w_{ij}^* > 1/m$ with certainty. Regression weighting can be used to preserve the marginal fractional weights. Because the values are all categorical, the raking ratio weighting method can be easily implemented.

Creating an imputed value $\tilde{y}_{ik}^*$ for y_{ik} is essentially creating an imputation cell for y_{ik}. Given the value of $\tilde{\mathbf{y}}_{i,\text{mis}}^{*(j)} = \tilde{\mathbf{y}}_{j,\text{mis}(i)}, (j = 1, 2, \ldots, m)$, we can perform a single hot deck imputation from the donor set of observed units with the same value of $\tilde{\mathbf{y}}_{j,\text{mis}(i)}$. Here, the value of $\tilde{\mathbf{y}}_{j,\text{mis}(i)}$ can be used as an imputation cell. If at least one donor is identified from the set of fully responding units with $\tilde{\mathbf{y}}_{k,\text{mis}(i)} = \tilde{\mathbf{y}}_{i,\text{mis}}^{*(k)}$, we can use the donor to obtain imputed values $\mathbf{y}_{i,\text{mis}}^{*(j)} = \mathbf{y}_{k,\text{mis}(i)}$ for missing $\mathbf{y}_{i,\text{mis}}$. If such a donor is not identifiable from the set of fully responding units, then we do hot deck imputation marginally using the marginal values of $\tilde{\mathbf{y}}_{j,\text{mis}(i)}$ for each item separately. The discrete version preserves most of the correlation structure in Y_k, and the marginal hot deck imputation will perform well if there is no systematic variation within categories of $\tilde{Y}$.

In summary, if $Y = (Y_1, Y_2, Y_3)$, the proposed fractional imputation method is performed by the following steps:

[Step 1] For each item k, transform Y_k into $\tilde{Y}_k$, a discrete version of Y_k. The value of $\tilde{Y}_k$ will serve the role of imputation cell for Y_k since the hot deck imputation for Y_k will be performed marginally within the cell value of $\tilde{Y}_k$.

[Step 2] Use the estimated joint probability to compute the fractional weights

$$\tilde{w}_{ij}^* = \frac{\tilde{\pi}(\tilde{\mathbf{y}}_{i,\text{obs}}, \tilde{\mathbf{y}}_{\text{mis}(i)}^{*(j)})}{\sum_j \tilde{\pi}(\tilde{\mathbf{y}}_{i,\text{obs}}, \tilde{\mathbf{y}}_{\text{mis}(i)}^{*(j)})}, \tag{10.59}$$

for fully efficient fractional imputation (FEFI), where $\tilde{\mathbf{y}}_{\text{mis}(i)}^{*(j)}$ is the j-th realization of the $\tilde{\mathbf{y}}_{\text{mis}(i)}$, the missing part of $\tilde{\mathbf{y}}$ for unit i. The FEFI assigns fractional weights for all possible combinations of the imputed values for missing $\tilde{\mathbf{y}}_{\text{mis}(i)}$. The joint probability $\tilde{\pi}$ can be estimated by a modified EM algorithm.

[Step 3] Among the possible values of $\tilde{\mathbf{y}}_{\text{mis}(i)}$, select a PPS sample of size m with probability proportional to the fractional weights in [Step 2]. The m imputed values of $\tilde{\mathbf{y}}_{\text{mis}(i)}$ will have equal initial fractional weights $w_{ij0}^* = 1/m$.

[Step 4] Using calibration weighting, create final fractional weights satisfying

$$\frac{1}{N} \sum_{i \in A} \sum_{j=1}^{m} w_i \tilde{w}_{ij}^* I\left(\tilde{y}_{Ii1}^{(j)} = a\right) = \tilde{\pi}_{a++},$$

$$\frac{1}{N} \sum_{i \in A} \sum_{j=1}^{m} w_i \tilde{w}_{ij}^* I\left(\tilde{y}_{Ii2}^{(j)} = b\right) = \tilde{\pi}_{+b+},$$

$$\frac{1}{N} \sum_{i \in A} \sum_{j=1}^{m} w_i \tilde{w}_{ij}^* I\left(\tilde{y}_{Ii3}^{(j)} = c\right) = \tilde{\pi}_{++c},$$

and

$$\sum_{j=1}^{m} \tilde{w}_{ij}^* = 1.$$

[Step 5] For each item k, use marginal hot deck imputation to select from the respondents with the same value of $\tilde{y}_k$. That is, each imputed value of $\tilde{y}_k$ can be used as an imputation cell for hot deck imputation for item k.

A version of fractional hot deck imputation method is now implemented in the R package FHDI (Im et al., 2018).

10.7 Imputation for Two-Phase Sampling

Two-phase sampling, sometimes called double sampling, is a cost-effective technique in survey sampling. By first selecting a large sample, observing cheap auxiliary variables and then incorporating the auxiliary variables into the second-phase sampling design, we can produce estimators with smaller variances than those based on a single-phase sampling design for the same cost. Two-phase sampling is also popular when the sampling frame for eligible elements do not exist. In the first-phase sampling, a large sample is drawn without considering eligibility. Next, the eligibility of the first-phase sample elements is examined. The second-phase sample will become a stratified sample with two strata where Stratum One is the eligibility stratum and Stratum Two is the noneligibility stratum. The second-phase sample size for the noneligible stratum will be zero because we are not interested in noneligible elements in the population.

Estimation for two-phase sampling deals with how to incorporate the auxiliary information collected in the first-phase sample. One approach for incorporating the partial information is through imputation, where the nonsampled part in the second-phase sampling is treated as missing data. In fact, two-phase sampling can be understood as a sampling design with planned missingness.

Imputation for the nonsampled part is often referred to as *mass imputation*. Mass imputation is particularly useful in creating efficient domain estimates. For example, the World Bank has been using a simulated census method, developed by Elbers et al. (2003) and Haslett and Jones (2005), to estimate small area poverty measures in Bangladesh, Nepal, and some other developing countries. Fuller (2003) discussed mass imputation to get improved estimates for domains.

In two-phase sampling, the sample selection is made in two different ways. In the first-phase sample, only x is measured. Based on the information of x, a second-phase sample is selected and (x, y) is measured. Let A_1 and A_2 be the set of indices for the first-phase and second-phase sample, respectively. The second-phase sample is not necessarily nested within the first-phase sample. Also, let w_{i1} and w_{i2} be the sampling weight of unit i for the first-phase and second-phase sample, respectively. An unbiased estimator of $Y = \sum_{i=1}^{N} y_i$ is $\hat{Y}_2 = \sum_{i \in A_2} w_{i2} y_i$, which does not use the observation x in the first-phase sample and thus is inefficient.

To incorporate x_i into the estimation of Y, we use the following estimator,

$$\hat{Y}_{tp} = \sum_{i \in A_1} w_{i1} m(x_i; \hat{\beta}) + \sum_{i \in A_2} w_{i2} \left\{ y_i - m(x_i; \hat{\beta}) \right\} \qquad (10.60)$$

for some $m(x_i; \hat{\beta})$. We call the estimator (10.60) a *two-phase regression esti-mator* with a corresponding subscript denotation. First, $m(x_i; \hat{\beta})$ represents a predictor of y_i from an implicit regression model

$$E(y_i \mid x_i) = m(x_i; \beta). \qquad (10.61)$$

The two-phase regression estimator (10.60) is expressed as a sum of two terms, a "projection term" and a "bias-correction term." The projection term is unbiased under model (10.61) but not otherwise. The bias correction term makes the resulting estimator (10.60) nearly unbiased regardless of whether the model (10.61) holds or not. The weights w_{i2} are constructed to satisfy $\sum_{i \in A_2} w_{i2} = \sum_{i \in A_1} w_{i1}$. Otherwise, it is better to use

$$\hat{Y}_{tp} = \sum_{i \in A_1} w_{i1} m(x_i; \hat{\beta}) + \left(\frac{\sum_{i \in A_1} w_{1i}}{\sum_{i \in A_2} w_{2i}} \right) \sum_{i \in A_2} w_{i2} \left\{ y_i - m(x_i; \hat{\beta}) \right\}, \qquad (10.62)$$

which essentially normalizes the second-phase sample weights.

Note that the two-phase regression estimator (10.62) can be written as

$$\hat{Y}_{FEFI} = \sum_{i \in A_2} w_{i1} y_i + \sum_{i \in A_1 / A_2} \sum_{j \in A_2} w_{i1} w_{ij}^* y_{ij}^*, \qquad (10.63)$$

where $y_{ij}^* = \hat{y}_i + \hat{e}_j$, $\hat{y}_i = m(\mathbf{x}_i; \hat{\beta})$, $\hat{e}_j = y_j - \hat{y}_j$, and

$$w_{ij}^* = \frac{w_{j2} - w_{j1}}{\sum_{k \in A_2} (w_{k2} - w_{k1})}.$$

The expression (10.63) implies that we impute all the elements in $A_1/A_2 = A_1 \cap A_2^c$. The estimator (10.63) is computed by augmenting the dataset with $(n_1 - n_2 + 1) \times n_2$ records, where n_1 and n_2 is the size of A_1 and A_2, respectively. For $i \in A_2$, element i has only one record with observation y_i and weight w_{1i}. For $i \in A_1/A_2$, element i has n_2 records where the j-th record has an imputed observation $y_{ij}^* = \hat{y}_i + \hat{e}_j$ with weight $w_{i1} w_{ij}^*$. The imputation method in (10.63) imputes all the elements in A_1/A_2 and is called the *fully efficient fractional imputation (FEFI) method*, as considered in Fuller and Kim (2005). The FEFI estimator is algebraically equivalent to the two-phase regression estimator, and can provide estimates for other parameters such as population quantiles.

If we want to limit the number of imputations, say m, fractional imputation using the regression weighting method of Fuller and Kim (2005) can be used. That is, we first select m values of $y_{ij}^* = \hat{y}_i + \hat{e}_j$ among the set of n_2 imputed values $\{y_{ij}^* : j \in A_2\}$ using an efficient sampling method. The new fractional weights $\tilde{w}_{ij}^*$ assigned to y_{ij}^* are determined so that

$$\sum_{j \in A_{I(i)}} \tilde{w}_{ij}^* \left(1, y_{ij}^*\right) = \sum_{j \in A_2} w_{ij}^* \left(1, y_{ij}^*\right), \tag{10.64}$$

where w_{ij}^* is the fractional weight in the FEFI estimator, $A_{I(i)}$ is the set of indices for the elements A_2, and $\hat{e}_j$ was selected for imputed value of unit $i \in A_1$. Condition (10.64) is equivalent to

$$\sum_{j \in A_{I(i)}} \tilde{w}_{ij}^* \left(1, \hat{e}_j\right) = (1, \bar{e}_{FEFI}), \tag{10.65}$$

where $\bar{e}_{FEFI} = \sum_{j \in A_2} w_{ij}^* \hat{e}_j$. The fractional weight satisfying (10.65) can be computed using the regression weighting method or the empirical likelihood method. Some additional constraints other than (10.65) can also be imposed. The fractional imputed data y_{ij}^* with weight $w_{i1} \tilde{w}_{ij}^*$ are constructed for the first-phase sample only. The resulting FI estimator of Y is then

$$\hat{Y}_{FI} = \sum_{i \in A_1} \sum_{j \in A_{I(i)}} w_i \tilde{w}_{ij}^* y_{ij}^*. \tag{10.66}$$

Replication variance estimation can be considered for estimating the variance of a FI estimator. For nested two-phase sampling, the k-th replicate of the two-phase regression estimator in (10.62) is computed by

$$\hat{Y}_{tp}^{(k)} = \sum_{i \in A_1} w_{i1}^{(k)} \left[m(x_i; \hat{\beta}) + \delta_i g^{(k)} \left(\frac{w_{2i}}{w_{1i}} \right) \left\{ y_i - m(x_i; \hat{\beta}) \right\} \right],$$

where $w_{i1}^{(k)}$ is the usual replication weights for w_{i1} in A_1, δ_i is the indicator function for the second-phase sample selection, and

$$g^{(k)} = \frac{\sum_{i \in A_1} w_{i1}^{(k)}}{\sum_{i \in A_2} w_{i1}^{(k)} (w_{i2}/w_{i1})}.$$

Using the reverse framework in (10.36), it can be shown that the replication variance estimator

$$\hat{V}_{rep} = \sum_{k=1}^{L} c_{1k} \left(\hat{Y}_{tp}^{(k)} - \hat{Y}_{tp} \right)^2$$

is consistent for the total variance if the sampling rate n/N is negligible. Note that the sampling variability of $\hat{\beta}$ can be safely ignored.

For the FEFI estimator in (10.63), the k-th replicate can be constructed by

$$\hat{Y}_{FEFI}^{(k)} = \sum_{i \in A_2} w_{i1}^{(k)} y_i + \sum_{i \in A_1/A_2} \sum_{j \in A_2} w_{i1}^{(k)} w_{ij}^{*(k)} y_{ij}^*,$$

where

$$w_{ij}^{*(k)} = \frac{w_{j1}^{(k)} (w_{j2}/w_{j1} - 1)}{\sum_{l \in A_2} w_{l1}^{(k)} (w_{l2}/w_{l1} - 1)}.$$

Also, for the FI estimator using (10.66), the same calibration can be used applying the replication weights. That is, we can find $\tilde{w}_{ij}^{*(k)}$ such that

$$\sum_{j \in A_{I(i)}} \tilde{w}_{ij}^{*(k)} (1, \hat{e}_j) = \left(1, \bar{e}_{FEFI}^{(k)} \right), \tag{10.67}$$

where $\bar{e}_{FEFI}^{(k)} = \sum_{j \in A_2} w_{ij}^{*(k)} \hat{e}_j$. Note that the effect of estimating β can be safely ignored. See Park and Kim (2019) for more theoretical details.

10.8   Synthetic Data Imputation

Synthetic data imputation is a technique of creating imputed values for items not observed in the current survey by incorporating information from other surveys. For example, suppose that there are two independent surveys, called Survey One and Survey Two, as in Table 10.1. Suppose that we observe x_i from Survey One and observe (x_i, y_i) from Survey Two. Let A_1 and A_2 be the index set of the sample elements in Survey One and Survey Two, respectively. In this case, we may want to create synthetic values of y_i in Survey One, so that inference about y can be made even in Survey One. This is particularly useful when Survey One is a large scale survey and item y is very expensive to measure. The setup of two independent samples with common items is often called non-nested two-phase sampling.

If a working regression model $E(Y \mid x) = m(x; \beta)$ is imposed, then the model parameter β can be estimated from Survey Two, and then synthetic values of y_i can be created by $y_i^* = \hat{y}_i \equiv m(x_i; \hat{\beta})$ or $y_i^* = \hat{y}_i + \hat{e}_i^*$, where $\hat{e}_i^*$

TABLE 10.1

Data Structure for Two Independent Survey Samples

Sample	Sample Size	X	Y
A_1	n_1: large	✓	
A_2	n_2: small	✓	✓

is randomly generated from the empirical distribution of the residuals in A_2. More generally, we can postulate a conditional distribution $f(y \mid x;\theta)$ and obtain a design-consistent estimator of θ by solving

$$\sum_{i \in A_2} w_{i2} S(\theta; x_i, y_i) = 0, \tag{10.68}$$

where w_{i2} is the sampling weights for the A_2 sample and $S(\theta; x, y) = \partial \log f(y \mid x;\theta)/\partial\theta$. Once $\hat{\theta}$ is obtained from (10.68), the imputed values are generated from $f(y \mid x; \hat{\theta})$. Let $y_{i1}^*, \ldots, y_{im}^*$ be m imputed values generated from $f(y \mid x_i; \hat{\theta})$. The synthetic estimator of $Y = \sum_{i=1}^{N} y_i$ obtained from sample A_1 is then given by

$$\hat{Y}_{FI} = \sum_{i \in A_1} w_{i1} \sum_{j=1}^{m} w_{ij}^* y_{ij}^*, \tag{10.69}$$

where w_{i1} is the sampling weight for unit i in A_1 sample and w_{ij}^* is the fractional weights. The fractional weights are computed by

$$w_{ij}^* = w_{ij0}^* + \hat{\lambda}' \left(S_{ij}^* - \bar{S}_i^* \right) w_{ij0}^*, \tag{10.70}$$

where $w_{ij0}^* = 1/m$ is the initial fractional weight, $\bar{S}_i^* = \sum_{j=1}^{m} w_{ij0}^* S_{ij}^*$, $S_{ij}^* = S(\hat{\theta}; x_i, y_{ij}^*)$, and

$$\hat{\lambda} = -\left\{ \sum_{i \in A_1} w_{i1} \sum_{j=1}^{m} w_{ij0}^* \left(S_{ij}^* - \bar{S}_i^* \right)^{\otimes 2} \right\}^{-1} \sum_{i \in A_1} w_{i1} \bar{S}_i^*. \tag{10.71}$$

Note that the fractional weights are constructed to satisfy

$$\sum_{i \in A_1} w_{i1} \sum_{j=1}^{m} w_{ij}^* S(\hat{\theta}; x_i, y_{ij}^*) = 0 \tag{10.72}$$

and $\sum_{j=1}^{m} w_{ij}^* = 1$. Condition (10.72) is used to guarantee that we obtain the same pseudo MLE $\hat{\theta}$ from the imputed values in A_1. If some of w_{ij}^* in (10.70) takes negative values, we may use

$$w_{ij}^* = \frac{w_{ij0}^* \exp\left(\hat{\lambda}' S_{ij}^* \right)}{\sum_{l=1}^{m} w_{il0}^* \exp\left(\hat{\lambda}' S_{il}^* \right)},$$

where $\hat{\lambda}$ is defined in (10.71).

The following theorem, originally proved by Kim and Rao (2012), shows that $\hat{Y}_{FI}$ in (10.69) is asymptotically unbiased under modest regularity conditions without requiring the imputation model $f(y \mid x) = f(y \mid x; \theta)$ to be correctly specified.

Theorem 10.2. *Assume that $\hat{\theta}$ computed from (10.68) satisfies*

$$\sum_{i \in A_2} w_{i2} \left\{ y_i - E(y_i \mid x_i; \hat{\theta}) \right\} = 0, \qquad (10.73)$$

where $E(y \mid x; \theta) = \int y f(y \mid x; \theta) dy$. Then, under some regularity conditions, for sufficiently large m, the imputed estimator in (10.69) is asymptotically equivalent to

$$\tilde{Y} = \sum_{i \in A_1} w_{i1} \tilde{y}_i + \sum_{i \in A_2} w_{i2} \left(y_i - \tilde{y}_i \right), \qquad (10.74)$$

where $\tilde{y}_i$ is the probability limit of $E(y \mid x_i; \hat{\theta})$ under the working model.

Condition (10.73) is satisfied in many cases when $\hat{\theta}$ is computed by (10.68). The result follows even when A_2 is a subset of A_1, which is the setup of classical two-phase sampling. Since $\tilde{Y}$ in (10.74) satisfies $E(\tilde{Y} \mid \mathcal{F}_N) = Y$, the synthetic estimator in (10.69) is asymptotically unbiased for Y, regardless of the value of $\tilde{y}_i$. Thus, the asymptotic unbiasedness does not require correct specification of the imputation model $f(y \mid x; \theta)$.

The asymptotic variance is

$$V\left(\tilde{Y} \mid \mathcal{F}_N\right) = V\left(\sum_{i \in A_1} w_{i1} \tilde{y}_i \mid \mathcal{F}_N \right) + V\left\{ \sum_{i \in A_2} w_{i2} \left(y_i - \tilde{y}_i \right) \mid \mathcal{F}_N \right\}. \quad (10.75)$$

The first term is the variance due to sampling in Survey One, and the second term is the variance due to sampling in Survey Two. The two terms are uncorrelated because of the assumption of independent sampling. The first term is small because n_1 is generally large. The second term is also small if the working model is good. For example, assume the simple random sampling in both surveys. Let n_1 be the sample size of Survey One and n_2 be the sample size of Survey Two. Then, (10.75) reduces to

$$V\left(\tilde{Y}\right) = \frac{1}{n_1}\left(1 - \frac{n_1}{N}\right)\frac{1}{N-1}\sum_{i=1}^{N}(m_i - \bar{m})^2 + \frac{1}{n_2}\left(1 - \frac{n_2}{N}\right)\frac{1}{N-1}\sum_{i=1}^{N}(\hat{e}_i - \bar{e})^2,$$

where $m_i = m(\mathbf{x}_i; \beta)$, $\bar{m} = N^{-1}\sum_{i=1}^{N} m_i$, $e_i = y_i - m(\mathbf{x}_i; \beta)$, and $\bar{e} = N^{-1}\sum_{i=1}^{N} e_i$. It follows from the last term in (10.75) that the bias-correction term cannot be ignored for variance estimation although it is zero under (10.73).

For variance estimation, we can use a replication method that incorporates variability in both sampling designs. Let L_1 be the number of replications for variance estimation for $\hat{X}_1 = \sum_{i \in A_1} w_{i1} x_i$. Assume that

$$\hat{V}_1(\hat{X}_1) = \sum_{k=1}^{L_1} c_{k1} \left(\hat{X}_1^{(k)} - \hat{X}_1 \right)^2,$$

where $\hat{X}_1^{(k)} = \sum_{i \in A_1} w_{i1}^{(k)} x_i$, is a consistent estimator of the variance of $\hat{X}_1$. It is possible to construct the same form of the replication variance estimator for estimating the variance of $\hat{Y}_2 = \sum_{i \in A_2} w_{i2} y_i$. That is, we can find $w_{i2}^{(k)}, k = 1, \dots, L_1$ such that

$$\hat{V}_1(\hat{Y}_2) = \sum_{k=1}^{L_1} c_{k1} \left(\hat{Y}_2^{(k)} - \hat{Y}_2 \right)^2, \tag{10.76}$$

where $\hat{Y}_2^{(k)} = \sum_{i \in A_2} w_{i2}^{(k)} y_i$ is consistent for the variance of $\hat{Y}_2$. Lemma A1 of Kim and Rao (2012) provided a way of constructing the L_1 replication weights $w_{i2}^{(k)}$ from A_2 sample.

The proposed replication variance estimation method can be described as follows:

[Step 1] For each $k = 1, \dots, L_1$, compute $\hat{\theta}^{(k)}$ by solving

$$\sum_{i \in A_2} w_{i2}^{(k)} S(\theta; x_i, y_i) = 0,$$

where $w_{i2}^{(k)}$ is the replication weights in (10.76).

[Step 2] Compute the initial replication fractional weights by

$$w_{ij0}^{*(k)} = \frac{f(y_{ij}^* \mid x_i; \hat{\theta}^{(k)}) / f(y_{ij}^* \mid x_i; \hat{\theta})}{\sum_{l=1}^{m} f(y_{il}^* \mid x_i; \hat{\theta}^{(k)}) / f(y_{il}^* \mid x_i; \hat{\theta})}. \tag{10.77}$$

Adjust the initial replication fractional weights to satisfy

$$\sum_{i \in A_1} w_{i1}^{(k)} \sum_{j=1}^{m} w_{ij}^{*(k)} S(\hat{\theta}^{(k)}; x_i, y_{ij}^*) = 0 \tag{10.78}$$

and $\sum_{j=1}^{m} w_{ij}^{*(k)} = 1$.

[Step 3] The k-th replicate of $\hat{Y}_{FI}$ is then given by

$$\hat{Y}_{FI}^{(k)} = \sum_{i \in A_1} w_{i1}^{(k)} \sum_{j=1}^{m} w_{ij}^{*(k)} y_{ij}^*$$

and the variance of $\hat{Y}_{FI}$ in (10.69) is estimated by

$$\hat{V}_1(\hat{Y}_{FI}) = \sum_{k=1}^{L_1} c_{k1} \left(\hat{Y}_{FI}^{(k)} - \hat{Y}_{FI} \right)^2.$$

Exercises

1. Using Cauchy–Schwartz inequality or something else, prove that q_i^* in (10.8) minimizes the variance term in (10.36).

2. Assume that the finite population values $\{(x_i, y_i) : i = 1, \ldots, N\}$ are generated independently with probability density $f(y_i \mid x_i; \theta) h(x_i)$, and the conditional distribution of y_i given x_i is in the exponential family

$$ f(y_i \mid x_i) = \exp\left\{ \frac{y_i \gamma_i - b(\gamma_i)}{\tau^2} - c(y_i, \tau) \right\}, $$

where γ_i is the canonical parameter and $g(\gamma_i) = x_i' \beta_0$. The marginal density of x, $h(x)$, is completely unspecified. Writing $\mu_i = E(y_i \mid x_i)$, we have $\mu_i = \partial b(\gamma_i)/\partial \gamma_i$. Note that the score function for

$$ S(\beta; x_i, y_i) = \frac{1}{\tau^2} (y_i - \mu_i) \{v(\mu_i) g_\mu(\mu_i)\}^{-1} x_i, $$

where $g_\mu(\mu_i) = \partial g(\mu_i)/\partial \mu_i$.

Now, under probability sampling design, we consider the problem of solving (10.4). Answer the following questions:

(a) Show that the solution $\hat{\beta}_q$ from (10.4) is asymptotically unbiased regardless of the choice of the $q(x_i)$ function in (10.4).

(b) Show that the asymptotic variance of the $\hat{\beta}_q$ obtained from (10.4) is

$$ J_q^{-1} V \left\{ \sum_{I_i=1} d_i e_i \{v(\mu_i) g_\mu(\mu_i)\}^{-1} x_i q(x_i) \right\} (J_q')^{-1}, \quad (10.79) $$

where $d_i = 1/\pi_i$ and $J_q = \sum_{i=1}^{N} \{v(\mu_i) g_\mu^2(\mu_i)\}^{-1} x_i x_i' q(x_i)$.

(c) Show that the optimal choice that minimizes (10.79), assuming independence between the terms in the summation in this expression, is

$$ q_i^* = v(\mu_i) \{E(d_i e_i^2 \mid x_i)\}^{-1} = E(e_i^2 \mid x_i) \{E(d_i e_i^2 \mid x_i)\}^{-1}. $$

3. Under the setup of Section 10.2, consider the following estimator

$$ \hat{Y}_{p,bc} = \sum_{i=1}^{N} \hat{m}_i + \sum_{i \in A} d_i (y_i - \hat{m}_i), $$

where $\hat{m}_i = m(\mathbf{x}_i; \hat{\beta})$ and $\hat{\beta}$ satisfies

$$ \sum_{i \in A} d_i \left\{ y_i - m(\mathbf{x}_i; \hat{\beta}) \right\} = 0. $$

(a) Show that, writing $\hat{Y}_{p,bc} = \hat{Y}_{p,bc}(\hat{\beta})$,

$$E\left\{\frac{\partial}{\partial \beta}\hat{Y}_{p,bc}(\beta)\right\} = 0.$$

(b) Argue that the effect of $\hat{\beta}$ in $\hat{Y}_{p,bc}$ can be safely ignored in variance estimation.

4. Under the setup of Section 10.2, consider the following instrumental variable estimator of $Y = \sum_{i=1}^{N} y_i$,

$$\hat{Y}_z = \sum_{i \in A} w_i y_i,$$

where

$$w_i = 1 + \left(\sum_{i \in A^c} \mathbf{x}_i'\right)\left(\sum_{i \in A} \mathbf{z}_i \mathbf{x}_i'\right)^{-1} \mathbf{z}_i.$$

(a) Show that $\hat{Y}_z$ satisfies the calibration condition (10.9).
(b) Show that if $\mathbf{z}_i'\mathbf{a} = d_i - 1$ for some $\mathbf{a}$, then $\hat{Y}_z$ is asymptotically design unbiased.
(c) Under the condition in (b), discuss how to estimate the variance of $\hat{Y}_z$.

5. Assume that a simple random sample of size n is obtained from a finite population of size N with auxiliary information $\bar{\mathbf{x}}_N = N^{-1}\sum_{i=1}^{N} \mathbf{x}_i$. Let the usual regression estimator be written in the form of $\bar{y}_{reg} = \sum_{i \in A} w_i y_i$, where

$$w_i = n^{-1} + (\bar{\mathbf{x}}_N - \bar{\mathbf{x}}_n)\left\{\sum_{i \in A}(\mathbf{x}_i - \bar{\mathbf{x}}_n)'(\mathbf{x}_i - \bar{\mathbf{x}}_n)\right\}^{-1}(\mathbf{x}_i - \bar{\mathbf{x}}_n)'.$$

Show that $\bar{y}_{reg,2} = \sum_{i \in A} w_{i,2} y_i$, where $w_{i,2} = \exp(nw_i)/\{\sum_{i \in A} \exp(nw_i)\}$, is asymptotically equivalent to the regression estimator. When is $w_{i,2}$ preferable to w_i in practice?

6. Consider the setup of two independence samples, A_1 and A_2, from the same finite population, where $\mathbf{x}_i$ is observed in both surveys and y_i are observed in Survey Two. Let d_{1i} and d_{2i} be the sampling weight of unit i in sample A_1 and sample A_2, respectively. Let $\hat{\mathbf{X}}_1 = \sum_{i \in A_1} d_{1i}\mathbf{x}_i$ and $\hat{\mathbf{X}}_2 = \sum_{i \in A_2} d_{2i}\mathbf{x}_i$ be unbiased estimators of $\mathbf{X} = \sum_{i=1}^{N} \mathbf{x}_i$ obtained from the two surveys and $\hat{Y}_2 = \sum_{i \in A_2} d_{2i}y_i$ be an unbiased estimator of $Y = \sum_{i=1}^{N} y_i$. We are interested in estimating Y combining the two surveys.

(a) Show that the optimal estimator of Y among the class of unbiased estimators of Y that are linear in $\hat{\mathbf{X}}_1$, $\hat{\mathbf{X}}_2$ and $\hat{Y}_2$ is

$$\hat{Y}_{opt} = \hat{Y}_2 + \left(\hat{\mathbf{X}}_{opt} - \hat{\mathbf{X}}_2\right)' \hat{B},$$

where

$$\hat{\mathbf{X}}_{opt} = \hat{\mathbf{K}}\hat{\mathbf{X}}_1 + (1 - \hat{\mathbf{K}})\hat{\mathbf{X}}_2,$$

$$\hat{\mathbf{K}} = \left\{\hat{V}(\hat{\mathbf{X}}_1) + \hat{V}(\hat{\mathbf{X}}_2)\right\}^{-1} \hat{V}(\hat{\mathbf{X}}_2),$$

and $\hat{B} = \{\hat{V}(\hat{\mathbf{X}}_2)\}^{-1}\widehat{Cov}(\hat{\mathbf{X}}_2, \hat{Y}_2)$.

(b) Let $\mathbf{x}_i' = (\mathbf{x}_{1i}', \mathbf{x}_{2i}')$ and assume that $\mathbf{X}_1 = \sum_{i=1}^{N}\mathbf{x}_{1i}$ is known. Find the optimal estimator of Y in this case.

7. Let A be the set of sample indices obtained from a probability sample of size n with the first-order inclusion probability π_i. Let $(\mathbf{x}_i, y_i)$ be the sample observations from A and $\bar{\mathbf{x}}_N = N^{-1}\sum_{i=1}^{N}\mathbf{x}_i$ is known. Let $\bar{y}_N = N^{-1}\sum_{i=1}^{N}y_i$ be the parameter of interest. Consider the following estimator:

$$\bar{y}_{reg1} = \bar{\mathbf{x}}_N \hat{\beta},$$

where

$$\hat{\beta} = \left(\sum_{i \in A}\pi_i^{-2}\mathbf{x}_i'\mathbf{x}_i\right)^{-1}\sum_{i \in A}\pi_i^{-2}\mathbf{x}_i'y_i.$$

(a) Find the conditions on $\mathbf{x}_i$ so that $\bar{y}_{reg1}$ is design consistent.

(b) Find a superpopulation model where $\bar{y}_{reg1}$ achieves the minimum model variance among the class of linear (in y) and model-unbiased estimators of $\bar{y}_N$.

(c) Let $y_i = \mathbf{x}_i'\beta + e_i$ be the superpopulation model with $e_i \sim (0, \gamma_{ii}\sigma^2)$ for some known $\gamma_{ii} = \gamma(\mathbf{x}_i)$. Find a set of conditions on $\mathbf{x}_i$ and π_i such that $\bar{y}_{reg1}$ is optimal in the sense that it minimizes the anticipated variance among the class of linear model-unbiased estimators and the class of fixed-sample size design-consistent estimators of $\bar{y}_n$ under a nonreplacement design with fixed probabilities.

8. Let A be the set of sample indices obtained from a probability sample of size n with the first-order inclusion probability π_i. The population size N is unknown and $\hat{N}_d = \sum_{i \in A}d_i$ is used to estimate N, where $d_i = 1/\pi_i$. For scalar x_i, consider the following regression estimator

$$\hat{Y}_{reg2} = \hat{Y}_d + \left(X - \hat{X}_d\right)\hat{B}_1,$$

where $X = \sum_{i=1}^{N} x_i$, $\left(\hat{X}_d, \hat{Y}_d\right) = \sum_{i \in A} d_i(x_i, y_i)$,

$$\hat{B}_1 = \left\{ \sum_{i \in A} d_i \left(x_i - \bar{x}_d \right)^2 \right\}^{-1} \sum_{i \in A} d_i \left(x_i - \bar{x}_d \right) y_i,$$

and $(\bar{x}_d, \bar{y}_d) = (\hat{X}_d, \hat{Y}_d)/\hat{N}_d$.

Answer the following questions:

(a) Show that $\hat{Y}_{reg2}$ is asymptotically unbiased for $Y = \sum_{i=1}^{N} y_i$.

(b) Derive the asymptotic variance of $\hat{Y}_{reg2}$.

(c) Compare the asymptotic variance of $\hat{Y}_{reg2}$ with the asymptotic variance of $\hat{Y}_{reg3}$, where

$$\hat{Y}_{reg3} = \hat{Y}_\pi + \left(X - \hat{X}_\pi \right) \hat{B}_1,$$

and $\left(\hat{X}_\pi, \hat{Y}_\pi \right) = N \left(\hat{X}_d, \hat{Y}_d \right)/\hat{N}_d$.

11

Data Integration

Statistical analysis of non-probability survey samples faces many challenges. Non-probability samples have unknown selection/inclusion mechanisms and are typically biased, and they do not represent the target population. A popular framework in dealing with the biased non-probability samples is to assume that auxiliary variable information on the same population is available from an existing probability survey sample. Combining the information from a non-probability sample and auxiliary information from a probability sample is called data integration.

One can view data integration as a missing data problem, and apply the imputation techniques or propensity score weights to combine information from several sources. We consider the following setup for data integration. Let A be a probability sample with observations on auxiliary variable $\mathbf{X}$; let B be the non-probability sample with information on both the study variable Y and the auxiliary variables $\mathbf{X}$. Table 11.1 presents the general setup of the two sample structure for data integration. As indicated in Table 11.1, sample B is not representative of the target population.

TABLE 11.1
Data Structure for Two Samples

Sample	Type	X	Y	Representative?
A	Probability Sample	✓		Yes
B	Non-probability Sample	✓	✓	No

Under the data structure in Table 11.1, we wish to develop methods for combining information from two samples. Note that y_is can be treated as missing values in sample A. Thus, roughly speaking, there are two different approaches for data integration. One approach is mass imputation, where we use sample B as a training sample for developing an suitable imputation model and use the trained model to impute missing values in sample A. The other approach is to use propensity score weighting, where the selection probability for sample B is trained using the observations in sample A and the inverse of the estimated selection probabilities are applied to the elements in sample B. The two approaches can be combined to develop doubly robust estimator, which is presented in Section 11.4.

DOI: 10.1201/9780429321740-11

11.1   Mass Imputation

To describe mass imputation formally, define δ to be the indicator variable for the unit being included in the non-probability sample B. We first assume that each unit in the population has a non-zero probability to be included in the sample B, i.e.,

$$P(\delta = 1 \mid \mathbf{x}) > 0 \tag{11.1}$$

for all $\mathbf{x}$ in the support of X. Positivity Assumption in (11.1) means that, for any possible value $\mathbf{x}$, there is a positive probability for this unit to be selected for sample B. Assumption (11.1) implies that the sample support of X in sample B coincides with the support of X in the population.

The prediction model $f(y \mid \mathbf{x})$ can be estimated by using observed (Y, X) from sample B if

$$f(y \mid \mathbf{x}, \delta = 1) = f(y \mid \mathbf{x}). \tag{11.2}$$

The prediction model $f(y \mid \mathbf{x})$ can then be used for creating mass imputation of y for the probability sample A. A sufficient condition for (11.2) is the missing-at-random (MAR) assumption for the sample B:

$$P(\delta = 1 \mid \mathbf{x}, y) = P(\delta = 1 \mid \mathbf{x}). \tag{11.3}$$

Assumption (11.3) is a strong assumption as it means that the sampling mechanism for sample B does not depend on Y after conditioning on $\mathbf{x}$. Given the data structure in Table 11.1, there is no way to test this assumption.

Under assumptions (11.1)–(11.2), it is possible to consider a mass imputation estimator based on nearest neighbor imputation as suggested by Rivers (2007). Nearest neighbor imputation is a nonparametric method that does not require any parametric model assumptions. While nonparametric imputation methods can provide robust estimation, it suffers from curse of dimensionality, and the asymptotic bias of the nearest neighbor imputation is not negligible if the dimension of $\mathbf{x}$ is greater than one (Yang and Kim, 2020).

Kim et al. (2021) considered a conditional mean model for sample B with the first moment specified as

$$E(Y \mid \mathbf{x}) = m(\mathbf{x}; \beta_0) \tag{11.4}$$

for some unknown $p \times 1$ vector β_0 and a known function $m(\cdot; \cdot)$. Let $(y_i, \mathbf{x}_i)$ be the observed values of (Y, X) for unit $i \in B$. Let $n_A = |A|$ and $n_B = |B|$ be the respective sample sizes. We assume that $\hat{\beta}$ is the unique solution to

$$\hat{U}(\beta) = \frac{1}{n_B} \sum_{i \in B} \{y_i - m(\mathbf{x}_i; \beta)\} \, \mathbf{h}(\mathbf{x}_i; \beta) = 0 \tag{11.5}$$

for some p-dimensional vector of functions $\mathbf{h}(\mathbf{x}_i; \beta)$. Note that $E\{\hat{U}(\beta_0)\} = 0$ under assumption (11.2) and model (11.4).

The estimator $\hat{\beta}$ is first obtained from sample B and then used to obtain the predicted value $\hat{y}_i = m(\mathbf{x}_i; \hat{\beta})$ for all $i \in A$. The mass imputation estimator for the finite population mean $\theta_N = N^{-1} \sum_{i=1}^{N} y_i$ is computed as

$$\hat{\theta}_I = \frac{1}{N} \sum_{i \in A} w_i \hat{y}_i, \qquad (11.6)$$

where w_i is the sampling weight of unit i.

To discuss asymptotic properties of the mass imputation estimator given in (11.6), define $\beta^* = \operatorname{plim} \hat{\beta}$, where the reference distribution is the sampling mechanism for sample B. Roughly speaking, if (11.3) holds, then $\beta^* = \beta_0$, where β_0 is the true model parameter in (11.4). Under some regularity conditions, we can show that the mass imputation estimator (11.6) satisfies

$$\hat{\theta}_I = \tilde{\theta}_I + o_p(n_B^{-1/2}), \qquad (11.7)$$

where

$$\tilde{\theta}_I = N^{-1} \sum_{i \in A} w_i m(\mathbf{x}_i; \beta^*) + N^{-1} \sum_{i \in B} \{y_i - m(\mathbf{x}_i; \beta^*)\} g(\mathbf{x}_i; \beta^*) \qquad (11.8)$$

with $g(\mathbf{x}_i; \beta^*) = \mathbf{c}' \mathbf{h}(\mathbf{x}_i; \beta^*)$ and $\mathbf{c}$ is the solution to

$$\sum_{i \in B} \dot{m}(\mathbf{x}_i; \beta^*) \mathbf{h}'(\mathbf{x}_i; \beta^*) \mathbf{c} = \sum_{i=1}^{N} \dot{m}(\mathbf{x}_i; \beta^*) \qquad (11.9)$$

and $\dot{m}(\mathbf{x}; \beta) = \partial m(\mathbf{x}; \beta) / \partial \beta$. Expression (11.8) is essentially the first-order Taylor linearization of $\hat{\theta}_I$ in (11.6), and the Taylor expansion is made around $\beta = \beta^*$, not around the true parameter value β_0. Note that we do not use the MAR assumption in (11.3) to obtain (11.7).

By (11.7), we can express

$$\tilde{\theta}_I - \theta_N = N^{-1} \left(\sum_{i \in A} w_i m_i^* - \sum_{i=1}^{N} m_i^* \right) + N^{-1} \left(\sum_{i=1}^{N} \delta_i g_i^* e_i^* - \sum_{i=1}^{N} e_i^* \right), \qquad (11.10)$$

where $m_i^* = m(\mathbf{x}_i; \beta^*)$, $e_i^* = y_i - m(\mathbf{x}_i; \beta^*)$ and $g_i^* = g(\mathbf{x}_i; \beta^*)$. Thus, ignoring the smaller order terms, the asymptotic bias is

$$
\begin{aligned}
E(\tilde{\theta}_I - \theta_N) &= E\left\{ N^{-1} \left(\sum_{i=1}^{N} \delta_i g_i^* e_i^* - \sum_{i=1}^{N} e_i^* \right) \right\} \\
&= -E\left(N^{-1} \sum_{i=1}^{N} e_i^* \right) \\
&= E\{Cov_N(\delta_i \pi_B^{-1}, e_i^*)\}, \qquad (11.11)
\end{aligned}
$$

where $\pi_B = n_B/N$, and the second equality follows from $E\left[\sum_{i=1}^{N} \delta_i \{y_i - m(\mathbf{x}_i; \beta^*)\} g_i^*\right] = 0$ by the definition of β^*. Here, we use notation $Cov_N(x_i, y_i) = N^{-1} \sum_{i=1}^{N} (x_i - \bar{x}_N)(y_i - \bar{y}_N)$. If $\beta^* = \beta_0$, then $e_i^* = e_i$ and the mass imputation estimator in (11.6) is unbiased. Otherwise, the bias is non-zero.

On the other hand, the asymptotic bias of $\bar{y}_B = n_B^{-1} \sum_{i \in B} y_i$ is

$$
\begin{aligned}
E(\bar{y}_B - \theta_N) &= E\{n_B^{-1} \sum_{i=1}^{N} \delta_i(y_i - \bar{Y}_N)\} \\
&= E\{Cov_N(\delta_i \pi_B^{-1}, y_i)\}.
\end{aligned} \tag{11.12}
$$

Thus, comparing (11.11) with (11.12), we find that the absolute value of the bias of $\tilde{\theta}_I$ is smaller than that of $\bar{y}_B$ as the variance of e_i^* will be smaller than the variance of y_i. Thus, the mass imputation estimator reduces the bias even when the sampling mechanism for sample B is non-ignorable.

Now, to investigate the variance, we now assume that the sampling mechanism for sample B is MAR as defined in (11.3). Thus, (11.10) reduces to

$$
\tilde{\theta}_I - \theta_N = N^{-1}\left(\sum_{i \in A} w_i m_i - \sum_{i=1}^{N} m_i\right) + N^{-1}\left(\sum_{i=1}^{N} \delta_i g_i e_i - \sum_{i=1}^{N} e_i\right), \tag{11.13}
$$

where $g_i = g(\mathbf{x}_i; \beta_0)$. Thus, we can obtain $V(\tilde{\theta}_I - \theta_N) = V_A + V_B$, where

$$
V_A = V\left(N^{-1} \sum_{i \in A} w_i m_i\right) \tag{11.14}
$$

is the design-based variance under the probability sampling design for sample A, and

$$
V_B = V\left\{N^{-1}\left(\sum_{i=1}^{N} \delta_i g_i e_i - \sum_{i=1}^{N} e_i\right)\right\} \tag{11.15}
$$

is the variance component for sample B under the superpopulation model. Furthermore, if $n_B/N = o(1)$ then we can express

$$
V_B = E\left\{N^{-2} \sum_{i \in B} (e_i^* g_i^*)^2\right\}. \tag{11.16}
$$

Note that we can easily estimate V_B in (11.16) even though the sampling mechanism for sample B is unknown. The asymptotic variance $V(\tilde{\theta}_I - \theta_N)$ consists of two parts. The first term V_A is of order $O(n_A^{-1})$, and the second term V_B is of order $O(n_B^{-1})$. If $n_A/n_B = o(1)$, i.e., the sample size n_B is much larger than n_A, the term V_B is of smaller order, and the leading term of the total variance is V_A. Otherwise, the two variance components both contribute to the total variance.

Under the linear regression model $y_i = \mathbf{x}_i'\beta + e_i$ with $e_i \sim (0, \sigma_e^2)$, independent among all i, we have $\hat{y}_i = \mathbf{x}_i'\hat{\beta}$, where $\hat{\beta} = \left(\sum_{i\in B}\mathbf{x}_i\mathbf{x}_i'\right)^{-1}\sum_{i\in B}\mathbf{x}_iy_i$. The mass imputation estimator of (11.6) under the regression model is given by $\hat{\theta}_{I,reg} = N^{-1}\sum_{i\in A}w_i\mathbf{x}_i'\hat{\beta}$. If the probability sample A is selected by simple random sampling, the asymptotic variance of $\hat{\theta}_{I,reg}$ is given by

$$V\left(\hat{\theta}_{I,reg} - \theta_N\right) \approx V\left(n_A^{-1}\sum_{i\in A}\mathbf{x}_i'\beta\right) + V\left(N^{-1}\sum_{i\in B}e_i\mathbf{x}_i'\mathbf{c}\right), \qquad (11.17)$$

where $\mathbf{c} = \left(N^{-1}\sum_{i\in B}\mathbf{x}_i\mathbf{x}_i'\right)^{-1}\bar{\mathbf{x}}_N$ and $\bar{\mathbf{x}}_N = N^{-1}\sum_{i=1}^{N}\mathbf{x}_i$. If $\mathbf{x}_i = (1, x_i)'$ and $\beta = (\beta_0, \beta_1)'$ and assuming $\pi_B = n_B/N$ is negligible, the asymptotic variance reduces to

$$V\left(\hat{\theta}_{I,reg} - \theta_N\right) \approx \frac{1}{n_A}(\beta_1)^2\sigma_x^2 + \frac{1}{n_B}\sigma_e^2 + E\left[\frac{(\bar{x}_N - \bar{x}_B)^2}{\sum_{i\in B}(x_i - \bar{x}_B)^2}\right]\sigma_e^2,$$

where $\sigma_x^2 = V(X)$ and $\bar{x}_B = n_B^{-1}\sum_{i\in B}x_i$. If sample B is a random sample from the population, then the third term is of order $O(n_B^{-2})$ and becomes negligible. However, since sample B is a non-probability sample, the third term might not be negligible.

Variance estimation for the mass imputation estimator (11.6) requires the estimation of the two components V_A and V_B. The first component can be estimated by

$$\hat{V}_A = \frac{1}{N^2}\sum_{i\in A}\sum_{j\in A}\frac{\pi_{ij} - \pi_i\pi_j}{\pi_{ij}}w_im(\mathbf{x}_i; \hat{\beta})w_jm(\mathbf{x}_j; \hat{\beta}),$$

where $\pi_{ij} = P(i, j \in A)$ are the joint inclusion probabilities and are assumed to be positive. The second component can be estimated by

$$\hat{V}_B = \frac{1}{N^2}\sum_{i\in B}\hat{e}_i^2\{\hat{\mathbf{c}}'\mathbf{h}(\mathbf{x}_i; \hat{\beta})\}^2, \qquad (11.18)$$

where $\hat{e}_i = y_i - m(\mathbf{x}_i; \hat{\beta})$ and $\hat{\mathbf{c}} = \left(N^{-1}\sum_{i\in B}\mathbf{x}_i\mathbf{x}_i'\right)^{-1}N^{-1}\sum_{i\in A}w_i\mathbf{x}_i$. The total variance of $\hat{\theta}_I$ can be estimated by $\hat{V}(\hat{\theta}_I - \theta_N) = \hat{V}_A + \hat{V}_B$.

11.2 Propensity Score Method

Under MAR, we can further build a model for $P(\delta = 1 \mid \mathbf{x})$ and use it to construct the propensity score weights for sample B. Suppose that $\pi(\mathbf{x}) = P(\delta = 1 \mid \mathbf{x})$ has a parametric form such that $\pi(\mathbf{x}) = \pi(\mathbf{x}; \phi)$ for some ϕ. The population log-likelihood function for ϕ can be written as

$$l(\phi) = \sum_{i=1}^{N} [\delta_i \log \pi(\mathbf{x}_i; \phi) + (1 - \delta_i) \log\{1 - \pi(\mathbf{x}_i; \phi)\}].$$

Thus, the (population-based) maximum likelihood estimator of ϕ can be obtained by solving

$$S_p(\phi) \equiv \sum_{i=1}^{N} \left\{ \frac{\delta_i}{\pi(\mathbf{x}_i; \phi)} - \frac{1 - \delta_i}{1 - \pi(\mathbf{x}_i; \phi)} \right\} \dot{\pi}(\mathbf{x}_i; \phi) = 0,$$

which is equivalent to solving

$$\sum_{i=1}^{N} \delta_i h(\mathbf{x}_i; \phi) = \sum_{i=1}^{N} \pi(\mathbf{x}_i; \phi) h(\mathbf{x}_i; \phi) \tag{11.19}$$

for ϕ, where

$$h(\mathbf{x}_i; \phi) = \frac{\dot{\pi}(\mathbf{x}_i; \phi)}{\pi(\mathbf{x}_i; \phi)\{1 - \pi(\mathbf{x}_i; \phi)\}}.$$

The left side of (11.19) can be constructed from sample B. Thus, we have only to estimate the right side of (11.19). Using the sampling weights, we can use

$$\sum_{i=1}^{N} \delta_i h(\mathbf{x}_i; \phi) = \sum_{i \in A} w_i \pi(\mathbf{x}_i; \phi) h(\mathbf{x}_i; \phi), \tag{11.20}$$

which does not require identification of the elements in both samples. Chen et al. (2020b) first proposed estimation using (11.20) for propensity score method for voluntary samples. The final propensity score estimator for θ_N is

$$\hat{\theta}_{PS} = \frac{\sum_{i \in B} \hat{\pi}_i^{-1} y_i}{\sum_{i \in B} \hat{\pi}_i^{-1}}, \tag{11.21}$$

where $\hat{\pi}_i = \pi(\mathbf{x}_i; \hat{\phi})$. Consistency and the asymptotic normality of the propensity score estimator in (11.21) can be established. See Chen et al. (2020b) for more details.

If $n_B = |B|$ is small compared with N, then the estimated probability $\hat{\pi}(\mathbf{x}_i)$ can take small values, and the resulting PS estimator in (11.21) can be unstable. We discuss an alternative estimation method as follows.

Let $f_k(\mathbf{x})$ be the density function of $\mathbf{x}$ given $\delta = k$, for $k = 0, 1$. Using Bayes formula,

$$\frac{1}{\pi(\mathbf{x})} = 1 + \frac{\pi_0}{\pi_1} r(\mathbf{x}),$$

where $r(\mathbf{x}) = f_0(\mathbf{x})/f_1(\mathbf{x})$.

Suppose that we employ a parametric model such as

$$\log\{r(\mathbf{x})\} = \phi_0 + \phi_1'\mathbf{x}. \tag{11.22}$$

Note that (11.22) is equivalent to

$$\pi(\mathbf{x}) = \frac{1}{1 + \exp(\phi_0^* + \phi_1'\mathbf{x})},$$

where $\pi_0^* = \log(\pi_0/\pi_1) + \phi_0$. To estimate the model parameter in (11.22), we can apply the maximum entropy method described in Section 7.5. Recall that the density ratio function can be understood as the maximizer of $Q(r)$ in (7.62).

Under model (11.22), the finite population version of the objective function for maximum entropy method is

$$
\begin{aligned}
Q(\phi) &= \frac{1}{N_0}\sum_{i=1}^{N} I(\delta_i = 0)\left(\phi_0 + \phi_1'\mathbf{x}_i\right) - \frac{1}{N_1}\sum_{i=1}^{N} I(\delta_i = 1)\exp(\phi_0 + \phi_1'\mathbf{x}_i) \\
&= \frac{1}{N_0}\sum_{i=1}^{N}\left(\phi_0 + \phi_1'\mathbf{x}_i\right) - \frac{1}{N_1}\sum_{i=1}^{N}\delta_i\left\{\exp(\phi_0 + \phi_1'\mathbf{x}_i) - \frac{N_1}{N_0}(\phi_0 + \phi_1'\mathbf{x}_i)\right\},
\end{aligned}
$$

where $N_k = \sum_{i=1}^{N} I(\delta_i = k)$ and ϕ_0 satisfies

$$N_1^{-1}\sum_{i=1}^{N}\delta_i\exp(\phi_0 + \phi_1'\mathbf{x}_i) = 1.$$

Using Horvitz-Thompson estimation, we can use

$$
\begin{aligned}
\hat{Q}(\phi) &= \frac{1}{\hat{N}_0}\sum_{i\in A} d_i\left(\phi_0 + \phi_1'\mathbf{x}_i\right) - \frac{1}{N_1}\sum_{i=1}^{N}\delta_i\left\{\exp(\phi_0 + \phi_1'\mathbf{x}_i) - \frac{N_1}{\hat{N}_0}(\phi_0 + \phi_1'\mathbf{x}_i)\right\} \\
&= \frac{1}{\hat{N}_0}\left\{\sum_{i\in A} d_i\left(\phi_0 + \phi_1'\mathbf{x}_i\right) - \sum_{i=1}^{N}\delta_i(\phi_0 + \phi_1'\mathbf{x}_i)\right\} - \frac{1}{N_1}\sum_{i=1}^{N}\delta_i\left\{\exp(\phi_0 + \phi_1'\mathbf{x}_i)\right\},
\end{aligned}
$$

as a sample-based objective function, where $\hat{N}_0 = \sum_{i\in A} d_i - N_1$. The maximizer of $\hat{Q}(\phi)$ is called the maximum entropy estimator. The maximum entropy estimator of ϕ is obtained by solving

$$\frac{1}{N_1}\sum_{i=1}^{N}\delta_i\exp(\phi_0 + \phi_1'\mathbf{x}_i)(1, \mathbf{x}_i') = \left(1, \hat{\bar{\mathbf{x}}}_0'\right),$$

where

$$\hat{\bar{\mathbf{x}}}_0 = \frac{1}{\hat{N}_0}\left\{\sum_{i\in A} d_i\mathbf{x}_i - \sum_{i=1}^{N}\delta_i\mathbf{x}_i\right\}.$$

We can use

$$U_1(\phi) := \sum_{i=1}^{N} \delta_i \exp(\phi_1' \mathbf{x}_i)(\mathbf{x}_i - \hat{\bar{\mathbf{x}}}_0) = \mathbf{0},$$

to obtain $\hat{\phi}_1$. We may use

$$\hat{\phi}_1^{(t+1)} = \hat{\phi}_1^{(t)} + \left\{ \sum_{i=1}^{N} \delta_i w_i^{(t)} (\mathbf{x}_i - \bar{\mathbf{x}}_w^{(t)})^{\otimes 2} \right\}^{-1} \left(\hat{\bar{\mathbf{x}}}_0 - \bar{\mathbf{x}}_w^{(t)} \right)$$

where $w_i^{(t)} = w_i(\hat{\phi}_1^{(t)})$ with $w_i(\phi_1) = \exp(\phi_1' \mathbf{x}_i)/\{\sum_{i=1}^{N} \delta_i \exp(\phi_1' \mathbf{x}_i)\}$ and $\bar{\mathbf{x}}_w^{(t)} = \sum_{i=1}^{N} \delta_i w_i^{(t)} \mathbf{x}_i$. The final propensity score estimator is

$$\hat{\theta}_{PS2} = \frac{1}{N} \sum_{i=1}^{N} \delta_i \left\{ 1 + \frac{\hat{N}_0}{N_1} \exp(\hat{\phi}_0 + \hat{\phi}_1' \mathbf{x}_i) \right\} y_i. \tag{11.23}$$

Note that the final propensity weight $\hat{\pi}_i^{-1} = 1 + (\hat{N}_0/N_1)\exp(\hat{\phi}_0 + \hat{\phi}_1' \mathbf{x}_i)$ satisfies

$$\sum_{i=1}^{N} \delta_i \hat{\pi}_i^{-1} \mathbf{x}_i = \sum_{i=1}^{N} \delta_i \mathbf{x}_i + \sum_{i \in A} d_i (1 - \delta_i) \mathbf{x}_i.$$

Thus, it satisfies the covariate-balancing property which is quite close to calibration property in survey sampling.

Elliott and Valliant (2017) proposed a different approach of propensity score method for data integration. Note that

$$P(\delta = 1 \mid X) \propto P(I_A = 1 \mid X) \cdot \frac{f(X \mid \delta = 1)}{f(X \mid I_A = 1)},$$

where I_A is the sample inclusion indicator function for sample A. Thus,

$$\frac{1}{P(\delta = 1 \mid X)} \propto \{P(I_A = 1 \mid X)\}^{-1} \cdot \frac{f(X \mid I_A = 1)}{f(X \mid \delta = 1)} := \tilde{w}(X) \cdot R(x).$$

Elliott and Valliant (2017) proposed estimating two terms separately. To estimate the first term $\tilde{w}(x_i)$, using

$$E(w_i \mid x_i, I_{A,i} = 1) = \frac{1}{P(I_{A,i} = 1 \mid x_i)},$$

one can apply regression of w_i on x_i from sample A. To estimate the second term, Elliott and Valliant (2017) proposed using

$$R(X) \equiv \frac{f(X \mid I_A = 1)}{f(X \mid \delta = 1)} \propto \frac{P(I_A = 1 \mid X, I_A + \delta \geq 1)}{P(\delta = 1 \mid X, I_A + \delta \geq 1)}.$$

One can apply a suitable classification method from the combined sample to estimate $R(x)$. The final pseudo weight for sample B is then

$$\hat{w}_i = \tilde{w}_i \hat{R}(x_i).$$

Rafei et al. (2020) uses Bayesian Additive Regression Trees (BART) to estimate the two components in the pseudo weights for voluntary big data sample.

Example 11.1. *A limited simulation study is performed to compare the data integration methods. The setup for the simulation study employed a 2×2 factorial structure with two factors. The first factor is the superpopulation model that generates the finite population. The second factor is the sample size for sample B. We generated two finite populations of $(\mathbf{x}'_i, y_i)$ of size $N = 100,000$, where $\mathbf{x} = (x_1, \ldots, x_{10})$ with each $x_k \sim N(2,1)$, independently and the study variable y_i are constructed differently for each model:*

1. Model I: $y_i = 1 + x_{1i} + x_{2i} + x_{3i} + e_i$.

2. Model II: $y_i = 1 + 0.5 x_{1i} x_{2i} + 0.5 x_{1i} x_{3i} + e_i$.

Here, $e_i \sim N(0,1)$.

From each of the two populations, we generated two independent samples. We use simple random sampling of size $n_A = 500$ to obtain sample A. In selecting sample B of size n_B, where $n_B \in \{500, 1000\}$, we create two strata where Stratum 1 consists of elements with $x_{1i} \leq 2$ and Stratum 2 consists of elements with $x_{1i} > 2$. Within each stratum, we select n_h elements by simple random sampling, independent between the two strata, where $n_1 = 0.7 n_B$ and $n_2 = 0.3 n_B$. We assume that the stratum information is unavailable at the time of data analysis. Using the two samples A and B, we compute five estimators of $\theta_N = N^{-1} \sum_{i=1}^{N} y_i$:

1. The simple mean (Simple Mean) from sample B: $\hat{\theta}_B = n_B^{-1} \sum_{i \in B} y_i$.

2. EV: The PS estimator using Elliott and Valliant (2017) method.

3. CLW: The PS estimator proposed by Chen et al. (2020b) using $\pi(\mathbf{x}_i; \phi) = \{1 + \exp(-\phi_0 - \sum_{k=1}^{p} \phi_k x_{ik})\}^{-1}$ as the working propensity score model.

4. New-1: The new PS estimator using the maximum entropy method using all covariates.

5. New-2: The new PS estimator using the maximum entropy method using the selected covariates from the working regression model (using SCAD method).

In Table 11.2, the summary of the performance of five estimators is presented. Simple mean estimator is severely biases because of the sampling mechanism for sample B depends on x variable. The proposed PS estimators (New-1 and New-2) perform better than the other PS estimators (EV, CLW) in terms of mean squared errors. The New-2 PS estimator is slightly more efficient than the New-1 PS estimator because New-2 method satisfies the covariate-balancing property for the selected covariates only. In New-1 method, the covariate-balancing property holds for all covariates which leads to an efficiency loss in the PS estimation in our simulation setup.

TABLE 11.2
Monte Carlo Bias, Monte Carlo Variance, and Mean Square Error (MSE) of
the Five Point Estimators, Based on 10,000 Monte Carlo Samples

Case	Estimator	Model I			Model II		
		Bias	Var $(\times 10^3)$	MSE $(\times 10^3)$	Bias	Var $(\times 10^3)$	MSE $(\times 10^3)$
	Simple Mean	-0.318	6.898	108.192	-0.637	8.898	415.283
Case 1	EV	0.003	9.385	9.392	0.004	19.663	19.683
	CLW	-0.007	11.390	11.440	-0.015	24.404	24.630
$(n_B = 500)$	New-1	0.001	8.338	8.339	0.001	15.663	15.665
	New-2	0.001	8.299	8.300	0.002	15.611	15.613
	Simple Mean	-0.319	3.410	105.102	-0.638	4.354	411.121
Case 2	EV	0.006	7.702	7.744	0.012	16.002	16.149
	CLW	-0.007	9.443	9.492	-0.015	20.153	20.369
$(n_B = 1,000)$	New-1	0.001	7.147	7.147	0.001	13.868	13.868
	New-2	0.001	7.128	7.129	0.001	13.846	13.846

11.3   Nonparametric Propensity Score Approach

We now consider an extension of the propensity score approach in the
previous section to the case when $\mathbf{x}$ is categorical. In the case of categorical $\mathbf{x}$,
we can still consider a parametric similar to (11.22) and apply the maximum
entropy method. To present the idea of the maximum entropy method, suppose
that $\mathbf{x} = (x_1,\ldots,x_K)$ and each x_k can take one of D_k values among the set
$\mathcal{X}_k = \{x_k^{(1)},\ldots,x_k^{(D_k)}\}$ with unknown probabilities. We assume that the density
ratio function $r(\mathbf{x}) = f_0(\mathbf{x})/f_1(\mathbf{x})$ satisfies

$$r(\mathbf{x}) = \prod_{k=1}^{K} r_k(x_k), \tag{11.24}$$

where

$$r_k(x_k) = \sum_{d=1}^{D_k} I(x_k = x_k^{(d)})r_{kd}$$

and

$$r_{kd} = \frac{P(x_k = x_k^{(d)} \mid \delta = 0)}{P(x_k = x_k^{(d)} \mid \delta = 1)}.$$

Thus, we have only to estimate the parameters r_{kd}. If we define

$$\gamma_{ik}^{(d)} = \begin{cases} 1 & \text{if } x_{ik} = x_k^{(d)} \\ 0 & \text{otherwise,} \end{cases}$$

we can express $r_k(x_k) = \sum_{d=1}^{D_k} \gamma_{ik}^{(d)} r_{kd}$. Writing $\mathbf{r}_k = (r_{k1}, \ldots, r_{kD_k})$, the finite population version of the objective function for maximum entropy estimation is

$$Q(\mathbf{r}_k) = \frac{1}{N_0} \sum_{i=1}^{N} (1 - \delta_i) \sum_{d=1}^{D_k} \gamma_{ik}^{(d)} \log(r_{kd})$$

subject to

$$\frac{1}{N_1} \sum_{i=1}^{N} \delta_i \sum_{d=1}^{D_k} \gamma_{ik}^{(d)} r_{kd} = 1. \tag{11.25}$$

Using Horvitz-Thompson estimation, we can use

$$\hat{Q}(\mathbf{r}_k) = \frac{1}{N_0} \sum_{i \in A} d_i \left\{ \sum_{d=1}^{D_k} \gamma_{ik}^{(d)} \log(r_{kd}) \right\} - \frac{1}{N_0} \sum_{i=1}^{N} \delta_i \left\{ \sum_{d=1}^{D_k} \gamma_{ik}^{(d)} \log(r_{kd}) \right\}$$

$$\tag{11.26}$$

subject to (11.25) as a sample-based objective function. However, such an approach can be problematic if there exist $x_k^{(d)}$ such that $\sum_{i=1}^{N} \delta_i I(x_{ik} = x_k^{(d)}) > 0$ but $\sum_{i \in A} I(x_{ik} = x_k^{(d)}) = 0$. That is, the sample support of B should be a subset of the sample support of sample A.

Another approach is based on the maximum likelihood method considered in Kim and Tam (2021). To formally describe the idea, recall that the finite population U is decomposed into two groups, $U = B \cup B^c$. We assume that $\pi = P(\delta = 1)$ is known and given by $\pi = N_1/N$. To discuss parameter estimation, we assume that

$$p(\mathbf{x} \mid \delta = 1) = \prod_{k=1}^{K} p_k(x_k \mid \delta = 1), \tag{11.27}$$

where $p_k(x_k \mid \delta = 1) = m_{kd}$ if $x_k = x_k^{(d)}$ and $\sum_{d=1}^{D_k} m_{kd} = 1$. Since we can observe $\mathbf{x}_i$ among $\delta_i = 1$, we can estimate m_{kd} using

$$\hat{m}_{kd} = \frac{1}{N_1} \sum_{i=1}^{N} \delta_i \gamma_{ik}^{(d)}. \tag{11.28}$$

Now, for the model for $p(\mathbf{x} \mid \delta = 0)$, we assume that

$$p(\mathbf{x} \mid \delta = 0) = \prod_{k=1}^{K} p_k(x_k \mid \delta = 0),$$

where $p_k(x_k \mid \delta = 0) = u_{kd}$ if $x_k = x_k^{(d)}$ and $\sum_{d=1}^{D_k} u_{kd} = 1$. Using $\gamma_{ik}^{(d)}$ notation, we can express $m_{kd} = P(\gamma_{ik}^{(d)} = 1 \mid \delta_i = 1)$ and $u_{kd} = P(\gamma_{ik}^{(d)} = 1 \mid \delta_i = 0)$.

To estimate u_{kd}, we use the following EM algorithm:

1. First note that, if δ_i were observed, the complete-sample pseudo log-likelihood would be

$$l_{com}(\mathbf{u} \mid \boldsymbol{\delta}, \gamma) = \sum_{i \in A} d_i \delta_i \log \left\{ \pi \prod_{k=1}^{K} m_{ik} \right\} + \sum_{i \in A} d_i (1 - \delta_i) \log \left\{ (1 - \pi) \prod_{k=1}^{K} u_{ik} \right\},$$

where

$$(m_{ik}, u_{ik}) = \sum_{d=1}^{D_k} \gamma_{ik}^{(d)} (m_{kd}, u_{kd}).$$

Note that m_{kd} are estimated by (11.28). Only u_{kd} are the parameters of interest.

2. In the E-step, we need to evaluate the conditional expectation of $l_{com}(\mathbf{u} \mid \boldsymbol{\delta}, \gamma)$ given the observed data. Thus, given the current parameters, we have only to compute

$$
\begin{aligned}
Q(\mathbf{u} \mid \mathbf{u}^{(t)}) &= E\left\{ l_{com}(\mathbf{u} \mid \boldsymbol{\delta}, \gamma) \mid \mathbf{u}^{(t)} \right\} \\
&= \sum_{i \in A} d_i \hat{p}_i^{(t)} \log \left\{ \pi \prod_{k=1}^{K} m_{ik} \right\} + \sum_{i \in A} d_i (1 - \hat{p}_i^{(t)}) \log \left\{ (1 - \pi) \prod_{k=1}^{K} u_{ik} \right\},
\end{aligned}
$$

where

$$
\begin{aligned}
\hat{p}_i^{(t)} &= E(\delta_i \mid \gamma_i; \hat{\mathbf{u}}^{(t)}) \qquad\qquad\qquad (11.29) \\
&= \frac{\pi \prod_{k=1}^{K} m_{ik}}{\pi \prod_{k=1}^{K} m_{ik} + (1 - \pi) \prod_{k=1}^{K} \hat{u}_{ik}^{(t)}}
\end{aligned}
$$

and $(m_{ik}, \hat{u}_{ik}^{(t)}) = \sum_{d=1}^{D_k} \gamma_{ik}^{(d)} (m_{kd}, \hat{u}_{kd}^{(t)})$.

3. The M-step is to maximize the Q over $\mathbf{u}$ to update the parameters. The updating formula is

$$\hat{u}_{kd}^{(t+1)} = \frac{\sum_{i \in A} d_i (1 - \hat{p}_i^{(t)}) \gamma_{ik}^{(d)}}{\sum_{i \in A} d_i (1 - \hat{p}_i^{(t)})}.$$

4. Set $t = t+1$ and go to Step 2. Continue until convergence.

Once $\hat{u}_{kd}$ are obtained, we can get

$$\hat{p}_i = \frac{\pi \prod_{k=1}^{K} \hat{m}_{ik}}{\pi \prod_{k=1}^{K} \hat{m}_{ik} + (1 - \pi) \prod_{k=1}^{K} \hat{u}_{ik}} \qquad\qquad (11.30)$$

and $(\hat{m}_{ik}, \hat{u}_{ik}^{(t)}) = \sum_{d=1}^{D_k} \gamma_{ik}^{(d)} (m_{kd}, \hat{u}_{kd}^{(t)})$. The final PS estimator is then obtained by

$$\hat{\theta}_{PS} = \frac{1}{\hat{N}} \sum_{i=1}^{N} \frac{\delta_i}{\hat{p}_i} y_i,$$

where $\hat{N} = \sum_{i=1}^{N} \delta_i / \hat{p}_i$.

11.4 Doubly Robust Method

The mass imputation using outcome regression model and the propensity score model can be combined to develop doubly robust estimation. Recently, Yang et al. (2020) developed a doubly robust method for data integration with high dimensional covariates.

They propose a two-step approach for variable selection and finite population inference. The two step approach can be described as follows:

1. Separate model selection step using penalized estimation equation

2. Joint Parameter estimation step using the selected covariates

In the model selection step, they employ the penalized estimating function (Johnson et al., 2008)

$$U^p(\theta) = U(\theta) - p_\lambda(\theta),$$

where $U(\theta)$ is the usual estimating equation for the full model and $p_\lambda(\theta)$ is the penalty function with tuning parameter λ. Table 11.3 presents some popular penalty function for the penalized estimation method.

TABLE 11.3
Popular Penalty Functions

Method	Penalty Function
LASSO	$p_\lambda(\theta) = \lambda \sum_{j=1}^{p} |\theta_j|$
A-LASSO	$p_\lambda(\theta) = \lambda \sum_{j=1}^{p} \left(|\theta_j| / \left| \hat{\theta}_j \right| \right)$
SCAD	$p'_\lambda(\theta) = \lambda \sum_{j=1}^{p} \left\{ I(|\theta_j| < \lambda) + \frac{(a\lambda - |\theta_j|)_+}{(a-1)\lambda} I(|\theta_j| \geq \lambda) \right\}$

In step one, two different models are selected separately using penalization method. The model selection step for the selection probability model can be described as finding $\mathbf{x}_1 \subset \mathbf{x}$, such that

$$P(\delta = 1 \mid \mathbf{x}) = P(\delta = 1 \mid \mathbf{x}_1),$$

by solving the penalized estimating equation for α

$$U_1^p(\alpha \mid \lambda_1) \equiv U(\alpha) - p_{\lambda_1}(\alpha) = 0 \tag{11.31}$$

where

$$U(\alpha) = N^{-1} \left\{ \sum_{i \in B} \frac{1}{\pi(\mathbf{x}_i; \alpha)} \mathbf{x}_i - \sum_{i \in A} w_i \mathbf{x}_i \right\}.$$

The tuning parameter λ_1 is determined by K-fold cross-validation using

$$Loss(\lambda_1) = \left\| \sum_{i \in B} \frac{1}{\pi\{\mathbf{x}_i; \hat{\alpha}(\lambda_1)\}} \mathbf{x}_i - \sum_{i \in A} w_i \mathbf{x}_i \right\|,$$

where $\hat{\alpha}(\lambda_1)$ is the solution to (11.31).

The model selection step for outcome regression model can be described as finding $\mathbf{x}_2 \subset \mathbf{x}$, such that

$$E(y \mid \mathbf{x}) = E(y \mid \mathbf{x}_2)$$

by solving the penalized estimating equation

$$U^P(\beta \mid \lambda_2) \equiv U(\beta) - p_{\lambda_2}(\beta) = 0, \qquad (11.32)$$

where

$$U(\beta) = N^{-1} \sum_{i \in B} \{y_i - m(\mathbf{x}_i; \beta)\} \mathbf{x}_i.$$

The tuning parameter λ_2 is determined by K-fold cross-validation using

$$Loss(\lambda_2) = \sum_{i \in B} \left[y_i - m\{\mathbf{x}_i; \hat{\beta}(\lambda_2)\} \right]^2,$$

where $\hat{\beta}(\lambda_2)$ is the solution to (11.32).

Once the models are selected from the above methods, we have the following second step for joint estimation.

1. Let $\mathbf{x}_c = \mathbf{x}_1 \cup \mathbf{x}_2$ be the combined covariates selected from Step 1.

2. Apply the working models to $\mathbf{x}_c$ to get

$$P(\delta = 1 \mid \mathbf{x}) = \pi(\mathbf{x}_c; \alpha) \quad \text{and} \quad E(Y \mid \mathbf{x}) = m(\mathbf{x}_c; \beta).$$

3. Express the DR estimator of $\mu = E(Y)$ as a function of (α, β):

$$\hat{\mu}_{DR}(\alpha, \beta) = \frac{1}{N} \left[\sum_{i \in A} w_i m(\mathbf{x}_{i,c}; \beta) + \sum_{i \in B} \frac{1}{\pi(\mathbf{x}_{i,c}; \alpha)} \{y_i - m(\mathbf{x}_{i,c}; \beta)\} \right].$$

4. Joint estimating equation is obtained by solving

$$\frac{\partial}{\partial \alpha} \hat{\mu}_{DR}(\alpha, \beta) = 0 \quad \text{and} \quad \frac{\partial}{\partial \beta} \hat{\mu}_{DR}(\alpha, \beta) = 0. \qquad (11.33)$$

Once the model parameters are estimated, we use the following doubly robust estiamtor

$$\hat{\theta}_{DR} = \sum_{i \in A} w_i \hat{y}_i + \sum_{i \in B} \hat{\pi}_i^{-1} (y_i - \hat{y}_i),$$

where $\hat{\pi}_i = \pi(\mathbf{x}_{i,c}; \hat{\alpha})$. This estimating strategy mitigates the possible first-step selection error and renders the doubly robust estimator root-n consistent if either the sampling probability or the outcome model is correctly specified. See Yang et al. (2020) for more details.

11.5 Statistical Matching

We now consider the problem of combining two samples with the non-monotone missingness structure in Table 11.4. Sample A and sample B are probability samples which were selected from the same finite population. In sample A, observe (X, Y_1) and in sample B, observe (X, Y_2). The question of interest is the associational relationship of Y_1 and (X, Y_2). If (X, Y_1, Y_2) were jointly observed, one can fit a simple regression model of Y_2 on (X, Y_1). However, based on the available data, Y_1 and Y_2 not available simultaneously.

TABLE 11.4
Missingness Patterns in the Combined Samples: "$\checkmark$" Means "Is Measured"

Monotone Missingness			
	d	X	Y
Sample A	$\checkmark$	$\checkmark$	$\checkmark$
Sample B	$\checkmark$	$\checkmark$	
Non-Monotone Missingness			
	d	X	Y_1	Y_2
Sample A	$\checkmark$	$\checkmark$	$\checkmark$	
Sample B	$\checkmark$	$\checkmark$		$\checkmark$

d is the design weight, where the subscript indicates the sample, X is the vector of auxiliary variables, Y, Y_1 and Y_2 are scalar outcome variables.

This problem fits into the statistical matching framework (D'Orazio et al., 2006). In statistical matching, the goal is to create Y_1 for each unit in sample B by finding a "statistical twin" from the sample A. Typically, one assumes the conditional independence assumption that Y_1 and Y_2 are conditionally independent given X, or equivalently,

$$f(Y_1 \mid X, Y_2) = f(Y_1 \mid X). \tag{11.34}$$

Then, the "statistical twin" is solely determined by "how close" they are in terms of X's. However, in a regression model of Y_1 on (X, Y_2), (11.34) sets the regression coefficient associated with Y_2 to be zero *a priori*, which is contrary to the study question of interest.

For a joint modeling of (X, Y_1, Y_2) without assuming (11.34), identification is an important issue. Consider the following joint model of (Y_1, Y_2) given X,

$$Y_1 = \alpha_0 + \alpha_1 X + e_1, \tag{11.35}$$
$$Y_2 = \beta_0 + \beta_1 X + \beta_2 Y_1 + e_2, \tag{11.36}$$

where e_1 and e_2 are mean zero and $\text{Cov}(e_1, e_2) = 0$. Because (X, Y_1) is observed in sample A, (α_0, α_1) is identifiable. Because (X, Y_2) is observed in sample B, $f(Y_2 \mid X)$ is identifiable.

Coupling (11.35) and (11.36) leads to

$$Y_2 = (\beta_0 + \alpha_0 \beta_2) + (\beta_1 + \alpha_1 \beta_2)X + \beta_2 e_1 + e_2.$$

Thus, only $\beta_0 + \alpha_0 \beta_2$ and $\beta_1 + \alpha_1 \beta_2$ are identifiable and $(\beta_0, \beta_1, \beta_2)$ is not.

In general, non-linear relationships can help achieve identification. For example, suppose that the linear relationship of X-Y_1 in (11.35) is changed to

$$Y_1 \quad = \quad \alpha_0 + \alpha_1 X + \alpha_2 X^2 + e_1. \tag{11.37}$$

Again, $(\alpha_0, \alpha_1, \alpha_2)$ is identifiable from sample A. Coupling (11.36) and (11.37) leads to

$$Y_2 = (\beta_0 + \alpha_0 \beta_2) + (\beta_1 + \alpha_1 \beta_2)X + (\alpha_2 \beta_2)X^2 + \beta_2 e_1 + e_2.$$

Thus, $\beta_0 + \alpha_0 \beta_2$, $\beta_1 + \alpha_1 \beta_2$ and $\alpha_2 \beta_2$ are identifiable from sample B. As long as $\alpha_2 \neq 0$, $(\beta_0, \beta_1, \beta_2)$ is then identifiable. For an identifiable model, parameter estimation can be implemented either using the EM algorithm or using GLS.

To illustrate the idea, let the joint model of (Y_1, Y_2) be parameterized as

$$f(y_1, y_2 \mid x) = f_1(y_1 \mid x; \alpha) f_2(y_2 \mid x, y_1; \beta)$$

for some α and β. Since (x_i, y_{1i}) are observed in sample A, we can obtain a consistent estimator of α by maximizing

$$l_A(\alpha) = \sum_{i \in A} d_i^{(A)} \log f_1(y_{1i} \mid x_i; \alpha)$$

with respect to α, where $d_i^{(A)}$ is the sampling weight of unit i in sample A. Once $\hat{\alpha}$ is obtained, a consistent estimator of β can be obtained by finding the maximizer of

$$l_B(\beta \mid \hat{\alpha}) = \sum_{i \in B} d_i^{(B)} \log \int f_2(y_{2i} \mid x_i, y_1; \beta) f_1(y_1 \mid x_i; \hat{\alpha}) d\mu(y_1), \tag{11.38}$$

where $d_i^{(B)}$ is the sample weight of unit i in sample B. As long as the joint model is identifiable, the maximizer of $l_B(\beta \mid \hat{\alpha})$ in (11.38) exists uniquely with the probability approaching to one. The following EM algorithm can be used to find the maximizer of (11.38).

[**E-step**] Given the current parameter values, compute the conditional expectation of the pseudo log-likelihood functions:

$$Q(\beta \mid \hat{\alpha}, \beta^{(t)}) = \sum_{i \in B} d_i^{(B)} E \left\{ \log f_2(y_{2i} \mid x_i, y_{1i}; \beta) \mid x_i, y_{2i}; \hat{\alpha}, \beta^{(t)} \right\}. \tag{11.39}$$

[M-step] Update the parameter β by maximizing $Q(\beta \mid \hat{\alpha}, \beta^{(t)})$ with respect to β.

$$\beta^{(t+1)} = \arg\max_{\beta} Q(\beta \mid \hat{\alpha}, \beta^{(t)}). \tag{11.40}$$

The E-step and M-step can be iteratively computed until convergence, leading to the pseudo maximum likelihood estimator $\widehat{\beta}$. The expectation in (11.39) can be taken with respect to

$$f(y_1 \mid x, y_2; \alpha, \beta) = \frac{f_1(y_1 \mid x; \alpha) f_2(y_2 \mid x, y_1; \beta)}{\int f_1(y_1 \mid x; \alpha) f_2(y_2 \mid x, y_1; \beta) d\mu(y_1)}. \tag{11.41}$$

To generate imputed values from (11.41), one may use Markov Chain Monte Carlo methods or the parametric fractional imputation of Kim (2011).

Other assumptions can be used to achieve model identification. Kim et al. (2016) used an instrumental variable assumption for model identification and developed fractional imputation methods for statistical matching. Park et al. (2016) presented an application of the statistical matching technique using fractional imputation in the context of handling mixed-mode surveys. Park et al. (2017) applied the above matching method to combine two surveys with measurement errors.

Example 11.2. *The non-monotone missingness setup in Table 11.4 can be used to illustrate handling measurement errors in mixed mode surveys. In Sample A, we use the survey mode A to obtain Y_1. In Sample B, we use the survey mode B to obtain Y_2. We may use the mode A as the gold standard and specify a measurement error model for Y_2 as*

$$Y_2 = \beta_0 + \beta_1 Y_1 + u$$

where $u \sim N(0, \sigma_u^2)$. In this case, we use

$$f(y_1, y_2 \mid \mathbf{x}) = f_1(y_1 \mid \mathbf{x}; \theta_1) f_2(y_2 \mid y_1; \theta_2),$$

where $\theta_2 = (\beta_0, \beta_1, \sigma_u^2)$, as the joint parametric distribution for applying the above matching method. Parameter estimation for α can be implemented using the observations in sample A only. Parameter estimation for θ_2 is obtained by the EM algorithm in (11.39)–(11.40).

11.6 Mass Imputation Using Multilevel Models

We now discuss mass imputation for data integration using multilevel models. Multilevel models, such as generalized linear mixed models (McCulloch et al., 2008), are widely used in the analysis of clustered data. When the sample contains area information such as counties, we can build a two-level model that borrows strength across other areas.

We now consider the problem of combining two samples with the monotone missingness structure in Table 11.4. Sample A and sample B are probability samples which were selected from the same finite population, and the area (or cluster) information is available from both samples. Let x_{ij} and y_{ij} be the realized value of the auxiliary variable and the study variable, respectively, for element j in cluster i. We observe x_{ij} for both samples but observe y_{ij} only from sample A. The study variable y_{ij} is completely unobserved in sample B, and our goal is to predict y_{ij} for sample B. This is essentially a small area estimation problem using unit level models (Rao and Molina, 2015).

For illustration, suppose that the following two-level models are specified for the clustered population:

$$y_{ij} \mid (x_{ij}, a_i) \sim f_1(y_{ij} \mid a_i, x_{ij}; \theta_1) \tag{11.42}$$

and

$$a_i \mid x_{ij} \sim f_2(a_i \mid x_{1i}; \theta_2). \tag{11.43}$$

Model (11.42) is the level one model, and model (11.43) is the level two model. The a_i represents the cluster-specific effects and x_{1i} is the cluster-level covariates. By allowing a_i random as in (11.43), we can borrow strength across clusters and obtain improved prediction.

Parameter estimation for two-level models under informative sampling has been discussed in Kim et al. (2017). To basic idea is to find $\hat{a}_i(\theta_1)$, the profile maximum likelihood estimator of a_i, by

$$\hat{a}_i(\theta_1) = \arg\max_{a_i} \sum_{(ij) \in A} d_{ij} \log f_1(y_{ij} \mid a_i, x_{ij}; \theta_1)$$

for given θ_1, where d_{ij} is the sampling weight of unit (ij) in the sample. We may assume that

$$\hat{a}_i(\theta_1) \mid (a_i, \theta_1) \sim N \left[a_i, \hat{V}(\hat{a}_i \mid a_i, \theta_1) \right], \tag{11.44}$$

where $\hat{V}(\hat{a}_i \mid a_i, \theta_1)$ is a design-unbiased variance estimator of $\hat{a}_i$ treating a_i as fixed. The normality approximation in (11.44) may be justified when n_i is large.

Thus, writing $\hat{V}_i(a_i, \theta_1) = \hat{V}(\hat{a}_i \mid a_i, \theta_1)$ and combining (11.43) with (11.44), we can define an approximated predictive distribution of v_i as

$$p(a_i \mid \theta; D) = \frac{\phi(\hat{a}_i; a_i, \hat{V}_i(a_i, \theta_1)) f_2(a_i \mid x_{1i}; \theta_2)}{\int \phi(\hat{a}_i; a_i, \hat{V}_i(a_i, \theta_1)) f_2(a_i \mid x_{1i}; \theta_2) da_i}, \tag{11.45}$$

where $\phi(x; \mu, \sigma^2)$ is the density function of $N(\mu, \sigma^2)$ and D is a collection of observations. Even if we use the normality in (11.44), the predictive distribution in (11.45) is not normal.

Instead of using (11.45), Kim et al. (2017) used a further approximation by using

$$p(a_i \mid \theta; D) = \frac{\phi(\hat{a}_i; a_i, \hat{V}_i(\hat{a}_i, \theta_1)) f_2(a_i \mid x_{1i}; \theta_2)}{\int \phi(\hat{a}_i; a_i, \hat{V}_i(\hat{a}_i, \theta_1)) f_2(a_i \mid x_{1i}; \theta_2) da_i}, \qquad (11.46)$$

which replaces $\hat{V}_i(a_i, \theta_1)$ with $\hat{V}_i(\hat{a}_i, \theta_1)$. In (11.46), the predictive distribution is normal, and the computation for E-step is much easier.

The approximate prediction model in (11.45) or (11.46) can be used to develop the E-step of the EM algorithm for parameter estimation. That is, we first compute

$$\hat{Q}_1(\theta_1 \mid \theta^{(t)}) = \sum_{(ij) \in A} d_{ij} E \left\{ \log f_1(y_{ij} \mid x_{ij}, a_i; \theta_1) \mid \hat{a}_i, \theta^{(t)} \right\}$$

and

$$\hat{Q}_2(\theta_2 \mid \theta^{(t)}) = \sum_{i \in A_I} d_i E \left\{ \log f_2(a_i; \theta_2) \mid \hat{a}_i, \theta^{(t)} \right\},$$

where A_I is the index set of clusters in sample A and d_i is the sampling weight associated with cluster i. In two-stage cluster sampling, d_i are known. The cluster weight reflects the representativeness of the sample clusters to the population of clusters.

Remark 11.1. *Another method is to use the profile score function directly. That is, instead of using (11.44), we can use*

$$\hat{S}_i(a_i; \theta_1) \mid (a_i, \theta_1) \sim N \left[0, \hat{V}(\hat{S}_i \mid a_i, \theta_1) \right], \qquad (11.47)$$

where

$$\hat{S}_i(a_i; \theta_1) \equiv \sum_{j \in A_i} d_{ij} S_a(a_i \mid x_{ij}, y_{ij}, \theta_1) = 0,$$

$$S_a(a_i \mid x_{ij}, y_{ij}, \theta_1) = \frac{\partial}{\partial a_i} \log f_1(y_{ij} \mid x_{ij}, a_i; \theta_1),$$

and $\hat{V}(\hat{S}_i \mid a_i, \theta_1)$ is a design-based variance estimator of $\hat{S}_i(a_i \mid \theta_1)$ treating a_i as fixed. Thus, the approximate predictive distribution is

$$p(a_i \mid \theta; D) = \frac{\phi\{\hat{S}_i(a_i; \theta_1); 0, \hat{V}_i(\hat{S}_i \mid a_i, \theta_1)\} f_2(a_i \mid x_{1i}; \theta_2)}{\int \phi\{\hat{S}_i(a_i \mid \theta_1); 0, \hat{V}(\hat{S}_i \mid a_i, \theta_1)\} f_2(a_i \mid x_{1i}; \theta_2) da_i}. \qquad (11.48)$$

Because the computation for $\hat{V}(\hat{S}_i \mid a_i, \theta_1)$ does not involve linearization, the approximate normality in (11.47) is likely to be more accurate than the approximate normality in (11.44). In this case, the predictive distribution using (11.48) may lead to more efficient estimation than the one using (11.46).

Once the parameters are estimated, the prediction of y_{ij} can be made in the following two-steps:

[Step 1] Generate M values of a_i, denoted by $a_i^{*(1)}, \ldots, a_i^{*(M)}$ from the predictive distribution in (11.45) or (11.48) evaluated at $\hat{\theta}$.

[Step 2] Generate M values of y_{ij}, denoted by $y_{ij}^{*(1)}, \ldots, y_{ij}^{*(M)}$, by

$$y_{ij}^{*(k)} \sim f_1(y_{ij} \mid x_{ij}, a_i^{*(k)}; \hat{\theta}_1),$$

for $k = 1, \ldots, M$. The final predictor of y_{ij} is

$$\hat{y}_{ij} = \frac{1}{M} \sum_{k=1}^{M} y_{ij}^{*(k)}. \tag{11.49}$$

The predictors in (11.49) can be used for mass imputation in sample B. The uncertainty associated with mass imputation estimator can be evaluated using the mean squared estimation formula based on Taylor linearization. See Berg and Kim (2020) for more details.

11.7 Data Integration for Regression Analysis

We now consider regression analysis combining partial information from external sources. To explain the idea, suppose that there are three data sources (A, B, C) with the data structure in Table 11.5, Sample A contains all the observations while samples B and C contain partial observations.

TABLE 11.5
Data Structure for Survey Integration

Sample	Sampling Weight	z	x_1	x_2	y
A	w_a	✓	✓	✓	✓
B	w_b	✓	✓		✓
C	w_c	✓		✓	✓

Suppose that we are interested in estimating the parameter in the regression model

$$E(Y \mid x_1, x_2) = m(\beta_0 + \beta_1 x_1 + \beta_2 x_2),$$

where $m(\cdot)$ is known but $\beta = (\beta_0, \beta_1, \beta_2)$ is unknown. The estimating equation for β using sample A can be written as

$$\hat{U}_a(\beta) \equiv \sum_{i \in A} w_{a,i}\{y_i - m(x_{1i}, x_{2i}; \beta)\}h(x_{1i}, x_{2i}; \beta) = 0, \tag{11.50}$$

for some $h(x_{1i},x_{2i};\beta)$ such that $\hat{U}_a(\beta)$ is linearly independent almost everywhere.

Now, we wish to incorporate the partial information from sample B. To do this, suppose that we have a "working" model for $E(Y \mid x_1,z)$:

$$E(Y \mid x_1,z) = m_1(x_1,z;\alpha) \tag{11.51}$$

for some α. Note that since (z_i,x_{1i},y_i) are observed, we can use sample B to estimate α by solving

$$\sum_{i \in B} w_{b,i} U_1(\alpha;x_{1i},z_i,y_i) = 0,$$

for some U_1 satisfying $E\{U_1(\alpha;x_1,z,Y) \mid x_1,z\} = 0$ under the working model (11.51).

Similarly, to incorporate the partial information from sample C, suppose that we have a "working" model for $E(Y \mid x_2,z)$:

$$E(Y \mid x_2,z) = m_2(x_2,z;\gamma) \tag{11.52}$$

for some γ. We can also construct an unbiased estimating equation

$$\sum_{i \in C} w_{c,i} U_2(\gamma;x_{2i},z_i,y_i) = 0$$

for some U_2 satisfying $E\{U_2(\gamma;x_2,z,Y) \mid x_2,z\} = 0$ under the working model (11.52). Once $\hat{\alpha}$ and $\hat{\gamma}$ are obtained, we can use this extra information to improve the efficiency of $\hat{\beta}$ in (11.50).

One approach is through empirical likelihood of Qin and Lawless (1994). We can find the maximizer of $Q(w_i,w_{ia}) = \sum_{i \in A} w_{ia} \log(w_i/w_{ia})$ subject to $\sum_{i \in A} w_i = N$ and

$$\sum_{i \in A} w_i [U_1(\hat{\alpha};x_{1i},z_i,y_i), U_2(\hat{\gamma};x_{2i},z_i,y_i)] = (0,0). \tag{11.53}$$

Constraint (11.53) incorporates the extra information. Once the solution $\hat{w}_i$ is obtained, we can use

$$\sum_{i \in A} \hat{w}_i \{y_i - m(x_{1i},x_{2i};\beta)\} h(x_{1i},x_{2i};\beta) = 0 \tag{11.54}$$

to estimate β. The idea is similar in spirit to model calibration idea discussed in Example 10.2.

The choice of the "working" conditional expectation model determines the efficiency of the resulting estimator. The following example illustrates the idea of the data integration method.

Example 11.3. *Suppose that we are interested in estimating the regression coefficients in the following linear regression model*

$$y_i = \beta_0 + \beta_1 x_{1i} + \beta_2 x_{2i} + e_i,$$

where $e_i \sim (0, \sigma^2)$. To incorporate the partial information, we can use sample A to obtain $\hat{x}_{1i} = \hat{E}(x_{1i} \mid z_i, x_{2i})$ and $\hat{x}_{2i} = \hat{E}(x_{2i} \mid z_i, x_{1i})$, where $\hat{E}(\cdot \mid \cdot)$ denotes the estimated conditional expectation. The conditional expectation requires a model specification of the conditional distribution, but the model does not have to be correctly specified.

Once these $\hat{x}_{1i}$ and $\hat{x}_{2i}$ are constructed, then we can construct the following working regression model

$$y_i = \alpha_0 + \alpha_1 x_{1i} + \alpha_2 \hat{x}_{2i} + e_{1i}, \tag{11.55}$$

where $e_{1i} \sim (0, \sigma_1^2)$, using sample B to obtain $\hat{\alpha}$. Thus, we use

$$U_1(\alpha; x_{1i}, \hat{x}_{2i}, y_i) = \{y_i - \alpha_0 - \alpha_1 x_{1i} - \alpha_2 \hat{x}_{2i}\}(1, x_{1i}, \hat{x}_{2i})'$$

as the estimating function under the working model in (11.55).

Similarly, we construct the following working regression model

$$y_i = \gamma_0 + \gamma_1 \hat{x}_{1i} + \gamma_2 x_{2i} + e_{2i}, \tag{11.56}$$

where $e_{2i} \sim (0, \sigma_2^2)$, using sample C to obtain $\hat{\gamma}$. Thus, we use

$$U_2(\gamma; \hat{x}_{1i}, x_{2i}, y_i) = \{y_i - \gamma_0 - \gamma_1 \hat{x}_{1i} - \gamma_2 x_{2i}\}(1, \hat{x}_{1i}, x_{2i})'$$

as the estimating function under the working model in (11.56).

Once $\hat{\alpha}$ and $\hat{\gamma}$ are constructed, we can use the same empirical likelihood method with

$$\sum_{i \in A} w_i \left[U_1(\hat{\alpha}; x_{1i}, \hat{x}_{2i}, y_i), U_2(\hat{\gamma}; \hat{x}_{1i}, x_{2i}, y_i) \right] = (0, 0),$$

instead of using (11.53), to combine all available information. Because the two working models, (11.55) and (11.56) resemble the original regression model that we are interested in, the parameter estimates for these working models are quite relevant to the parameters in the original regression model.

11.8   Record Linkage

To combine information from different sources, one assume that it is possible to identify the records with the same entity, which is not always the case in practice. If the data do not contain unique identification number, identifying records from the same entity becomes a challenging problem. *Record linkage* is the term describing the process of joining records that are believed to be related to the same entity. Suppose that we have two data files A and B that are believed to have some common entities. Our goal is to find the true matches among all possible pairs of the two data files. Let the bipartite *comparison space* $\Omega = A \times B = M \cup U$ consist of *matches* M and *non-matches* U between the records in files A and B. For any pair of records $(a,b) \in \Omega$, let γ_{ab} be the *comparison vector* between a set of *key* variables associated with $a \in A$ and $b \in B$, respectively, such as name, sex, birthdate. The key variables and the comparison vector γ_{ab} are fully observed over Ω. In cases where the key variables may be affected by measurement errors, a match (a,b) may not have complete agreement in terms of γ_{ab}, and a non-match (a,b) can nevertheless agree on some (even all) of the key variables.

In the classical approach of Fellegi and Sunter (1969), one recognises the probabilistic nature of γ_{ab} due to the perturbations that cause key-variable errors. The related methods are referred to as *probabilistic record linkage*. To explain the probabilistic record linkage method of Fellegi and Sunter (1969), let $m_{ab} = f(\gamma_{ab} \mid (a,b) \in M)$ be the probability mass function of the discrete values γ_{ab} can take given $(a,b) \in M$. Similarly, we can define $u_{ab} = f(\gamma_{ab} \mid (a,b) \in U)$. The ratio

$$r_{ab} = \frac{m_{ab}}{u_{ab}}$$

is then the basis of the likelihood ratio test (LRT) for $H_0 : (a,b) \in M$ vs. $H_1 : (a,b) \in U$. Let $M^* = \{(a,b) : r_{ab} > c_M\}$ be the pairs classified as matches and $U^* = \{(a,b) : r_{ab} < c_U\}$ the non-matches, the remaining pairs are classified by clerical review, where (c_M, c_U) are related to the probabilities of false links (of pairs in U) and false non-links (of pairs in M), respectively, defined as

$$\mu = \sum_{\gamma} u(\gamma)\delta(M^*;\gamma) \quad \text{and} \quad \lambda = \sum_{\gamma} m(\gamma)\delta(U^*;\gamma), \quad (11.57)$$

where $\delta(M^*;\gamma) = 1$ if $\gamma_{ab} = \gamma$ means $(a,b) \in M^*$ and 0 otherwise, similarly for $\delta(U^*;\gamma)$.

In reality m_{ab} and u_{ab} are unknown. Nor is the *prevalence* $\pi = |M|/|\Omega| := n_M/n$. Let $\boldsymbol{\eta}$ contain π and the unknown parameters of $m(\gamma)$ and $u(\gamma)$. Let $g_{ab} = 1$ if $(a,b) \in M$ and 0 if $(a,b) \in U$. Given the complete data $\{(g_{ab}, \gamma_{ab}) : (a,b) \in \Omega\}$, Winkler (1988) and Jaro (1989) assume the log-likelihood to be

$$h(\boldsymbol{\eta}) = \sum_{(a,b)\in\Omega} g_{ab} \log(\pi m_{ab}) + \sum_{(a,b)\in\Omega} (1 - g_{ab}) \log((1-\pi)u_{ab}). \quad (11.58)$$

An EM-algorithm follows by treating $g_{\Omega} = \{g_{ab} : (a,b) \in \Omega\}$ as the missing data. In reality the comparison vectors of any two pairs are not independent, as long as they share a record. Thus, $h(\boldsymbol{\eta})$ given by (11.58) cannot be the complete-data log-likelihood.

Nevertheless, the ranking of the record pairs in Ω by the estimated $\hat{r}_{ab}$ may not be heavily affected by this problem, and the algorithm of Winkler (1988) and Jaro (1989) is still widely used in practice. Steorts et al. (2016) developed a Bayesian approach to record linkage using a graphical representation of the link patterns. Also, Lee et al. (2021) applied the maximum entropy method to record linkage problems.

Statistical analysis with linked data under linkage error is also important area of research. See Di Consiglio and Tuoto (2018) and Han and Lihiri (2019) for comprehensive overviews of the existing methods on this topic. Salvati et al. (2021) developed a framework for small area estimation with linked data.

12

Advanced Topics

In this chapter, we introduce some advanced topics for handling missing data. Basically, modern nonparametric function estimation techniques can be used to develop imputation and propensity score weighting. Specifically, the reproducing kernel Hilbert space theory can be used for function estimation. We are particularly interested in making statistical inference after the nonparametric imputation with modern function estimation techniques.

12.1 Smoothing Spline Imputation

Smoothing spline is a classical method of function estimation. Without missing data, the smoothing spline estimator for $m(x) = E(Y \mid x)$ for scalar $x \in [a,b]$ with $a,b < \infty$ arises as a solution to the minimization problem

$$\min_{m \in W^2[a,b]} \left[\sum_{i=1}^{n} \{y_i - m(x_i)\}^2 + \lambda \int_a^b \{m^{(2)}(x)\}^2 dx \right] \qquad (12.1)$$

where $\lambda > 0$ is a fixed smoothing parameter and $W^q[a,b]$ denotes the Sobolev space of order q. That is,

$$W^q[a,b] = \{f : f \text{ has } q-1 \text{ absolutely continuous derivatives,}$$

$$\int_a^b \{f^{(q)}(x)\}^2 dx < \infty\}.$$

According to Schoenberg (1964), the optimization problem in (12.1) has an explicit, unique minimizer which satisfies the following conditions:

1. $\hat{m}(x)$ has continuous $(2q-2)$ derivatives on $(-\infty, +\infty)$.
2. $\hat{m}$ is $(2q-1)$-th polynomial in each of the intervals (x_i, x_{i+1})
3. $\hat{m}$ is $(q-1)$-th polynomial in $(-\infty, x_1)$ and also in (x_n, ∞).

For $q = 2$, the solution $\hat{m}(x)$ is called a natural cubic splines with knots at the unique values of x_i, $i = 1,\ldots,n$. The first term of (12.1) measures closeness to the data and the second term penalizes curvature in the function. Large

---

DOI: 10.1201/9780429321740-12

values of λ produce smoother curves while smaller values produce more wiggly curves.

If the solution is a natural cubic spline, we can write $m \in W^2[a,b]$ as

$$m(x) = \sum_{i=1}^{n} \alpha_i B_i(x) \tag{12.2}$$

where $\{B_1(x), \ldots, B_n(x)\}$ is the B-spline basis functions for the natural cubic splines. See Section 5.2.1 of Hastie et al. (2009) for details of the B-spline basis functions. Reinsch (1967) proposed a regularized regression among the class of the natural spline function in (12.2), called the smoothing spline regression. The solution to the smoothing spline regression is obtained by minimizing

$$Q_\lambda(\alpha) = \left[\sum_{i=1}^{n} \left\{ y_i - \sum_{j=1}^{n} \alpha_j B_j(x_i) \right\}^2 + \lambda \sum_{i=1}^{n} \sum_{j=1}^{n} \alpha_i \alpha_j \Omega_{ij} \right] \tag{12.3}$$

where

$$\Omega_{ij} = \int_a^b B_i^{(2)}(x) B_j^{(2)}(x) dx$$

is the (i,j)-th element in the penalty matrix Ω.

Writing $\hat{\mathbf{m}} = (\hat{m}(x_1), \ldots, \hat{m}(x_n))'$ with $\hat{m}(x) = \sum_{i=1}^{n} \hat{\alpha}_i B_i(x)$, we can express

$$\begin{aligned} \hat{\mathbf{m}} &= B \left(B'B + \lambda\Omega \right)^{-1} B'\mathbf{y}_n \\ &= (I + \lambda D)^{-1} \mathbf{y}_n, \end{aligned} \tag{12.4}$$

where $\mathbf{y}_n = (y_1, \ldots, y_n)'$, B is $n \times n$ matrix of $B_{ik} = B_k(x_i)$, and $D = (B')^{-1}\Omega B^{-1}$. By (12.4), we can express

$$\hat{\mathbf{m}} = S_\lambda \mathbf{y}_n,$$

where $S_\lambda = (I + \lambda D)^{-1}$ is known as the smoother matrix. Using the eigen-decomposition of D, we can write $D = \sum_{k=1}^{n} d_k \mathbf{u}_k \mathbf{u}_k'$ which leads to

$$S_\lambda = \sum_{k=1}^{n} \rho_k(\lambda) \mathbf{u}_k \mathbf{u}_k',$$

where $\rho_k(\lambda) = (1 + \lambda d_k)^{-1}$. The effective degrees of freedom of the linear smoother S_λ is defined as

$$\mathrm{df}_\lambda = \mathrm{trace}(S_\lambda) = \sum_{k=1}^{n} \rho_k(\lambda),$$

which is roughly the dimension of the number of effective basis functions in the projection space. Note that a larger value of λ gives a smaller value of $\rho(\lambda)$, which means higher level of smoothing.

Now, in the missing data setup, we can use

$$\min_{m \in W^2[a,b]} \left[\sum_{i=1}^{n} \delta_i \{y_i - m(x_i)\}^2 + \lambda \int_a^b \{m^{(2)}(x)\}^2 dx \right] \qquad (12.5)$$

to obtain $\hat{y}_i = \hat{m}(x_i)$ as a smoothing spline imputation for y_i. The resulting imputation estimator of $\theta = E(Y)$ is

$$\hat{\theta}_I = \frac{1}{n} \sum_{i=1}^{n} \{\delta_i y_i + (1 - \delta_i)\hat{m}(x_i)\}.$$

Smoothing spline is a theoretically elegant and it is a special case of the Kernel Ridge Regression (KRR) in Sobolev space (Wahba, 1990). Because smoothing spline is a special case of KRR, the asymptotic properties of the smoothing spline imputation can be obtained from the asymptotic properties of the KRR imputation, which will be covered in Section 12.2.

12.2   Kernel Ridge Regression Imputation

Kernel ridge regression is a modern regression technique using the reproducing kernel Hilbert space (RKHS) theory. Given the input space $\mathcal{X}$, a kernel function K is a bilinear function

$$K : \mathcal{X} \times \mathcal{X} \mapsto \mathbb{R}$$

satisfying $K(x,x') = K(x',x)$. Kernel K is positive semidefinite if and only if every Gram matrix $K_{ij} = K(x_i, x_j)$ is positive semidefinite. That is,

$$\sum_i \sum_j a_i a_j K_{ij} \geq 0$$

for all $a_i \in \mathbb{R}$.

According to Aronszajn (1950), K is a positive definite kernel on the set $\mathcal{X}$ if and only if there exists a Hilbert function space $\mathcal{H}$ and a mapping

$$\Phi(x) = K(\cdot, x) \in \mathcal{H}$$

such that

$$K(x,x') = \big\langle \Phi(x), \Phi(x') \big\rangle_{\mathcal{H}}$$

holds for any $x, x' \in \mathcal{X}$. See Scholkopf and Smola (2002) for more details about the theory for the reproducing kernel Hilbert space.

In the context of nonparametric regression, the KRR can be described as finding

$$\hat{m} = \arg\min_{m \in \mathcal{H}} \left[\sum_{i=1}^{n} \{y_i - m(\mathbf{x}_i)\}^2 + \lambda \|m\|_{\mathcal{H}}^2 \right], \tag{12.6}$$

where $\|m\|_{\mathcal{H}}^2$ is the norm of m in the Hilbert space $\mathcal{H}$ generated by a positive definite kernel function K.

By the representer theorem for RKHS (Wahba, 1990), the KRR estimator can be obtained by solving the finite-dimensional optimization problem

$$Q_\lambda(\alpha) = \left[(\mathbf{y} - K\alpha)'(\mathbf{y} - K\alpha) + \lambda\alpha' K\alpha \right].$$

The solution is

$$\hat{\mathbf{m}} = (K'K + \lambda K)^{-1} K'\mathbf{y}_n = \left(I_n + \lambda K^{-1} \right)^{-1} \mathbf{y}_n, \tag{12.7}$$

where $K = (K(\mathbf{x}_i, \mathbf{x}_j))_{ij} \in \mathbb{R}^{n \times n}$ and $\mathbf{y} = (y_1, \ldots, y_n)'$.

To discuss the asymptotic properties of the KRR estimator in (12.7), we first introduce Marcer's theorem: For any positive definite kerel function K, we can express

$$K(x, x') = \sum_{j=1}^{\infty} \mu_j \phi_j(x) \phi_j(x'), \tag{12.8}$$

where $\mu_1 \geq \mu_2 \geq \cdots \geq 0$ are eigenvalues and $\{\phi_j\}_{j=1}^{\infty}$ is an orthonomal basis for $L^2(P)$.

Now, the average mean squared error of $\hat{m}(x)$ is

$$\mathrm{AMSE}(\hat{m}) = \frac{1}{n} \sum_{i=1}^{n} \{\mathrm{Bias}(\hat{m}(x_i))\}^2 + \frac{1}{n} \sum_{i=1}^{n} \mathrm{V}\{\hat{m}(x_i)\}. \tag{12.9}$$

For linear smoothers, $\hat{\mathbf{m}} = S_\lambda \mathbf{y}$, we have

$$\frac{1}{n} \sum_{i=1}^{n} \mathrm{V}\{\hat{m}(x_i)\} = \frac{\mathrm{tr}(S_\lambda S_\lambda')}{n} \sigma^2.$$

Using $\mathrm{tr}(S_\lambda S_\lambda) \leq \mathrm{tr}(S_\lambda)$, we have

$$\frac{1}{n} \sum_{i=1}^{n} \mathrm{V}\{\hat{m}(x_i)\} \leq \frac{\mathrm{df}_\lambda}{n} \sigma^2,$$

where

$$\mathrm{df}_\lambda = \mathrm{tr}(S_\lambda) = \sum_{j=1}^{\infty} \frac{\mu_j}{\mu_j + \lambda} := \gamma(\lambda)$$

and μ_j are the eigenvalues in (12.8).

For the squared bias part in (12.9), we make the following assumptions.

[A1] For some $k \geq 2$, there is a constant $\rho < \infty$ such that $\mathrm{E}[\phi_j(X)^{2k}] \leq \rho^{2k}$ for all $j \in \mathbb{N}$, where $\{\phi_j\}_{j=1}^{\infty}$ are orthonormal basis by expansion from Mercer's theorem in (12.8).

[A2] The function $m \in \mathcal{H}$, and for $\mathbf{x} \in \mathcal{X}$, we have $\mathrm{E}[\{Y - m(\mathbf{x})\}^2] \leq \sigma^2$, for some $\sigma^2 < \infty$.

Under [A1] and [A2], Zhang et al. (2013) showed that

$$\{\mathrm{Bias}(\hat{m}(x))\}^2 \leq C \cdot \lambda \|m\|_{\mathcal{H}}^2 .$$

Therefore, we have

$$\mathrm{AMSE}(\hat{m}) = O(1) \times \left\{ \lambda \|m\|_{\mathcal{H}}^2 + \frac{\gamma(\lambda)}{n} \right\}. \tag{12.10}$$

For the ℓ-th order of Sobolev space, we have $\mu_j \leq Cj^{-2l}$ and

$$\gamma(\lambda) = \sum_{j=1}^{\infty} \frac{1}{1 + j^{2\ell}\lambda} \leq O\left(\lambda^{-1/(2\ell)} \right). \tag{12.11}$$

Note that (12.10) is minimized when

$$\lambda = \frac{\gamma(\lambda)}{n}$$

which is equivalent to $\lambda \asymp n^{-\frac{2\ell}{2\ell+1}}$ under (12.11). The optimal rate $\lambda \asymp n^{-\frac{2\ell}{2\ell+1}}$ leads to

$$\mathrm{AMSE}(\hat{m}) = O(n^{-\frac{2\ell}{2\ell+1}}) \tag{12.12}$$

which is the optimal rate in Sobolev space, discussed by Stone (1982).

In the context of imputation, we can consider the KRR imputation as the solution to the following optimization problem.

$$\hat{m} = \arg\min_{m \in \mathcal{H}} \left[\sum_{i=1}^{n} \delta_i \{y_i - m(\mathbf{x}_i)\}^2 + \lambda \|m\|_{\mathcal{H}}^2 \right], \tag{12.13}$$

where $\mathcal{H}$ is the reproducing kernel Hilbert space generated by kernel K. Using the KRR imputation in (12.13), we can obtain the imputed estimator of $\theta = \mathrm{E}(Y)$ as follows:

$$\hat{\theta}_I = \frac{1}{n} \sum_{i=1}^{n} \{\delta_i y_i + (1 - \delta_i)\hat{m}(\mathbf{x}_i)\}, \tag{12.14}$$

where $\hat{m}(\cdot)$ is defined in (12.13). Because $\hat{m}(\mathbf{x})$ is a nonparametric regression estimator of $m(\mathbf{x}) = \mathrm{E}(Y \mid \mathbf{x})$, we can expect that the KRR imputation estimator in (12.14) is consistent for $\theta = E(Y)$ under MAR, as long as $\hat{m}(\mathbf{x})$ is a consistent estimator of $m(\mathbf{x})$. Wang and Kim (2021c) established the following theorem.

Theorem 12.1. *Suppose Assumption [A1] and [A2] hold for a Sobolev kernel of order ℓ. As long as λ satisfies*

$$n\lambda \to 0 \quad and \quad n\lambda^{1/2\ell} \to \infty, \tag{12.15}$$

we have, under MAR,

$$\sqrt{n}\left(\hat{\theta}_I - \theta\right) \xrightarrow{\mathcal{L}} N(0, \sigma^2),$$

where $\sigma^2 = V(\eta)$ with

$$\eta = m(\mathbf{x}) + \delta \frac{1}{\pi(\mathbf{x})}\{y - m(\mathbf{x})\} \tag{12.16}$$

and $\pi(\mathbf{x}) = E(\delta \mid \mathbf{x})$.

The proof of Theorem 12.1 can be found in Wang and Kim (2021c). Note that the optimal rate $\lambda \asymp n^{-\frac{2\ell}{2\ell+1}}$ does not satisfy (12.15). To control the bias part, we need smaller λ such as $\lambda = n^{-\kappa}$ with $\kappa > 1$. Similar conditions are used for bandwidth selection for nonparametric Kernel regression in Theorem 6.2:

$$nh \to \infty \text{ and } nh^4 \to 0,$$

which is derived under $\dim(\mathbf{x}) = 1$.

To discuss variance estimation of $\hat{\theta}_I$, we need to find a nonparametric estimator of $\pi(\mathbf{x}_i)$ to get

$$\hat{\eta}_i = \hat{m}(\mathbf{x}_i) + \frac{\delta_i}{\hat{\pi}(\mathbf{x}_i)}\{y_i - \hat{m}(\mathbf{x}_i)\}$$

as the linearized estimator of η_i in (12.16). Once $\hat{\eta}_i$ are obtained, standard variance estimator can be applied to $\hat{\eta}_i$. A nonparametric estimation of $\pi(\mathbf{x})$ is discussed in the next section.

12.3   KRR-Based Propensity Score Estimation

In this section, we discuss how to develop the propensity score estimation based on RKHS theory in Section 12.2. Recall that the density ratio function $r(\mathbf{x}) = f(\mathbf{x} \mid \delta = 0)/f(\mathbf{x} \mid \delta = 1)$ can be understood as the minimizer of $Q(r)$ in (7.62). Thus, the maximum entropy estimator of $r(\mathbf{x})$ can be obtained by maximizing

$$\hat{Q}(r) = \frac{1}{n_1}\sum_{i=1}^{n}\delta_i r(\mathbf{x}_i)\left[\log\{r(\mathbf{x}_i)\} - 1\right] \tag{12.17}$$

subject to calibration constraint. We may use

$$\frac{1}{n}\sum_{i=1}^{n}\delta_i\left\{1+\frac{n_0}{n_1}r(\mathbf{x}_i)\right\}[h_1(\mathbf{x}_i),\dots,h_L(\mathbf{x}_i)] = \frac{1}{n}\sum_{i=1}^{n}[h_1(\mathbf{x}_i),\dots,h_L(\mathbf{x}_i)]$$

(12.18)

for the calibration constraint. The resulting PS estimator of $\theta = E(Y)$ is in the form of

$$\hat{\theta}_{PS} = \frac{1}{n}\sum_{i=1}^{n}\delta_i\left\{1+\frac{n_0}{n_1}\hat{r}(\mathbf{x}_i)\right\}y_i.$$

(12.19)

The following theorem presents the asymptotic properties of the PS estimator constructed from the above optimization problem.

Lemma 12.1. *Assume that the response mechanism is MAR. If*

$$m(\mathbf{x}) = \mathbb{E}(Y \mid \mathbf{x}) \in \mathcal{H} = span\{h_1(\mathbf{x}),\dots,h_L(\mathbf{x})\}$$

(12.20)

and $\hat{r}(\mathbf{x})$ satisfies (12.18), then $\hat{\theta}_{PS}$ is unbiased for θ.

Proof. Writing $\hat{\theta}_n = n^{-1}\sum_{i=1}^{n}y_i$, we have

$$\begin{aligned}
E\left(\hat{\theta}_{PS}-\hat{\theta}_n \mid x_1,\dots,x_n\right) &= \frac{1}{n}\sum_{i=1}^{n}\delta_i\left\{1+\frac{n_0}{n_1}\hat{r}(\mathbf{x}_i)\right\}m(\mathbf{x}_i)-\frac{1}{n}\sum_{i=1}^{n}m(\mathbf{x}_i)\\
&= 0,
\end{aligned}$$

where the second equality follows from (12.18) and (12.20). Thus, because $\hat{\theta}_n$ is unbiased for θ, $\hat{\theta}_{PS}$ is also unbiased. □

We solve the optimization problem using $\hat{Q}(r)$ as the objective function with constraint (12.18). Since the primal problem is a convex optimization problem with linear constraints, we can obtain the dual problem with an unconstrained concave maximization. That is, using Lagrange multiplier method, the solution belongs to the class

$$\log\{r(\mathbf{x})\} = \phi_0 + \sum_{k=1}^{L}\phi_k h_k(\mathbf{x})$$

and the optimization problem can be expressed as solving

$$\min_{\phi_1\in\mathbb{R}^L}\left\{\frac{1}{n_1}\sum_{i=1}^{n}\delta_i r(\mathbf{x}_i;\phi)-\frac{1}{n_0}\sum_{i=1}^{n}(1-\delta_i)\log\{r(\mathbf{x}_i;\phi)\}\right\}$$

(12.21)

and ϕ_0 is a normalizing constant satisfying

$$n_1 = \sum_{i=1}^{n}\delta_i\exp\{\phi_0 + \sum_{k=1}^{L}\hat{\phi}_k h_k(x_i)\}.$$

More generally, if

$$f(y \mid \mathbf{x}) = f(y \mid h_1(\mathbf{x}), \ldots, h_L(\mathbf{x})) \tag{12.22}$$

then the MAR reduces to

$$Y \perp \delta \mid \mathcal{H}, \tag{12.23}$$

where $\mathcal{H} = \text{span}\{h_1(\mathbf{x}), \ldots, h_L(\mathbf{x})\}$. Under (12.23), the PS estimator using the basis functions of $\mathcal{H}$ is unbiased. The basis function $\{h_1(\mathbf{x}), \ldots, h_L(\mathbf{x})\}$ satisfying (12.23) is called the balancing score (Rosenbaum and Rubin, 1983). By constructing the propensity score function using balanced score, efficient propensity score estimation can be achieved. See Wang and Kim (2021a) for more details.

We now extend the idea to a more general function space $\mathcal{H}$. Instead of a fixed-dimensional function space, we consider a reproducing kernel Hilbert space generated by a kernel K. Now, as we have $m \in \mathcal{H}$, and by the representer theorem in KRR, we know that $m \in \text{span}\{K(\cdot, \mathbf{x}_1), \ldots, K(\cdot, \mathbf{x}_n)\}$. Thus, the calibration constraint is

$$\frac{1}{n_1} \sum_{i=1}^{n} \delta_i r(\mathbf{x}_i)(K(\cdot, \mathbf{x}_1), \ldots, K(\cdot, \mathbf{x}_n)) \cong \frac{1}{n_0} \sum_{i=1}^{n} (1 - \delta_i)(K(\cdot, \mathbf{x}_1), \ldots, K(\cdot, \mathbf{x}_n)). \tag{12.24}$$

Further, we want to incorporate with the normalization constraint $\sum_{i=1}^{n} \delta_i \omega(\mathbf{x}_i) = n$, i.e.,

$$\frac{1}{n_1} \sum_{i=1}^{n} \delta_i r(\mathbf{x}_i) = \frac{1}{n_0} \sum_{i=1}^{n} (1 - \delta_i). \tag{12.25}$$

Minimizing $\hat{Q}(g)$ subject to (12.24) and (12.25) is the maximum entropy method considered in Nguyen et al. (2010). Using Lagrangian multiplier method, the solution to this optimization problem can be written as

$$\log\{r(\mathbf{x})\} \equiv \log\{r(\mathbf{x}; \phi)\} = \phi_0 + \sum_{i=1}^{n} \phi_i K(\mathbf{x}, \mathbf{x}_i) \tag{12.26}$$

for some $\phi = (\phi_0, \ldots, \phi_n)' \in \mathbb{R}^{n+1}$.

Hence, using the representer theorem again, the maximum entropy estimator of $r(\mathbf{x})$ can be obtained as

$$\min_{\phi \in \mathbb{R}^n} \left\{ \frac{1}{n_1} \sum_{i=1}^{n} \delta_i r(\mathbf{x}_i; \phi) - \frac{1}{n_0} \sum_{i=1}^{n} (1 - \delta_i) \log\{r(\mathbf{x}_i; \phi)\} + \tau \phi' \mathbf{K} \phi \right\} \tag{12.27}$$

and ϕ_0 is a normalizing constant satisfying

$$n_1 = \sum_{i=1}^{n} \delta_i \exp\left\{ \phi_0 + \sum_{j=1}^{n} \hat{\phi}_j K(\mathbf{x}_i, \mathbf{x}_j) \right\}. \tag{12.28}$$

The tuning parameter τ is chosen to minimize

$$D(\tau) = \left\| \frac{1}{n} \sum_{i=1}^{n} \delta_i \left\{ 1 + \frac{n_0}{n_1} \cdot \hat{r}_\tau(x_i) \right\} \hat{m}(x_i) - \frac{1}{n} \sum_{i=1}^{n} \hat{m}(x_i) \right\|,$$

where $\hat{m}(x)$ is determined by KRR estimation. Thus, we can use the following two-step procedure to determine the tuning parameter τ.

1. Use KRR estimation to obtain $\hat{m}(\mathbf{x})$.

2. Given $\hat{m}(\mathbf{x})$, find $\hat{\tau}$ that minimizes $D(\tau)$.

As the objective function in (12.27) is convex, we apply the limited-memory Broyden-Fletcher-Goldfarb-Shanno (L-BFGS) algorithm to solve the optimization problem with the following first order partial derivatives:

$$\frac{\partial U}{\partial \phi_0} = \frac{1}{n_1} \sum_{i=1}^{n} \delta_i \exp\left(\phi_0 + \sum_{j=1}^{n} \phi_j K(\mathbf{x}_i, \mathbf{x}_j) \right) - 1,$$

$$\frac{\partial U}{\partial \phi_k} = \frac{1}{n_1} \sum_{i=1}^{n} \delta_i K(\mathbf{x}_i, \mathbf{x}_k) \exp\left(\phi_0 + \sum_{j=1}^{n} \phi_j K(\mathbf{x}_i, \mathbf{x}_j) \right) - \frac{1}{n_0} \sum_{i=1}^{n} (1 - \delta_i) K(\mathbf{x}_i, \mathbf{x}_k)$$

$$+ 2\tau \sum_{i=1}^{n} K(\mathbf{x}_i, \mathbf{x}_k) \phi_i, k = 1, \ldots, n,$$

where U is the objective function in (12.27).

Now, to discuss the asymptotic properties of the PS estimator (12.19) using the maximum entropy method in (12.27). We now establish the following theorem. The proof can be found in Wang and Kim (2021c).

Theorem 12.2. *Under some regularity conditions, we can establish that*

$$\sqrt{n}\left(\hat{\theta}_{PS} - \theta \right) \xrightarrow{\mathcal{L}} N(0, \sigma^2), \tag{12.29}$$

where $\sigma^2 = V(\eta)$ with η in (12.16).

By Theorem 12.2, the KRR-based PS estimator is asymptotically equivalent to the KRR imputation estimator that we discussed in Section 12.2. Thus, for variance estimation, we can use the asymptotic equivalence to obtain consistent variance estimator of $\hat{\theta}_{PS}$. That is, we use

$$\hat{V}(\hat{\theta}_{PS}) = \frac{1}{n(n-1)} \sum_{i=1}^{n} (\hat{\eta}_i - \bar{\eta}_n)^2,$$

where

$$\hat{\eta}_i = \hat{m}(\mathbf{x}_i) + \delta_i \hat{\omega}(\mathbf{x}_i) \{ y_i - \hat{m}(\mathbf{x}_i) \}$$

$\hat{\omega}(\mathbf{x}) = 1 + (n_0/n_1)\hat{r}(\mathbf{x})$ and $\bar{\eta}_n = n^{-1} \sum_{i=1}^{n} \hat{\eta}_i$.

12.4 Soft Calibration

We now consider the problem of propensity weighting with high dimensional covariates. When the dimension of the covariates is high, imposing the calibration constraint for all covariates can lead to inefficient estimation. To handle this problem, we can either use variable selection technique or use a shrinkage method. We will discuss shrinkage method for calibration in this section.

Let $\mathbf{x} = (\mathbf{x}_1, \mathbf{x}_2)$ and $\dim(\mathbf{x}_2) = q$. For calibration, we may wish to achieve

$$\frac{1}{n}\sum_{i=1}^{n}\delta_i\left\{1 + \frac{n_0}{n_1}r(\mathbf{x}_i)\right\}\mathbf{x}_{i1} = \frac{1}{n}\sum_{i=1}^{n}\mathbf{x}_{i1} \tag{12.30}$$

exactly but

$$\frac{1}{n}\sum_{i=1}^{n}\delta_i\left\{1 + \frac{n_0}{n_1}r(\mathbf{x}_i)\right\}\mathbf{x}_{i2} \cong \frac{1}{n}\sum_{i=1}^{n}\mathbf{x}_{i2} \tag{12.31}$$

approximately. If we are certain that $\mathbf{x}_1$ are important covariates for $\mathbb{E}(Y \mid \mathbf{x})$ but less certain about $\mathbf{x}_2$, then we can impose *hard calibration* for $\mathbf{x}_1$ and *soft calibration* for $\mathbf{x}_2$. Zubizarreta (2015) also considered an algorithm for soft calibration using some prespecified tolerance level.

To handle this problem, it is helpful to use the mixed-model representation

$$\mathbf{y} \mid \mathbf{u} \sim N(X_1\beta + X_2\mathbf{u}, I_n\sigma_e^2), \quad \mathbf{u} \sim N\left(\mathbf{0}, I_q\sigma_u^2\right), \tag{12.32}$$

where $\mathbf{y} = (y_1, \ldots, y_n)'$, $\mathbf{X}_1$ is $n \times p$ matrix with $\mathbf{x}'_{i1}$ in the i-th row, and $\mathbf{X}_2$ is $n \times q$ matrix with $\mathbf{x}'_{i2}$ in the i-th row. Under (12.32) and MAR, the best linear unbiased predictor of y_i is

$$\hat{y}_i = \mathbf{x}'_{1i}\hat{\beta} + \mathbf{x}'_{2i}\hat{\mathbf{u}} \tag{12.33}$$

where $\hat{\beta}$ and $\hat{\mathbf{u}}$ are the solution to

$$\begin{pmatrix} X_1'\Delta_n X_1 & X_1'\Delta_n X_2 \\ X_2'\Delta_n X_1 & X_2'\Delta_n X_2 + \lambda I_q \end{pmatrix}\begin{pmatrix}\beta \\ \mathbf{u}\end{pmatrix} = \begin{pmatrix}X_1'\Delta_n\mathbf{y} \\ X_2'\Delta_n\mathbf{y}\end{pmatrix}$$

where $\Delta_n = \mathrm{diag}\{\delta_1, \ldots, \delta_n\}$ and $\lambda = \sigma_e^2/\sigma_u^2$.

In the spirit of Lemma 12.1, it is reasonable to construct the propensity weights to satisfy

$$\frac{1}{n}\sum_{i=1}^{n}\delta_i\left\{1 + \frac{n_0}{n_1}r(\mathbf{x}_i)\right\}y_i = \frac{1}{n}\sum_{i=1}^{n}\hat{y}_i, \tag{12.34}$$

where $\hat{y}_i$ is defined in (12.33). Note that (12.34) implies that

$$\frac{1}{n}\sum_{i=1}^{n}\delta_i\left\{1 + \frac{n_0}{n_1}r(\mathbf{x}_i)\right\}E(y_i \mid \mathbf{x}_i, \mathbf{u}) = \frac{1}{n}\sum_{i=1}^{n}E(\hat{y}_i \mid \mathbf{x}_i, \mathbf{u}). \tag{12.35}$$

Now, note that

$$
\begin{aligned}
E(\hat{\beta} \mid X, \mathbf{u}) &= \beta \\
E(\hat{\mathbf{u}} \mid X, \mathbf{u}) &= (\hat{\Sigma}_{22 \cdot 1} + \lambda I_q)^{-1} \hat{\Sigma}_{22 \cdot 1} \mathbf{u} \\
&= \left\{ I_q - \lambda (\hat{\Sigma}_{22 \cdot 1} + \lambda I_q)^{-1} \right\} \mathbf{u}.
\end{aligned}
$$

where $\hat{\Sigma}_{22 \cdot 1} = X_2' \Delta_n X_2 - (X_2' \Delta_n X_1)(X_1' \Delta_n X_1)^{-1}(X_1' \Delta_n X_2)$.

Thus, (12.35) implies that

$$
n^{-1} \sum_{i=1}^{n} \delta_i \left\{ 1 + \frac{n_0}{n_1} r(\mathbf{x}_i) \right\} \mathbf{x}_{1i} = n^{-1} \sum_{i=1}^{n} \mathbf{x}_{1i} \tag{12.36}
$$

$$
n^{-1} \sum_{i=1}^{n} \delta_i \left\{ 1 + \frac{n_0}{n_1} r(\mathbf{x}_i) \right\} \mathbf{x}_{2i} = n^{-1} \sum_{i=1}^{n} \mathbf{x}_{2i} \{1 + R(\lambda)\}, \tag{12.37}
$$

where

$$
R(\lambda) = -\lambda (\hat{\Sigma}_{22 \cdot 1} + \lambda I_q)^{-1}.
$$

Thus, if $\lambda = o(n^{1/2})$, then $R(\lambda) = o(n^{-1/2})$ and soft calibration holds for $\mathbf{x}_2$.

The random effect interpretation of the soft calibration is described in Park and Fuller (2009) in the context of survey sampling.

12.5 Penalized Regression Imputation

We now briefly discuss penalized regression approach to imputation. To explain the idea of penalized regression, consider the linear regression model

$$
Y_i = \mathbf{x}_i' \beta + \epsilon_i, \quad i = 1, 2, \ldots, n,
$$

where Y_i is the response variable, $\mathbf{x}_i$ is a d-dimensional vector of predictors for the i-th subject, β is a d-dimensional vector of regression coefficients, and $(\epsilon_1, \ldots, \epsilon_n)$ are independent and identically distributed errors with mean zero. The penalized least squares estimator of β is the minimizer of the objective function

$$
Q_p(\beta) = \sum_{i=1}^{n} \left(y_i - \mathbf{x}_i' \beta \right)^2 + n \sum_{j=1}^{d} p_\lambda(|\beta_j|), \tag{12.38}
$$

where $p_\lambda(\cdot)$ is the penalty function. Table 12.1 presents popular penalty functions for penalized regression.

We are interested in investigating the asymptotic properties of the penalized regression imputation estimator of $\theta = E(Y)$ given by

$$
\hat{\theta}_I = \frac{1}{n} \sum_{i=1}^{n} \left\{ \delta_i y_i + (1 - \delta_i) \mathbf{x}_i' \hat{\beta}_\lambda \right\}, \tag{12.39}
$$

TABLE 12.1

Some Popular Penalty Functions

Method	Reference	Penalty Function
Ridge	Hoerl and Kennard (1970)	$p_\lambda(\beta) = \lambda \sum_{j=1}^{p} \beta_j^2$
LASSO	Tibshirani (1996)	$p_\lambda(\beta) = \lambda \sum_{j=1}^{p} |\beta_j|$
Ada. LASSO	Zou (2006)	$p_\lambda(\beta) = \lambda \sum_{j=1}^{p} \left(|\beta_j| / \left|\hat{\beta}_j\right| \right)$
SCAD	Fan and Li (2001)	$p'_\lambda(\beta) = q_\lambda(\beta)$ in (12.42)

where $\hat{\beta}_\lambda$ is the minimizer of $Q_\lambda(\beta)$ in (12.38). Fan and Li (2001) presents some asymptotic properties of $\hat{\beta}_\lambda$ and discuss sufficient conditions for the $\sqrt{n}$-consistency and the sparsity of $\hat{\beta}_\lambda$. Johnson et al. (2008) extended the idea further to develop variable selection using penalized estimating functions.

In the context of imputation, we may use

$$Q_p(\beta) = \frac{1}{n} \sum_{i=1}^{n} \delta_i \left(y_i - \mathbf{x}_i'\beta \right)^2 + \sum_{j=1}^{d} p_\lambda(|\beta_j|). \tag{12.40}$$

Without loss of generality, we assume that the true model is

$$Y_i = \mathbf{x}_{1i}'\beta_1 + \mathbf{x}_{2i}'\beta_2 + \epsilon_i, \quad i = 1, 2, \ldots, n,$$

where β_1 is a s-dimensional vector of non-zero parameter values and $\beta_2 = 0$.

If we define

$$\hat{U}_p(\beta) = \frac{\partial}{\partial \beta} Q_p(\beta),$$

we can express

$$\hat{U}_p(\beta) = U(\beta) - \mathbf{q}_\lambda(|\beta|)\mathrm{sgn}(\beta),$$

where $\hat{U}(\beta) = n^{-1} \sum_{i=1}^{n} (y_i - \mathbf{x}_i'\beta)\mathbf{x}_i$, $\mathbf{q}_\lambda(\beta)$ is a d-dimensional vector of $q_\lambda(|\beta_j|) = p'_\lambda(|\beta_j|)$, and $\mathrm{sgn}(a) = a/|a|$ for $a \neq 0$ and $\mathrm{sgn}(a) = 0$ for $a = 0$. Let $\beta^{(0)}$ be the true value of β. Johnson et al. (2008) made the following two assumptions on the penalty function:

[C1] For nonzero fixed β,

$$\lim_n n^{1/2} q_{\lambda_n}(|\beta|) = 0 \quad \text{and} \quad \lim_n q'_{\lambda_n}(|\beta|) = 0 \tag{12.41}$$

[C2] For any $M > 0$,

$$\lim_{|\beta| \leq Mn^{-1/2}} \inf n^{1/2} q_{\lambda_n}(|\beta|) \to \infty.$$

The first condition [C1] prevents the j-th element of the penalized estimating function from being dominated by the penalty term for $\beta_j^{(0)} \neq 0$. The second condition [C2] implies that the j-th element of the penalized estimating function is dominated by the penalty term if $\beta_j^{(0)} = 0$, so that any consistent solution to $\hat{U}_p(\beta) = 0$ must satisfy $\hat{\beta}_j = 0$ with the probability converging to one, as $n \to \infty$. Under the Smoothly Clipped Absolute Deviation (SCAD) penalty of Fan and Li (2001), the two conditions are satisfied if we choose $\lambda_n \to 0$ and $\sqrt{n}\lambda_n \to \infty$. Since the SCAD penalty is

$$q_\lambda(\beta_j) = \lambda \left\{ I(|\beta_j| \le \lambda) + \frac{(a\lambda - |\beta_j|)_+}{(a-1)\lambda} I(|\beta_j| > \lambda) \right\}, \tag{12.42}$$

with $a > 2$, we can obtain $\sqrt{n}q_\lambda(|\beta_j|) = q_\lambda'(|\beta_j|) = 0$ for $\beta_j \neq 0$ and $\sqrt{n}\inf_{|\beta_j| \le Mn^{-1/2}} q_{\lambda_n}(|\beta_j|) = \sqrt{n}\lambda_n$.

We now briefly discuss statistical inference with the penalized regression imputation estimator in (12.39). Note that we can express $\hat{\theta}_I(\hat{\beta}_\lambda) = n^{-1}\sum_{i=1}^n \mathbf{x}_i'\hat{\beta}_\lambda$ and apply Taylor linearization around $\beta^{(0)}$, the true parameter values, to obtain

$$\begin{aligned}
\hat{\theta}_I &= \hat{\theta}_I(\beta^{(0)}) + E\left\{ \frac{\partial}{\partial \beta_1} \hat{\theta}_I(\beta^{(0)}) \right\} \left(\hat{\beta}_1 - \beta_1^{(0)} \right) \\
&\quad + E\left\{ \frac{\partial}{\partial \beta_2} \hat{\theta}_I(\beta^{(0)}) \right\} \left(\hat{\beta}_2 - \beta_2^{(0)} \right) + o_p(n^{-1/2}).
\end{aligned}$$

Because $\hat{\beta}_2 = 0$ with the probability converging to one, as $n \to \infty$ and $\beta_2^{(0)} = 0$, we can obtain

$$\hat{\theta}_I = \hat{\theta}_I(\beta^{(0)}) + E\left\{ \frac{\partial}{\partial \beta_1} \hat{\theta}_I(\beta^{(0)}) \right\} \left(\hat{\beta}_1 - \beta_1^{(0)} \right) + o_p(n^{-1/2}). \tag{12.43}$$

Thus, the uncertainty associated with $\hat{\beta}_2$ can be safely ignored for the oracle estimator. Furthermore, we can apply the same Taylor expansion to $\hat{U}_p(\hat{\beta}_\lambda) = 0$ around $\beta^{(0)}$ to get

$$0 = \hat{U}_p(\beta^{(0)}) + E\left\{ \frac{\partial}{\partial \beta_1} \hat{U}_p(\beta^{(0)}) \right\} \left(\hat{\beta}_1 - \beta_1^{(0)} \right) + o_p(n^{-1/2}). \tag{12.44}$$

Here, the uncertainty associated with $\hat{\beta}_2$ is also ignored by the oracle property. Thus, combining (12.43) and (12.44), we obtain

$$\begin{aligned}
\hat{\theta}_I &= n^{-1}\sum_{i=1}^n \mathbf{x}_{1i}'\beta_1 \\
&\quad + \{E(n^{-1}\sum_{i=1}^n \mathbf{x}_{1i}')\}(\Sigma_{11} + \Lambda_1)^{-1}\left\{ n^{-1}\sum_{i=1}^n \delta_i\mathbf{x}_{1i}\left(y_i - \mathbf{x}_{1i}'\beta_1\right) - \mathbf{q}_\lambda(|\beta_1|)\mathrm{sgn}(\beta_1) \right\} \\
&\quad + o_p(n^{-1/2}),
\end{aligned}$$

where $\Sigma_{11} = E\{n^{-1}\sum_{i=1}^{n}\delta_i\mathbf{x}_{1i}\mathbf{x}_{1i}'\}$ and $\Lambda_1 = \text{diag}\{q_\lambda'(|\beta_1|),\ldots,q_\lambda'(|\beta_s|)\}$. Under assumption (12.41), we have $\Lambda_1 = o(1)$ and $\sqrt{n}\mathbf{q}_\lambda(|\beta_1|) = o(1)$. Thus, we obtain, ignoring the smaller order terms,

$$\hat{\theta}_I \cong n^{-1}\sum_{i=1}^{n}\left\{\mathbf{x}_{1i}'\beta_1 + \kappa_1'\mathbf{x}_{1i}\delta_i\left(y_i - \mathbf{x}_{1i}'\beta_1\right)\right\}, \tag{12.45}$$

where

$$\kappa_1 = \left\{E(\sum_{i=1}^{n}\delta_i\mathbf{x}_{1i}\mathbf{x}_{1i}')\right\}^{-1}E\left(\sum_{i=1}^{n}\mathbf{x}_{1i}\right).$$

Note that (12.45) is equivalent to the standard linearization formula using the selected covariates only in the regression imputation.

The Lasso penalty does not satisfy the oracle property. In this case, we may consider the double machine learning estimator of Chernozhukov et al. (2018). The basic idea is to randomly split the sample to separately perform model selection and parameter estimation.

12.6 Sufficient Dimension Reduction

Sufficient dimension reduction (SDR) is also a popular technique of finding the space of covariates in the regression model. There is a large literature on sufficient dimension reduction involving a response Y and a covariate vector X, where the distribution of (Y,X) is nonparametric. Suppose that B is a fixed matrix with the smallest possible row dimension such that Y and X are independent conditional on BX, i.e., $Y \perp X \mid BX$. The linear function BX carries all information contained in X about Y so that statistical inference on unknown quantities in the distribution of Y or conditional distribution of $Y \mid X$ can be made using (Y,BX) instead of (Y,X). Although B is not unique, the linear space generated by the rows of B, denoted by $\mathcal{L}(B)$, is unique and called central subspace. The matrix B having the smallest row dimension and satisfying $\mathscr{Y} \perp \mathscr{X} \mid B\mathscr{X}$ is denoted as $B_{\mathscr{Y}|\mathscr{X}}$. The goal of SDR is to estimate nonparametrically the typically unknown $B_{Y|X}$ based on data observed from (Y,X). The most popular sufficient dimension reduction methods are the sliced inverse regression (SIR) in Li (1991), the sliced average variance estimation (SAVE) in Cook and Weisberg (1991), the directional regression in Li and Wang (2007), the minimum average variance estimation in Xia and Zhu (2002), the average partial mean estimation in ?, and the semiparametric method in Ma and Zhu (2012).

In the context of missing data, SDR can be used under the situation where the response Y has missing values and the covariate vector X is fully observed. Under MAR, we can simply apply any available SDR method using observed data from $(Y, X, \delta = 1)$, because $Y \perp X \mid B_{Y|X} X$ is equivalent to $Y \perp X \mid (B_{Y|X}, \delta = 1)$ when missing data are ignorable.

If the dimension of the SDR space is l, our goal is to find $B_{Y|X} \in \mathbb{R}^{l \times d}$ such that

$$Y \perp \delta \mid B_{Y|X} \mathbf{x},$$

where $l < d$ and $d = \dim(\mathbf{x})$. To find B in a function space, we now introduce the SDR method of Stojanov et al. (2019). For a positive semi-definite kernel function k, let $\mathcal{H}_k$ denote the induced reproducing kernel Hilbert space (RKHS). In particular, there exists a feature map $\psi : \mathbb{R}^d \to \mathcal{H}_k$, which maps $\mathbf{x}$ to an abstract space. On the other hand, the kernel t may induce a map $\mathbb{R} \to \mathcal{H}_t$ for the response variable. Further, let u induces a map $\mathbb{R}^l \to \mathcal{H}_u$ for sufficient dimension reduction variable $\mathbf{b}(\mathbf{x}) = B_{Y|X} \mathbf{x}$. For $\xi \in \mathcal{H}_u$, $\zeta \in \mathcal{H}_t$, the cross-covariance operator from $\mathcal{H}_u$ to $\mathcal{H}_t$ is defined as

$$\langle \zeta, \mathcal{C}_{Y, \mathbf{b}(\mathbf{x})} \xi \rangle_{\mathcal{H}_t} \equiv \mathbb{E}_{Y, \mathbf{b}(\mathbf{x})} \left\{ \xi(\mathbf{b}(\mathbf{x})) \zeta(Y) \right\} - \mathbb{E}_{\mathbf{b}(\mathbf{x})} \left\{ \xi(\mathbf{b}(\mathbf{x})) \right\} \mathbb{E}_Y \left\{ \zeta(Y) \right\}.$$

In addition, the conditional covariance operator can be defined as

$$\mathcal{C}_{YY|\mathbf{b}(\mathbf{x})} \equiv \mathcal{C}_{YY} - \mathcal{C}_{Y, \mathbf{b}(\mathbf{x})} \mathcal{C}_{\mathbf{b}(\mathbf{x}), \mathbf{b}(\mathbf{x})}^{-1} \mathcal{C}_{\mathbf{b}(\mathbf{x}), Y}.$$

According to Stojanov et al. (2019), the SDR problem can be transformed into the following optimization problem on Grassmann manifold for the data with $\delta_i = 1$:

$$\underset{B}{\arg\min} \quad \mathrm{Trace} \left\{ \widehat{\mathcal{C}}_{YY|B\mathbf{x}} \right\}$$

$$s.t. \quad BB' = I. \tag{12.46}$$

When missing data are nonignorable, however, we cannot carry out sufficient dimension reduction using data from $(Y, X, \delta = 1)$, since $Y \perp X \mid B_{Y|X} X, \delta = 1$ is not the same as $Y \perp X \mid B_{Y|X} X$. To the best of our knowledge, the sufficient dimension reduction technique has not been extended to any general nonignorable nonresponse problem. Furthermore, in the presence of nonresponse, $B_{Y|X} X$ alone is not enough for analysis, i.e., we need to find $B_{Y, \delta|X}$ such that

$$(Y, \delta) \perp X \mid B_{Y, \delta|X} X. \tag{12.47}$$

Recently, Zhao et al. (2021) proposed SDR method under nonignorable nonresponse. To explain the idea, let $f(\cdot \mid \cdot)$ be a generic notation for conditional probability density. Since $f(Y, \delta \mid X) = f(\delta \mid Y, X) f(Y \mid X)$, to find $B_{Y, \delta|X}$ in (12.47), it suffices to find $B_{Y|X}$ and $B_{\delta|Y, X}$, respectively, for

$$Y \perp X \mid B_{Y|X} X \quad \text{and} \quad \delta \perp (Y, X) \mid Y, B_{\delta|Y, X} X. \tag{12.48}$$

Let $\mathcal{L}(M_1, ..., M_k)$ denote the linear space generated by the rows of matrices $M_1, ..., M_k$. Then $\mathcal{L}(B_{Y, \delta|X}) = \mathcal{L}(B_{Y|X}, B_{\delta|Y, X})$, where $B_{Y, \delta|X}$ is in (12.47). However, $B_{Y|X}$ or $B_{\delta|Y, X}$ in (12.48) cannot be directly estimated because Y

is subject to nonignorable nonresponse. To overcome this difficulty, we utilize the identity

$$f(Y \mid X)P(\delta = 1 \mid Y, X) = P(\delta = 1 \mid X)f(Y \mid X, \delta = 1)$$

and conclude that $\mathcal{L}(B_{Y,\delta|X}) = \mathcal{L}(B_{Y|X,\delta=1}, B_{\delta|X})$, where $B_{Y|X,\delta=1}$ and $B_{\delta|X}$ are matrices for

$$Y \perp X \mid B_{Y|X,\delta=1}X, \ \delta = 1 \qquad \text{and} \qquad \delta \perp X \mid B_{\delta|X}X. \qquad (12.49)$$

Based on (12.49), one can apply an available sufficient dimension reduction method to estimate $B_{Y|X,\delta=1}$ using data (Y_i, X_i) with $\delta_i = 1$, $i = 1, ..., n$, and $B_{\delta|X}$ using data (X_i, δ_i), $i = 1, ..., n$. Note that, although $\mathcal{L}(B_{Y|X,\delta=1}, B_{\delta|X}) = \mathcal{L}(B_{Y|X}, B_{\delta|Y,X}) = \mathcal{L}(B_{Y,\delta|X})$, we are not able to identify individual $B_{Y|X}$ or $B_{\delta|Y,X}$ under this approach.

We now establish a result that can be used to combine the two searches of $B_{Y|X,\delta=1}$ and $B_{\delta|X}$ in (12.49) into a single search. For any $t \geq 0$,

$$
\begin{aligned}
P(\delta e^Y \leq t \mid X) &= P(\delta = 1, e^Y \leq t \mid X) + P(\delta = 0 \mid X) \\
&= P(Y \leq \log t \mid X, \delta = 1)P(\delta = 1 \mid X) + P(\delta = 0 \mid X),
\end{aligned}
$$
$$(12.50)$$

where $\log t$ is defined to be $-\infty$ when $t = 0$. If the rows of a matrix C span $\mathcal{L}(B_{Y|X,\delta=1}, B_{\delta|X})$, then it follows from (12.49)–(12.50) that $P(\delta e^Y \leq t \mid X) = \text{pr}(\delta e^Y \leq t \mid CX)$. This means that $\mathcal{L}(B_{\delta e^Y|X}) \subseteq \mathcal{L}(B_{Y|X,\delta=1}, B_{\delta|X})$. On the other hand, if

$$\delta e^Y \perp X \mid B_{\delta e^Y|X}X, \qquad (12.51)$$

then

$$P(\delta = 0 \mid X) = P(\delta e^Y = 0 \mid X) = P(\delta e^Y = 0 \mid B_{\delta e^Y|X}X) = P(\delta = 0 \mid B_{\delta e^Y|X}X),$$

which implies $\mathcal{L}(B_{\delta|X}) \subseteq \mathcal{L}(B_{\delta e^Y|X})$ because $B_{\delta|X}$ has the smallest row dimension. Zhao et al. (2021) proved the following result.

Theorem 12.3. *Let* $B_{Y,\delta|X}$, $B_{Y|X,\delta=1}$, $B_{\delta|X}$, *and* $B_{\delta e^Y|X}$ *be defined in* (12.47), (12.49), *and* (12.51). *Then*

$$\mathcal{L}(B_{Y,\delta|X}) = \mathcal{L}(B_{\delta e^Y|X}) = \mathcal{L}(B_{Y|X,\delta=1}, B_{\delta|X}).$$

Theorem 12.3 layouts a way to carry out sufficient dimension reduction when nonignorable nonresponse exists. To obtain an estimator of $B_{\delta e^Y|X}$ in (12.51), we can apply any available method in the literature of sufficient dimension reduction, using data $\delta_i e^{Y_i}$, $i = 1, ..., n$, in which any nonrespondent Y_i does not cause any problem since $\delta_i e^{Y_i} = 0$ when $\delta_i = 0$. The dimension of $\mathcal{L}(B_{\delta e^Y|X})$ can be estimated by the modified BIC-type criterion proposed by Zhu et al. (2010).

12.7   Neural Network Model

We now consider using neural network models for imputation. The neural network model has recently attracted a lot of attention in the machine learning literature as a general function approximator, due to its ability to capture a complex nonlinear trend in data (Goodfellow et al., 2016). To explain the model with one hidden layer, we use

$$y \mid \Lambda \sim p(y \mid \Lambda), \tag{12.52}$$

$$\Lambda = g^{(y)}\left(\beta_0 + \sum_{j=1}^{d} \beta_j h_j(\mathbf{x}; \alpha_j)\right), \tag{12.53}$$

$$h_j(\mathbf{x}; \alpha_j) = g^{(1)}\left(\alpha_{j0} + \sum_{r=1}^{p} \alpha_{jr} x_r\right), \; j = 1, \ldots, d, \tag{12.54}$$

$$\alpha_{10} < \alpha_{20} < \cdots < \alpha_{d0} \quad \text{and} \quad \sum_{r=1}^{p} \alpha_{jr}^2 = 1, j = 1, \ldots, d. \tag{12.55}$$

where $p(y|\Lambda)$ in (12.52) is the probability density for y with a parameter Λ; Equation (12.53) is the 'output layer' that specifies a generalized linear model for Λ with an inverse link function $g^{(y)}(\cdot)$ and coefficients $\beta = [\beta_0, \beta_1, \ldots, \beta_d]'$; Equation (12.54) is the 'hidden layer' that defines the basis functions $h_1(\mathbf{x}; \alpha_1), \ldots, h_d(\mathbf{x}; \alpha_d)$, with an 'activation function' $g^{(1)}(\cdot)$ and another set of coefficients $\alpha_j = [\alpha_{j0}, \ldots, \alpha_{jp}]'$ for $j = 1, \ldots, d$. The coefficients β_j and α_{jr} are the parameters to be estimated from the sample and the size of hidden layers (d) plays the role of model complexity parameter. The inverse link function $g^{(y)}(\cdot)$ is determined by the data type of y and the activation function $g^{(1)}(\cdot)$ is chosen among standard choices such as the sigmoid function and the rectified linear unit (ReLU) (Goodfellow et al., 2016). The constraints in (12.55) are required to overcome unindentifiability of the parameter $\gamma = [\beta', \alpha']'$, where $\alpha = [\alpha'_1, \ldots, \alpha'_d]'$. Figure 12.1 gives a graphical illustration of the single hidden-layer neural network model with $p = 3$ and $d = 4$. Variable $h_j = h_j(\mathbf{x})$ denote the j-th element of the hidden layer in the neural network model.

Assuming continuous y, the link function $g^{(y)}(\cdot)$ is defined as an identify function, which leads to the following output layer:

$$\Lambda(\mathbf{x}) = m(\mathbf{x}; \gamma) = \beta_0 + \sum_{j=1}^{d} \beta_j h_j(\mathbf{x}; \alpha_j).$$

For the activation function $g^{(1)}(\cdot)$ we use ReLU, i.e.,

$$g^{(1)}(w) = \max(0, w)$$

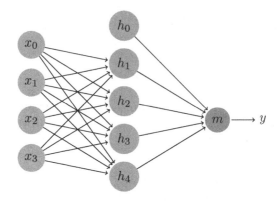

FIGURE 12.1
A graphical illustration of the single hidden-layer neural network model with $p = 3$ and $d = 4$.

for $-\infty < w < \infty$, which is known for a better computational stability without sacrificing model flexibility and hence used as the standard choice for activation function in the modern neural network literature (see, e.g., Chapter 6.3.1 in Goodfellow et al., 2016). Based on the fact that the basis functions $h_j(\cdot)$ are functions of α_j, i.e., $h_j(\mathbf{x}) = h_j(\mathbf{x}; \alpha_j)$, we estimate the model parameters $\gamma = [\beta', \alpha']'$ by minimizing the following cost function

$$Q(\gamma) = \sum_{i=1}^{n} \delta_i \left\{ y_i - \beta_0 - \sum_{j=1}^{d} \beta_j h_j(\mathbf{x}_i; \alpha_j) \right\}^2 \qquad (12.56)$$

with respect to γ. Once the minimizer $\widehat{\gamma}$ of (12.56) is obtained, the imputed estimator of $\theta = E(Y)$ is given by

$$\widehat{\theta}_I = \frac{1}{n} \sum_{i=1}^{n} \{ \delta_i y_i + (1 - \delta_i) m(\mathbf{x}_i; \widehat{\gamma}) \}, \qquad (12.57)$$

where

$$m(\mathbf{x}_i; \widehat{\gamma}) = \widehat{\beta}_0 + \sum_{j=1}^{d} \widehat{\beta}_j h_j(\mathbf{x}_i, \widehat{\alpha}_j). \qquad (12.58)$$

To discuss variance estimation of $\widehat{\theta}_I$ in (12.57), we first ignore the uncertainty associated with estimating α_j in constructing $h_j(\mathbf{x}; \widehat{\alpha}_j)$. In this case, we can apply the standard linearization method of Kim and Rao (2009) to $\widehat{\theta}_I$.

That is, using linearization with respect to β only, we obtain

$$
\widehat{\theta}_I \cong \frac{1}{n} \sum_{i=1}^{n} [m(\mathbf{x}_i; \gamma^*) + \delta_i \widehat{w}_i(\alpha^*) \{y_i - m(\mathbf{x}_i; \gamma^*)\}]
$$

$$
:= \frac{1}{n} \sum_{i=1}^{n} \eta_i(\gamma^*) = \tilde{\theta}_I(\gamma^*),
$$

where $\widehat{w}_i(\alpha^*) = \widehat{k}' \mathbf{h}(\mathbf{x}_i; \alpha^*)$ and

$$
\widehat{\kappa} = \left\{ \sum_{i=1}^{n} \delta_i \mathbf{h}(\mathbf{x}_i; \alpha^*) \mathbf{h}'(\mathbf{x}_i; \alpha^*) \right\}^{-1} \sum_{i=1}^{n} \mathbf{h}(\mathbf{x}_i; \alpha^*).
$$

By construction, we have

$$
E\left\{ \frac{\partial \tilde{\theta}_I(\gamma^*)}{\partial \beta} \right\} = 0. \tag{12.59}
$$

Furthermore, Chang et al. (2021) show that, under some general conditions, we also obtain

$$
E\left\{ \frac{\partial \tilde{\theta}_I(\gamma^*)}{\partial \alpha} \right\} = 0. \tag{12.60}
$$

By (12.59) and (12.60), the effect of estimating β as well as the effect of estimating α in the imputation procedure can be safely ignored following the argument in Randles (1982):

$$
\sqrt{n}\{\widehat{\theta}_I(\widehat{\gamma}) - \theta\} = \sqrt{n}\{\tilde{\theta}_I(\gamma^*) - \theta\} + \sqrt{n}(\widehat{\beta} - \beta^*) E\left\{ \frac{\partial \tilde{\theta}_I(\gamma^*)}{\partial \alpha} \right\}
$$

$$
+ \sqrt{n}(\widehat{\alpha} - \alpha^*) E\left\{ \frac{\partial \tilde{\theta}_I(\gamma^*)}{\partial \beta} \right\} + o_p(1)
$$

$$
= \sqrt{n}\{\tilde{\theta}_I(\gamma^*) - \theta\} + o_p(1).
$$

Thus, the asymptotic distributions of $\widehat{\theta}_I$ and $\tilde{\theta}_I(\gamma^*)$ are identical and consequently the asymptotic variance of $\tilde{\theta}_I(\gamma^*)$ is equivalent to that of $\widehat{\theta}_I$. In addition, the continuity of $m(\mathbf{x}, \gamma)$ and the consistency of $\widehat{\gamma}$, we have only to use $\eta_i(\widehat{\gamma}) = m(\mathbf{x}_i; \widehat{\gamma}) + \delta_i \widehat{w}_i(\widehat{\alpha}) \{y_i - m(\mathbf{x}_i; \widehat{\gamma})\}$ in the variance estimation formula. That is, we can use

$$
\widehat{V}_I = \frac{1}{n} s_\eta^2
$$

to estimate the variance of $\widehat{\theta}_I$, where

$$
s_\eta^2 = \frac{1}{n-1} \sum_{i=1}^{n} (\eta_i(\widehat{\gamma}) - \bar{\eta}_n)^2
$$

and $\bar{\eta}_n = n^{-1} \sum_{i=1}^{n} \eta_i(\widehat{\gamma})$.

Bibliography

Akaike, H. (1974). A new look at the statistical model identification. *IEEE Transactions on Automatic Control*, 19:716–723.

Albert, P. S. and Follmann, D. A. (2000). Modeling repeated count data subject to informative dropout. *Biometrics*, 56:667–677.

Alho, J. M. (1990). Adjusting for nonresponse bias using logistic regression. *Biometrika*, 77:617–624.

Allen, M. B. and Issacson, E. L. (1998). *Numerical Analysis for Applied Science*. John Wily & Sons, New York.

Amemiya, T. (1985). *Advanced Econometrics*. Harvard University Press, Cambridge, MA.

Amstrup, S. C., McDonald, T. L., and Manly, B. F. J. (2005). *Handbook of Capture–Recapture Analysis*. Princeton University Press, Princeton, NJ.

Anderson, R. L. (1957). Maximum likelihood estimates for the multivariate normal distribution when some observations are missing. *Journal of the American Statistical Association*, 52:200–203.

Aronszajn, N. (1950). Theory of reproducing kernels. *Transactions of the American mathematical society*, 68(3):337–404.

Bacharoglou, A. (2010). Approximation of probability distributions by convex mixtures of Gaussian measures. *Proceedings of the American Mathematical Society*, 138(7):2619–2628.

Baker, S. G. and Laird, N. M. (1988). Regression analysis for categorical variables with outcome subject to nonignorable nonresponse. *Journal of the American Statistical Association*, 83:62–69.

Bang, H. and Robins, J. M. (2005). Doubly robust estimation in missing data and causal inference models. *Biometrics*, 61:962–973.

Berg, E. and Kim, J. K. (2020). An approximate best prediction approach to small area estimation for sheet and rill erosion under informative sampling. *Annals of Applied Statistics*, 15:102–125.

Berg, E., Kim, J. K., and Skinner, C. J. (2016). Imputation under informative sampling. *Journal of Survey Statistics and Methodology*, 4(4):436–462.

Billingsley, P. (1986). *Probability and Measure (2nd edition)*. John Wiley & Sons, New York, NY.

Binder, D. A. (1983). On the variances of asymptotically normal estimators from complex surveys. *International Statistical Reviews*, 51:279–292.

Bishop, Y. M. M., Fienberg, S. E., and Holland, P. W. (1975). *Discrete Multivariate Analysis: Theory and Practice*. MIT Press, Cambridge, MA.

Booth, J. G. and Hobert, J. P. (1999). Maximizing generalized linear models with an automated Monte Carlo EM algorithm. *Journal of the Royal Statistical Society: Series B*, 61:625–685.

Breidt, F. J. and Opsomer, J. D. (2017). Model-assisted survey estimation with modern prediction techniques. *Statistical Science*, 32:190–205.

Cameron, A. C. and Trivedi, P. K. (2005). *Microeconometrics: Methods and Applications*. Cambridge University Press, New York, NY.

Cao, W., Tsiatis, A. A., and Davidian, M. (2009). Improving efficiency and robustness of the doubly robust estimator for a population mean with incomplete data. *Biometrika*, 96:723–734.

Chambers, R. L., Steel, D. G., Wang, S., and Welsh, A. (2012). *Maximum Likelihood Estimation for Sample Surveys*. Chapman & Hall / CRC, Boca Raton, FL.

Chan, K. C. G., Yam, S. C. P., and Zhang, Z. (2016). Globally efficient nonparametric inference of average treatment effects by empirical balancing calibration weighting. *Journal of the Royal Statistical Society: Series B*, 78:673–700.

Chang, W., Kim, S., and Kim, J. K. (2021). Statistical inference with neural network imputation for item nonresponse. Submitted.

Chatterjee, N., Chen, Y.-H., Maas, P., and Carroll, R. (2016). Constrained maximum likelihood estimation for model calibration using summary-level information from external big data sources. *Journal of the American Statistical Association*, 111:107–117.

Chen, J., Shao, J., and Fang, F. (2020a). Instrument search in pseudo likelihood approach for nonignorable nonresponse. *Annals of the Institute of Statistical Mathematics*. 73: 519–533.

Chen, Q. and Ibrahim, J. G. (2006). Semiparametric models for missing covariate and response data in regression models. *Biometrics*, 62:177–184.

Chen, S. and Haziza, D. (2017). Multiply robust imputation procedures for the treatment of item nonresponse in surveys. *Biometrika*, 104:439–453.

Chen, S. and Kim, J. K. (2017). Semiparametric fractional imputation using empirical likelihood in survey sampling. *Statistical Theory and Related Fields*, 1:69–81.

Chen, S. X., Qin, J., and Tang, C. Y. (2013). Mann-Whitney test with adjustments to pretreatment variables for missing values and observational study. *Journal of the Royal Statistical Society: Series B*, 75:81–102.

Chen, Y., Li, P., and Wu, C. (2020b). Doubly robust inference with non-probability survey samples. *Journal of the American Statistical Association*, 115:2011–2021.

Chen, Y. H. and Chen, H. (2000). A unified approach to regression analysis under double-sampling designs. *Journal of the Royal Statistical Society: Series B*, 62:449–460.

Cheng, P. E. (1994). Nonparametric estimation of mean functionals with data missing at random. *Journal of the American Statistical Association*, 89:81–87.

Chernozhukov, V., Chetverikov, D., Demirer, M., Duflo, E., Hansen, C., Newey, W., and Robins, J. M. (2018). Double/debiased machine learning for treatment and structural parameters. *The Econometrics Journal*, 21(1):C1–C68.

Cook, R. D. and Weisberg, S. (1991). Discussion of "Sliced inverse regression for dimension reduction". *Journal of the American Statistical Association*, 86:28–33.

Copas, A. and Farewell, V. (1998). Dealing with non-ignorable non-response by using an 'enthusiasm-to-respond' variable. *Journal of the Royal Statistical Society: Series A*, 161:385–396.

Copas, J. B. and Eguchi, S. (2001). Local sensitivity approximations for selectivity bias. *Journal of the Royal Statistical Society: Series B*, 63:871–895.

Copas, J. B. and Eguchi, S. (2005). Local model uncertainty and incomplete-data bias (with discussion). *Journal of the Royal Statistical Society: Series B*, 67:459–513.

Copas, J. B. and Li, H. G. (1997). Inference for non-random samples. *Journal of the Royal Statistical Society: Series B*, 59:55–95.

Cox, D. R. (1972). Regression models and life-tables. *Journal of the Royal Statistical Society: Series B*, 34:187–220.

Cox, D. R. and Reid, N. (1987). Parameter orthogonality and approximate conditional inference. *Journal of the Royal Statistical Society: Series B*, 49:1–39.

Davidian, M. and Caroll, R. J. (1987). Variance function estimation. *Journal of the American Statistical Association*, 82:1079–1091.

Dawid, A. (1979). Conditional independence in statistical theory. *Journal of the Royal Statistical Society: Series B*, 41:1–31.

Dempster, A. P., Laird, N. M., and Rubin, D. B. (1977). Maximum likelihood from incomplete data via the EM algorithm. *Journal of the Royal Statistical Society: Series B*, 39:1–37.

Deville, J. C. and Särndal, C. E. (1992). Calibration estimators in survey sampling. *Journal of the American Statistical Association*, 87:376–382.

Di Consiglio, L. and Tuoto, T. (2018). When adjusting for the bias due to linkage errors: A sensitivity analysis. *Statistical Journal of the IAOS*, 34:589–597.

D'Orazio, M., Zio, M. D., and Scanu, M. (2006). *Statistical Matching: Theory and Practice*. Wiley: Chichester, UK.

Drew, J. H. and Fuller, W. A. (1980). Modeling nonresponse in surveys with callbacks. In *Proceedings of the Survey Resaerch Methods Section*, pages 639–642, Washington, DC. American Statistical Association.

Efron, B. and Hinkley, D. V. (1978). Assessing the accuracy of the maximum likelihood estimator: Observed versus expected Fisher information. *Biometrika*, 65:457–487.

Elbers, C., Lanjouw, J. O., and Lanjouw, P. (2003). Micro-level estimation of poverty and inequality. *Econometrica*, 71:355–364.

Elliott, M. and Valliant, R. (2017). Inference for nonprobability samples. *Statistical Science*, 32(2):249–264.

Fan, J. and Li, R. (2001). Variable selection via nonconcave penalized likelihood and its Oracle properties. *Journal of the American Statistical Association*, 96:1348–1360.

Fang, F. and Shao, J. (2016). Model selection with nonignorable nonresponse. *Biometrika*, 103:861–874.

Fay, R. E. (1992). When are inferences from multiple imputation valid ? In *Proceedings of the Survey Research Methods Section*, pages 227–232, Washington, DC. American Statistical Association.

Fellegi, I. P. and Sunter, A. B. (1969). A theory for record linkage. *Journal of the American Statistical Association*, 64(328):1183–1210.

Fisher, R. A. (1922). On the mathematical foundations of theoretical statistics. *Philosophical Transactions of the Royal Society of London A*, 222:309–368.

Follmann, D. A. and Wu, M. (1995). An approximate generalized linear model with random effects for informative missing data. *Biometrics*, 51:151–168.

Fuller, W. A. (2003). Estimation for multiple phase samples. In Chambers, R. L. and Skinner, C. J., editors, *Analysis of Survey Data*, pages 307–322. Wiley: Chichester, England.

Fuller, W. A. (2009). *Sampling Statistics*. John Wiley & Sons, Inc., Hoboken, NJ.

Fuller, W. A. and Battese, G. E. (1973). Transformations for estimation of linear models with nested-error structure. *Journal of the American Statistical Association*, 68:626–632.

Fuller, W. A. and Kim, J. K. (2005). Hot deck imputation for the response model. *Survey Methodology*, 31:139–149.

Fuller, W. A., Loughin, M. M., and Baker, H. D. (1994). Regression weighting in the presence of nonresponse with application to the 1987-1988 Nationwide Food Consumption Survey. *Survey Methodology*, 20:75–85.

Geman, S. and Geman, D. (1984). Stochastic relaxation, Gibbs distributions, and the Bayeisan restoration of images. *IEEE Transactions on Pattern Analysis and Machine Intelligence*, 6:721–741.

Gilks, W. R. and Wild, P. (1992). Adaptive rejection sampling for Gibbs sampling. *Applied Statistics*, 41:337–348.

Givens, G. H. and Hoeting, J. A. (2005). *Computational Statistics*. John Wiley & Sons, Inc., Hoboken, NJ.

Godambe, V. P. and Joshi, V. M. (1965). Admissibility and Bayes estimation in sampling finite populations. I. *Annals of Mathematical Statistics*, 36:1707–1722.

Godambe, V. P. and Thompson, M. E. (1986). Parameters of superpopulation and survey population: their relationships and estimation. *International Statistical Review*, 54:127–138.

Goh, G. and Kim, J. K. (2021). Accounting for model uncertainty in multiple imputation under informative sampling. *Scandinavian Journal of Statistics*. 48: 930–949.

Gong, G. and Samaniego, F. J. (1981). Pseudo maximum likelihood estimation: Theory and applications. *Annals of Statistics*, 9:861–869.

Goodfellow, I., Bengio, Y., Courville, A., and Bengio, Y. (2016). *Deep Learning*, volume 1. MIT Press, Cambridge.

Graham, B. S., Pinto, C., and Egel, D. (2012). Inverse probability tilting for moment condition models with missing data. *Review of Economic Studies*, 79:1053–1079.

Guo, Y. and Little, R. J. A. (2011). Regression analysis with covariates that have heteroscedastic measurement error. *Statistics in Medicine*, 30:2278–2294.

Hainmueller, J. (2012). Entropy balancing for causal effects: A multivariate reweighting method to produce balanced samples in observational studies. *Political Analysis*, 20:25–46.

Hampel, F. R. (1974). The influence curve and its role in robust estimation. *Journal of the American Statistical Association*, 69:383–393.

Han, P. and Wang, L. (2013). Estimation with missing data: Beyond double robustness. *Biometrika*, 100:417–430.

Han, Q. and Wellner, J. A. (2021). Complex sampling designs: Uniform limit theorems and applications. *Annals of Statistics*, 49:459–485.

Han, Y. and Lihiri, P. (2019). Statistical analysis with linked data. *International Statistical Review*, 87:S139–S157.

Haslett, S. J. and Jones, G. (2005). Small area estimation using surveys and censuses: Some practical and statistical issues. *Statistics in Transition*, 7:541–556.

Hastie, T., Tibshirai, R., and Friedman, J. (2009). *The Elements of Statistical Learning*. Springer, 2nd edition.

Hastings, W. K. (1970). Monte Carlo sampling methods using Markov chains and their applications. *Biometrika*, 57:97–109.

Heckman, J. J. (1979). Sample selection as a specification error. *Econometrica*, 47:153–161.

Heitjan, D. F. and Rubin, D. B. (1991). Ignorability and coarse data. *Annals of Statistics*, 19:2244–2253.

Henmi, M. and Eguchi, S. (2004). A paradox concerning nuisance parameters and projected estimating functions. *Biometrika*, 91:929–941.

Hirano, K., Imbens, G., and Ridder, G. (2003). Efficient estimation of average treatment effects using the estimated propensity score. *Econometrica*, 71:1161–1189.

Hoerl, E. and Kennard, R. W. (1970). Ridge regression: Biased estimation for nonorthogonal problems. *Technometrics*, 12:55–67.

Hoeting, J. A., Madigan, D., Raftery, A. E., and Volinsky, C. T. (1999). Bayesian model averaging: a tutorial. *Statistical Science*, 14:382–417.

Huang, A. (2014). Joint estimation of the mean and error distribution in generalized linear models. *Journal of the American Statistical Association*, 109:186–196.

Huggins, R. and Hwang, W. H. (2011). A review of the use of conditional likelihood in capture-recapture experiments. *International Statistical Review*, 79:385–400.

Ibrahim, J. G. (1990). Incomplete data in generalized linear models. *Journal of the American Statistical Association*, 85:765–769.

Ibrahim, J. G., Lipsitz, S. R., and Chen, M. H. (1999). Missing covariates in generalized linear models when the missing data mechanism is nonignorable. *Journal of the Royal Statistical Society: Series B*, 61:173–190.

Im, J., Cho, I. H., and Kim, J. K. (2018). FHDI: An R package for fractional hot deck imputation. *The R journal*, 10:140–154.

Imai, K. and Ratkovic, M. (2014). Covariate balancing propensity score. *Journal of the Royal Statistical Society: Series B*, 76:243–263.

Isaki, C. T. and Fuller, W. A. (1982). Survey design under the regression superpopulation model. *Journal of the American Statistical Association*, 77:89–96.

Ishwaran, H. and James, L. F. (2011). Gibbs sampling methods for stick-breaking priors. *Journal of the American Statistical Association*, 96:161–173.

Jaro, M. A. (1989). Advances in record-linkage methodology as applied to matching the 1985 census of Tampa, Florida. *Journal of the American Statistical Association*, 84(406):414–420.

Jiang, D. and Shao, J. (2012). Semiparametric pseudo likelihood for longitudinal data with outcome-dependent nonmonotone nonresponse. *Statistica Sinica*, 22:1103–1121.

Johnson, B. A., Lin, D., and Zeng, D. (2008). Penalized estimating functions and variable selection in semiparametric regression models. *Journal of the American Statistical Association*, 103:672–680.

Kalton, G. and Kish, L. (1984). Some efficient random imputation methods. *Communications in Statistics: Series A*, 13:1919–1939.

Kang, J. D. Y. and Schafer, J. L. (2007). Demystifying double robustness: A comparison of alternative strategies for estimating a population mean from incomplete data. *Statistical Science*, 22:523–529.

Kenward, M. G. (1998). Selection models for repeated measurements with non-random dropout: An illustration of sensitivity. *Statistics in Medicine*, 17:2723–2732.

Kim, H. J., Reiter, J. P., Wang, Q., Cox, L. H., and Karr, A. F. (2014). Multiple imputation of missing or faulty values under linear constraints. *Journal of Business and Economic Statistics*, 32:375–386.

Kim, J. K., Park, S., and Lee, Y. (2017). Statistical inference using generalized linear mixed models under informative cluster sampling. *Canadian Journal of Statistics*, 45:479–497.

Kim, J. K. and Tam, S. (2021). Data integration by combining big data and survey sample data for finite population inference. *International Statistical Review*. 89: 382–401.

Kim, J. K. and Yang, S. (2017). A note on multiple imputation under informative sampling. *Biometrika*, 104:221–228.

Kim, J. K. (2004). Finite sample properties of multiple imputation estimators. *Annals of Statistics*, 32:766–783.

Kim, J. K. (2010). Calibration estimation using exponential tilting in sample surveys. *Survey Methodology*, 36:145–155.

Kim, J. K. (2011). Parametric fractional imputation for missing data analysis. *Biometrika*, 98:119–132.

Kim, J. K., Berg, E., and Park, T. (2016). Statistical matching using fractional imputation. *Survey Methodology*, 42:19–40.

Kim, J. K., Brick, M. J., Fuller, W. A., and Kalton, G. (2006a). On the bias of the multiple imputation variance estimator in survey sampling. *Journal of the Royal Statistical Society: Series B*, 68:509–521.

Kim, J. K. and Fuller, W. A. (2004). Fractional hot deck imputation. *Biometrika*, 91:559–578.

Kim, J. K. and Haziza, D. (2014). Doubly robust inference with missing data in survey sampling. *Statistica Sinica*, 24:375–394.

Kim, J. K. and Im, J. (2012). Propensity score adjustment with several follow-ups. *Biometrika*, 101:439–448.

Kim, J. K. and Kim, J. J. (2007). Nonresponse weighting adjustment using estimated response probability. *Canadian Journal of Statistics*, 35:501–514.

Kim, J. K., Navarro, A., and Fuller, W. A. (2006b). Replicate variance estimation after multi-phase stratified sampling. *Journal of the American Statistical Association*, 101:312–320.

Kim, J. K. and Park, H. A. (2006). Imputation using response probability. *Canadian Journal of Statistics*, 34:171–182.

Kim, J. K. and Park, M. (2010). Calibration estimation in survey sampling. *International Statistical Review*, 78:21–39.

Kim, J. K., Park, S., Chen, Y., and Wu, C. (2021). Combining non-probability and probability survey samples through mass imputation. *Journal of the Royal Statistical Society: Series A*. 184: 941–963.

Kim, J. K., Park, S., and Kim, K. (2019). A note on propensity score weighting method using paradata in survey sampling. *Survey Methodology*, 45:451–463.

Kim, J. K. and Rao, J. N. K. (2009). Unified approach to linearization variance estimation from survey data after imputation for item nonresponse. *Biometrika*, 96:917–932.

Kim, J. K. and Rao, J. N. K. (2012). Combining data from two independent surveys: A model-assisted approach. *Biometrika*, 99:85–100.

Kim, J. K. and Riddles, M. K. (2012). Some theory for propensity-score-adjustment estimators in survey sampling. *Survey Methodology*, 38:157–165.

Kim, J. K. and Shin, D. W. (2012). The factoring likelihood method for non-monotone missing data. *Journal of the Korean Statistical Society*, 41:375–386.

Kim, J. K. and Yu, C. L. (2011). A semi-parametric estimation of mean functionals with non-ignorable missing data. *Journal of the American Statistical Association*, 106:157–165.

Kott, P. S. and Chang, T. (2010). Using calibration weighting to adjust for nonignorable unit nonresponse. *Journal of the American Statistical Association*, 105:1265–1275.

Lahiri, P. and Larsen, M. D. (2005). Regression analysis with linked data. *Journal of the American Statistical Association*, 100:222–230.

Lee, D. and Kim, J. K. (2021). Semiparametric imputation using conditional Gaussian mixture models under item nonresponse. *Biometrics*. https://doi.org/10.1111/biom.13410.

Lee, D., Zhang, L.-C., and Kim, J. K. (2021). Maximum entropy classification for record linkage. Submitted.

Lesage, E., Haziza, D., and D'Haultfoeuille, X. (2019). A cautionary tale on instumental calibration for the treatment of nonignorable nonresponse in surveys. *Journal of the American Statistical Association*, 114:906–915.

Li, B. and Wang, S. (2007). On directional regression for dimension reduction. *Journal of the American Statistical Association*, 102:997–1008.

Li, K. C. (1991). Sliced inverse regression for dimension reduction. *Journal of the American Statistical Association*, 86:316–327.

Lin, N. X., Shi, J. Q., and Henderson, R. (2012). Doubly misspecified models. *Biometrika*, 99:285–298.

Little, R. J. A. (1982). Models for nonresponse in sample surveys. *Journal of the American Statistical Association*, 77:237–250.

Little, R. J. A. (1995). Modeling the drop-out mechanism in longitudinal studies. *Journal of the American Statistical Association*, 90:1112–1121.

Little, R. J. A. and Rubin, D. B. (2002). *Statistical Analysis with Missing Data* (2nd Ed.). John Wiley & Sons, Hoboken, NJ.

Loh, W. Y. (2002). Regression trees with unbiased variable selection and interaction detection. *Statistica Sinica*, 12:361–386.

Louis, T. A. (1982). Finding the observed information matrix when using the EM algorithm. *Journal of the Royal Statistical Society: Series B*, 44:226–233.

Ma, Y. and Zhu, L. P. (2012). A semiparametric approach to dimension reduction. *Journal of the American Statistical Association*, 107:168–179.

Madigan, D. and Raftery, A. E. (1994). Model selection and accounting for model uncertainty in graphical models using Occam's window. *Journal of the American Statistical Association*, 89(428):1535–1546.

Magee, L. (1998). Improving survey-weighted least squares regression. *Journal of the Royal Statistical Society: Series B*, 60:115–126.

Matei, A. and Ranalli, M. G. (2015). Dealing with non-ignorable nonresponse in survey sampling: a latent modeling approach. *Survey Methodology* 41:145–164.

McCulloch, C. E., Searle, S. R., and Neuhaus, J. M. (2008). *Generalized, Linear, and Mixed models*. Wiley, 2nd edition.

McLachlan, G. J. and Krishnan, T. (2008). *The EM Algorithm and Extensions*. John Wiley & Sons, Hoboken, NJ.

Meilijson, I. (1989). A fast improvement to the EM algorithm on its own terms. *Journal of the Royal Statistical Society: Series B*, 51:127–138.

Meng, X. L. (1994). Multiple-imputation inferences with uncongenial sources of input (with discussion). *Statistical Science*, 9:538–573.

Metropolis, N., Rosenbluth, A. W., Rosenbluth, M. N., Teller, A. H., and Teller, E. (1953). Equations of state calculations by fast computing machines. *Journal of Chemical Physics*, 21:1087–1091.

Miao, W., Ding, P., and Geng, Z. (2016). Identifiability of normal and normal mixture models with nonignorable missing data. *Journal of the American Statistical Association*, 111:1673–1683.

Molenberghs, G., Beunckens, C., and Kenward, M. G. (2008). Every missingness not at random has a missingness at random counterpart with equal fit. *Journal of the Royal Statistical Society: Series B*, 70:371–388.

Morikawa, K. and Kim, J. K. (2021). Semiparametric optimal estimation with nonignorable nonresponse data. *Annals of Statistics*. Accepted for publication.

Morikawa, K., Kim, J. K., and Kano, Y. (2017). Semiparametric maximum likelihood estimation under nonignorable nonresponse. *Canadian Journal of Statistics*, 45:393–409.

Moustaki, I. and Knott, M. (2000). Weighting for item non-response in attitude scales using latent variable models with covariates. *Journal of the Royal Statistical Society: Series A*, 163:445–459.

Müller, U. U. (2009). Estimating linear functionals in nonlinear regression with response missing at random. *Annals of Statistics*, 98:2245–2277.

Murray, J. S. and Reiter, J. P. (2016). Multiple imputation of missing categorical and continuous values via bayesian mixture models with local dependence. *Journal of the American Statistical Association*, 111:1466–1479.

Navidi, W. (1997). A graphical illustration of the EM algorithm. *American Statistician*, 51:29–31.

Nawata, K. and Nagase, N. (1996). Estimation of sample-selection bias models. *Econometric Letters*, 42:387–400.

Nguyen, X., Wainwright, M. J., and Jordan, M. I. (2010). Estimating divergence functionals and the likelihood ratio by convex risk minimization. *IEEE Transactions on Information Theory*, 56(11):5847–5861.

Nielsen, S. F. (2003). Proper and improper multiple imputation. *International Statistical Review*, 71:593–607.

Oakes, D. (1999). Direct calculation of the information matrix via the EM algorithm. *Journal of the Royal Statistical Society: Series B*, 61:479–482.

O'Muircheartaigh, C. and Moustaki, I. (1999). Symmetric pattern models: A latent variable approach to item non-response in attitude scales. *Journal of the Royal Statistical Society: Series A*, 162:177–194.

Orchard, T. and Woodbury, M. (1972). A missing information principle: Theory and applications. In *Proceedings of the 6th Berkeley Symposium on Mathematical Statistics and Probability*, volume 1, pages 695–715, Berkeley, California. University of California Press.

Owen, A. B. (1988). Empirical likelihood ratio confidence intervals for a single functional. *Biometrika*, 75:237–249.

Owen, A. B. (2001). *Empirical Likelihood*. Chapman and Hall / CRC, New York, NY.

Paik, M. C. (1997). The generalized estimating equaiton approach when data are not missing completely at random. *Journal of the American Statistical Association*, 92:1320–1329.

Park, M. and Fuller, W. A. (2009). The mixed model for survey regression estimation. *Journal of Statistical Planning and Inference*, 139:1320–1331.

Park, S. and Kim, J. K. (2019). Mass imputation for two-phase sampling. *Journal of the Korean Statistical Society*, 48:578–592.

Park, S., Kim, J. K., and Park, S. (2016). An imputation approach for handling mixed-mode surveys. *Annals of Applied Statistics*, 10(2):1063–1085.

Park, S., Kim, J. K., and Stukel, D. (2017). A measurement error model for survey data integration: combining information from two surveys. *Metron*, 75:345–357.

Park, T. and Brown, M. (1994). Models for categorical data with nonignorable nonresponse. *Journal of the American Statistical Association*, 89:44–52.

Park, T. and Casella, G. (2008). The Bayesian Lasso. *Journal of the American Statistical Association*, 103:681–686.

Pawitan, Y. (2001). *In All Likelihood: Statistical Modelling and Inference Using Likelihood*. Oxford University Press.

Qin, J. and Lawless, J. (1994). Empirical likelihood and general estimating equations. *Annals of Statistics*, 22:300–325.

Qin, J., Leung, D., and Shao, J. (2002). Estimation with survey data under non-ignorable nonresponse or informative sampling. *Journal of the American Statistical Association*, 97:193–200.

Qin, J., Zhang, B., and Leung, D. (2009). Empirical likelihood in missing data problems. *Journal of the American Statistical Association*, 104:1492–1503.

Rafei, A., Flannagan, C. A. C., and Elliott, M. R. (2020). Big data for finite population inference: Applying quasi-random approaches to naturalistic driving data using bayesian additive regression trees. *Journal of Survey Statistics and Methodology*, 8:148–180.

Randles, R. H. (1982). On the asymptotic normality of statistics with estimated parameters. *Annals of Statistics*, 10:462–474.

Rao, J. N. K. and Molina, I. (2015). *Small Area Estimation*, 2nd edition. Wiley.

Rao, J. N. K. and Shao, J. (1992). Jackknife variance estimation with survey data under hot deck imputation. *Biometrika*, 79:811–822.

Rao, J. N. K. and Sitter, R. R. (1995). Variance estimation under two-phase sampling with application to imputation for missing data. *Biometrika*, 82:453–460.

Rathouz, P. J. and Gao, L. (2009). Generalized linear models with unspecified reference distribution. *Biostatistics*, 10:205–218.

Redner, R. A. and Walker, H. F. (1984). Mixture densities, maximum likelihood and the EM algorithm. *SIAM Review*, 26:195–239.

Reinsch, C. (1967). Smoothing by spline functions. *Numerische Mathematik*, 10:177–183.

Riddles, M., Kim, J. K., and Im, J. (2016). Propensity score adjustment method for nonignorable nonresponse. *Journal of Survey Statistics and Methodology*, 4:215–245.

Rivers, D. (2007). Sampling for web surveys. In *Proceedings of the Survey Research Methods Section of the American Statistical Association*, pages 1–26.

Robert, C. P. and Casella, G. (1999). *Monte Carlo Statistical Methods*. Springer, New York.

Robins, J. M., Rotnitzky, A., and Zhao, L. P. (1994). Estimation of regression coefficients when some regressors are not always observed. *Journal of the American Statistical Association*, 89:846–866.

Robins, J. M. and Wang, N. (2000). Inference for imputation estimators. *Biometrika*, 87:113–124.

Rosenbaum, P. R. (1987). Model-based direct adjustment. *Journal of the American Statistical Association*, 82:387–394.

Rosenbaum, P. R. and Rubin, D. B. (1983). The central role of the propensity score in observational studies for causal effects. *Biometrika*, 70:41–55.

Rotnitzky, A. and Robins, J. M. (1997). Analysis of semi-parametric regression models with non-ignorable non-response. *Statistics in Medicine*, 6:81–102.

Rubin, D. B. (1974). Characterizing the estimation of parameters in incomplete data problems. *Journal of the American Statistical Association*, 69:467–474.

Rubin, D. B. (1976). Inference and missing data. *Biometrika*, 63:581–590.

Rubin, D. B. (1978). Multiple imputation in sample surveys - a phenomenological Bayesian approach to nonresponse. In *Proceedings of the Survey Research Methods Section*, pages 20–34, Washington, DC. American Statistical Association.

Rubin, D. B. (1981). The Bayesian bootstrap. *Annals of Statistics*, 9:130–134.

Rubin, D. B. (1987). *Multiple Imputation for Nonresponse in Surveys*. John Wiley & Sons, New York.

Rubin, D. B. and van der Laan, M.J. (2008). Empirical efficiency maximization: Improved locally efficient covariate adjustment in randomized experiments and survival analysis. *International Journal of Biostatistics*, 4(1):Article 5.

Rubin, D. B. and Schenker, N. (1986). Multiple imputation for interval estimation from simple random samples with ignorable nonresponse. *Journal of the American Statistical Association*, 81:366–374.

Rubin-Bleuer, S. and Schiopu-Kratina, I. (2005). On the two-phase framework for joint model and design-based inference. *Annals of Statistics*, 33(6):2789–2810.

Salvati, N., Fabrizi, E., Ranalli, M. G., and Chambers, R. L. (2021). Small area estimation with linked data. *Journal of the Royal Statistical Society: Series B*, 83:78–107.

Sang, H., Kim, J. K., and Lee, D. (2020). Semiparametric fractional imputation using Gaussian mixture models for multivariate missing data. *Journal of the American Statistical Association*. https://doi.org/10.1080/01621459.2020.1796358.

Scharfstein, D., Rotnizky, A., and Robins, J. M. (1999). Adjusting for nonignorable dropout using semi-parametric models. *Journal of the American Statistical Association*, 94:1096–1146.

Schoenberg, I. (1964). Spline functions and the problem of graduation. *Proceedings of the National Academy of Science*, 52:947–950.

Scholkopf, B. and Smola, A. J. (2002). *Learning with Kernels*. The MIT Press.

Schwarz, E. (1978). Estimating the dimension of a model. *Annals of Statistics*, 6:461–464.

Scott, A. J. and Wild, C. J. (1997). Fitting regression models to case-control data by maximum likelihood. *Biometrika*, 84:57–71.

Seber, G. A. F. and Wild, C. J. (1989). *Nonlinear Regression*. John Wiley & Sons, New York, NY.

Sethuraman, J. (1994). A constructive definition of Dirichlet priors. *Statistica Sinica*, 4:639–650.

Shao, J. (2018). Semiparametric propensity weighting for nonignorable non-response: A discussion of "Satistical inference for nonignorable missing data problems: a selective review" by Niansheng Tang and Yuanyuan Ju. *Statistical Theory and Related Fields*, 2:141–142.

Shao, J. and Steel, P. (1999). Variance estimation for survey data with composite imputation and nonnegligible sampling fraction. *Journal of the American Statistical Association*, 94:254–265.

Shao, J. and Wang, L. (2016). Semiparametric inverse propensity weighting for nonignorable missing data. *Biometrika*, 103:175–187.

Shao, J. and Zhao, J. (2013). Estimation in longitudinal studies with nonignorable dropout. *Statistics and Its Interface*, 6:303–313.

Shortreed, S. M. and Ertefaie, A. (2017). Outcome-adaptive Lasso: Variable selection for causal inference. *Biometrics*, 73:1111–1122.

Steorts, R. C., Hall, R., and Fienberg, S. E. (2016). A Bayesian approach to graphical record linkage and deduplication. *Journal of the American Statistical Association*, 111(516):1660–1672.

Stojanov, P., Gong, M., Carbonell, J. G., and Zhang, K. (2019). Low-dimensional density ratio estimation for covariate shift correction. *Proceedings of machine learning research*, 89:3449.

Stone, C. (1982). Optimal global rates of converence for nonparametric regression. *Annals of Statistics*, 10:1040–1053.

Takeuchi, K. (1976). Distribution of informational statistics and a criterion of model fitting. *SuriKagaku (Mathematical Sciences)*, 153:12–18.

Tan, Z. (2006). A distributional appproach for causal inference using propensity scores. *Journal of the American Statistical Association*, 101:1619–1637.

Tang, G., Little, R. J. A., and Raghunathan, T. E. (2003). Analysis of multivariate missing data with nonignorable nonresponse. *Biometrika*, 90:747–764.

Tanner, M. A. and Wong, W. H. (1987). The calculation of posterior distribution by data augmentation. *Journal of the American Statistical Association*, 82:528–540.

Tibshirani, R. (1996). Regression shrinkage and selection via the Lasso. *Journal of the Royal Statistical Society: Series B*, 58:267–288.

Tobin, J. (1958). Estimation of relationships for limited dependent variables. *Econometrica*, 26:24–36.

Uehara, M. and Kim, J. K. (2020). Semiparametric response model with nonignorable nonresponse. Submitted.

van der Vaart, A. W. (1998). *Asymptotic Statistics*. Cambridge University Press, Cambridge.

Wahba, G. (1990). *Spline Models for Observational Data*, volume 59. SIAM.

Wang, D. and Chen, S. X. (2009). Empirical likelihood for estimating equations with missing values. *Annals of Statistics*, 37:490–517.

Wang, H. and Kim, J. K. (2021a). Propensity score estimation using density ratio model under item nonresponse. Submitted.

Wang, H. and Kim, J. K. (2021b). Maximum sampled conditional likelihood for informative subsampling. Submitted.

Wang, H. and Kim, J. K. (2021c). Statistical inference using regularized M-estimation in the reproducing kernel Hilbert space for handling missing data. Submitted.

Wang, L., Shao, J., and Fang, F. (2021). Propensity model selection with nonignorable nonresponse and instrumental variable. *Statistica Sinica*, 31: 647–672.

Wang, N. and Robins, J. M. (1998). Large-sample theory for parametric multiple imputation procedures. *Biometrika*, 85:935–948.

Wang, S., Shao, J., and Kim, J. K. (2014). An instrumental variable approach for identification and estimation with nonignorable nonresponse. *Statistica Sinica*, 24:1097–1116.

Wang, Z., Kim, J. K., and Yang, S. (2018). Approximate Bayesian inference under informative sampling. *Biometrika*, 105:91–102.

Wei, G. C. and Tanner, M. A. (1990). A Monte Carlo implementation of the EM algorithm and the poor man's data augmentation algorithms. *Journal of the American Statistical Association*, 85:699–704.

Winkler, W. E. (1988). Using the EM algorithm for weight computation in the Fellegi-Sunter model of record linkage. In *Proceedings of the Section on Survey Research Methods*, pages 667–671. American Statistical Association.

Wong, R. K. W. and Chan, K. C. G. (2018). Kernel-based covariate functional balancing for observational studies. *Biometrika*, 105:199–213.

Wu, C. and Sitter, R. R. (2001). A model-calibration approach to using complete auxiliary information from survey data. *Journal of the American Statistical Association*, 96(453):185–193.

Wu, C. F. J. (1983). On the convergence properties of the EM algorithm. *Annals of Statistics*, 11:95–103.

Wu, M. and Follmann, D. A. (1999). Use of summary measures to adjust for informative missingness in repeated measures with random effects. *Biometrics*, 55:75–84.

Xia, Y., Tong, H., Li, W.K., and Zhu, L.X. and Zhu, L. X. (2002). An adaptive estimation of dimension reduction space (with discussion). *Journal of the Royal Statistical Society: Series B*, 64:363–410.

Xie, X. and Meng, X.-L. (2017). Dissecting multiple imputation from a multiphase inference perspective: What happens when god's, imputer's and analyst's models are uncongenial? *Statistica Sinica*, 27:1485–1594.

Xu, J. (2007). *Methods for intermittent missing responses in longitudinal data.* PhD thesis, University of Wisconsin, Madison.

Xu, L. and Shao, J. (2009). Estimation in longitudinal or panel data models with random-effect-based missing responses. *Biometrics*, 65:1175–1183.

Yang, S., Kim, J. K., and Song, R. (2020). Doubly robust inference when combining probability and non-probability samples with high dimensional data. *Journal of the Royal Statistical Society: Series B*, 82:445–465.

Yang, S. and Kim, J. K. (2014). Fractional hot deck imputation for robust inference under item nonresponse in survey sampling. *Survey Methodology*, 40:211–230.

Yang, S. and Kim, J. K. (2016). A note on multiple imputation for method of moments estimation. *Biometrika*, 103:244–251.

Yang, S. and Kim, J. K. (2020). Predictive mean matching imputation in survey sampling. *Scandinavian Journal of Statistics*, 47:839–861.

Yang, S., Kim, J. K., and Zhu, Z. (2013). Parametric fractional imputation for mixed models with nonignorable missing data. *Statistics and Its Interface*, 6:339–347.

Yuan, K.-H. and Jennrich, R. I. (2000). Estimating equations with nuisance parameters: Theory and applications. 52:343–350.

Zhang, Y., Duchi, J., and Jou Wainwright, M. (2015). Divide and conquer kernel ridge regression. *Journal of Machine Learning Research*, 16: 3299–3340.

Zhao, J. and Ma, Y. (2018). Optimal pseudolikelihood estimation in the analysis of multivariate missing data with nonignorable nonresponse. *Biometrika*, 105:479–486.

Zhao, J. and Shao, J. (2015). Semiparametric pseudo-likelihoods in generalized linear models with nonignorable missing data. *Journal of the American Statistical Association*, 110:1577–1590.

Zhao, P., Haziza, D., and Wu, C. (2020). Survey weighted estimating equation inference with nuisance functional. *Journal of Econometrics*, 216:516–536.

Zhao, P., Wang, L., and Shao, J. (2021). Sufficient dimension reduction and instrument search for data with nonignorable nonresponse. *Bernoulli*, 27:930–945.

Zhao, Q. (2019). Covariate balancing propensity score by tailored loss functions. *Annals of Statistics*, 47:965–993.

Zhou, M. and Kim, J. K. (2012). An efficient method of estimation for longitudinal surveys with monotone missing data. *Biometrika*, 99:631–648.

Zhu, L. P., Zhu, L. X., and Feng, Z. H. (2010). Dimension reduction in regressions through cumulative slicing estimation. *Journal of the American Statistical Association*, 105:1455–1466.

Zou, H. (2006). The adaptive Lasso and its oracle properties. *Journal of the American Statistical Association*, 101:1418–1429.

Zubizarreta, J. R. (2015). Stable weights that balance covariates for estimation with incomplete outcome data. *Journal of the American Statistical Association*, 110:910–922.

Index

$O_p(1)$, 59
$\sqrt{n}$-consistency, 178

AIC, 158
Aitken acceleration, 58
Akaike Information Criterion, 158
Approximate Bayesian bootstrap
    (ABB), 118
Approximate conditional model
    (ACM), 256
Ascent method, 32
Asymptotically linear, 84

Balanced score, 330
Bartlett identity, 10
Bayes formula, 14, 29, 57, 65, 74, 208
Bayes theorem, *see* Bayes formula
Bayesian bootstrap, 117
Bayesian Information Criterion
    (BIC), 159
Bisection method, 31
Bounded in probability, 59, 178

Calibration, 275, 306
Calibration estimator, 224, 268
    Instrumental variable, 270
Calibration sample, *see* Validation
    sample
Callbacks, 221
Capture-recapture sampling, 225
Cauchy distribution, 62
Cauchy–Schwartz inequality, 294
CC method, *see* Complete case
    method
CDF, 141
Cell mean model, 94
Censoring at the first missing, 242
Complete Case (CC) analysis, 16

Complete case method, 162
Conditional independence, 15, 28,
    65, 208
Conditional likelihood, 203, 223, 225
Congeniality, 121, 129
Conjugate prior, 107
Convergence in distribution, 205
Convergence in probability, 80
Covariate balancing, 275, 306
Covariate-dependent missing, 232

Data augmentation, 67
Density ratio function, 165, 273
Dirichlet distribution, 117
double machine learning, 336
Double sampling, *see* Two-phase
    sampling
Doubly protected estimator, *see*
    Doubly robust estimator
Doubly robust estimator, 187
    Fractional imputation, 190
    Local efficiency, 189
DR estimator, *see* Doubly robust
    estimator

Effective degrees of freedom, 325
EM algorithm, 47, 216
    Convergence of, 48
    Convergence rate, 51
    for fractional imputation, 139
    Monotone likelihood increase
        property, 47
    Observed information, 49
EM by weighting, 55, 285
Empirical Fisher information, 37
Empirical likelihood,151, 191
Ergodic, 60

361